ADVANCES IN ENERGY, ENVIRONMENT AND MATERIALS SCIENCE

T0139026

PROCEEDINGS OF THE 2ND INTERNATIONAL CONFERENCE ON ENERGY, ENVIRONMENT AND MATERIALS SCIENCE (EEMS 2016), JULY 29–31, 2016, SINGAPORE

Advances in Energy, Environment and Materials Science

Editors

Yeping Wang
Tongji University, China

Shiquan Zhou
Huazhong University of Science and Technology, China

CRC Press
Taylor & Francis Group
Boca Raton London New York

CRC Press is an imprint of the
Taylor & Francis Group, an **informa** business

A BALKEMA BOOK

Published by:
CRC Press/Balkema
P.O. Box 447, 2300 AK Leiden, The Netherlands
e-mail: Pub.NL@taylorandfrancis.com
www.crcpress.com – www.taylorandfrancis.com

First issued in paperback 2020

© 2017 by Taylor & Francis Group, LLC
CRC Press/Balkema is an imprint of the Taylor & Francis Group, an informa business

No claim to original U.S. Government works

Typeset by V Publishing Solutions Pvt Ltd., Chennai, India

ISBN 13: 978-0-367-73660-6 (pbk)
ISBN 13: 978-1-138-03600-0 (hbk)

Visit the Taylor & Francis Web site at
http://www.taylorandfrancis.com

and the CRC Press Web site at
http://www.crcpress.com

Table of contents

Material science and engineering

Preface

The previous First International Conference on Energy, Environment and Materials Science (EEMS 2015) has successfully taken place on August 25th–26th, 2015 in Guangzhou, China. All accepted papers were published by CRC Press/Balkema (Taylor & Francis Group) and have been indexed by Ei Compendex and Scopus and CPCI.

The 2016 2nd International Conference on Energy, Environment and Materials Science (EEMS 2016) was held on July 29–31, 2016 in Singapore. We have invited scholars and researchers from all over the world to present their research at the conference.

EEMS 2016 is dedicated to presenting and publishing novel and fundamental advances in energy and environment research fields. Scholars and specialists at EEMS 2016 will share their knowledge and research results. During the conference, an international stage is prepared for the participants to present their study theories. They will also find practical applications for these theories in related fields.

In order to organize EEMS 2016, we have sent our invitation to scholars and researchers all around the world. Eventually, over 150 research papers were submitted. These papers were subject to a strict review process performed by a team of international reviewers. All the submissions were double-blind reviewed, both the reviewers and the authors remaining anonymous. First, all the submissions were divided into several sections based on the topics, and the information of the authors, including name, affiliation, email and so on, removed. Then the editors assigned the submissions to reviewers according to their research interests. Each submission was reviewed by two reviewers. The review results were requested to be sent to chairs on time. If two reviewers had conflicting opinions, the paper would be transmitted to a third reviewer assigned by the chairs. Only papers which were approved by all reviewers were accepted for publication.

With the hard work of the reviewers, only 65 papers were finally accepted for publication. These papers were divided into three sections:

– Energy Science and Energy Technology
– Environmental Science and Environmental Engineering
– Material Science and Engineering

To prepare EEMS 2016, we have received a lot help from many people.

We thank all the contributors for their interest and support to EEMS 2016. We also feel honored to have the support from our international reviewers and committee members. Moreover, the support from CRC Press/Balkema (Taylor & Francis Group) is also deeply appreciated; without their effort, it would not have been possible to publish this book.

The Organizing Committee of EEMS 2016

Organizing committees

CONFERENCE CHAIRS

Prof. Yeping Wang, *Tongji University, China*
Prof. Li Xie, *Tongji University, China*

TECHNICAL PROGRAM COMMITTEES

Prof. Buxing Han, *Institute of Chemistry, Chinese Academy of Sciences, China*
MSc. Aline Albuquerque Chemin, *University of Sao Paulo, Brazil*
Dr. Rosinei Batista Ribeiro, *Superior Institute of Research and Scientific Initiation, Brazil*
Dr. Hao Wang, *Material Research Institute, Technical Center for China Tobacco Yunnan, China*
A.Prof. Mohamed Ramadan Eid, *Hail University, Saudi Arabia*
A.Prof. Chi-wai Kan, *Hongkong Polytechnic University, Hong Kong*
Prof. Bharat Bhushan, *The Ohio State University, USA*
Prof. Ping Chen, *Dalian University of Technology, China*
Prof. Antonio Messineo, *University of Enna Kore, Italy*
Prof. Gunhee Jang, *Hanyang University, Korea*
Prof. Dr. Aidy Ali, *Department of Mechanical Engineering, Universiti Pertahanan Nasional, Malaysia*
A.Prof. Gajendra Sharma, *Kathmandu University, Department of Computer Science and Engineering, Nepal*
Dr. Harish Kumar Sahoo, *International Institute of Information Technology (IIIT), Bhubaneswar, India*
Prof. Alaimo Andrea, *Faculty of Engineering and Architecture, Italy*
Prof. Govind Sharan Dangayach, *Malaviya National Institute of Technology, India*
Prof. Antonio Gil, *Public University of Navarra, Spain*
Prof. Dzintra Atstaja, *BA School of Business and Finance, Latvia*
Dr. Asimananda Khandual, *Biju Pattanaik University of Technology, India*
Dr. Fábio Robereto Chavarette, *Department of Mathematics, Brasil*
Prof. Heyong He, *Fudan University, China*
Prof. Leonid Getsov, *The Polzunov Central Boiler and Turbine Inst, Russia*
Prof. Loganina, *Penza State University of Architecture and Construction, Russia*
Prof. Maroš Soldán, *Slovak technical university, Slovak Republic*
Dr. Saima Shabbir, *Department of Materials Science and Engineering, Institute of Space Technology, Pakistan*
Prof. Saffi Mohamed, *Mohamed V University in Rabat (ESTS), Morocco*
Prof. Ahmad Rezaeian, *Isfahan University of Technology, Iran*
Prof. Yingkui Yang, *Hubei University, China*
Prof. Jie Du, *Hainan University, China*
Dr. Cheng Yang, *Shenzhen, Tsinghua University, China*
Dr. Calogero Orlando, *Kore University of Enna, Italy*
A.Prof. Haitao Wu, *University of Jinan, China*
Prof. Guang Xu, *Wuhan University of Science and Technology, China*
Dr. Daohai Zhang, *Guizhou University, China*
Prof. Weifeng Yao, *Shanghai University of Electric Power, China*

Energy science and energy technology

Advances in Energy, Environment and Materials Science – Wang & Zhou (Eds)
© 2017 Taylor & Francis Group, London, ISBN 978-1-138-03600-0

A study on the new energy-saving logistics container and its calculation of cooling capacity

Guo-Feng Xu, Nan-Nan Zhang, Li Wang & Hai-Feng Yu
Harbin University of Commerce, China, Heilongjiang Province, Harbin

ABSTRACT: In this paper, the technical characteristics of the logistics container which is based on online fresh refrigeration, and calculates its heat load and refrigerating capacity is introduced. The logistics container adopts an advanced technology of cold storage, by using removable cooling plates to refrigerate, charging cold for cooling plates by using a special apparatus at night, and then by reducing the peak of power load during the day. Parting it as module, the goods in different temperature range placed respectively, not only make full use of cold, but also ensure the freshness of goods. The logistics container is combined with RFID technology, monitoring temperature, and humidity conditions of the goods at any time, and it has the characteristics of energy conservation, high efficiency, and safety.

1 INTRODUCTION

Energy conservation is an eternal topic, in the traditional cold-chain transportation, consuming large quantity resource and high investment, violating the principle of energy conservation (Yan, 2015). However, with the rapid development of economy and improvement of living standards, there is a higher demand for cold-chain transportation. Cold storage technology is a refrigeration energy-saving technology praised highly in the present society, combining it with logistics container, not only to save energy effectively, but also to improve the efficiency of logistics operation (Huang, 2010). Classifying it by modules, goods of different species and different temperature ranges are kept in different containers, distinguishing it by different colors and different sizes. According to the different requirements of temperature, the different cooling plates are chosen and cold energy is used effectively.

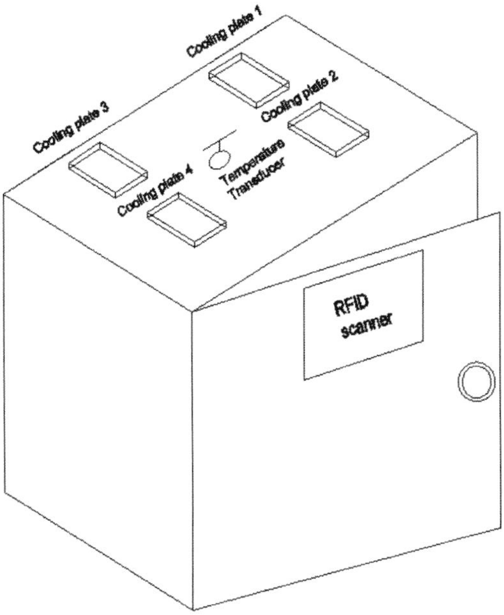

RFID scanners and temperature transducers are installed in the logistics container and each item is labeled with RFID tags. From the goods sent to be picked by customers, by using the RFID memory, the temperature and logistics information, customers can inquire about their goods at any time. The memory function ends after customers pickup their goods. Figure 1 shows the schematic diagram of the cabinet structure.

Figure 1. Schematic diagram of the cabinet structure.

2 CHARACTERISTICS OF THE LOGISTICS CONTAINER

The logistics container is different from the traditional refrigerated container (Paster, 2011). (1) It depends on the cool storage medium in which in the cooling plate melting is used to maintain low temperature and constant temperature is advantageous to the preservation of fresh food. (2) There is no need to provide refrigeration machines and power sources, thereby reducing the weight and saving the investment. (3) The logistics container adopts the advanced cold storage technology, charging chillness for the cooling plate at night when the electricity trough, and releasing the chillness when the heat is at its peak during the day, to achieve the goal of peak shift ultimately and save a great amount of energy. (4) The logistics container is classified by modules, different kinds of fresh food, and place, respectively, and then, making full use of the chillness and avoiding the freshness food tainted by other odor, thereby ensuring the safety and freshness of food. (5) It is the free convection of the container and the outside world instead of being forced by the fan, the cooling loss is less relatively when the door opened. (6) The logistics containers have a fixed site, just the normal temperature distributes it, thereby reducing the waste caused by vehicles (Zhang, 2007). (7) Divide the container into several modules, according to the different temperature requirements of food, choosing the different location. The fresh food stored in the same container with the same temperature range has different demands for low temperature. For the same container, due to the heat transfer of the wall and outside world, the middle temperature is lower than the surrounding temperature.

The cooling plate is the core component of the container; according to the different goods and distance, different cooling plates are chosen. Some goods need to be refrigerated, and some need to frozen, their demands for low temperature are various. In order to reach the best logistics efficiency, the goods of different temperature ranges are placed in different containers, and choosing the different cooling plates. Different low temperature environment needs the different storage medium filling in the cooling plate. The cool storage medium is the material of storing cold energy, storage and release of cold energy both by its phase change (Chen, 2005). The cool storage medium is mixed low-melting-point inorganic salt and the lowest refrigeration temperature can reach −32°C (Ren, 2010).

The thermal insulation material and structure design of the container directly affect the heat load, and then affect the heat preservation activity. The vesicant is injected between the outer skin and inner tank that is rigid polyurethane, and is foaming on-site (Xiao, 2011). Rigid polyurethane is a thermal insulation material that helps in both environmental protection and energy conservation; its thermal conductivity is only 0.020 W/(m·K)~0.023 W/(m·K) (Wu, 2005), its thickness is commonly 40~60 mm, and then it reduces the loss of cooling capacity. Polyurethane foam has a good heat preservation effect and a high adiabatic

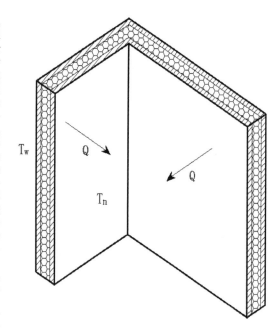

Figure 2. Schematic of the logistics container's heat transfer.

constant. However, the adiabatic constant of the thermal insulation material is not invariable; it will decrease with the change of refrigeration time, the change of the adiabatic constant could be ignored when calculating the amount of chillness (Liang, 2001). The outer skin and inner tank of the logistics container use the Glass Fiber Reinforced Plastics (GFRP), GFRP has good physical properties, such as high toughness, no pollution, recycling ability, strong carrying capacity, and so on (Xue, 2015) and meets the requirement of health, structural strength, and thermal performance (Wang, 2008). The thermal conductivity of the GFRP is 0.348 W/(m·K)~0.464 W/(m·K), defining its thickness is 25 mm.

3 CALCULATION OF COOLING CAPACITY

For the purpose of retaining the freshness of goods, we need to balance the heat load in the cabinet firstly, making the cold quantity of cold storage plate is bigger than the heat load of the cabinet. There are two specific technical parameters such as the leakage rate of heat and air, with the change of refrigeration time and constant opening of the door, there will be a leakage phenomenon, which leads to the loss of cooling capacity. And refrigerated food should consider its respiratory heat (frozen food without respiratory heat). In addition, solar radiation and cabinet lighting will also produce some heat load. In our experiment, a small container is chosen to calculate and its size is $291 \times 223 \times 257$ mm.

Respectively to hold 5 kg spinach, 5 kg pork, and 5 kg grape as an example, the calculation of heat load and cooling capacity is performed. The container's total heat transfer coefficient is calculated as follows:

$$K = \frac{1}{\frac{1}{a_1} + \frac{\delta_1}{\lambda_1} + \frac{\delta_2}{\lambda_2} + \frac{\delta_3}{\lambda_3} + \frac{1}{a_2}} \quad (1)$$

In the formula, a_1 and a_2 are the heat transfer coefficients of the external surface and inner surface, the heat transfer coefficient of gas natural convection is $1\sim10$ W/(m²·K) (Yang, 2006), ordering $a_1 = a_2 = 7$ W/(m²·K).

λ_1 and λ_3 are the thermal conductivities of the cabinet shell of the inter tank, ordering $\lambda_1 = \lambda_3 = 0.4$ W/(m·K).

λ_2 is the thermal conductivity of thermal insulation materials and its value is 0.021 W/(m·K);

δ_1 and δ_3 are the thicknesses of the cabinet shell of the inter tank;

δ_2 is the thickness of the thermal insulation materials, which is 0.5 m. Substituting the above data in the equation (1), the total heat transfer coefficient $K = 0.039$ W/(m²·K).

The heat load calculation of the container is as follows:

(1) The heat load by heat conduction into the cabinet is as follows:

$$Q_1 = KA(T_w - T_n) \quad (2)$$

In the above-stated equation, A is the heat transfer area of the cabinet;

T_w is the external temperature, ordering it is the average maximum temperature of 27°C in summer in Harbin;

T_n is the minimum temperature inside the cabinet, for spinach is 0°C, for pork is −18°C, for grape is −3°C. Substituting the above data in equation (2), we can get $Q_{1Spinach} = 17.562$ W, $Q_{1Pork} = 29.270$ W, and $Q_{1Grape} = 19.513$ W.

(2) The heat of air leakage is given by the following equation (Xiao, 2011):

$$Q_2 = \alpha Q_1 \quad (3)$$

In the equation, α is the air leakage coefficient of the cabinet, whose value is $\alpha = 0.1\sim0.2$. If $\alpha = 0.1$, $Q_{2Spinach} = 1.756$ W, $Q_{2Pork} = 2.927$ W, and $Q_{2Grape} = 1.951$ W.

(3) The heat of solar radiation is given by the following equation:

$$Q_3 = \gamma KA (T_y - T_w)t_y/24 \quad (4)$$

In the equation, γ is the percentage of the solar irradiation surface accounted for the cabinet's heat transfer surface as follows: $\gamma = 0.35$;

T_y is the average temperature of the cabinet radiated by solar energy, where $T_y = T_w + 20°C = 47°C$

t_y is the time of the cabinet radiated by solar radiation, is 12 h. Substituting the above data in equation (4), we can get $Q_{3Spinach} = Q_{3Pork} = Q_{3Grape} = 2.277$ W.

(4) The heat of opening the door into the cabinet is given by the following equation:

$$Q_4 = f(Q_1 + Q_3) \quad (5)$$

In the equation, f is the heat load coefficient for opening the door, f is 0.25 when the door is not opened in the transport, f is 0.5 when the door is opened $1\sim5$ times, f is 0.75 when the door is opened $6\sim10$ times, f is 1 when the door is opened $11\sim15$ times. The value of f is 1, $Q_{4Spinach} = 19.839$ W, $Q_{4Pork} = 31.547$ W, and $Q_{4Grape} = 21.790$ W.

(5) The heat generated by internal lighting of the cabinet is calculated by using the following equation:

$$Q_5 = P_d \bullet t_d/24 \quad (6)$$

In the above-mentioned equation, P_d is the power of internal lighting in the cabinet, and its value is 40 W;

t_d is the lighting time of the cabinet, and is about 4h. Substituting the above data in the equation (6), we can get $Q_{5Spinach} = Q_{5Pork} = Q_{5Grape} = 6.667$ W.

(6) The heat load caused by respiratory heat of fresh food in the cabinet can be calculated by using the following equation:

$$Q_6 = \Sigma m_i q_i \quad (7)$$

In the equation, m_i is the quality of fresh food in the cabinet;

q_i is the respiratory heat of fresh food (frozen food does not possess respiratory heat). The respiratory heat of spinach is 0.5 W/kg, grapes is 0.009 W/kg, and the heat load caused by respiration is as follws: $Q_{6Spinach} = 2.5$ W and $Q_{6Grape} = 0.045$ W; pork has no respiratory heat.

In summary, the heat load generated per hour in the cabinet is calculated by using the following equation:

$$Q_0 = Q_1 + Q_2 + Q_3 + Q_4 + Q_5 + Q_6 \quad (8)$$

Table 1. Comparison of storage conditions and capacity for spinach and pork.

Category	T_w	T_n	Q_0	Q_z
	°C	°C	W	W
Spinach	27	0	50.601	607.212
Pork	27	−18	72.688	872.256
Grape	27	−3	52.243	626.916

Substituting the above data in equation (8), we can get the heat load generated per hour in the cabinet $Q_{0Spinach} = 50.601$ W, $Q_{0Pork} = 72.688$ W, and $Q_{0Grape} = 52.243$ W.

The total cooling capacity required in the use of the cold storage plate can be calculated as follows:

$$Q_z = Q_0 \times t_z \tag{9}$$

In the equation, t_z is the use time of the cabinet, which is 12 h. $Q_{zSpinach} = 607.212$ W, $Q_{zPork} = 872.256$ W, and $Q_{zGrape} = 626.916$ W.

The temperature of the cold storage plate should meet the following requirements:

$$Q_d \geq Q_z \tag{10}$$

Through the above data analysis, for the same size cabinet, storage of goods with different temperature, the heat load is different, and the amount of cold storage needed for the container is also different. As a result, the container that is used for the storage of different goods should be installed with different specifications or different numbers of cold storage plate.

4 CONCLUSION

The cold storage container is constructed by combining the cold storage technology and heat insulation technology. The storage temperature is relatively stable and it has the following characteristics: good thermal insulation performance, low investment cost, and energy conservation. According to the different requirements such as low temperature and different sizes of a cabinet, different storage cold media and different numbers of cool storage plates are chosen and the amount of cold waste used is minimized. By calculating the heat load and cooling capacity and by optimizing the cabinet constantly, the cooling capacity is fully utilized. In the container, refrigerated, frozen, and normal temperature food can be stored, only requires an ordinary truck to transport and distribute, reduce the cost of transportation, so

as to make the logistics efficiency optimization. It also meet the needs of the modern society for low temperature transport that refrigerating food multi-temperature.

However, the cooling time is limited, the cooling speed is less, and it needs further improvement. For the problem of limited refrigeration time, the standby cold storage plate can be set up in the mobile station to prevent the chillness of the cold storage plate not being sufficient. For the problem of the slow cooling speed, we can precool the cabinet before placing the fresh food, and make the temperature reach the required temperature in advance, to remove the heat stored in the cabinet.

REFERENCES

Chen Huanxin, Zhu Xianfeng, Liu Guofeng. Numerical Simulation of Eutectic Solution During Freezing Period in the Cold-over Plate. Cryogenics. 2005, 2:24~29.

Huang Xuelian, Yu Xin, Ma Yongquan. Study and application of cool storage technology in fruits and vegetables preservation [J]. Journal of zhongkai University of Agriculture and Engineering, 2010, 2(23):67~70.

Liang, D.Q., Guo, K.H., Fan, S.S. Prediction of Refrigerant Gas Hydrates Formation Condition. J. Thermal Sci. 2001, 10(1):64~68.

Paster, M.D., Ahluwalia, R.K., Berry, G. et al. Hydrogen storage technology options for fuel cell vehicleσ:well-to-wheel costs, energy efficiencies, and greenhouse gas emissions [J]. International Journal of Hydrogen Energy. 2011, 36(22):14534~14551.

Ren Junsheng, Li Aimin. Characteristics and cooling capacity calculation of cold storage plate refrigerated vehicle [J]. Technical Forum, 2010, 8:47~49.

Wang Chuancheng. Lining plate substitute material for cold storage container [J]. Container Industry, 2008 (6):32~34.

Wu Lin, Yang He, Yi d-elian. Technical status and development trend of thermal insulation materials [J]. Shanxi Architecture, 2005, 31 (19):1~2.

Xiao Ying, Liu Dejun. Research on the technology and equipment of cold storage and transportation of agricultural products [J]. Journal of Agricultural Mechanization Research, 2011, 1:57~60.

Xue Zhongmin. Review and Prospect of the development of glass fiber reinforced plastics and composite materials in China [J]. Fiber Reinforced, 2015, (1): 5~12.

Yan Can, Liu Sheng, Jia Lie et al. Research progress on cold chain logistics technique of vegetables[J]. Food and Machinery, 2015, 31 (4): 260~265.

Yang Shiming, Tao Wenquan. Heat transfer [M]. Beijing: Higher Education Press, 2006:4~15.

Zhang Xuelai, Zhang Juvying, Lv Leilei, et al. Design of ice storage cold storage transport container[A]. Transaction of China Refrigeration Association Annual Conference on cold storage transportation Specialized Committee [C]. 2007.

Advances in Energy, Environment and Materials Science – Wang & Zhou (Eds)
© *2017 Taylor & Francis Group, London, ISBN 978-1-138-03600-0*

Empirical analysis of the Compressed Natural Gas (CNG) refueling behaviors

Yang Li
College of Engineering and Technology, Tianjin Agricultural University, Tianjin, China
Tianjin Huabei Gas and Heating Engineering Design Co. Ltd., Tianjin, China

Dengchao Jin, Zhenbo Bao, Hao Jin, Junwang Guo & Yulian Zhao
College of Engineering and Technology, Tianjin Agricultural University, Tianjin, China

Jian Shao & Di Yang
Tianjin Huabei Gas and Heating Engineering Design Co. Ltd., Tianjin, China

ABSTRACT: We empirically study refueling behaviors in a CNG secondary filling station. To get the central tendency and dispersion of normal refueling behaviors, abnormal behaviors or outliers are detected by the boxplot method and kernel density estimation. We find that refueling behaviors with initial pressure larger than 15.55 MPa be outliers which account for about 3.17% of the whole data set analyzed. Most normal refueling actions with the value of initial pressure around 6.0 MPa, 12.4 m³ in volume, and final pressure with 19.7 MPa. To improve the refueling effectiveness and reduce operating costs, the causes of outliers are needed to be found out.

1 INTRODUCTION

For its almost zero evaporative emissions, Compressed Natural Gas (CNG) is broadly benefited from the ongoing intensification of environmental policy. Additionally, its lower price has the potential to turn CNG into a clean alternative to the ordinary automobile fuels such as gasoline (petrol) and diesel. The Natural Gas Vehicles (NGV) usually receive natural gas from high-pressure reservoirs at filling station during refueling process. For some reasons, natural gas has not been widely accepted as an alternative fuel to gasoline. One of the problems (M. Farzaneh-Gord, 2011; 2013) with CNG secondary filling stations is the amount of time used to refuel a NGV. The other problem (M. Farzaneh-Gord, 2011; 2013) to the wide spread marketing of CNG is the on board storage capacity of NGV. In order to provide a system to refuel an NGV in a time comparable to that of a gasoline dispenser, a lot of studies have been done (M. Farzaneh-Gord, 2011; 2013; M. Baratta, 2015; N. N. Nahavandi, 2014; M. Saadat-Targhi, 2016). A bibliography (M. I. Khan, 2016) reviews the potential of CNG as transportion fuel, at the end of which containing 822 references from 111 scientific journals to papers, conference proceedings and dissertations on the subject that are published between 2001 and 2015. As far as we known, these articles about CNG and NGV are mainly based on theoretical or numerical study. With the rapid development of refueling technolo—gy, it is able to record a CNG secondary filling station's refueling activities. In this study, the refueling behaviors of a CNG secondary filling station, characterized by the amount of gas filled (in volume), the initial and final pressure of the refueled NGV, are empirically studied. Specially, basic statistical descriptions of the central tendency and dispersion of the data set are studied by Python language. Outlier behaviors are also studied based on statistical approaches. At the same time, graphic displays of the basic statistical descriptions, including quantile plots, histograms, and scatter plots, are used in this paper to visualize data.

The paper is arranged as follows: In Sec. 2 we briefly introduce the data studied in this paper, and related methods are also introduced. In Sec. 3 we present the empirical results. In Sec. 4 we conclude our paper.

2 DATA AND METHODS

2.1 *Data description*

The data set used in this study contains 125,021 records from a CNG secondary filling station in Tianjin, over a period for 3 months from May to

Date	Volume	Initial Pressure	Final Pressure
2014-5-1 0:01	14.11	4.5	18.6
2014-5-1 0:01	18.95	0.6	18.9
2014-5-1 0:01	17.89	1.5	18.7
2014-5-1 0:01	14.12	2.4	18.8
2014-5-1 0:01	15	4.4	18.7
2014-5-1 0:04	12.66	6.4	19
2014-5-1 0:04	18.66	1.6	19.5
2014-5-1 0:04	13.12	1.5	19.2
2014-5-1 0:04	15.34	4.3	19.4
2014-5-1 0:05	11.59	7.6	19.9

Figure 1.   Part of the data set used in this paper.

July in 2014. Each refueling record logs the refueling behavior's starting time, the initial and final pressure of the NGV, and the volume added into the NGV in this refueling process. Fig. 1 shows the CNG's refueling record table with the first ten records (the names of attributes are in Chinese). The first column of the table is the refueling time, which is referenced as t in this paper. The second column of the table is the total volume of CNG injected into the NGV during a refueling process, which is referenced as V. The third and final columns record the initial and final pressure of an NGV during its refueling process, and are referenced as $P_{initial}$, P_{final} respectively.

2.2   Central tendency and dispersion of data

The central tendency (J. W. Han, 2012; G. Casella, 2002; X. R. Chen, 2000) measures the location of the middle or center of a data distribution, which can given an attribution where do most of the data's values fall. In this paper, three measures of central tendency, i.e. mean, median, mode, are used to present an overall picture of the CNG refueling records. The mean and median of a set of data can be directly calculated from functions in Python language. The mode, which is defined as the value that appears most frequently in a set of data, is another measure of central tendency. No function could be used directly to get the mode of a continuous numeric attribute. Alternatively, a short Python program (see Appendix A) is written for this calculation in this paper.

In addition to assessing the central tendency of the CNG refueling record, it is essential to have an idea of the dispersion of the data set. The measures of dispersion of a data distribution, such as quartiles, standard deviation, can give a overall picture about how are the data spread out (J. W. Han, 2012; G. Casella, 2002; X. R. Chen, 2000). The first, second, and third quartile is the 25th, 50th, and 75th percentile, which denoted by Q_1, Q_2, and Q_3 respectively. Q_2 is really the median of the data set. Quartiles and standard deviation can be easily calculated by corresponding functions in Python language.

2.3   Outlier detection and graphic displays of basic statistical descriptions

In this paper, a common rule of thumb for identifying suspected outliers (J. W. Han, 2012)[7] is used to select out values not falling in the following range

$$(Q_1 - 1.5 \times IQR, Q_3 + 1.5 \times IQR) \tag{1}$$

where Inerquartile Range (IQR) is defined as

$$IQR = Q_3 - Q_1 \tag{2}$$

$Q_1 - 1.5 \times IQR$ is referenced as Min, while $Q_3 - 1.5 \times IQR$ referenced as Max. As discussed in reference (J. W. Han, 2012), this outlier detection method is like to using 3σ as the threshold for normal distribution.

Rather than assuming the data set is normal, kernel density estimation, a nonparametric method, estimates the probability density distribution from the input data only. The probability density function of a distribution is estimated by kde. KDE1d(), the kde function from Python, with a way taking into account the boundary condition by 'renormalizing' the kernel.

To make the central tendency, dispersion, and outlier easy to understand, graphics are helpful for visualizing inspection of data. Quantile plot and histograms are used to show univariate distributions. A quantile plot is an effective method to show a data distribution for one attribute, such as X. Suppose xi (i = 1, 2, ..., N) sorted in increasing order, and let each observation, xi, is paired with a percentage, fi, which is defined as

$$f_i = \frac{i - 0.5}{N}, i = 1, 2, ..., N \tag{3}$$

On a quantile plot, x_i is plotted against f_i, which allows us to see their quantiles, such as Q_1, median (Q_2), Q_3 in this paper.

3   DATA ANALYSIS AND RESULTS

3.1   Outlier detection and data visualization

Since we want to get the central tendency and dispersion of normal refueling activities, the outliers of this data set need to be selected out firstly. For each record, there are three attributes, V, $P_{initial}$, P_{final}.

In other words, we need to detect outliers in a three-dimensional data. The approach for outlier detection in this three-dimensional data is to search for outliers in various subspaces (J. W. Han, 2012).

By the methods introduced in Sec.2, we firstly detect the outliers in only one dimension. If only considering the attribute of initial pressure $P_{initial}$, 3709 records are abnormal based on initial pressure larger than 16.15 MPa. If only considering the attribute of volume V, 3422 records are abnormal based on volume lesser than 0.645 m³. And 8409 records are outliers based only on final pressure P_{final} lesser than 18.15 MPa.

Secondly, we consider the outliers in two-dimensional subspaces of the three attributes. There are 2879 records being outliers if considering both initial pressure and volume. On the other side, there are 242 or 726 records being outliers if considering final pressure and inital pressures, or volume separately. The largest final pressure is 19.9 and it is needed not much volume of CNG to increase the initial pressure from 16.15 to the final pressure of 19.9. So the result that lesser volume and high initial pressure are highly correlated is reasonable. And that high initial pressure is lowly correlated with low final pressure is also understandable in that a high initial pressure needs a more higher final pressure.

Thirdly, the ideal case is that a outlier is an activity with the above three abnormal values. However, there are only 236 records being outliers if all the three attributes considered at the same time.

Therefore, a simple and reasonable approach to detect the three-dimensional data is to search for outliers by the initial pressure only. If a refueling action with a initial pressure larger than the largest nonoutlier value (Max), this action is outlier. And its volume and final pressure are removed in calculating their statistics. The outlier searching process is realized by a recursive algorithm, which is stopped when there is no outlier left. The recursive process and measures of the central tendency and dispersion of the three attributive are listed in Table 1.

Table 1 contains 8 measures of three attributes, i.e. V, $P_{initial}$, P_{final}. These measures describe the data set's central tendency, such as mean, median and mode, and dispersion of each attribute, such as Min, Q_1, Q_3, Max, and σ. For each attribute, there are three sets of values. The first set is calculated without removing any record. The second set is obtained by removing the behaviors with initial pressure bigger than the Max, which taking as this attribute's biggest nonoutlier value. The third set is the final results, because there is no outlier left for attribute of initial pressure.

Figure 2 includes the three attributes' time series, quantile plot and histograms and their estimated density functions. The red sites in subfigures of the first column represents the outliers. While the red lines in other columns represent the statistical results of the whole data set without removing outliers. The blue lines are results of normal refueling behaviors without outliers.

3.2 Results and discussion

From Table 1, as expected, the three attributes' modes do not change by removing outliers. The other two statistics of central tendency, i.e. the mean and median, change in different directions for $P_{initial}$, V and P_{final}. As the outliers removed, the central tendency of $P_{initial}$ decreases. The distribution of $P_{initial}$ is lightly negatively skewed for its mode is slightly greater than the median. And the central tendency of V increases as expected because the outliers are actions with lesser volume. The distribution of V is positively skewed since its mode is slightly lesser than the median. But the mean of P_{final} increases little, and its median doesn't change. This results are understandable from the distribution of P_{final} in Figure 2, which shows that its value mainly falls at 19.9, the maximum of P_{final}.

The dispersion of both initial pressure and volume narrowed as expected because the outliers are sites lie in their boundary. Where the dispersion of final pressure has different behavior for the outliers is not locating in its boundary.

Table 1. The statistical values of central tendency and dispersion of the CNG refueling behaviors.

Measures	Mean	Median	Mod	Min	Q1	Q3	Max	Sigma
P_initial	6.251	5.9	6.0	X	3.4	8.5	16.15	3.885
	5.884	5.7	6.0	X	3.3	8.2	15.55	3.317
	5.863	5.7	6.0	X	3.3	8.2	X	3.288
V	11.865	12.25	12.0	0.645	9.24	14.97	23.565	4.300
	12.212	12.41	12.0	1.284	9.55	15.06	23.325	3.863
	12.235	12.41	12.0	1.36	9.58	15.06	23.28	3.835
P_final	19.323	19.7	19.9	18.15	19.2	19.9	X	1.258
	19.334	19.7	19.9	18.4	19.3	19.9	X	1.262
	19.335	19.7	19.9	18.4	19.3	19.9	X	1.261

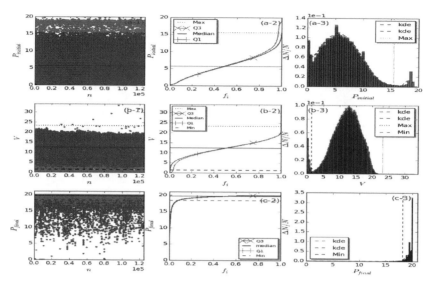

Figure 2. (Color online) Scatter, q-quartile and histograms and kernetical distribution estimation functioins of volume, initial and finial pressure of the CNG refueling behaviors. Red sites represents the outliers.

4 CONCLUSIONS

By analyzing the 125,021 records from a CNG secondary filling station in Tianjin, we get the first empirical results of the CNG refueling behaviors. Firstly, we find that a reasonable way of finding out the refueling behaviors' outliers is to detect which by initial pressure. Secondly, the central tendency and dispersion of the data set are analyzed and results are explained. Refueling behaviors with higher value of NGV's initial pressure than 15.55 MPa may be taken as outliers. As expected, this behaviors also with less volume, which means that little CNG is added into the NGV. But the causes of these outliers are still needed to be found out.

ACKNOWLEDGMENTS

This work is supported by the Innovation Fund for Small Technology-based Firms under Grant No. 14ZXCXGX00392.

APPENDIX A

```
import pandas
import numpy as np
ser = pandas.Series(x)
ax = ser.plot(kind = 'kde')
kdex = ax.get_children()[2]._x
kdey = ax.get_children()[2]._y
kdeymax = np.max(kdey)
for i in range(len(kdey)):
    if np.abs((kdey[i]-kdeymax) < = 0.000001):
        print kdex[i]
```

REFERENCES

Baratta M., H. Kheshtinejad, D. Laurenzano, D. Misul, S. Brunetti, Modeling aspects of a CNG injection system to predict its behavior under steady state conditions and throughout driving cycle simulations[J], Journal of natural gas science and engineering, 24 (2015):52–63.

Casella G., R. L. Berger, Statistical Inference (2nd edition), Wadsworth Group, 2002.

Chen X. R., Probability and mathematical statistics, Science Press, Beijing, 2000. https://pythonhosted.org/PyQt-Fit/KDE_tut.html

Farzaneh-Gord M., M. Deymi-Dashtebayaz, Optimizing natural gas fuelling station reservoirs pressure based on ideal gas model [J], Polish journal of chemical technology, 15(1)(2013):88–96.

Farzaneh-Gord M., M. Deymi-Dashtebayaz, H. R. Rahbari, Studying effects of storage types on performance of CNG filling stations [J], Journal of natural gas science and engineering, 3(2011):334–340.

Han J. W., M. Kamber, J. Pei, Data mining: concepts and techniques (3rd edition), Elsevier Pte Ltd, Singapre, 2012.

Khan M. I., T. Yasmeen, A. Shakoor, N. B. Khan, M. Wakeel, B. Chen, Exploring the potential of compressed natural gas as a viable fuel option to sustainable transport: a bibliography (2001–2015)[J], Journal of natural gas science and engineering, 31(2016):352–381.

Nahavandi N. N., M. Farzaneh-Gord, Numerical simulation of filling process of natural gas onboard vehicle cylinder[J], Journal of Brazilian society of mechanical science and engineering, 36(4)(2014): 837–846.

Saadat-Targhi M., J. Khadem, M. Farzaneh-Gord, Themodynamic analysis of a CNG refueling station considering the reciprocating compressor[J], Journal of natural gas science and engineering, 29(2016):453–461.

An overview of the inspection technique and requirement of the nuclear power plant's high energy piping system

Guo-Zheng Zhou

Suzhou Nuclear Power Research Institute Co., Ltd., Suzhou, China

ABSTRACT: From the inspection history of high energy piping in nuclear power plants, the harm of high energy piping breaking and the necessity of inspection are introduced in this paper. The damage mechanisms, material selection considerations, and the break design principle of total load reduction are mainly presented. Moreover, the application, advantage, disadvantage, acceptant criteria, and implementation considerations of different non-destructive testing methods are analyzed at length in this paper; the aim is to discuss the design optimization and inspection technology application of high energy piping.

1 INTRODUCTION

NRC defines the high energy piping of nuclear power plants in NUREG 0800 as fluid systems that, under normal plant conditions that are either in operation or maintained pressurized under conditions where either or both of the following are met:

a. Maximum operating temperature exceeds 95°C (200°F), or
b. Maximum operating pressure exceeds 1900 kPa (275 psig).

But if the operating time of the piping system does not exceed or is less than two percent of this system's operating time under the temperature exceeds 95°C (200°F), or the pressure exceeds 1900 kPa (275 psig), or the operating time is less than one percent of the system's operating time under the high temperature or pressure, these pipings cannot be considered as high energy piping. Due to the high hammer energy after the high energy piping breaking, which may cause great harm, and even affect nuclear safety, high energy piping break must be considered during the design of the nuclear power plant. These systems can be subjected to a number of damage mechanisms, including creep, fatigue, microstructural instability, and Flow-Accelerated Corrosion (FAC). An effective In-Service Inspection (ISI) program anticipates the occurrence of damage and provides for a cost-effective program to identify this damage during an early stage of development in order to allow for budgeted repair or replacement.

The high energy piping described in this paper follows NUREG 0800-3.6.1, as NRC's technology documentation "PLANT DESIGN FOR PROTECTION AGAINST POSTULATED PIPING FAILURES IN FLUID SYSTEMS OUTSIDE CONTAINMENT", this piping is mainly present in main steam system and feed water system outside of the containment, as established in Rochester Gas and Electric Corporation's Report "Effect of Postulated Pipe Breaks Outside the Containment Building", dated October 29, 1973, though this inspection is not forced in ASME XI, it is ruled in NRC's technology documentation, and included in augmented ISI program in America. Based on the harmfulness of the high energy piping break and drawing on the experience, the damage mechanisms, inspection technologies, and evaluation ways of high energy piping are analyzed in this paper.

2 DAMAGE MECHANISMS

In general, a high energy piping examination experience indicates that damage development and cracking have been encountered at specific base metal locations and weldments. In some cases, damages have been ascribed to deficiencies during manufacture or welding. However, damages are developed as a result of long-term and in-service mechanisms in most cases.

Damage mechanisms mainly depend on the component, operating conditions, materials, and methods used in manufacture. Frequently, two or more damage mechanisms are at work simultaneously on a component, e.g. pre-crack, creep cracking, fatigue cracking, combined creep-fatigue cracking, corrosion-fatigue cracking, and so on and therefore, each unit and its components must be treated individually; damage depends on factors specific to a generating unit (design, operation, and chemistry).

Damage primarily occurs in very thin section components, in highly stressed components, or in materials that have rapid crack growth rates. Techniques used for these conditions include calculations based on history, extrapolations of failure statistics, strain measurements, accelerated mechanical testing, microstructural evaluations, oxide scale growth, hardness measurements, and advanced NDT techniques. In most instances of cracking in high energy piping sections, the critical crack size at failure is sufficiently large that NDT for crack sizing along with crack growth analysis for life assessment is appropriate.

Determining the degree of damage to a component typically requires the application of one or more inspection methods. Assessing the rate of damage accumulation implies that the process of monitoring damage will be performed repeatedly or is ongoing. Determining the damage degree required to cause failure requires the use of one or more analytical or experimental techniques. In some cases, it is advisable to adopt a conservative approach. Depending on the active damage mechanism, several techniques and software programs can be used to predict the remaining life for a component. The lifetime of a high energy piping component is a complex function of factors such as operating conditions (stresses and environment), geometry (piping layout, support placement, and wall thickness), material, and damage type.

3 MATERIAL SELECTION CONSIDERATIONS

The four primary damage mechanisms that are most likely to affect high energy piping systems are creep, fatigue, creep-fatigue, and FAC. In each case, proper steel selection can minimize the probability of damage. Based on the relevant damage mechanism, Table 1 lists the type of steel that is most susceptible to damage and the preferred selection to resist that type of damage in the nuclear power plant. Different steels have different levels of damage resistance, depending on the damage mechanism and the preferred steel for resistance to one form of damage can have higher susceptibility to another form of damage. Therefore, optimal steel selection for a given application depends on a thorough understanding of the potential application's operating environment.

Higher-strength steels are being selected more frequently for power plant applications, so that a thinner cross section can be used for piping that holds the same pressure. Thinner cross sections lead to easier, faster, and significantly less expensive construction and repair. However, if a crack is initiated, then component life is necessarily shorter for a thinner cross section than a thicker one. Although newly developed steels with higher strengths often exhibit higher resistance to crack extension, the thinner cross section is a major factor in the overall component life after damage has been initiated.

Table 1. Steel selection guide based on damage mechanism susceptibility.

Damage mechanism	Necessary conditions for damage	Most susceptible steel	Preferred steel selection
Creep	High-temperature exposure under sufficient stress for an extended period of time	Brittle low-alloy steels	higher alloy content steels
Fatigue	Cyclical stress of sufficient amplitude to lie above the material's endurance limit; presence of stress risers	Pearlitic Martensitic plain carbon steels	Higher alloy steels with high tensile strengths and ductility
Creep-fatigue	Combination of high-temperature exposure under a cyclical stress of sufficient amplitude and duration to lie above the material endurance limit	Lower alloy content steels	Higher alloy content steels including stainless steel and Incoloy
FAC	Exposure to turbulent water flow at temperatures below 662°F (350°C), and a pH<9	Carbon and low-alloy steels	Carbon and low-alloy steels
Corrosion (pitting)	Presence of a low-temperature, stagnant and oxygenated aqueous solution for an extended period of time	Plain carbon or low-alloy structural steel	316 stainless steel with 2%–3% molybdenum
Erosion	Exposure to a high-velocity aqueous solution with hard insoluble particle entrainment	Plain carbon or low-alloy structural steel	Plain carbon or low-alloy structural steel

4 BREAK ANALYSES

The imaginary fracture position of high energy piping follows the ANSI/ANS–58.2. It is a must to assume that there is break in the end of piping, concerning the high energy piping without piping system, the middle break should be in the welding part of piping during stress calculation, e.g. the connection parts of valve, tee and elbow, concerning the high energy piping with piping system, the middle break should be the place where the calculation stress of which is over the value in regulation.

The high energy piping design has high requirements for piping layout, besides the technology systems; the piping stress should be reduced, for the target to reduce the middle break. Piping loads consist of persisted loads and dynamic loads. The gravity of the piping part and piping media, thermal loads led by pressure and temperature are persisted loads; the thermal loads of piping is generated when the piping suffers constraint from the support during operation. The way to reduce the thermal loads is by reducing the piping support during piping layout, so that the thermal expansion displacement can be relaxed by piping deformation. The dynamic loads are generated for the piping shaking leaded by earthquakes, increasing the supports reasonably can reduce the response of earthquake.

During the safety design of piping, obviously its a contradiction between the thermal loads and dynamic loads for the supports; the supports should be reduced for thermal loads, but increased for dynamic loads. Therefore, these factors should be considered reasonably, thermal loads and dynamic loads should be balancing in this contradiction by adjusting the position of supports, changing the style of supports, or even changing the piping run sometimes, so that the total loads can be the lowest.

That no imaginary middle breaks in high energy piping can be achieved by reasonable piping layout, detail stress calculation, and repeated iteration analysis.

5 NON-DESTRUCTIVE TESTING

The NDT decision is based on manufacture history, operating history, specific damage mechanisms of the component to be inspected, and risk-benefit.

The applications, advantages, and disadvantages of common NDT methods used for high energy piping are as follows:

- Visual: VT can be applied to any component in which the damage manifests itself at accessible surfaces to a degree that permits detection by visual means. Considerable ingenuity can often

be exercised to produce a viewing and transport system for most plant components. VT should always be the first method applied.
- Liquid Penetrant: PT can be applied to essentially any exposed surface, but it requires a more extensive surface preparation than MT. More commonly, PT is used by default for surface inspections of non-ferromagnetic materials for piping systems, such as austenitic stainless steels, where it is not possible to apply MT.
- Magnetic Particle: MT is one of the most frequently used NDT methods and is applied in power stations for routine inspection. MT can be applied essentially to any exposed surface of any ferromagnetic material. Surface preparation is not as demanding as for dye penetrant inspection and reasonable sensitivity can be achieved, particularly by using the wet fluorescent method.
- Ultrasonic: probably 75% of all UT performed in power plants involves conventional S-wave or L-wave techniques. The term conventional is used as a descriptive term for pulse-echo or pitch-catch testing using broad beams whose beam characteristics are controlled only by transducer size and frequency and the inherent material considerations. TOFD is generally applicable to inspect a relatively simple and uniform geometry. However, it is primarily used to inspect the seam welds and circumferential welds in piping systems.
- Radiography: generally, radiography is used to inspect welds. The most common use in power plants is for inspecting repair welds in boiler tubes. It is ideal for detecting macroscopic defects that are common to welds, such as porosity or slag inclusion.
- Acoustic Emission: currently, acoustic emission is used primarily to monitor seam-welded lines. But, it is sometimes difficult to separate the signal noise from the background.

In the ISI program of the nuclear power plant in America, ASME XI IWC-3514 is recommended as the acceptance criteria, the content of which is the same as IWB-3514 which includes the acceptance criteria of different materials and defects.

6 APPROACHES

These experiences demonstrate the need for a comprehensive understanding of the design and manufacture of high energy piping systems, the identification of probable damage mechanisms, and the tools available for developing and implementing a comprehensive high energy piping program with the following objectives and benefits:

- Safety: personnel safety is the highest priority. Although it is impossible to remove all risks, the goal of an effective high energy piping program should be to lower the risk of failure to a level that the employees readily accept.

- System reliability: an additional objective of the high energy piping program is to reduce the financial impact associated with forced outages resulting from failure. This can be accomplished by using state-of-the-art inspection techniques to identify the damage in sufficient time to effect mitigation or any required repair and replacement during scheduled maintenance overhauls.
- Optimized inspection costs: an effective high energy piping program can minimize inspection costs through the proper documentation and optimal inspection timing.
- Outage planning: program costs can be minimized through complete and effective forecasting of inspection and repair requirements.

7 CONCLUSIONS

The risk of high energy piping is mainly related to the design, material, and building. Since the early 1970s, studies of many institutes show that the probability of DEGB (Double Ended Guillotine Break) accidents is very low in the high energy piping of nuclear plants. Based on these studies, NRC revised the 10 CRF 50, Appendix A, GDC-4 in nuclear safety regulations, and the LBB (Leak Before Break) technology can be used in the piping design under some conditions. With an increase in the operation experience and improvement of analyzing technology, we suggest paying more attention on the design, material, and building of high energy piping, and then deciding whether inspection should be performed according to the operation conditions, mechanics analysis, and life analysis.

REFERENCES

ASME Boiler and Pressure Vessel Code: Section XI, Rules for In-service Inspection. American Society of Mechanical Engineers, New York, 2001.

EPRI TR-1012201, "Fossil Plant High-Energy Piping Damage: Theory and Practice Volume 1: Piping Fundamentals," June, 1997.

Kai Ding, Gang LI, and Bing-bing Liang, "Study on Pipe Whip Analysis of High Energy Pipe Break," Nuclear Power Engineering. Chengdu, vol. 32 S1, pp. 13–17, January 2008.

NUREG-0800, "Protection against Postulated Piping Failures in Fluid Systems Outside Containment," March, 2007.

Qiang Li, Peng Ceng, Hong-dong Zhen, "Overview of LBB Analysis Technologies for High Power Lines in Nuclear Power Plant," Nuclear power Engineering, Chengdu, vol. 32 S1, pp. 189–191, June 2011.

Advances in Energy, Environment and Materials Science – Wang & Zhou (Eds)
© 2017 Taylor & Francis Group, London, ISBN 978-1-138-03600-0

A discussion on the inspection and test technology of the post-irradiation fuel assembly

Yu-Chun Tao & Guo-Zheng Zhou
Suzhou Nuclear Power Research Institute Co., Ltd., Suzhou, China

ABSTRACT: Based on the manufacture and use of fuel assemblies of a Pressurized Water Reactor (PWR) nuclear power unit and the rapid growth of nuclear power in China, inspection and testing technology of the fuel assembly of the nuclear power plant are combined in this paper according to the practice and feedback in the field of inspection and testing of the fuel assembly at home and abroad, and the test results and acceptance criteria of the fuel assembly are discussed. Finally, the idea of development and construction of China's independent fuel system inspection and testing technology is proposed in this paper.

1 INTRODUCTION

Nuclear fuel assemblies are the first safety barrier of the PWR nuclear power plant, UO_2 pellets are sealed in zircaloy cladding to prevent the radioactive material from leaking and the integrity of fuel assemblies is directly related to the safe operation of the nuclear plant. With regard to the PWR, fuel assemblies are in the center of the reactor and suffer high temperature, high pressure, and intense radiation. Moreover, many factors, such as water impact, foreign equipment, vibration, corrosion, heat transfer, radiation, etc. will influence fuel assemblies during the operation. It is really challenging for fuel assemblies to keep all the performance indicators and integrity in such an operation environment. Cracks, wear, deformation, oxide impurity, precipitation accumulation, and other phenomena may occur in fuel assemblies while operating in a core environment, it will affect the integrity of fuel assemblies and the safety of the reactor when the problem is serious.

There was a statistics of PWR fuel assemblies from 1986 to 1996, which states that 825 fuel assemblies had failed. Though the structure and materials of the PWR fuel assembly were improved several times, and the failure rates have been greatly reduced, fuel assembly failure still occurs sometimes during long operations. According to the public reports of WANO, a lot of failure still occurs abroad every year. Besides the wastage of remaining fuel, it will influence the operation of the reactor and adjustment of the non-conventional plan, and lead to a more indirect economic loss.

In light of this, an effective inspection and test should be performed on fuel assemblies to find the potential factors, which will influence the function and integrity of fuel assemblies, and then the damaged fuel assemblies could be repaired in time, which will ensure the safety of reload fuel assemblies and the efficiency will be improved. Moreover, a root cause analysis can be performed according to the result of inspection and test, which can improve the design and manufacture process of fuel assemblies and help in analyzing higher quality fuel assemblies.

2 INSPECTION AND TEST TECHNOLOGY

2.1 *Current situation in China and Abroad*

Because of the nuclear power industry was started initially in Europe and America, they have deeply studied the inspection and test technology of the fuel assembly for many years and the industrial chain of nuclear power operation and maintenance technology is integral. Many companies, such as Westing-

Table 1. Sort of fuel assembly inspection and test technology.

Type	Function
Character measurement of the fuel assembly	Peripheral fuel rods' oxide layer thickness measurement of the fuel assembly
	Deformation measurement of the fuel assembly
	Oxide layer thickness and defect measurement of the fuel rod
Leakage detection and repair of the fuel assembly	On-line sipping and off-line sipping
	Leaking rods detection of the fuel assembly
	Repair technology of the damaged fuel assembly

house, AREVA, and ENUSA, have entire technology and equipment, but China is still in its infancy. It is deficient to improve the design of the fuel assembly and deal with leakage in the fuel assembly.

According to the using direction, the inspection and test technology of the fuel assembly can be divided into two types, as shown in Table 1.

Currently, the main management process of the fuel assembly of PWRs in China is as follows:

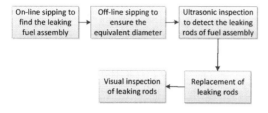

As shown in the management process depicted above, current management only focuses on the detection and replacement of leaking rods and there is no progress in the inspection and testing of fuel performance and characteristic parameters, e.g. oxide thickness measurement and deformation measurement, which are important for the design, modification, and optimization of the fuel assembly.

2.2 *Oxide layer thickness measurement of the fuel assembly*

Because of the reaction between the coolant and fuel cladding during the operation of the reactor, there will be less than 100 micros of zirconium oxides in the fuel. During the heat conduction process between fuel rod cladding and coolant, the negative influence of zirconium oxides and the loss of metal will lead to fuel cladding thickness reduction and therefore, there is a limit for the oxide layer thickness of fuel rods, the rod must be replaced once the oxide layer thickness reaches the limit, or the fuel assembly cannot be used. It is very important to evaluate the performance, operation limit, and lifetime of the fuel assembly by measuring the oxide layer's thickness.

A method for measuring the thickness of an oxide layer is eddy currents. Since the oxide layer is a non-conductor, when the oxide layer becomes thicker, the value of the eddy current will become lower and this is called as the lift-off of eddy currents. The oxide layer thickness of the fuel rod can be measured by measuring the lift-off value. Currently, the measurement accuracy of the mainstream device can reach 6 micros and only peripheral rods of the fuel assembly can be measured if the fuel assembly is not disassembled.

2.3 *Oxide layer thickness measurement and defect detection of fuel rods*

As described in the previous section, if the inner rods of the fuel assembly should be measured, the fuel assembly must be disassembled and the rod to be measured should be drawn out. For the high risk and cost, this measurement is only performed when there is a leaking rod in the fuel assembly. Leaking rod detection and drawing will be introduced in the following section. The mechanism of oxide layer thickness measurement of the fuel rod is the same as that of the fuel assembly. The method of crack measurement is also by using eddy currents, but the mechanism is different from that of oxide layer thickness measurement; when there is a crack on the fuel rod, the eddy current signal will be twisted and the defect status and size can be measured by using the twisted signal. Oxide layer thickness and defect measurement play an important role in the root cause analysis of the damaged fuel assembly.

Bobbin probe and rotation probe are used for defect inspection and measurement. A Bobbin probe is used to locate the defect in the axial direction and the accuracy should reach 1 mm. The rotation probe is used to locate the defect in the circumference and measure the length of the defect and the uncertainty of length measurement should reach 1 mm.

2.4 *Deformation measurement of the fuel assembly*

Fuel assemblies operate in high temperature, high pressure and high radiation environment, fuel rod will deform partly due to the high radiation and the fuel assembly will bend and twist because of the assembly stress and thermal stress. If the deformation is serious, it will influence the water gap of the fuel assembly and normal insertion of control rods, which will lead to rod drop due to difficult and out-of-tolerance circumstances; it is a threat to the safe operation of the reactor. Therefore, the deformation of the fuel assembly must be measured at regular intervals, including the full length; twist; tilt; bending; and grid width of fuel assembly. According to the measurement type, the deformation measurement of the fuel assembly can be divided into contact measurement and non-contact measurement. LVDT (Linear Variable Differential Transformer) is a typical contact measurement technology; the non-contact measurement technology includes laser, ultrasonic, and video image-sensors; however, there is a difference in the parameters that can measured by using different technologies.

2.5 Sipping

Sipping is a method to judge whether the fuel assembly is leaking or not by detecting the gas or solid radioisotope from the fuel rod. The method of detection is to make the fission product migrate due to pressure difference and mechanical scrubbing, that is the fuel assembly to be inspected is isolated, and then increase the internal pressure or reduce the external pressure of fuel rod, which will speed up the fission gas or water solubility fission product leaking from leaking rods, this fission product will be released into the water and air around, and a leaking fuel assembly can be detected by using the existing radioisotope from the sample of water or air.

Sipping can be divided into on-line sipping and off-line sipping according to the detection way. On-line sipping is performed at the time of fuel upload; the flowing media will be introduced into the measuring chamber, and the activity of γ or β is directly and continually measured. The fuel assembly will be fixed in the chamber and it only takes about two minutes to perform the inspection. An off-line sipping device was installed in the spent fuel pool, the fuel assembly will be isolated in the chamber, and the inner pressure of fuel rods is increased by heating. If there are leaking rods, the fission product will be released with flow media, we can authenticate whether there are leaking rods by comparing the activity of γ of flow media before and after heating. Moreover, we can calculate the equivalent diameter of the leaking hole. The fuel assembly can be operated if the equivalent value is less than the limit value, but the leaking fuel assembly is forbidden to load again for the conservative consideration of the nuclear power plant.

2.6 Detection of leaking fuel rods

A leaking fuel assembly can be authenticated by using the sipping technology, but the leaking rods must be detected by using the ultrasonic technology, for it is a critical step in leaking rod replacement. The detection mechanism is as follows: when ultrasonic signals are sent to the fuel rod cladding and is received from the other side, whether the fuel rod is leaking or not can be judged by the intensity and wave form of ultrasonic signals received. If the fuel rod was leaking, water will be released into the fuel rod cladding and some energy will be absorbed and the signal received will be very weak; if perfect, the signal received will be relatively strong.

As shown in Figure 1, when the leaking rod is full of water, the ultrasonic signal declines by up to 30 percent, it is easy to distinguish the leaking rods of the fuel assembly.

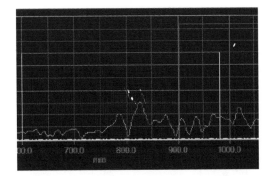

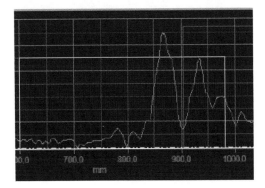

Figure 1. Ultrasonic signals of perfect and leaking rods.

2.7 Fuel assembly repair

Damage of the location grid may occur in the post-irradiation fuel assembly, including grid tear, outward, and deformation, which will lead to hardness of fuel loading and uploading and wear of the fuel rod during operation. Moreover, the fission product be released into the primary coolant if the fuel rod is leaking, which will lead to an increase in the radiation level and increase in the collective dose. If the fuel assemblies had not undergone any process, they are forbidden to load again for the consideration of safe operation.

If the damaged fuel assembly only operated for one cycle, it would arouse loss of assets directly and the cost of dealing with spent fuel would increase, and according to the symmetrical requirement of the refueling scheme, the other three fuel assemblies are also unavailable for the unavailability of this fuel assembly, it really leads to a large waste of nuclear fuel and influences the following refueling scheme. On this account, it is important to repair the damaged fuel assembly; fuel assembly repair includes skeleton repair and leaking rod replacement.

The skeleton of the fuel assembly comprises location grid, up and bottom nozzles, guide tube, neutron flux measuring tube. The key of skeleton repair is to replace the damaged skeleton and keep perfect fuel rods and the common way is to draw out fuel rods from the bottom nozzle after fuel assembly and disassembly and then insert perfect fuel rods into a new skeleton. The process of leaking rod replacement is to remove the top nozzle of the fuel assembly and pull out a leaking rod by the fall of a new fuel elevator, and finally, insert stainless steel rods by the rising of the new fuel elevator.

3 SUMMARY

Inspection and test technology of the post-irradiation fuel assembly is necessary for the safe operation, design, update, and optimization of the fuel assembly; it can provide critical inspection and verification ways to the fuel assembly. Some theoretical research has been carried out in China, but systematic solutions and the capability of equipment supply are not arrived at yet. Some of the nuclear power plants purchase inspection and test equipment in foreign markets, but it is difficult and expensive to purchase spare parts; moreover, when the structure of the fuel assembly is updated, it will cost a lot to update the inspection and test equipment. China exhibits rapid nuclear power development, but the capability of inspection and test technology of the post-irradiation fuel assembly is hysteretic seriously, as it has been a stone of the development of the fuel area. Therefore, it is necessary to speed up the self-research and self-design of inspection and test technology of the post-irradiation fuel assembly. It can meet the requirement of nuclear plants for the purpose of fuel safety; meanwhile, it can also improve the capability of self-design and self-manufacture of the fuel assembly.

REFERENCES

Jin-lu Liu, Yin Wang, and Deng-fei Xian, "Evaluation of nuclear fuel assembly deformation," Mechanical & Electrical Engineering Technology. Guangzhou, vol.31 No.1, pp. 70–72, January 2008.

Jun-xian Deng, Xi-juan Zhao and Ye-xiao Li, "Inspection, reparation and reconstitution of PWR," Nuclear Physics Review, Shanghai, vol.16 No.2, pp.126–130, June 1999.

Qi-zhen Ye, Xiao-ming Li and Zhong-de Yu. China Electrical Engineering Canon. Beijing, CEPP, 2009.

Qiu-ping Shen, Zhi-qing Chen and Dao-ping Xu, "Design of fuel subassembly repairing device," Nuclear Techniques, Shanghai, vol.33 No.2, pp.148–151, February 2010.

Yong-ming Gao, Sheng Li and Li-dan Li, "A Method for Survey of Underwater Non-contact Deformation of fuel assemblies," Nuclear Power Engineering, Chengdu, vol.31 No.4, pp.87–90, August 2010.

Zhi-xin Deng, Ling-bin Wang, "Testing, Analysis and Countermeasurement Study for Nuclear Fuel Radioactive Distortion of Qinshan II," China Nuclear Power, vol.6 No.4, pp.307–311, December 2013.

Advances in Energy, Environment and Materials Science – Wang & Zhou (Eds)
© 2017 Taylor & Francis Group, London, ISBN 978-1-138-03600-0

Research on nonlinear control for bi-directional DC/DC converters in energy storage systems

S.L. Ma, M.X. Chen, L. Huang & J.W. Wu
School of Automation Science and Electrical Engineering, Beihang University, Beijing, China

P. Wang
Changping Power Supply Company, Beijing, China

ABSTRACT: In recent years, a distributed optical power generation system has been developed and the power of the power storage tracking source has become a hot research topic. In this paper, we first establish an energy storage system in a non-isolated bidirectional DC/DC converter based on a nonlinear model and obtain the large signal modeling of nonlinear state space description; and then, according to the state feedback linearization conditions, the nonlinear mapping of nonlinear space to linear space is used and the state feedback coefficient is calculated. Finally, verified by using digital simulation, the control method proposed in this paper can accurately and quickly realize the power command tracking function, and in the voltage disturbance, load mutation, inaccurate parameters etc. the case reflects strong robustness. The simulation results verify the correctness and validity of the control method proposed.

1 INTRODUCTION

Distributed generation can make full use of clean and renewable energy sources and centralized power generation complementing each other is an effective means to solve the problem of energy and environment [1–3]. Under the premise of ensuring the safe and stable operation of the power grid, it is required to develop the permeability of the distributed generation. At present, the effective technical scheme is to join the energy storage system and reasonable and effective control of its balance of distributed new energy power generation, large power grid, and user load between the power demands of the three. The design and control of the power electronic converter in the energy storage system has become the key technology of the development of the distributed generation system [4–6].

2 BIDIRECTIONAL DC/DC CONVERTER

2.1 Bidirectional DC/DC topological structure

The bidirectional DC/DC converter is divided into two categories according to whether they are segregated. The utility model of a non-isolated bidirectional DC/DC converter, as shown in Figure 2, has the advantages of simple structure, less number of switches, and simple control, but the output power and voltage level cannot be too high, and is more suitable for low voltage and low power level of distributed generation applications.

2.2 Copying old text onto a new file

In Figure 1, S1 and S2 are switching devices, l is the inductance, C is the capacitance of Ub said energy storage voltage between the two poles of the battery, the UDC said the voltage across the capacitor, namely the DC bus voltage, within the dotted box with a DC voltage source and a series resistance to simulate high voltage side power supply, comprising a photovoltaic electricity pool DC bus access and inverter DC bus output in two parts. Ueq is the simulation of the power supply voltage on the high-voltage side and R is the simulation of the high voltage side equivalent resistance. The switch devices S1 and S2 are complementary, and when the S1 works in the Buck mode, the system can charge the energy storage battery; the switch device S2 works in the Boost mode and the energy storage battery discharges in this mode.

2.3 Large signal modeling and analysis

For the pulse width modulation type DC/DC converters, the most common modeling method is the state space averaging method. The traditional state space average model of the switching converter is proved to be the Euler Lagrange (Euler-Lagrange, EL) dynamic system in the literature. According to

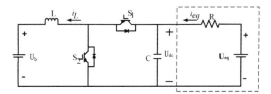

Figure 1. Main circuit of Bi-directional DC/DC converter.

the bidirectional DC/DC circuit topology and the state space averaging method, the nonlinear model of the non-isolated bidirectional DC/DC converter can be established:

$$\begin{cases} L\dfrac{di_L}{dt} = (1-d) \times U_{dc} - U_b \\ C\dfrac{dU_{dc}}{dt} = (d-1) \times i_L + \dfrac{U_{eq} - U_{dc}}{R} \end{cases} \quad (1)$$

where Udc and iL, respectively are the average value of the inductor current and capacitor voltage, the duty cycle of the switch D and the duty cycle of the switch S1 (1-D).

If [iL, Udc] is selected as the state variable, the single input and single output affining the nonlinear system of the non-isolated bidirectional DC/DC converter based on the differential geometry method can be sorted out according to equation (1). However, in the distributed generation system, the function of the bidirectional DC/DC converter is to track the parameters of a given power instruction, which is not reflected by the structure of the PI control and the control effect is limited. When the power command Pref is positive, the energy storage system is charged; when Pref is negative, the energy storage system is discharged. According to the 1.2 topological analysis, we can see that the tracking of Pref can be transformed into the tracking output voltage Udc

$$\begin{cases} \dot{e} = f(e) + g(e)u \\ y = h(e) = e_2 \end{cases} \quad (2)$$

By using the nonlinear control theory and equation (2) system description, the following equation can be obtained:

$$\begin{aligned} L_g \omega(e) &= \dfrac{\partial \omega}{\partial e} g(e) = \dfrac{\partial \omega}{\partial e_1}\left(\dfrac{1}{L}(e_2 + U_{ref})\right) \\ &+ \dfrac{\partial \omega}{\partial e_2}\left(-\dfrac{1}{C}(e_1 + I_{ref})\right) = 0 \end{aligned} \quad (3)$$

Solving equation (3) can be a solution to yield the following equation:

$$\begin{cases} \dot{e} = f(e) + g(e)u \\ \bar{y} = \omega(e) = Le_1^2 + Ce_2^2 + 2Le_1 I_{ref} + 2Ce_2 U_{ref} \end{cases} \quad (4)$$

On the new state space description of the output equation, coordinate transformation, as in equation (4), occurs with nonlinear mapping system and linear system under the new coordinate system, as shown in equation (5), Z for linear systems with state variables and V for the linear system as the control variables.

$$\begin{aligned} z = \phi(e) &= \begin{bmatrix} \omega(e) \\ L_f \omega(e) \end{bmatrix} \\ &= \begin{bmatrix} Le_1^2 + Ce_2^2 + 2Le_1 I_{ref} + 2Ce_2 U_{ref} \\ -2U_b(e_1 + I_{ref}) + 2(e_2 + U_{ref})(U_{eq} - e_2 - U_{ref})/R \end{bmatrix} \end{aligned} \quad (5)$$

$$\dot{z} = \begin{bmatrix} 0 & 1 \\ 0 & 0 \end{bmatrix} z + \begin{bmatrix} 0 \\ 1 \end{bmatrix} v \quad (6)$$

3 SIMULATION ANALYSIS

3.1 Simulation parameters

In order to verify the state feedback linearization method, control the storage can power command by tracking the superiority, based on the MATLAB/Simulink simulation platform. First, verify that control parameter K and zeta influence on control effect are in line with the analysis of 2.3. Then, assuming the control parameter k = 20, zeta = 5, according to the system start process, high side voltage fluctuation and load change are analyzed (i.e. power flow direction and amplitude changes and changes of system parameters of the battery charging and discharging process in simulation are analyzed, in order to illustrate the effectiveness and robustness of the method). The solid line represents a given target, the dotted line represents the tracking curve, and the system parameters are shown in Table 1.

Table 1. Simulation parameters.

Setting	Value
Udc (V)	650
Ub (V)	216
Pb (kW)	5
Qb (Ah)	1.5
L (mH)	2
C (uF)	200
Req (Ω)	10
Frequency (kHz)	10

3.2 Simulation result

According to the simulation conditions of Table 1, different control parameters are set up to simulate and the power tracking command is +5 kW when the time is assumed to be 0.5 s. The control parameters affect the response performance of the system, as shown in Figure 2.

From Figures 2 and 3, it can be seen that, with the increase of K rise time and decrease of adjusting time, the steady-state accuracy is improved; when k is fixed, with the decrease of zeta, the system overshoot and oscillation response are significantly faster; and when k = 0, the system oscillation of linear periodic oscillation T = 1.15 ms. Analysis conclusion is consistent with that of the simulation results with the previous control parameters

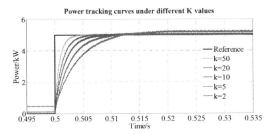

Figure 2. The impact contrast diagram of control parameters (K = 5).

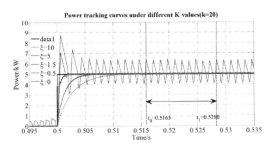

Figure 3. The impact contrast diagram of control parameters (K = 5).

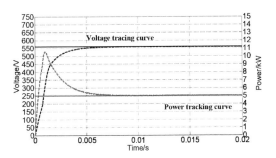

Figure 4. Voltage tracking, power tracking, and battery SOC in the starting process.

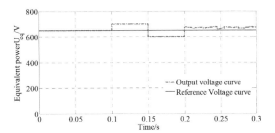

Figure 5. Voltage tracking, power tracking, and battery SOC in DC voltage fluctuation.

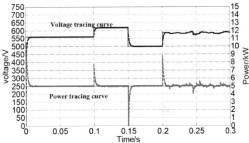

Figure 6. Voltage tracking, power tracking, and battery SOC in DC voltage fluctuation (voltage and power).

and can be combined with control parameters that affect the conclusions and the actual needs of the reasonable choice of control parameters K and zeta.

According to the simulation condition, by assuming power by follow the instructions for +5 kW, initial battery State of Charge (SOC) SOE = 50%, the battery discharge is on the DC side, the control effects are as shown in Figure 5 and 6. The solid lines represent given values and dotted lines represent actual values.

From Figure 5 can be seen that during system boot, does not exceed the tone or shock, resulting in start voltage being too high or volatile, and in the 4~6 ms time reached with the command tracking; the tracking precision is high, can meet the rapid and steady state performance of the system requirements, and is in line with distributed generation system of the energy storage system to track the response of requirements.

4 CONCLUSIONS

In this paper, a nonlinear control method for a non-isolated bidirectional DC/DC converter for a distributed photovoltaic energy storage power generation system is presented. The following conclusions are drawn from this work:

1. For non-isolated bidirectional DC/DC converter nonlinear systems and the state feedback linearization, the large signal modeling and the state feedback controller of the static working point of the system are not required and used in energy storage systems frequently under charge and discharge conditions.
2. The controller design presented in this paper is not sensitive to the system parameters; it has strong robustness, strong anti-interference ability to the DC high voltage and side voltage disturbance, load mutation, and so on;
3. The control method used in this paper has good starting performance, can quickly and accurately exhibit control, shows no adjustment of tracking power command, obtains a good control effect, can effectively achieve an energy storage unit and a DC high voltage side of the bidirectional energy flow, and has a good application prospect to the power system in distributed photovoltaic storage.

ACKNOWLEDGMENT

This paper is supported by the National Natural Science Foundation of China (51377007) and the Specialized Research Fund for the Doctoral Program of Higher Education of China (20131102130006).

REFERENCES

Kimjy, Jeoh J H, Kim S K. Cooperative control strategy of energy storage system and microsources for stabilizing the microgrid during islanded operation [J]. IEEE Trans Power Electronics, 2010, 25(12): 3037–3048.

LI Peng, Lehman B.A Simple Design for Paralleling Current-Mode Controlled DC-DC Converters [J]. IEEE APEC, 2003: 898–904.

Liang Liang, Li Jianlin. Operating modes of photovoltaic/energy-storage hybrid system and its control strategy [J]. Electronic Power Automation Equipment, 2011, 31(8): 20–23.

Paraskevadaki E V, Papathanassiou S A. Evaluation of MPP voltage and power of mc-Si PV modules in partial shading conditions [J]. IEEE Transactions on Energy Conversion, 2011, 26(3): 923–932.

Patel H, Agarwal V. Maximum power point tracking scheme for PV systems operating under partially shaded conditions [J]. IEEE Trans. on Industrial Electronics, 2008, 55(4): 1689–1698.

Wong Pit-leong, Lee F C, JIA Xiao-chuan, et al. A Novel Modeling Concept for Multi—coupling Core Structures [J]. IEEE APEC,2001.102–108.

Advances in Energy, Environment and Materials Science – Wang & Zhou (Eds)
© *2017 Taylor & Francis Group, London, ISBN 978-1-138-03600-0*

System analysis of scale pig breeding and biogas energy development

Bi-Bin Leng
*School of Economics and Management, Jiangxi Science and Technology Normal University,
Nanchang, Jiangxi, China*
*Center for Central China Economic Development Research, Nanchang University, Nanchang,
Jiangxi, China*

Cong Chen
*School of Economics and Management, Jiangxi Science and Technology Normal University,
Nanchang, Jiangxi, China*

Qiao Hu
*Center for Central China Economic Development Research, Nanchang University, Nanchang,
Jiangxi, China*

ABSTRACT: To explore the effective pattern of waste disposal in scale pig breeding, on the basis of analyzing waste pollution and secondary pollution in scale pig breeding through SD, the biogas energy shortage in households and lack of a technical service system, a scale pig breeding and household biogas resources development mode is built. The results of the SD system archetype analysis showed that the development model realized the anti-constraint transformation for the constraint of negative feedback in the problems of new pollutants of scale pig breeding, the shortage of household biogas resources, and the lack of a technological service system.

1 INTRODUCTION

Breeding industry is one of the most important industries in agriculture. Pig breeding and sustainable development is of great significance for the development of the rural economy in China. Since the reform and opening up, with the development of the national economy, the improvement of living standards and the increase of population, the residents' demand of pork consumption have been expanded. At present, China is the largest producer of pork and its pork production has been accounted for about 50% in the world's total output. In 2014, in order to achieve the appropriate scale of operation of open space for development, Document NO.1 of the central government separated the management right from the managerial right, thereby achieving a major breakthrough in theory and practice. Nowadays, large-scale and concentrated degree of big breeding has become more and more popular. In 2010, the proportion of pig farms increased from 23.2% in 1998 to 64.51%, and the scale of evolution has deepened gradually. Research also pointed out that the scale of farming adopted feeds in factory to concentrated raise, which has greatly saved the cost of work force, field, etc. and drew a conclusion that the scale breeding is superior to the backyard breeding. With the deepening of the evolution in scale pig breeding, scale breeding produces a lot of highly concentrated and uneasily digested and absorbed fecal pollution, which leads to the problem of environmental pollution. Therefore, the environmental problems in scale pig breeding became the new problem in scale pig breeding.

A large number of scholars have carried out a lot of research toward pig breeding waste, but from the current situation of the process of waste disposal in scale breeding, pollution is still a serious problem. The System Dynamics (SD) is a kind of method based on the feedback control theory and it is the analysis method of the complex social economy and ecological environment system; since the 1950s, the United States Massachusetts Institute of Technology Forester (Jay Forrester W) Professor founded many strategies and decisive analysis etc. These have been successfully applied in countries and regions and even in the world scale, which is known as the "strategy and decision lab". Pig breeding and the problem of waste disposal are exactly a complex social, economic, and ecological environment system. The SD's in-depth analysis of scale pig farming waste pollution, secondary pollution, households on the basis of deficiency of the biogas energy shortage, and the technical service system is used to explore an effective mode of scale pig breeding waste processing and the development of students, thereby promoting the sustainable development of pig breeding.

2 NEW ENVIRONMENTAL POLLUTANTS IN THE SCALE PIG BREEDING RESTRICTIVE ARCHETYPE

2.1 *The pollution problems from excess biogas*

According to the problems, such as the emission of pollutants and the pollution of excess biogas in scale pig breeding, a biogas excess growth ceiling archetype is built in this paper, as shown in Figure 1.

Figure 1 shows the following positive feedback loop "breeding scale $\xrightarrow{+}$ the government support $\xrightarrow{+}$ biogas construction scale $\xrightarrow{+}$ scale breeding benefit $\xrightarrow{+}$ breeding scale", which revealed that as countries increasingly support scale pig breeding, scale breeding has developed rapidly. The scale breeding enterprises used policies supported by the government and through the large and medium-sized methane project construction and constantly expanding farm scale, many large farms and the farming villages were formed. The effectiveness of the biogas engineering benefit growth positive feedback loop is shown in Figure 1.

The following negative feedback loop in Figure 1 "breeding scale $\xrightarrow{+}$ Pig urine amount $\xrightarrow{+}$ degree of pollution of the environment $\xrightarrow{+}$ breeding scale" revealed that as countries continually support the growing scale breeding pigs, scale aquaculture has developed rapidly. The scale breeding enterprises used government-supported policies, through large and medium-sized methane project construction and constantly expand the scale of farming, thereby leading to the formation of many large farms and the farm-ing community. Although the construction of a biogas engineering unit can eradicate pollution and dispose part of the farm manure, due to the constant expansion of the scale, the ineffective disposal of most pig manure, and the limitation of the breeding area of the arable land, farm manure cannot be digested fully, which causes a lot of excess pig urine, enable an unbalance in the nutrients in the farming area and severely pollute soil and water in the scale breeding area, thus forming the negative feedback loop caused by pig's breeding scale and biogas excess growth ceiling, as shown in Figure 1.

2.2 *The secondary pollution problems caused by Biogas slurry and biogas utilization ability are insufficient*

According to the discharge of wastes and the secondary pollution of the biogas engineering problem in the scale of pig breeding, the secondary pollution scale biogas slurry for large and medium-sized methane project biogas secondary pollution growth ceiling archetype is built in this paper, as shown in Figure 2.

The following negative feedback loop "breeding scale $\xrightarrow{+}$ The government support $\xrightarrow{+}$ biogas construction scale $\xrightarrow{+}$ the biogas production of biogas slurry $\xrightarrow{+}$ the remaining amount of biogas slurry $\xrightarrow{+}$ degree of environmental pollution $\xrightarrow{}$ scale breeding efficiency $\xrightarrow{+}$ breeding scale" revealed that through the construction of a large and medium-sized methane project and the anaerobic fermentation to partial feces caused by scale pig breeding, which reduces the pollution of direct emissions in farm manure, at the same time, produce a lot of biogas slurry and biogas. The scale enterprises utilization ability of biogas slurry and biogas is insufficient, leading to the secondary pollution of biogas slurry and biogas, which causes the second-pollution scale, as shown in Figure 2.

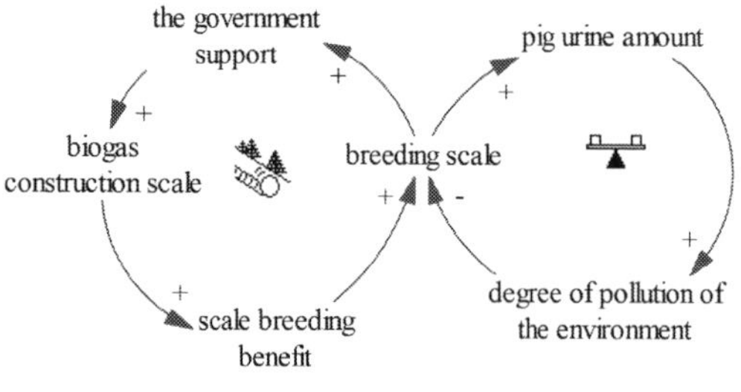

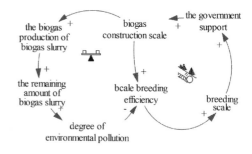

Figure 1.   Schematic of the scale pig breeding biogas feedstock excess growth ceiling archetype.

Figure 2.   Schematic of the scale biogas engineering and biogas slurry methane's secondary pollution growth limit negative feedback loop.

3 HOUSEHOLD BIOGAS FEEDSTOCK SHORTAGES AND INADEQUATE TECHNICAL SERVICE SYSTEM DYNAMICS CONSTRAINTS ARCHETYPES

China has paid continuous attention to and support the development of clean energy sources in rural areas and household biogas project construction. The rural biogas engineering construction has become an important national strategic move and people's livelihood project for changing the Chinese rural energy structure, thereby developing low-carbon agriculture, reducing the rural non-point source pollution, and protecting the ecological environment. However, with the expansion of household biogas construction, a fair amount of household biogas failed to exhibit its real effectiveness, and even appeared as a waste phenomenon of the household biogas tank in a short period of time. The main reasons are firstly due to the growing number of migrant farmers, the shortage of supply of pig labor in rural households, and the lack of raw materials for biogas; secondly, the lack of technical services. Most biogas farmers do not acquire a comprehensive utilization technology and the household biogas possesses low benefits. The archetype is as shown in Figure 3.

As can be seen from the above, the following positive feedback loop "household biogas construction scale $\xrightarrow{+}$ household biogas efficiency $\xrightarrow{+}$ government performance $\xrightarrow{+}$ the government supports household biogas $\xrightarrow{+}$ household biogas construction scale" revealed that with policy supporting national rural biogas construction land, rural household biogas developed rapidly. As an important livelihood project, the household biogas engineering construction to improve the efficiency of agricultural production and improving farmers' quality of life played a huge role in the development of rural courtyard ecological economy and has played a positive propulsion, prompting further development of the rural household biogas.

The following negative feedback loop "household biogas construction scale $\xrightarrow{+}$ demand of technical support service $\xrightarrow{+}$ technical service demand shortage $\xrightarrow{-}$ biogas use effect $\xrightarrow{+}$ farmers' enthusiasm for construction $\xrightarrow{+}$ household biogas construction scale" and the negative feedback loop "household biogas construction scale $\xrightarrow{+}$ biogas manure feedstock demand $\xrightarrow{+}$ biogas manure feedstock shortfall $\xrightarrow{-}$ biogas use effect $\xrightarrow{+}$ farmers' enthusiasm for construction $\xrightarrow{+}$ household biogas construction scale" reveal that with the growing number of migrant farmers, rural households without pig labor is in short supply and the lack of raw materials for the biogas; secondly, the lack of technical services and most of the comprehensive utilization of biogas farmers do not have technology; household biogas is with low benefit. In the expansion of the household biogas construction unit at the same time, due to the universal problems constructed quite a number of household biogas production units which failed to play their real benefits.

4 SCALE PIG FARMING AND ANTI-CONSTRAINTS TRANSFORM IN THE ARCHETYPES OF HOUSEHOLD BIOGAS DEVELOPMENT SYSTEMS

With the development of scale pig breeding, pig breeding has become the main mode of pig breeding, but the scale of pig breeding by using pig urine as the main raw material of rural biogas fermentation engineering as the huge potential for biogas production. In 2010, more than 50 years market scale cultivation of marketable fattened stock up to 602.504 million head, according to the V (biogas annual output) = X (pig manure and urine) * TS (the rate of dry gas is 257.3 m3/t) * A (the rate of dry matter and pig manure is 18%, and the rate of pig urine is 3%) computing (1.38 kg per emission of pigs manure, and pig urine is 2.12 kg), 2010 (more than 50 heads) tier of pig scale cultivation biogas from the pig urine volume of 17654164915 cubic meters, 0.714 kg with 1 cubic meter of methane of standard coal equivalent calculation, 2010 pigs scale cultivation waste produced biogas is equivalent to 12605073.75 tons of standard coal; visible pig breeding scale waste disposal possesses

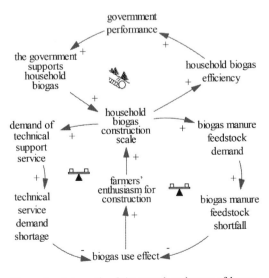

Figure 3. Schematic of the growth archetype of household biogas manure urine material shortage and technical service shortage.

huge potential for biogas production. It shows that the scale pig breeding and the household biogas development system provides realistic conditions.

4.1 *Innovation mode of cooperation between pig scale cultivation and household biogas production*

The pig breeding scale environmental pollution system dynamics restricting the base model reflects the pig breeding size restriction system of environmental pollution. The shortage of household biogas and insufficient technical service reflect the household biogas shortage of raw materials and the technical service system, the comprehensive scale pig breeding waste disposal and household biogas engineering problem. We found that new problems in the disposal of scale pig breeding waste provide household biogas resources development. Pig breeding scale enterprises can provide the conditions for household biogas resource development. The scale farming enterprises can provide household biogas with waste which can not be included by farming enterprises and can be used as the raw materials in the household biogas tank. And these enterprises have the ability to provide technical and capital support for the perfection of the biogas service system; but the household biogas tank provides the processing places for the increasing amount of waste of the pig breeding scale.

The scale of pig breeding and household biogas resources cooperation development mode are a complex system, and then, we apply the SD model to analyze the pig breeding scale environmental pollution and the shortages of the household biogas system as well as the technical service system of the merger and anti-constraint transformation.

4.2 *Pig breeding scale and household biogas cooperation development system of the reverse restraint transformation archetype*

In this paper, the pig breeding scale environmental pollution is combined with the shortage of the household biogas system and the technical service system. Figures 1–3 were combined to produce the model of the scale breeding pigs and household biogas development system archetype, as shown in Figure 4.

According to the above analysis and the actual research scenario, one is the scale pig farming in the construction of the large and medium-sized biogas project which is performed after the existing farming manure biogas large surplus is subjected to harmless treatment, a large number of organic fertilizers and energy- efficient biogas produce large and medium-sized biogas projects which cannot be used effectively, thereby posing a serious threat to the ecological environment; but on the

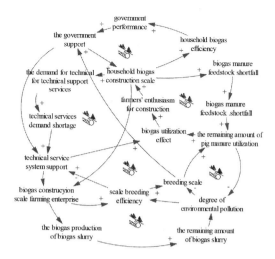

Figure 4. Schematic of the pig breeding scale cooperation with household biogas development system archetype.

other hand, the lack of household biogas appeared due to the shortage of raw materials and technical services. A crucial reason for this contradiction is the lack of effective cooperation in the development of pig breeding enterprises and households among biogas farmers, resource development, and utilization mode feedback. Due to the shortage of current scale pig farming enterprises and household biogas farmers cooperation of development, as a result of which it is a harm to the interests of the development of the large-scale breeding process, breeding companies, and local farmers. And the goal of government's environmental protection is faced with the challenge, which has seriously affected the development of the local rural economy and the construction of a harmonious society. From the scale of pig farming and household biogas cooperation development system dynamics, we can make the following observations:

1. The following original scale pig breeding environment pollution negative feedback loop system dynamics constraints archetype "breeding scale $\xrightarrow{+}$ biogas manure feedstock shortfall $\xrightarrow{+}$ degree of environmental pollution $\xrightarrow{+}$ breeding scale" now into the positive feedback loop "breeding scale $\xrightarrow{+}$ the government support $\xrightarrow{+}$ household biogas construction scale $\xrightarrow{+}$ biogas manure feedstock demand $\xrightarrow{+}$ biogas manure feed stock shortfall $\xrightarrow{+}$ the remaining amount of pig manure utilization $\longrightarrow$ degree of environmental pollution $\longrightarrow$ breeding scale", the evolution of the transformation process is as follows:

The transformation revealed that in scale pig breeding and household biogas development

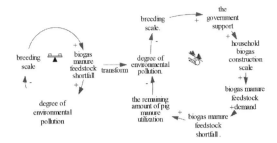 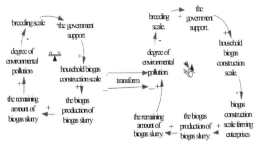

Figure 5. Pollution control and transformation of pig scale farming and the household biogas cooperative development system.

Figure 6. The secondary pollution control conversion of pig scale cultivation and the household biogas cooperative development system.

cooperation system, under the support of government policy, the remaining amount of pig manure produced by the scale of the breeding household biogas project has been absorbed, thus reducing the pollution of the environment and promoting the continuous growth of pig breeding. Pig breeding and biogas development cooperation system has its anti-constraint effect on scale pig farming waste disposal thereby causing excess pollution.

2. The following negative feedback loop "breeding scale $\xrightarrow{+}$ the government support $\xrightarrow{+}$ household biogas construction scale $\xrightarrow{+}$ the biogas production of biogas slurry $\xrightarrow{+}$ the remaining amount of biogas slurry $\xrightarrow{+}$ degree of environmental pollution $\xrightarrow{-}$ breeding scale" now converted into the following positive feedback loop "breeding scale $\xrightarrow{+}$ the government support $\xrightarrow{+}$ household biogas construction scale $\xrightarrow{-}$ biogas construction scale farming enterprises $\xrightarrow{+}$ the biogas production of biogas slurry $\xrightarrow{+}$ the remaining amount of biogas slurry $\xrightarrow{-}$ breeding scale" the evolution of the transformation process is as follows:

Fig. 6 revealed that in the scale pig farming and household biogas development cooperation system, under the support of the government policy, households with biogas development the diversion of farming enterprises, the expansion of biogas engineering scale, thus reduce the farming enterprises biogas utilization remaining amount and avoid the secondary pollution of the pig farm scale biogas engineering; scale pig farming and the household biogas development cooperation system of farming enterprises large and medium-sized biogas project methane biogas slurry pollution with anti-constraint.

3. The original household biogas material shortage and technical service shortage system dynamics restriction on the following negative feedback loop "household biogas construction scale $\xrightarrow{+}$ biogas

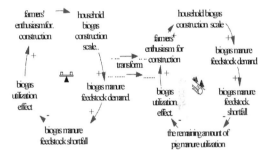

Figure 7. The anti-constraint transformation of the household biogas energy and technical shortage of the pig scale farming and household biogas cooperative development system.

manure feedstock demand $\xrightarrow{+}$ biogas manure feedstock shortfall $\xrightarrow{-}$ biogas utilization effect $\xrightarrow{+}$ farmers' enthusiasm for construction $\xrightarrow{+}$ household biogas construction scale" is transformed into a positive feedback loop "household biogas construction scale $\xrightarrow{+}$ biogas manure feedstock demand $\xrightarrow{+}$ biogas manure feedstock shortfall $\xrightarrow{+}$ the remaining amount of pig manure utilization $\xrightarrow{+}$ biogas utilization effect $\xrightarrow{+}$ farmers' enthusiasm for construction $\xrightarrow{+}$ household biogas construction scale". The evolution of the transformation process is as follows:

The cooperative development system revealed that in the scale pig farming and household biogas development cooperation system, the government meanwhile has obtained the political achievements, making the scale pig farming and household biogas system developed supported by policies, with an increase of breeding in scale pig breeding enterprises and constructive intensity of the construction of the biogas technology service system, the two common problems of the existing energy shortages and the technical demand in household biogas has been effectively solved, which promotes the utilization efficiency and continuous development in household biogas; the scale pig breeding and household biogas development cooperation system of households with biogas energy and technology shortage has its anti-restriction. In summary, the pig breeding and household archetype with the methane cooperative development system revealed that the scale pig farming households with biogas development cooperation can not only solve the scale pig farming pollution problem, but also improve production efficiency, increase the income of the farmers, and realize the goal of the project of improving the quality of life of the rural life of the important people's livelihood project.

5 CONCLUSION AND PROSPECTS

In 2014, the "Document NO.1" separates the right to the contracted management, which achieved a major breakthrough in theory and practice as well as further expanding the development space for the realization of the moderate scale management. The scale of pig breeding as a kind of inevitable trend of pig breeding, with the expansion of pig breeding and the increase of scale farms, the environmental problems in pig scale breeding are increasingly prominent. Through SD archetype and scale pig breeding waste disposal through gas resources development mode innovation, the scale pig farming and household biogas development innovation mode of cooperation was proposed in this paper, which is an effective mode of waste disposal and resource circulation development.

The development mode innovation in the scale of pig breeding waste disposal by using gas resources is a new exploration for the scale of pig breeding waste management mode. It is based on the SD archetype of pig breeding and household with analysis of the corporation pattern of biogas, which mainly demonstrates the effectiveness of the development mode from the model of complex systems of mutual feedback effects; but the empirical research of the stability and influential elements

in the proceeding of the scale pig breeding and household biogas resources cooperation development mode needs a further in-depth investigation, which is also the direction and focus in the next step research.

ACKNOWLEDGMENTS

The authors gratefully acknowledge the grant of project Scale Pig Breeding Ecological Energy System Stability Feedback Simulation Study (71501085) supported by the National Natural Science Foundation of China and the Project of Humanities and Social Sciences in Colleges and Universities by Jiangxi Province (GL1536).

REFERENCES

Bhattacharya, S.C., P. Abdul Salam. Low Greenhouse Gas Biomass Option for Cooking in the Developing countries [J], Biomass and Bio Energy, 2002, 22(24).

Chenglin, Ma, Zhou Deyi, The Empirical Study Toward The Growth of Scale Pig Breeding and The Changes in Breeding Technology [J], Journal of Hua Zhong Agricultural University, 2014, (4).

Feng Yonghui, The Scale Pig Breeding Pig and the Area Lay out Trend in Our Country [J], Chinese Journal of Animal Husbandry 2006, (4).

Guoping Tu, Leng Bibin, Jia Renan. System Archetype of the "Company & Farmer & Futures, Options" Basing on the Core of Archetype Generating Set. [J]. System Engineering Theory and Practice, 2011, 31 (5).

Leng Bibin, Tu Guoping, Jia Renan. Scale Pig Breeding and Household Biogas Development System Dynamic Stability Basing on SD Evolutionary Game Model [J], System Engineering, 2014, (3).

Wang Chen, He Zhongwei, Gaoran, etc. Analysis of the Technical Efficiency in Pig Production in Our Country: The Empirical Research Based on DEA Model [J], Agricultural Outlook 2012, (2).

Wang Cuixia, The Feedback Analysis of Circular Economy Growth Ceiling System in Scale Breeding [J], System Engineering, 2007, 25(5).

Zhang Xingyue, The Influence Factor Analysis of Evolution of Scale Pig Breeding in China [D]. Hangzhou: Zhejiang University, 2013.

Advances in Energy, Environment and Materials Science – Wang & Zhou (Eds)
© *2017 Taylor & Francis Group, London, ISBN 978-1-138-03600-0*

Experimental research on the influence of the excess air coefficient on boiler heat loss

Qin Cai

Chongqing Special Equipment Inspection and Research Institute, Chongqing, China

ABSTRACT: The influence of excess air coefficient on heat loss from exhaust gas, heat loss from gas incomplete combustion, and heat loss from solid incomplete combustion was analyzed by carrying out the thermal balance experiment of the boiler. The results showed that heat loss from exhaust gas increased with an increase in the excess air coefficient. Heat loss from gas incomplete combustion and heat loss from solid incomplete combustion first decreased and then increased. There existed an optimum excess air coefficient corresponding to the minimum heat loss of the boiler. As a result, the excess air coefficient has an important effect on the thermal efficiency of the boiler. Under the condition of complete combustion of adequate oxygen and fuel as much as possible, the excess air coefficient is lower and the combustion is more economical.

1 INTRODUCTION

Industrial boilers are important thermal power equipment, mainly used in the field of factory power, building heating, and so on (Sandro et al. 2015). As heating equipment, energy consumption of industrial boilers accounts for a large proportion, which is only less than that of power station boilers. At present, more than six hundred thousand industrial boilers exist in China. Every year, about six hundred million tons of standard coal is consumed by industrial boilers. Due to various reasons, in China, the average operating efficiency of the industrial boiler is mostly between 60% and 65%. When compared with developed countries, the average operating efficiency of the coal boiler is commonly 10~15% which is below the value of efficiency (Saidur et al. 2010). Therefore, it has become an important problem to improve the energy efficiency and reduce the heat loss of industrial boilers (Wang et al. 2013, Shyan et al. 2010).

Excess air coefficient is an important index to evaluate the efficiency of boilers (Zhao et al. 2015). In the process of fuel combustion, adequate air is used to complete the violent oxidation reaction. If the value of excessive air coefficient is not controlled properly, the combustion condition of the boiler's furnace will be poor (Li et al. 2013). Such energy will be greatly wasted and the atmospheric environment will be polluted. Therefore, the excess air coefficient has an important influence on the combustion condition and economic operation of the boiler (Vladimir 2005).

2 HEAT LOSS OF THE BOILER

In order to ensure a good combustion condition, the actual amount of air present in the boiler furnace is much more than the theoretical amount. The amount of excess air is measured by the ratio of the actual air volume and the theoretical air volume, which is called the excess air coefficient (Li et al. 2015).

For coal boilers, the excess air coefficient described by the compositions of the exhaust gas is given by the following equation:

$$\alpha_{py} = \frac{21}{21-79\dfrac{O_2'-0.5CO'}{100-RO_2'-O_2'-CO'}} \tag{1}$$

where O_2', CO', and RO_2' are O_2, CO, and RO_2 in the component content of exhaust gas, respectively.

Most of the heat absorbed by the boiler is used to produce steam, which is the effective heat. The rest is the heat loss of the boiler (Seung et al. 2016). The thermal balance equation is expressed as follows:

$$q_1 + q_2 + q_3 + q_4 + q_5 + q_6 = 100\% \tag{2}$$

where q_1 is the percentage of effective utilization heat in the boiler, q_2 is the heat loss from exhaust gas, q_3 is the heat loss from gas incomplete combustion, q_4 is the heat loss from solid incomplete combustion, q_5 is the surface heat loss, and q_6 is the heat loss from coal slag and ash.

q_2 is the percentage of the heat that is taken away by the exhaust gas in total input heat. It is calculated by using the following equation:

$$q_2 = \frac{K_{q4}}{Q_r}(h_{py} - h_{lk}) \times 100 \qquad (3)$$

where K_{q4} is the correction factor. Q_r is the total heat released into the boiler, h_{py} is the enthalpy of exhaust gas, and h_{lk} is the enthalpy of air into the furnace.

The heat loss from gas incomplete combustion exists because the heat of combustible gas composition is not released completely. It is obtained from the following equation:

$$q_3 = \frac{V_{gy}K_{q4}}{Q_r} \times \left(126.36CO' + 107.98H_2'\right.$$
$$\left. + 358.18C_mH_n'\right) \times 100 \qquad (4)$$

where V_{gy} is the volume of dry flue gas and CO', H_2' and C_mH' are CO, H_2, C_mH_n in the component content of exhaust gas, respectively.

The heat loss from solid incomplete combustion exists because the combustible component of the coal slag and ash is not burned completely (Wu et al. 2015). It is expressed by the following equation:

$$q_4 = \left(\alpha_{lz}\frac{C_{lz}}{100 - C_{lz}} + \alpha_{lm}\frac{C_{lm}}{100 - C_{lm}}\right.$$
$$\left. + \alpha_{fh}\frac{C_{fh}}{100 - C_{fh}}\right) \times \frac{328.664A_{ar}}{Q_r} \qquad (5)$$

where α_{lz}, α_{lm}, and α_{fh} are the percentages of coal slag, coal leakage, and fly ash in the total amount of coal ash, respectively; C_{lz}, C_{lm}, and C_{fh} are the combustible contents of the coal slag, coal leakage, and fly ash, respectively; and A_{ar} is the ash content of the coal as received.

For running boilers, the thermal efficiency of the boilers could be obtained through the thermal balance experiment. And then, the heat loss of the boiler is determined. An effective way could be obtained to improve the operating conditions and the thermal efficiency of the boiler (Brundaban 2016).

3 EXPERIMENTAL SYSTEM

In this paper, the influence of excess air coefficient on heat loss was analyzed by carrying out the thermal balance experiment of the coal boiler. The test items mainly included evaporation capacity

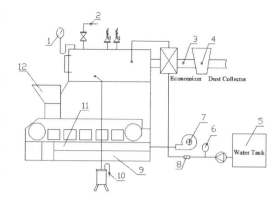

Figure 1. Measuring point diagram of the thermal balance experiment.
1-Steam pressure, 2-Steam sampling, 3-Temperature and composition of the exhaust gas, 4-Fly ash sampling, 5-Feed water temperature, 6-Feed water pressure, 7-air temperature, 8-Feed water flow, 9-Coal slag sampling and weighing, 10-Boiler water sampling, 11-Coal leakage sampling, 12-Fuel consumption.

of the boiler, the compositions of the exhaust gas, the temperature of the exhaust gas, the excess air coefficient, and the thermal efficiency of the boiler. Evaporation capacity of the boiler was measured by using an ultrasonic flowmeter. In order to carry out the ash balance calculation, the coal slag and coal leakage was weighed and sampled. Fly ash was only needed to be sampled. During the test the steam pressure of the boiler is 1.0 MPa and the average evaporation capacity is 3200 kg/h.

The arrangement of the measuring points in the thermal balance experiment is shown in Figure 1. Measuring points of the exhaust gas temperature and compositions were arranged behind the economizer. The gas sample points should be arranged in the straight section of the flow. In the process of the experiment, the combustion conditions of the boiler should be approximately consistent. The safety valve and blowdown valve of the boiler were not allowed to be opened. At the end of the experiment, the level of water and coal scuttle should be consistent with the beginning of the experiment.

4 EXPERIMENTAL RESULTS AND ANALYSIS

4.1 The influence of excess air coefficient on heat loss from exhaust gas

Figure 2 shows the influence of excess air coefficient on heat loss from exhaust gas under different exhaust gas temperatures. When the exhaust gas temperature was constant, heat loss from the exhaust gas increased with an increase in the excess

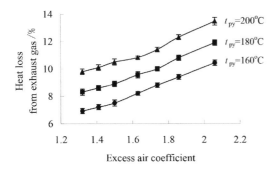

Figure 2. The relation between the excess air coefficient and heat loss from exhaust gas.

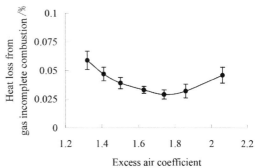

Figure 3. The relation between excess air coefficient and heat loss from gas incomplete combustion.

air coefficient. The higher the exhaust gas temperature, the greater the heat loss from exhaust gas. The heat loss from exhaust gas increased by about 1%, when the exhaust gas temperature increased by 10°C. The heat loss from exhaust gas mainly depends on the exhaust gas temperature and the excess air coefficient. Much leakage and unreasonable distribution of the air in the boiler furnace caused the excess air coefficient to increase. An increase in the flue gas volume leaded to an increase of heat loss from exhaust gas. Therefore, air leakage of the boiler must be minimized in order to reduce heat loss from exhaust gas.

4.2 *The influence of excess air coefficient on heat loss from gas incomplete combustion*

The relationship between the excess air coefficient and heat loss from gas incomplete combustion is shown in Figure 3. A too low or high value of excess air coefficient caused heat loss from gas incomplete combustion to increase. When the value of excess air coefficient was too low, the amount of air was not sufficient for combustion. An increase in the combustible component resulted in the increase of heat loss from gas incomplete combustion. When the value of excess air coefficient was too high, the temperature of the boiler furnace was decreased. Heat loss from gas incomplete combustion increased because of the poor combustion conditions of the boiler furnace.

4.3 *The influence of excess air coefficient on heat loss from solid incomplete combustion*

The influence of excess air coefficient on heat loss from solid incomplete combustion is shown in Figure 4. When the value of excess air coefficient increased, the heat loss from the solid incomplete combustion first decreased and then increased. The combustion condition of the boiler furnace

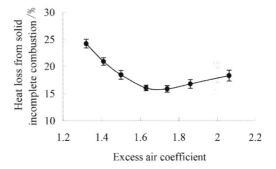

Figure 4. The relation between the excess air coefficient and heat loss from solid incomplete combustion.

was poor because of low excess air coefficient. The carbon content of the coal slag increased and then the heat loss from solid incomplete combustion increased. When the value of excess air coefficient was too high, the velocity of gas increased. The staying time of the coal in the boiler furnace was shortened. As a result, the heat loss from solid incomplete combustion increased.

It can be seen from Figures 2–4 that heat loss from solid incomplete combustion accounted for the largest proportion of the total heat loss, followed by heat loss from exhaust gas. Both of them accounted for more than 80% of the total heat loss.

4.4 *Optimum excess air coefficient*

According to the previous analysis, heat loss from exhaust gas, heat loss from gas incomplete combustion, and heat loss from solid incomplete combustion were all influenced by the excess air coefficient. As a result, the excess air coefficient has an important effect on the thermal efficiency of the boiler.

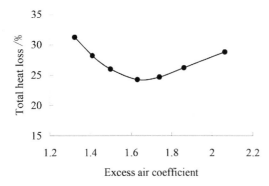

Figure 5. The relation between the excess air coefficient and heat loss.

The relationship between the total heat loss and excess air coefficient is shown in Figure 5. When the value of excess air coefficient increased, the total heat loss first decreased and then increased. Because the sum of the heat loss from exhaust gas, heat loss from gas incomplete combustion, and heat loss from solid incomplete combustion accounted for more than 80% of the total heat loss. There existed an optimum excess air coefficient corresponding to the minimum heat loss of the boiler. The higher the excess air coefficient, the greater the heat loss from the exhaust gas will be. The combustion condition of the boiler furnace was very poor when excessive air coefficient was not controlled properly. The energy resource was greatly wasted and the atmospheric environment was heavily polluted.

When the excess air coefficient exceeded 1.63, the total heat loss of the boiler increased with an increase of the excess air coefficient. As a result, under the condition of complete combustion of adequate oxygen and fuel as much as possible, the value of the excess air coefficient is lower and the combustion is more economical.

5 CONCLUSIONS

In this paper, the influence of excess air coefficient on heat loss was analyzed by performing a thermal balance experiment of the coal boiler. The experimental results showed that the heat loss from exhaust gas increased with an increase in the excess air coefficient. Heat loss from gas incomplete combustion and heat loss from solid incomplete combustion first decreased and then increased. Heat loss from solid incomplete combustion accounted for the largest proportion of the total heat loss, followed by the heat loss from exhaust gas. Both of them accounted for more than 80% of the total heat loss.

When the value of excess air coefficient increased, the total heat loss first decreased and

then increased. There existed an optimum excess air coefficient corresponding to the minimum heat loss. As a result, the excess air coefficient has an important effect on the thermal efficiency of the boiler. The excess air coefficient should be adjusted as far as possible in the optimum value range to ensure the efficient operation of the boiler. When the value of the excess air coefficient exceeded 1.63, the total heat loss of the boiler increased with an increase of excess air coefficient. Under the condition of complete combustion of adequate oxygen and fuel as much as possible, the value of the excess air coefficient is lower and the combustion is more economical.

REFERENCES

Brundaban Patro (2016). Efficiency studies of combination tube boilers. *Alexandria Engineering Journal* 55(1), 193–202.

Li Debo, Xu Qisheng & Shen Yueliang (2013). Numerical Simulation on Combustion Characteristics of Tangentially-fired Boilers at Different Air Speeds. *Chinese Journal of Power Engineering 33(3)*, 172–177.

Li Xuesong, Jin Xiuzhang & Han Chao (2015). The Research of Boiler Optimum Excess Air Coefifcient Based on PSO. *Instrumentation Customer 4*, 15–17.

Saidur. R, Ahamed J. U & Masjuki H. H. (2010). Energy and economic analysis of industrial boilers. *Energy Policy 38(5)*, 2188–2197.

Sandro Dal Secco, Olivier Juan & Myriam Louis-Louisy (2015). Using a genetic algorithm and CFD to identify low NOx configurations in an industrial boiler. *Fuel 158*, 672–683.

Seung Hee Euh, Sagar Kafle & Yun Sung Choi (2016). A study on the effect of tar fouled on thermal efficiency of a wood pellet boiler: A performance analysis and simulation using Computation Fluid Dynamics. *Energy 103*, 305–312.

Shyan-Shu Shieh, Yi-Hsin Chang & Shi-Shang Jiang (2010). Statistical key variable analysis and model-based control for the improvement of thermal efficiency of a multi-fuel boiler. *Fuel 89(5)*, 1141–1149.

Vladimir I. Kuprianov (2005). Applications of a cost-based method of excess air optimization for the improvement of thermal efficiency and environmental performance of steam boilers. *Renewable and Sustainable Energy Reviews 9(5)*, 474–498.

Wang Jianguo, Shyan-Shu Shieh & Shi-Shang Jang (2013). A two-tier approach to the data-driven modeling on thermal efficiency of a BFG/coal co-firing boiler. *Fuel 111*, 528–534.

Wu Jiangquan, Meng Jianqiang & Yan Taisen (2015). Effect of excess air ratio on slagging characteristics of Zhundong coal. *Journal of Harbin Institute of Technology 47(7)*, 78–83.

Zhao Junjie, Luo Liquan & Wu Hao (2015). Effect of the Excess Air Coefficient on Boiler Thermal Coefficient and Denitrification in Coal-Fired Power Plants. *Boiler Technology 46(3)*, 30–34.

Advances in Energy, Environment and Materials Science – Wang & Zhou (Eds)
© 2017 Taylor & Francis Group, London, ISBN 978-1-138-03600-0

Dynamics analysis on bearingless motorized spindle based on transfer matrix method

J. Meng, J.J. Tian, Z.L. Li & Z.Z. Lei
College of Mechanical and Power Engineering, Chongqing University of Science and Technology, Chongqing, China

ABSTRACT: The structure characteristics of the bearingless motorized spindle are introduced and analyzed. Its dynamics analysis is done by the transfer matrix method. Firstly, the transfer matrix suitable for bearingless motorized spindle is generated. Then transfer matrix equation of the bearingless motorized spindle, which is also the frequency equation of the shafting, is educed. Taking a certain bearingless motorized spindle for example, the process of calculating the natural frequency, critical speed and corresponding mode shape by the transfer matrix method are given. The first three orders' natural frequency, critical speed and mode shape are gained. This kind of calculation method and process provides foundation for further research on the bearingless motorized spindle's dynamic characteristic.

1 INTRODUCTION

It is well recognized that motorized spindle which integrates the rotated shaft with the motor is a promising technology. It is also critical parts of high speed machining systems and has been widely used in numerical control machine tools (Meng & Chen 2008). Bearingless motorized spindle is a recently arisen new technology. It solves the supporting problem of the shaft and can make the machine tools achieve higher speed. Being the core component of high speed machining systems, its performance decides the developmental level of the machine tools directly and its dynamic characteristic also affects the machine tools running safely and credibility.

Transfer matrix method is the easiest and most universal way of calculating the dynamic characteristic of the shaft. It simplified the shaft with mass continuous distributing into a series of rigid discrete disks with lumped mass. Rigid discrete disks are connected with massless elastic shaft elements. In this way, the whole shaft is dispersed to many discrete models. Usually, discrete points are chosen at the place of bearings and cross sections which changes suddenly on the shaft. In terms of the lever principle, the mass of the shaft segment is predigested to its two ends' cross sections. Every shaft segment couples with each other through the transfer matrix. Then the state equation of the whole shafting is available with considering the boundary condition (Zhong et al. 1987,

Wu et al. 2014). Solving the equation will obtain the natural frequency and the dynamic response to the system.

2 TRANSFER MATRIX FOR BEARINGLESS MOTORIZED SPINDLE

2.1 Structure characteristics of the bearingless motorized spindle

The bearingless motorized spindle is composed of the bearingless motor, radial magnetic bearing, axial magnetic bearing, and the rotated shaft etc. Besides the advantage of common motorized spindle, due to bearingless motorized spindle is a non contact type support, its structure is simple, volume is small, the speed and the torque can be increased at the same time, and the service life is longer (Gourc et al. 2011, Xie et al. 2014, Bu et al. 2014). The rotors of bearingless motor, radial magnetic bearing and axial magnetic bearing are incorporated with the shaft in the motorized spindle, as Figure 1 shown. When making the dynamics analysis, the rotor and the shaft should be treated as one. The bearingless motorized spindle is an elastic stepped shaft with mass continuous distributing. It has infinite degrees of freedom, namely it has infinite natural frequencies and vibration modes. When analyzing its dynamic characteristic, the whole shafting is simplified as multi-degree of freedom system generally, which has some lumped masses.

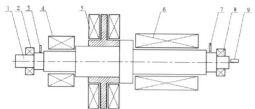

1-shaft 2,8-assistant bearing 3,7-radial displacement sensor
4-radial magnetic bearing 5-axial magnetic bearing
6-bearing-less motor 9-axial displacement sensor

Figure 1. Structure of bearingless motorized spindle.

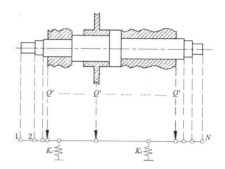

Figure 2. Simplified model of bearingless motorized spindle.

2.2 Model simplification

Founded on transfer matrix method and considering structure characteristics, the bearingless motorized spindle is simplified as follow:

1. The bearingless motorized spindle is dispersed to N shaft segments. According to the centroid immovability principle, the mass of each shaft segment is converted to its two ends' cross sections. Therefore, the mass and inertia are concentrated to the end of the shaft segment and constitute rigid discrete disks. Meantime, the shaft segment is predigested into the massless elastic shaft element with equal cross section.
2. The shaft is supported by bearingless motor and radial magnetic bearing with the rigidity K_1 and K_2, respectively.
3. The natural frequency and the vibration mode of the shaft are related to the distribution of shaft's mass and the stiffness. In order to make the mass and the stiffness of the simplified model distribute more approached to the real shaft, the more it is segmented the closer it to the fact. But the computational error will be increased at the same time. Accordingly, when predigesting the entire shafting, the segment containing rotor core who is heavier and longer than other parts should be separated to several shaft segments again. Due to some parameters (e.g. density) of the rotor and the shaft are different, it will bring several troubles and errors when calculating J_p, J_d, EI of the transfer matrix. The weight of the rotor is considered to be equivalent force, while establishing the simplified model. The equivalent force averagely acts on disks which belong to the shaft segment containing rotor, and each is denoted by Q'.

Figure 2 shows the simplified model of the bearingless motorized spindle. The Arabic numerals in the figure are a set of serial numbers indicating shaft segments after being simplified. The state of every rigid discrete disk can be expressed as $\{Z\} = [Q\ M\ Y\ \theta]$. Here Q, M, Y and θ represent the shearing force, moment, deflection and rotation angle, respectively.

2.3 The transfer matrix of bearingless motorized spindle

The discrete motorized spindle is simplified as the model with lumped masses. The disk of No.k ($k = 1, 2, ..., N-1$) with elastic supporting and the shaft element of No.k are taken to be analyzed, as shown in Figure 3. According to the classical dynamics and Figure 3, Formula (1) and (2) can be gained. Matrix $[B]_k$ is the transfer matrix of the shaft element and Matrix $[D]_k$ is the transfer matrix of the rigid disk.

$$\{Z\}_k^R = [D]_k \{Z\}_k^L = \begin{Bmatrix} y \\ \theta \\ M \\ Q \\ 1 \end{Bmatrix}_k^R$$

$$= \begin{bmatrix} 1 & 0 & 0 & 0 & 0 \\ 0 & 1 & 0 & 0 & 0 \\ 0 & \left(J_p\dfrac{\Omega}{\omega} - J_d\right)\omega^2 + K & 1 & 0 & 0 \\ m\omega^2 - K & 0 & 0 & 1 & Q' \\ 0 & 0 & 0 & 0 & 1 \end{bmatrix}_k \begin{Bmatrix} y \\ \theta \\ M \\ Q \\ 1 \end{Bmatrix}_k^L$$

(1)

$$\{\bar{Z}\}_{k+1}^L = [\bar{B}]_k \{\bar{Z}\}_k^R = [B]_k$$

$$= \begin{bmatrix} 1 & l & \dfrac{l^2}{2EI} & \dfrac{l^3}{6EI}(1-v) & 0 \\ 0 & 1 & \dfrac{l}{EI} & \dfrac{l^2}{2EI} & 0 \\ 0 & 0 & 1 & l & 0 \\ 0 & 0 & 0 & 1 & 0 \\ 0 & 0 & 0 & 0 & 1 \end{bmatrix}_k$$

(2)

34

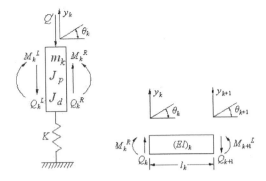

Figure 3. Dynamics analysis of No.k disk and shaft element.

where l is the length of the shaft element (mm); m is the mass of the rigid disk (kg); ω is angular velocity of precession (r/min); Ω is angular velocity of rotation (r/min); J_d is diametral moment of inertia (kg · mm^2); J_p is polar moment of inertia (kg · mm^2); K is the bearings stiffness (N/mm); EI is the flexural rigidity of the cross section (N · mm^2), E represents the Young's modulus of the material (N/mm^2), I represents the cross sectional moment of inertia (mm^4); $v = (6EI/\alpha GAl^2)$, α is the coefficient which is decided by the shape of the section, G is shear modulus of elasticity (N/mm^2), A is the area of the cross section (mm^2).

If the disk has no supporting, K is equal to zero. When the disk of the shaft that doesn't contain rotor, $Q' = 0$.

Therefore, the state vector of the shaft's typical element is $\{Z\}_{k+1} = [B]_k [D]_k \{Z\}_k$.

For making the analysis and calculation facility, the rigid disk and the shaft element are coupled into a typical element. The transfer matrix of the typical element is $[T]_k = [B]_k [D]_k$. The state vector can be written as $\{Z\}_{k+1} = [T]_k \{Z\}_k$. The transfer matrix equation of the entire shafting is expressed as $\{Z\}_N = [T]_{N-1} [T]_{N-2} \cdots [T]_1 \{Z\}_1$. When the motor drives the motorized spindle rotated freely, two ends' cross sections of the spindle are freedom, i.e. $Q_1 = 0$, $M_1 = 0$, $Q_N = 0$, $M_N = 0$. Hence, $\{Z\}_1 = [y, \theta, 0, 0, 1]_1^T$, $\{Z\}_N = [y, \theta, 0, 0, 1]_N^T$. Introducing these boundary conditions into the transfer matrix equation, the frequency equation of the whole shafting is formed as $[y, \theta, 0, 0, 1]_N^T = [T]_{N-1} \cdots [T]_1 [y, \theta, 0, 0, 1]_1^T$.

3 DYNAMICS ANALYSIS ON BEARINGLESS MOTORIZED SPINDLE

Calculating the transfer matrix equation, the frequency and the rotate speed can be obtained under the circumstances of Ω/ω is given. When $\Omega/\omega = 1$, the natural frequency and critical speed can be worked out.

Taking self designed bearingless motorized spindle for example, which is dispersed model has shown in Figure 4. The shaft containing rotors including bearingless motor, radial magnetic bearing and axial magnetic bearing will divided into several segments as shown in Figure 4. For example, to the rotor of radial magnetic bearing, it is obvious that Q_1 is half of rotor's weight and acts on the 4th disk, 5th disk and 6th disk. Due to the 4th disk and 6th disk are the two end points of this shaft segment, the both force are $Q'/2$.

MATLAB is used to calculate and simulate. The process is illustrated in Figure 5.

Based on the characteristic of the example, the interested rotate speed range is selected at 40000~ 200000 r/min. And the cross section is selected the left end of the shafting just the initial cross section. After other parameters are put into the transfer matrix, the left end's natural frequency is received by computing along the Figure 5, such as Table 1.

Then also using the transfer matrix equation, the state of every typical element can be figured out at the given speed. The course is just like Figure 6 has expressed. The mode shape curves of first three orders are shown in Figure 7.

It is can be seen that for the bearingless motorized spindle the amplitude of the left end is larger than the right side. This is due to the right support is provided by bearingless motor's rotor whose stiffness and supporting range are both higher than the radial magnetic bearing which at the left

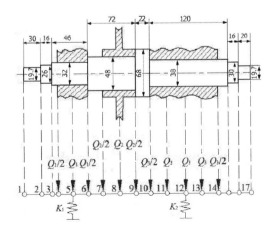

Figure 4. The dispersed model of bearingless motorized spindle.

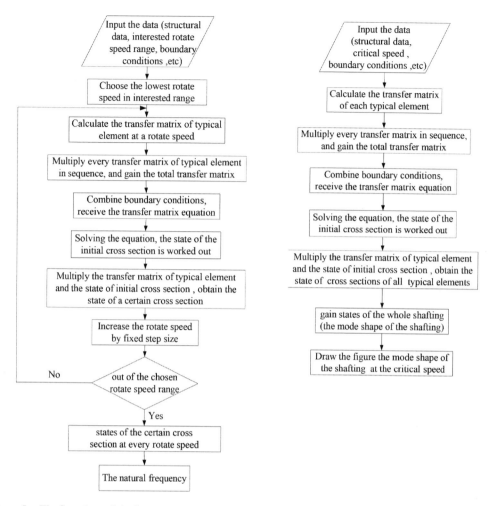

Figure 5.   The flow chart of the frequency response curve solution.

Table 1.   Frequency and critical speed of the first three orders.

Order	Natural frequency Hz	Critical speed r/min
1	758.95	45537
2	1359.8	81588
3	3019.9	181194

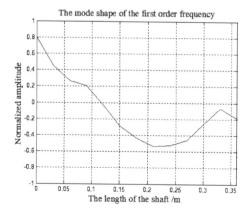

Figure 6.   The flow chart of the mode shape curve solution.

side of the shaft. There are some inflection point in the vibration shape curve, most of which is appear at the point where the section changes or near the spindle supporting.

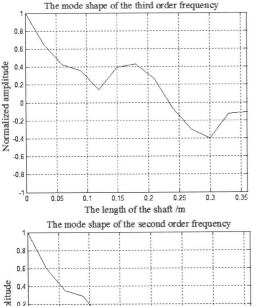

The mode shape of the third order frequency

The mode shape of the second order frequency

Figure 7. The mode shape of the shaft at the critical speed.

4 CONCLUSIONS

1. Based on the transfer matrix method and the structural characteristic of the bearingless motorized spindle, the typical elements of the bearingless motorized spindle is set up which integrate the rotated shaft with the rotor of bearingless motor, radial magnetic bearing and axial magnetic bearing, and the transfer matrix equation of the whole shafting is also established.
2. Using the transfer matrix equation to calculate a certain motorized spindle, solving processes are described and results of the critical speed, natural frequency and mode shape of the entire shafting are obtained.
3. The dynamic analysis of bearingless motorized spindle has been done on basis of the transfer matrix method. From the analysis results, it is known the vibration amplitude of the left end of bearingless motorized spindle can be improved by increase the thickness of radial magnetic bearings or shorten the extended length of the shaft at the left end, appropriately.

ACKNOWLEDGEMENTS

This paper is supported by National Natural Science Foundation of China (No.51505049), Scientific and Technological Research Program of Chongqing Municipal Education Commission (No.KJ1501314) and Chongqing Research Program of Basic Research and Frontier Technology (No. cstc2013 jcyjA70004).

REFERENCES

Bu, Wenshao., Lu, Chunxiao., Zu, Conglin. & Niu, Xinwen (2014). Research on Dynamic Decoupling Control Method of Three-phase Bearingless Induction Motor. *International Journal of Control and Automation. 7(5)*, 77–86.
Gourc, E., Seguy, S. & Arnaud, L. (2011). Chatter milling modeling of active magnetic bearing spindle in high-speed domain. *International Journal of Machine Tools and Manufacture. 51(12)*, 928–936.
Meng, Jie. & Chen, Xiaoan (2008). The Transfer Matrix Method of Dynamics Analysis on Motorized Spindle. *Journal of Machine Deign. 25(7)*, 37–40.
Wu, Yuhou., Zhang, Jiao., Zhang, Lixiu. & Shi, Huaitao (2014). Analysis of Ceramic Motorized Spingdle Rotor Dynamic Characteristic Based on Transfer Matrix and Finite Element Method. *Journal of Shenyang University (Natural Science). 30(3)*, 510–515.
Xie, Zhenyu., Yu, Kun. & Wen, Liangtang (2014). Characteristics of motorized spindle supported by active magnetic bearings. *Chinese Journal of Aeronautics. 27(6)*, 1619–1624.
Zhong,Yie., He, Yanzhong. & Wang, Zheng (1987). *Dynamics of rotor.* Beijing: Tsinghua University Press.

Advances in Energy, Environment and Materials Science – Wang & Zhou (Eds)
© 2017 Taylor & Francis Group, London, ISBN 978-1-138-03600-0

Short-term load forecasting model for electric vehicles considering the forgetting function and cross entropy

Rishang Long, Wenxia Liu, Jianhua Zhang, Yang Liu, Hao Hu, Tao Tan & Yuan Meng
State Key Laboratory of New Energy Power System, North China Electric Power University, Beijing, China

ABSTRACT: A new combined forecasting model based on data effectiveness and cross entropy is proposed to study the short-term load forecasting method of bus charging station. Firstly, the load characteristics of the bus charging station are analyzed, and the daily-charging load is characterized as large fluctuations, periodic, and closely related to meteorological conditions (temperature, rainfall, etc.). Secondly, aiming at the historical error accumulation problem, the combination predicting model is improved as follows: 1) Considering the prediction accuracy and stability of the single model, the combined forecasting model would adjust weights dynamically based on Cross Entropy (CE) algorithm and Normal Distribution probability density function; 2) Considering the time validity of the data, the concept of the forgetting function is proposed, and the prediction accuracy of the combined model is further improved. Finally, we construct training samples and testing samples based on history charging data at one bus station in Beijing city and compare the predicting results of several models to demonstrate the effectiveness of the combined forecasting model stated in this paper.

1 INTRODUCTION

In order to reduce the emissions of CO_2 and alleviate the energy crisis, the world has become increasingly concerned about the development of Electric Vehicles (EV) (Razeghi, 2014). Through the electric vehicle charging technology, the safety and economic operation of the grid is ensured by reducing the network loss and improving the voltage level (Clement, 2010, Zhang, 2011; Masoum, 2011). However, due to the large fluctuations in the charging load and close correlation to the day's temperature, weather and other factors, these give a certain impact on the order of the charging. Therefore, it is necessary to study the load forecasting method of EV charging station, which is a powerful support for the orderly charging of EV.

At present, the charging mode of EV mainly consists of AC slow charging, DC fast charging and charge of the replaced battery (Liu, 2014). Different charging methods have different characteristics in terms of charging starting time, charging efficiency, charging frequency, charging scale, which lead to different load characteristics of EV charging stations. References (Liu, 2014; Yuan, 2013; Liu, 2015; Zhang, 2014) consider the different models and charging methods, using the Monte Carlo method to draw EV initial charging time and mileage, then accumulate each EV charging load to obtain total short-term load simulation value. However, it cannot conduct as the load prediction

method for the prediction and forecasting. Reference (Hu, 2012) discusses different charging characteristics of the battery and establishes a rapid charging probability distribution model to predict total load of the rapid charging station, but fails to take full account for the effectiveness of the historical charge data. References (Chang, 2014; Liu, 2014) consider the influence of temperature, weather and day types on load, and make a short-term load forecasting of exchange type electric bus charging station on the foundation of selection of similar days. This method uses the historical data in a more efficient way, but the large fluctuation in the EV charging station load results in large forecast errors at certain moments. Due to exchange type electric bus charging station can interact with the grid through the load regulation, it has been developing rapidly in several big cities in China. In order to achieve better orderly charging, this paper will study its load characteristics and short-term load forecasting method.

At present, the short-term load-forecasting model is divided into two categories. One is traditional method based on historical data, such as regression model, time series model, etc. The other is artificial intelligence methods based on the historical data and load factors, such as neural network (Tascikaraoglu, 2014; Zhang, 2012), support vector machine (Liu, 2014; Jin, 2009; Xiao, 2015), fuzzy reasoning (Vapnik, 1996), wavelet analysis technology (Hinojosa, 2010), etc. In order to reduce the risk, in 1969, Bates and Granger put forward combined forecasting method based on the weights (Bates, 1969) which can combine the data characteristics of different methods and effectively improve the prediction accuracy. The key problem of combined forecasting is the weight coefficient. In recent years, many scholars have carried on the research to the weight coefficient, and proposed the improved methods (Chen, 2008; Fan, 2009; Velásquez, 2014; Su, 2005; He, 2013; Gao, 2008; Zhao, 2008). Cross Entropy (CE) is a Probability Density Function (PDF), which is a method to calculate the information difference between two random vectors. The CE is introduced into the combination forecasting by Li (2014) and Chen (2012), and the stability of the forecast results is improved. It overcomes the instability of the data on single algorithm, and improves the stability of prediction results. However, the PDF of Li (2014) adapts the failure rate function, which is not suitable for short-term load forecasting; Chen (2012) proposes the prediction method based on normal distribution of wind load, but do not take full account of the time characteristics of historical data, as well the method for solving is more complex.

In this paper, a new combined forecasting model based on forgetting function and CE is proposed to study the short-term load forecasting of the bus charging station in Beijing city. The combined forecasting model is improved as follows: 1) Considering precision and stability of each single model in the process of prediction, based on CE algorithm, we dynamically adjust weight in the combined forecasting model; 2) the time characteristic of the data is considered, and the concept of the forgetting function is proposed to improve the accuracy of the model.

2 LOAD CHARACTERISTICS ANALYSIS OF BUS CHARGING STATION

The data is taken from a bus charging station in Beijing city, which includes charging load values from July to September in 2012. Fig. 1 shows the comparison of week load (normalized values) among the second week (8–14), third week (15–21) and fourth week (22–28), where 8, 9, 15, 16 and 22, 23 are the weekends. The sampling interval is 15 minutes, a total of 96 points per day.

From Fig. 1, we can see that the daily charging fluctuations of the load is very large and have an obvious day cycle. From 8:00 to 10:00, the load starts to rise from zero, and then reaches many different peaks, mostly distributed between 12:00 to 14:00 and 17:00 to 20:00, and then the load gradually decreases to zero. The load value and day types are closely related. The load curves of days in different weeks are similar.

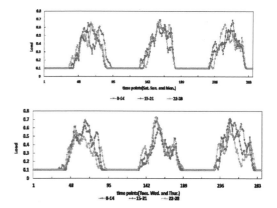

Figure 1. Weekly load comparison.

Therefore, the reasonable selection of similar days and the time characteristic of historical data can improve the accuracy of prediction. Now there are many methods of selecting similar days, the ideal methods include evidence theory, clustering analysis, trend similarity method, grey correlation method, etc.

3 CHARGING LOAD FORECASTING BASED ON CE MODEL

3.1 *Selection of single model*

There is no existing uniform rule to choose the single model for combined forecasting. The factors considered in this paper include: independence, diversity and accuracy of the algorithm.

In this paper, we carry out co-integration test and tolerance test to single methods, please refer to the Jiang (2014). We use a variety of methods to predict and screening according to the precision and stability of results. Finally we choose 4 models: Fuzzy neural network (FNN), Support Vector Machine (SVR), Grey Model (GM), Wavelet Transform (WT) model.

3.2 *Combined forecasting model based on CE and forgetting function*

The combined forecasting model is composed of m single forecasting model, and the relative effectiveness of each model is determined by historical data. If the combined forecast value at time t is y_t, ω_{it} is the weight of the ith model at time t, $\hat{y}_{it}$ is the prediction value of ith model at time t. The problem of combination forecasting is described as follows:

$$y_t = \sum_{i=1}^{m} \omega_{it} \hat{y}_{it} \qquad (1)$$

where $\sum_{i=1}^{m} \omega_{it} = 1$. From (1) we can know that there are two factors which influence the result of the final combined forecasting: One is a single model; The other is the weight of a single forecasting model. This paper focuses on the latter, through the CE model for weight optimization.

1. The CE Model

According to the definition of entropy, a method to calculate the difference of information between two random vectors is defined as Cross Entropy. The CE model can determine the extent of the mutual support degree through the judgment of intersection degree between different information sources. Also, using the mutual support degree to determine the weight of information sources, the higher mutual support, the greater the weight [28]. It is also called *Kullback-Leibler(K-L)* distance. The cross entropy of the two probability distribution is expressed as $D(f \parallel g)$.

For discrete case,

$$D(g \parallel f) = \sum_{1}^{n} g_i \ln \frac{g_i}{f_i} \qquad (2)$$

For continuous case,

$$D(g \parallel f) = \int g(x) \ln \frac{g(x)}{f(x)} dx$$
$$= \int g(x) \ln g(x) dx - \int g(x) \ln f(x) dx \qquad (3)$$

f and g denote the probability vector in the discrete case, respectively, the PDF in the continuous case.

CE model quantifies the "distance" between the amount of information. Nevertheless, the *K-L* distance is not the real length distance, but the difference between two probability distributions. When the two probability distributions are more uniform, the CE value is smaller. For the combined forecasting model based on CE, the CE model is the support of the combined forecasting. Therefore, the objective function of this paper is to find a single algorithm with maximum degree of support, which is also the maximum of the corresponding weights.

Using CE model should solve two major problems: the establishment of PDF and the formation of cross entropy objective function; the solution of weight coefficient through iteration.

2. The PDF of prediction model considering Forgetting Function

In general combined model, the effects of prediction error at different historical moment to the weight is regarded the same, but according to the principle of load development near the small, the prediction error should be treated differently in different historical moments. In the past literature, it is lack of scientific basis to determine subjectively the weight of data. Thus this paper introduces the forgetting function, and compares the influence of several forms of forecasting error. The forgetting function reflects the importance of the data in predicting the future. The data more closely to the test day, the more useful information it contained, the higher the value it is, which is in line with the idea of information theory. From the mentioned above, the forgetting function is a non-decreasing time function, the concrete form is shown as follows:

$$F(j) = (j)^q \qquad (4)$$

where j is the days before the forecasting day. The selection of forgetting function is not determined by the rules, the value of q can be 0.5, 1, 2, etc. This article will analyze it in the next section.

For the PDF of predicting power $f(x)$, it can be considered that $f(x)$ is the PDF $fi(x)$ of the single forecasting method multiplied by the corresponding weight. From central limit theorem [29], if a variable is influenced by many small independent random factors, then we can think the variable subjects to normal distribution, therefore the value of charging load power at a certain time can be considered to meet the normal distribution. The minimum CE is used to determine the probability distribution of the different forecasting methods, then the combined probability distribution of the charging power is obtained.

The PDF for method i is $(i = 1, 2, ..., m)$

$$f_i(x) = \frac{1}{\sqrt{2\pi}\sigma_i} e^{-(x-\mu_i)^2/(2\sigma_i^2)} \qquad (5)$$

where μ_i is mean value, σ_i is variance.

Making prediction using single methods based on similar days, obtain the forecast power value P_{ijt} at time t $(t = 1, 2, ..., T)$ in the jth day by the ith single method.

Based on forgetting function, the mean value and variance at time t by the ith single method is calculate as follows:

$$\mu_{it} = \frac{\sum_{j=1}^{n} F_j P_{ijt}}{n \sum_{j=1}^{n} F_j} \qquad (6)$$

41

$$\sigma_{it} = \frac{\sum_{j=1}^{n} F_j \left(P_{ijt} - \mu_{it} \right)^2}{n \sum_{j=1}^{n} F_j} \qquad (7)$$

3. Weight optimization and solving method of combination forecasting

The combination of the PDF of the predicted power can be obtained by the PDF of the single prediction method:

$$f(x) = \omega_{it} f_i(x) \qquad (8)$$

Therefore,

$$f(x) \sim N\left(\sum_{i}^{m} \omega_{it} \mu_{it}, \sum_{i}^{m} (\omega_{it} \sigma_{it})^2 \right) \qquad (9)$$

From (9), the objective function of minimum CE optimization problem is described as:

$$\min F = \min D\left(f_i(x) \| f(x) \right) \qquad (10)$$

$$\text{s.t.} \quad \begin{cases} 0 \le \omega_{it} \le 1 \\ \sum_{i=1}^{m} \omega_{it} = 1 \end{cases}$$

To select the appropriate weight vector to achieve the minimum F is to determine the support of different algorithms. The deduction of weight coefficients is shown in Appendix. Finally the expression of the weight coefficient is obtained as follows:

$$\omega_{it} = \frac{\sum_{k=1}^{N} I_{\{S(x) > \gamma\}} \dfrac{(x_k - \mu_{it})^2}{\sigma_{it}}}{\sum_{i=1}^{m} \sum_{k=1}^{N} I_{\{S(x) > \gamma\}} \dfrac{(x_k - \mu_{it})^2}{\sigma_{it}}} \qquad (11)$$

4 CHARGING LOAD FORECASTING BASED ON CE MODEL

The example data is from a bus charging station charging load data in Beijing city from July to September. The selection of single method has been introduced before. Algorithm is achieved in Matlab 2012a version, running on PC, 2.5GHz, 4GB. In this paper, the Root Mean Square Error (RMSE), Mean Relative Percent Error (MRPE), the Maximum Percent Error (MRE) are used to characterize the performance of the different combined forecasting model.

To illustrate the improvement of cross entropy combination algorithm in prediction accuracy, this paper selects the other two common combination forecasting methods, such as similar weight combination forecasting method (SA), GR regression model. The forecast results are shown in Fig. 3 (September 18th's forecast results) and Table 1:

From Table 1, Fig. 2 can be seen that:

1. The combined forecasting may not be optimal in certain moments or in one day prediction results, but from the point of view of long-term prediction, a single method in certain moments of forecast errors might be very big, the combined forecasting can significantly reduce the risk. The reason is that CE algorithm seeks the maximum degree of support of single method for the PDF, and not simply determine the weight on the basis of historical prediction error, making a single method of prediction result more suitable for combined forecasting and the weight more reasonable and credible.
2. Forecasting precision of single method can affect the accuracy of the combined forecasting. Therefore, to improve the prediction accuracy of single method can also for combination forecasting accuracy.
3. SA and GR prediction results are not able to achieve good accuracy. For MPE index, SA and GR prediction results in Sep.18, Sep.19 have large deviation, the CE model has higher accuracy than SA model GR even in the maximum error

Table 1. Error analysis of each methods.

	CE			SA			GR		
	MAPE	MPE	RMSE	MAPE	MPE	RMSE	MAPE	MPE	RMSE
Sep. 15	10.62%	84.95%	3.07%	11.03%	96.10%	3.93%	11.75%	96.30%	4.12%
Sep. 16	6.50%	44.00%	5.06%	8.40%	64.10%	6.03%	7.54%	78.02%	6.04%
Sep. 17	6.66%	67.37%	3.33%	6.52%	66.23%	3.25%	8.60%	69.21%	4.10%
Sep. 18	11.70%	133.18%	7.45%	11.91%	174.64%	8.02%	13.00%	164.86%	8.38%
Sep. 19	14.21%	93.92%	6.96%	14.49%	105.39%	7.98%	15.50%	94.77%	8.25%
Sep. 20	8.46%	38.16%	5.59%	8.79%	48.28%	5.94%	9.79%	47.50%	6.70%
Sep. 21	10.54%	71.08%	4.56%	11.78%	92.80%	5.36%	12.01%	89.53%	5.55%
Average	9.81%	76.09%	5.14%	10.42%	92.51%	5.79%	11.17%	91.46%	6.16%

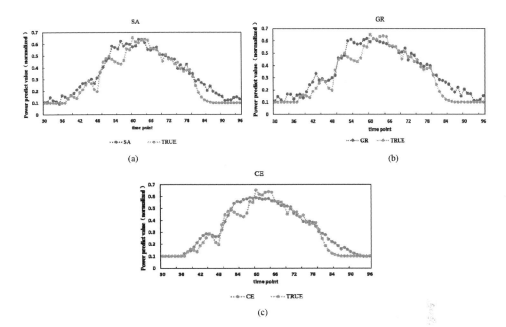

Figure 2. Prediction results of each methods.

in the Sep. 18, which reduces the error of about 56% than the other two methods, the stability of the CE model is better.

5 CONCLUSION

This paper analyses the EV charging station load characteristics, and on the basis of reasonable selection of similar days and full consideration to the historical data of time and effectiveness, the combination forecast model is improved by using the CE algorithm and the forgetting function. The experimental results show that the prediction accuracy of the combined forecasting model can meet the requirements of the bus charging station, and greatly improve the stability of the forecast.

At present, the size of the bus charging station is not large, the charging station load of has feature of random and volatile, the forecast accuracy cannot reach the traditional power load forecasting accuracy. In the future, the forecast accuracy can be improved by the following ways: 1) to improve the prediction accuracy of the single forecasting model. 2) to study the new fresh degree function and the new method to evaluate the single forecasting model.

REFERENCES

Bates J., C. Granger, "The combination of forecast," Operations Research Quarterly, vol. 20, pp. 451–468, 1969.

Boer P.D., D.P. Kroese, S. Mannor et al., "A Tutorial on the Cross-Entropy Method," Analysis of Operations Research. vol. 134, pp. 19–67, 2005.

Chang D.Z., J. Ren, J.W. Zhao, et al, "Research of Short-Term Load Forecasting Model for Electrical Vehicle Charging Station Based on RBF-NN," Journal of Qingdao University (Engineering & Technology Edition), vol. 12, no. 29, pp. 45–51, 2014.

Chen H.Y., "The combination forecasting method effective theory and its application," Beijing: Science Press, pp. 40–98, 113–125, 224–237, 2008.

Chen N., Q. Sha, Y. Tang, et al., "A Combination M ethod for Wind Power Predication Based on Cross Entropy Theory," Proceedings of the CSEE, vol.32, pp. 29–34, 2012.

Clement K., E. Haesen, et al., "The impact of charging plug-in hybrid electric vehicle on a residential distribution grid," IEEE Transaction on Power Systems, vol. 25, no.1, pp. 371–380, 2010.

Energy saving and new energy automobile industry planning[Online]. Available:http://www.gov.cn/zwgk/2012-07/09/content_2179032.htm

EVI, "Globe EV Outlook:Understanding the Electric Vehicle Landscape to 2020," EVI, 2013.

Fan S., L.N. Chen, and W.J. Lee, "Short-Term Load Forecasting Using Comprehensive Combination Based on Multimeteorological Information," IEEE Transactions on Industry Applications, vol. 45, no. 4, pp. 55–59, 2009.

Gao S., L. Mei, "Combined forecasting model of electricity price based on support vector machine," Electric Power Automation Equipment, vol. 11, no. 28, pp. 50–55, 2008.

He X.Q., N. Cai, "Research on Combined Estimation Model Construction of Economic Time Series Based on the Method of Fuzzy Adaptive Variable Weight," Soft Science, vol. 1, no. 27, pp. 141–145, 2013.

Hinojosa V.H., A. Hoese, "Short-Term Load Forecasting Using Fuzzy Inductive Reasoning and Evolutionary Algorithms," IEEE Transactions on Power Systems, vol.25, no.1, pp. 565–574, 2010.

Hu R., P.X. Zhao, " Load Characteristics of Fast Charging Station of Electric Vehicles," Journal of Shanghai University of Electric Power, vol. 4, no. 28, pp. 156–160, 2012.

Jiang C.J., "The Research of Adaptive Combination Forecasting Based on Rule Guildline of Selecting Models," Southeast University, 2014.

Jin Y.L., M.Q. Zhou, X.S. Wang, "Research on text classification method of SVM and K-means," Computer Technology and Development, vol. 11, no. 19, pp. 35–41, 2009.

Li R., H.L. Liu, Y. Lu, "HAN Biao.A combination method for distribution transformer life prediction based on cross entropy theory,"Power System Protection and Control, vol. 42, pp. 97–101, 2014.

Liu Q., Z.Y. Qi, "Electric vehicles Load Forcasting Model Based on Monte Carlo Simulation," Electric Power Science and Engineering, vol. 30, no. 10, pp. 14–21, 2014.

Liu Q., Z.Y. Qi, "The Load Forecast Model for Power Grid with the Accessing of Large-scale Electric Vehicles by Considering Spatial Motion Characteristics," Modern Electric Power, vol. 32, no. 1, pp. 76–87, 2015.

Liu W.X., X.B. Xu, X. Zhou, "Daily load forecasting based on SVM for electric bus charging station," Electric Power Automation Equipment, vol. 34, pp. 41–48, 2014.

Masoum A.S., S. Deilami, et al., "Smart load management of plug-in electric vehicles in distribution and residential networks with charging stations for peak shaving and loss minimisation considering voltage regulation," IET Generation, Transmission & Distribution, vol. 5, no. 8, pp. 877–888, 2011.

Razeghi G., L. Zhang, et al., "Impacts of plug-in hybrid electric vehicles on a residential transformer using stochastic and empirical analysis," Journal of Power Sources, vol. 252, pp. 277–285, 2014.

Su X.H. "Study on gas short term load forecasting based on artificial neural network," Chongqing University, 2005.

Tascikaraoglu A., M. Uzunoglu, "A review of combined approaches for prediction of short-term wind speed and power," Renewable and Sustainable Energy Reviews, vol. 34, pp. 243–254, 2014.

Vapnik V., S. Golowich, A. Smola, "Support vector machine for function approximation, regression estimation and signal processing," Adv Neural Inform Process Syst, vol. 9, pp. 281–287, 1996.

Velásquez J.D., C. Zambrano,C. J. Franco, "Forecast Combining Using a Generalized Single Multiplicative Neuron," IEEE Latin America Transactions, vol. 12, no. 4, pp. 101–108, 2014.

Xiao B., P. Nie, G. Mu, et al., "A Spatial Load Forecasting Method Based on Multilevel Clustering Analysis and Support Vector Machine,"Automation of Electric Power Systems, vol. 39, no. 12, pp. 56–61, 2015.

Yuan Z.P., W. Zhou, W.B. Wang, "Charging Load Forecasting Method for Electric Vehicles," East China Electric Power, vol. 12, no. 41, pp. 2657–2665, 2013.

Zhang C.H., Q. Huang, et al., "Smart grid facing the new challenge: the management of electric vehicle charging loads," Energy Procedia, vol. 12, pp. 98–103, 2011.

Zhang H.C., Z.C. Hu, Y.H. Song, et al., "A Prediction Method for Electric Vehicle Charging Load Considering Spatial and Temporal Distribution," Automation of Electric Power Systems, vol. 38, no. 1, pp. 13–20, 2014.

Zhang W.Y., W.C. Hong, et al., "Application of SVR with chaotic GASA algorithm in cyclic electric load forecasting," Energy, vol. 45, pp. 850–858, 2012.

Zhao W.Q., Y.L. Zhu, X.Q. Zhang, "Combinational Forecast for Transformer Faults Based on Support Vector Machine," Proceedings of the Chinese Society for Electrical Engineering, vol. 9, no. 28, pp. 14–21, 2008.

Advances in Energy, Environment and Materials Science – Wang & Zhou (Eds)
© 2017 Taylor & Francis Group, London, ISBN 978-1-138-03600-0

A quick and practical method for history matching of numerical simulation

Jing Xia
Department of Oil and Gas Field Development, PetroChina Research Institute of Petroleum Exploration and Development, Beijing City, China

Li Kuang & Lingyun Zheng
College of Petroleum Engineering, Northeast Petroleum University, Daqing City, Heilongjiang Province, China

ABSTRACT: This paper proposes an approach for reservoir simulation, which substantially simplifies numerical simulation. It shows that end point matching leads to identical prediction result by a thorough history matching and presents a method to attain the same. According to this research, when we perform a reservoir simulation work, we need not to do thorough history matching. We can get the same prediction result just by matching the indices such as water cut, pressure, residual oil saturation, and recover degree at the end point of the whole history. This improves the efficiency of reservoir simulation significantly.

1 INTRODUCTION

Reservoir numerical simulation is a new method that uses computers to study the multiphase flow in reservoirs. It is widely used in reservoir development and serves as an important technique to facilitate decision-making for field development (Wang et al. 2014, Watts 1997). Reservoir numerical simulation can solve many problems in oilfield, such as dynamic index prediction in the process of development program, infill adjusting, layer adjusting, programming of injection/production system, fracturing, water plugging, profile control, injection rate, injection pressure, flowing pressure, EOR technique evaluation, and driving mechanism (WAN et al. 2016, Souza & Abreu 2014). Black oil simulator is a numerical simulation model for nonvolatile oil. At present, it is the most appropriate, mature, and widely used reservoir simulation model. Furthermore, it is capable of simulating all types of normal development problems. This study is performed with black oil simulator (Møyner & Lie 2016, Huan 1986). It is very crucial to construct the input data stream during numerical simulation of a reservoir. In the process of building the input data, because many items are time-variant, it is necessary to first divide the whole history into hundreds of time section with different injection/production rate and then performs the history matching (Alireza & Karl 2012, Shahkarami et al. 2014). This makes the system very complex and heavy, thereby limiting the development and spreading of the numerical simulation technique. This paper aims at addressing the aforementioned problem, makes the numerical simulation using black oil model, contrasts the result of index prediction of thorough history matching and that of end point matching, and proves that their index prediction results are identical. This paper also provides a simple approach to achieve end point matching, that is, replacing the real history of every well with a smooth history whose injection/production rate equals the average of the real. The index prediction results of the real case and smoothed case are very similar, and the relative differences in residual oil saturation, recovery degree, water cut, and pressure are all less than 2%. This study is of great significance for improving the efficiency of reservoir numerical simulation.

2 BASIC MODEL

The model is square shaped with dimensions 1100×1100 m. It is divided into $11 \times 11 \times 2$ grids and there are a total of 242 blocks whose step is 100 m in X direction and 100 m in Y direction and equivalent to the thickness of the simulated layers in Z direction. Fig. 1 shows the distribution of water wells and oil wells and all wells are perforated at all of the two layers. There are 13 wells, among which 1, 2, 3, and 4 are water injection wells and 5, 6, 7, 8, 9, 10, 11, 12, and 13 are oil production wells. Figs. 2 to 9 show the distribution of effective thickness, permeability, and original oil saturation.

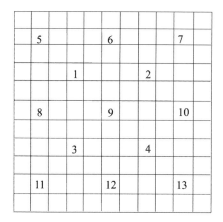

Figure 1.   Distribution of wells.

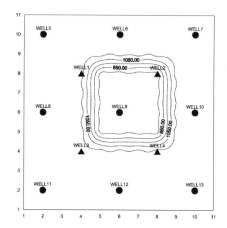

Figure 4.   Permeability in X direction (K = 1).

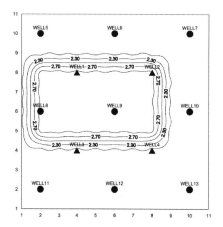

Figure 2.   Distribution of effective thickness (K = 1).

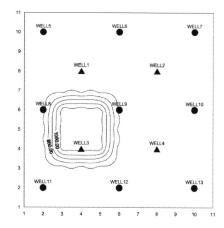

Figure 5.   Permeability in X direction (K = 2).

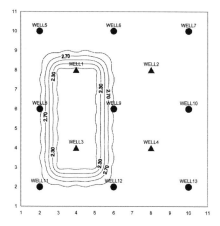

Figure 3.   Distribution of effective thickness (K = 1).

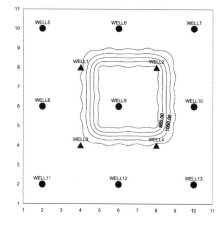

Figure 6.   Permeability in Y direction (K = 1).

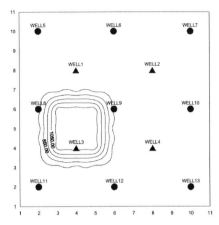

Figure 7.   Permeability in Y direction (K = 2).

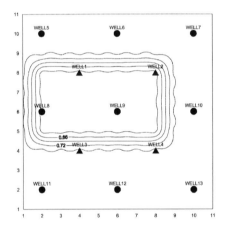

Figure 8.   Original oil saturation (K = 1).

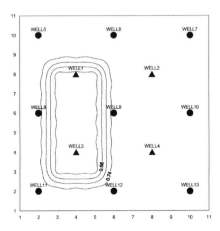

Figure 9.   Original oil saturation (K = 2).

3   HISTORY MATCHING

3.1   *Thorough history matching*

Thorough history matching means accurate matching of indices such as pressure, ware cut, production rate, and recovery degree at every time point of the history. End point matching means matching of indices such as pressure, ware cut, production rate, and recovery degree at the end point of the history. Table 1 shows the real production history of the wells.

Simulation is done for 3600 days. The calculation shows that, at the end of the 3600th day, the water cut of the produced fluid is 95.8% and the recovery degree is 13.168%. The distributions of pressure and residual oil at the end of the 3600th day are shown in Figs. 10 to 13.

3.2   *End point matching*

The injection/production rate of each well is constant and equal to "cumulative injected water or cumulative produced fluid ÷the simulation time (here, it is 3600 days)", as shown in Table 2.

The simulation result shows that, at the end of the 3600th day, the water cut of the produced fluid is 95.8% and the recovery degree is 13.204%. The distributions of pressure and residual oil at the end of the 3600th day are shown in Figs. 14 to 17.

At the end of the 3600th day, thorough history matching and end point matching obtain identical results. The water cut value is 95.8%, the recovery degrees are 13.168% and 13.204%, respectively, and the relative difference is only 0.273%. The relative difference for the pressure distribution and that for the residual oil are less than 2%. All these facts prove that averaging the injection/production history can achieve end point matching and thus simplify data processing substantially.

4   INDEX PREDICTION

In the following paragraphs, we will prove that thorough history matching and end point matching can lead to the same index prediction result in the forecasting period.

On the basis of the aforementioned history matching, index prediction is conducted. The injection/production rates used in this period are listed in Table 3.

4.1   *Thorough history matching*

The prediction result shows that, on the 3900th day, the comprehensive water cut is 95.4%, the

Table 1. Injection/production rate of the wells used in thorough history matching unit: m³/day.

Time	1–700 (day)												
Well name	1	2	3	4	5	6	7	8	9	10	11	12	13
Inje./prod. Rate	200	150	120	100	60	70	80	50	70	70	60	50	60
Time	701–1400 (day)												
Well name	1	2	3	4	5	6	7	8	9	10	11	12	13
Inje./prod. Rate	150	170	130	120	80	80	60	90	40	100	40	30	50
Time	1401–2100 (day)												
Well name	1	2	3	4	5	6	7	8	9	10	11	12	13
Inje./prod. Rate	130	190	140	110	50	90	50	60	60	60	50	90	60
Time	2101–2800 (day)												
Well name	1	2	3	4	5	6	7	8	9	10	11	12	13
Inje./prod. Rate	160	170	110	130	70	100	70	50	40	50	50	70	70
Time	2801–3500 (day)												
Well name	1	2	3	4	5	6	7	8	9	10	11	12	13
Inje./prod. Rate	120	150	150	150	60	60	50	70	70	50	70	60	80
Time	3501–3600 (day)												
Well name	1	2	3	4	5	6	7	8	9	10	11	12	13
Inje./prod. Rate	100	130	180	160	50	80	80	50	60	70	50	80	50

Table 2. Injection/production rate of the wells used in end point matching unit: m³/day.

Time	1–3600 (day)						
Well name	1	2	3	4	5	6	7
Inje./prod. rate	150.555	165.0	131.385	123.06	63.61	80.0	62.5
Well name	8	9	10	11	12	13	
Inje./prod. rate	63.61	56.11	66.11	53.89	60.56	63.61	

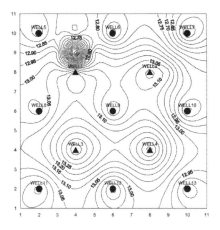

Figure 10. Distribution of pressure on the 3600th day for thorough history matching (K = 1).

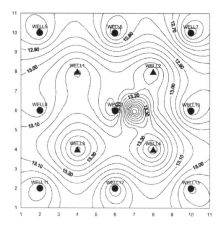

Figure 11. Distribution of pressure on the 3600th day for thorough history matching (K = 2).

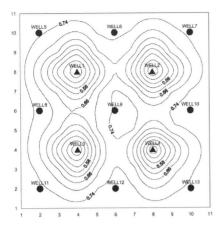

Figure 12. Distribution of residual oil on the 3600th day for thorough history matching (K = 1).

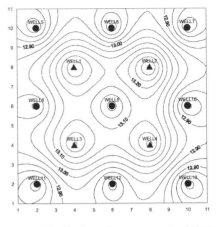

Figure 15. Distribution of pressure on the 3600th day for end point matching (K = 2).

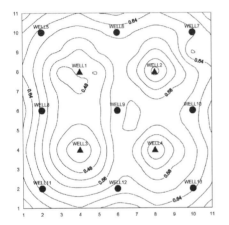

Figure 13. Distribution of residual oil on the 3600th day for thorough history matching (K = 2).

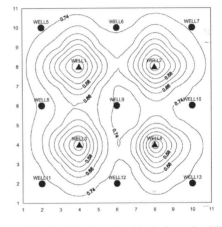

Figure 16. Distribution of residual oil on the 3600th day for end point matching (K = 1).

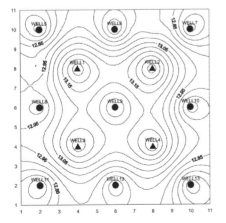

Figure 14. Distribution of pressure on the 3600th day for end point matching (K = 1).

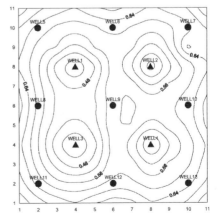

Figure 17. Distribution of residual oil on the 3600th day for end point matching (K = 2).

recovery degree is 13.495%, and the pressure distribution and residual oil distribution are shown in Figs. 18 to 21.

4.2 End point matching

In the case of end point matching, on the 3900th day, the comprehensive water cut is 95.4%, the recovery degree is 13.514%, and the pressure distribution and residual oil distribution are shown in Figs. 22 to 25.

4.3 Comparison of the two cases

At the end of the 3900th day, both thorough history matching and end point matching obtain the same water cut (95.4%). The recovery degrees are 13.495% and 13.514% for thorough history matching and end point matching, respectively, and their relative difference is 0.141%. The relative difference in pressure distribution for the two cases is less than 2% and that in residual oil distribution is less than 1%. This fact shows that end point matching and thorough history matching have identical function for reservoir simulation.

Table 3. Injection/prediction rate used in prediction period unit: m³/day.

Time	3601–3900 (day)												
Well name	1	2	3	4	5	6	7	8	9	10	11	12	13
Inje./prod. rate	50	30	80	60	30	20	20	20	40	30	10	30	20

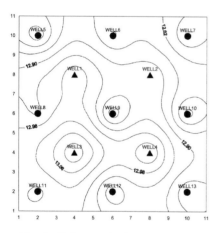

Figure 18. Prediction result of pressure on the 3900th day for thorough history matching (K = 1).

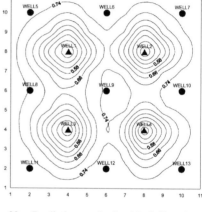

Figure 20. Prediction result of residual oil on the 3900th day for thorough history matching (K = 1).

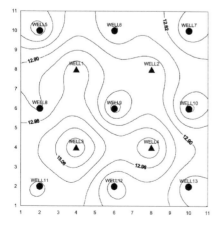

Figure 19. Prediction result of pressure on the 3900th day for thorough history matching (K = 2).

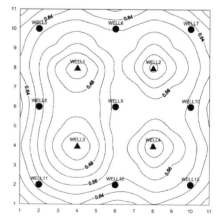

Figure 21. Prediction result of residual oil on the 3900th day for thorough history matching (K = 2).

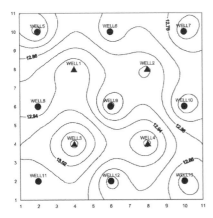

Figure 22. Prediction result of pressure on the 3900th day for end point matching (K = 1).

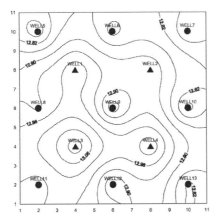

Figure 23. Prediction result of pressure on the 3900th day for end point matching (K = 2).

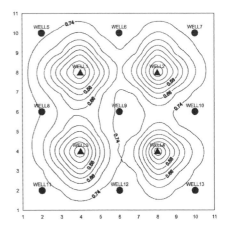

Figure 24. Prediction result of residual oil on the 3900th day for end point matching (K = 1).

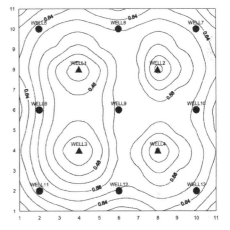

Figure 25. Prediction result of residual oil on the 3900th day for end point matching (K = 2).

5 CONCLUSIONS

The average injection/production history can attain end point matching.

End point matching and thorough history matching have identical function for reservoir simulation. They lead to the same prediction result for indices such as pressure, residual oil saturation, water cut, and recovery degree.

The method of end point matching is very significant to simplify the data processing of numerical simulation and improve the efficiency of reservoir simulation.

ACKNOWLEDGMENT

This paper is supported by science and technology research project of PetroChina Company Limited, Key technology research on gas field development (No.2016B-1504).

REFERENCES

De Souza, G., Pires, A. P., & de Abreu, E. (2014). Well-Reservoir Coupling on the Numerical Simulation of Horizontal Wells in Gas Reservoirs. Society of Petroleum Engineers.doi:10.2118/169386-MS

Huan Guanren (1986). The Black Oil Model for a Heat Oil Reservoir. SPE14853.

Kazemi Alireza & Stephen Karl D (2012). Schemes for automatic history matching of reservoir modeling: A case of Nelson oil field in the UK. J. Petroleum Exploration and Development. 03:326–337.

Møyner, O., & Lie, K.-A. (2016). A Multiscale Restriction-Smoothed Basis Method for Compressible Black-Oil Models. Society of Petroleum Engineers. doi:10.2118/173265-PA

Shahkarami, A., Mohaghegh, S. D., Gholami, V., & Haghighat, S. A. (2014). Artificial Intelligence Assisted History Matching. Society of Petroleum Engineers. doi: 10.2118/169507-MS

Watts J W (1997). Reservoir Simulation: Past, Present and Future. SPE38441.

Wang Yan-Feng, Liu Zhi-Feng & Wang Xiao-Hong (2014). Finite analytic numerical method for three-dimensional fluid flow in heterogeneous porous media. J. Journal of Computational Physics.

Wan Yizhao, Liu Yuewu, Ouyang Weiping, Han Guofeng & Liu Wenchao (2016). Numerical investigation of dual-porosity model with transient transfer function based on discrete-fracture model. J. Applied Mathematics and Mechanics (English Edition). (05)

Advances in Energy, Environment and Materials Science – Wang & Zhou (Eds)
© 2017 Taylor & Francis Group, London, ISBN 978-1-138-03600-0

Direct radiation prediction method for the power generation capacity simulation of an optical thermal power station

Bingxia Yu, Huang Ding & Zhijia Wang
China Electric Power Research Institute (CEPRI), Beijing, China

ABSTRACT: This paper introduces a direct radiation prediction method using conventional weather forecast, which uses clustering, classification, and regression method to process historical data, obtains the next-day prediction model, and then inputs meteorological forecast data such as direct radiation data to the simulation model using SAM software. Finally, we propose a power generation capacity simulation method based on conventional weather radiation forecast. The proposed method is conducive to the simulation of power generation capacity in optical thermal power station, which lacks actual meteorological data and numerical weather forecast data.

1 INTRODUCTION

Direct Solar Radiation (DNI) is the source of thermal power generation. Therefore, DNI data are the basis of power generation prediction in solar thermal power plant. Accurate solar radiation data are essential to analyze the function of a thermal power generation project. It is essential to evaluate the economic benefits of thermal power station in future. Solar direct radiation distribution determines the maximum potential of electricity generation, which is mainly affected by the influence of astronomical radiation, atmospheric absorption, and terrain shielding. Direct solar radiation is mainly calculated by the formula calculation method, numerical weather forecast method, and so on. The formula calculation method can only be used to calculate direct radiation of the sun qualitatively. Its accuracy is lower than others. Numerical weather forecast method is more accurate, but not without difficulties.

In order to solve the problem of direct radiation data loss and obtain more accurate direct radiation data, we propose a direct radiation prediction method using conventional weather forecast result. Using this method, we can forecast the direct radiation of the next day.

2 DIRECT RADIATION SIMULATION

We first divide the direct radiation data into three types by cluster analysis method, sunny, sunny cloudy, and cloudy, and then we fit the three types of curves using loess regression and create weather-type curve model. Finally, according to

the results of the conventional weather of the next day, we choose different weather type curve models to obtain the next 0–24 h direct radiation data. Its specific model is shown in Figure 1:

1. Get historical measured direct radiation data yesterday, and according to the weather type classification algorithm, divide the data type into sunny, sunny cloudy, and cloudy.
2. Fit direct radiation curve based on the weather type.

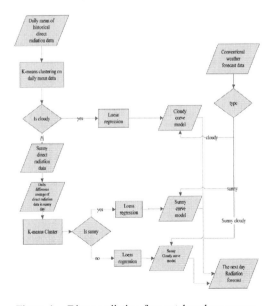

Figure 1. Direct radiation forecast based on conventional weather forecast.

3. Input the conventional weather forecast, choosing the curve, and obtain the next 24 h direct radiation.

The model is described in detail from three aspects, such as weather type classification, direct radiation model, and case analysis.

3 WEATHER TYPE CLASSIFICATION

According to the size and trend of direct radiation, we can divide the historical data into three types, sunny, cloudy and sunny, and cloudy day, by K-means clustering method. The specific procedure is as follows: first, weather is divided into two types, sunny and cloudy, by the daily average cluster, and then based on the difference of daily average data, it is further divided into sunny and sunny cloudy.

3.1 Classification algorithm for sunny and cloudy days

K-means clustering is the most famous clustering algorithm, whose purpose is to make the data points into different clusters, and a cluster is the composition of the closest-distance points. K-means algorithm chose Euclidean distance as the similarity measure method, which is used to find the corresponding initial cluster center vector V optimal classification, making the evaluation target smallest. The error square sum criterion function is used as the clustering criterion function.

The main difference between sunny and cloudy weather is the size of radiation values; therefore, by taking the radiation data between the daily sunrise and sunset to obtain the daily mean and using K-means clustering method, we obtain two categories sunny and cloudy. On the basis of the minimum mean of the sunny sample, we obtain the extreme data of cloudy and sunny classifications. If the radiation value is less than the extreme value, then that day is cloudy. Otherwise, it is a sunny day.

For example, when the data is divided into K (here is 2) clusters, we measure the daily radiation mean sample sequence $\{x^{(1)},...,x^{(m)}\}$, and the specific algorithm is described as follows:

1. Randomly select K cluster centroid point.
2. Repeat the following process until it converges:

For each sample i, using formula 1, calculate the class that it should belong to:

$$c^{(i)} := \mathbf{argmin} \| x^{(i)} - \mu_j \|^2 \tag{1}$$

For each class j, re-compute the centroid of the class using formula 2:

$$\mu_j := \frac{\sum_{i=1}^{m} 1\{c^{(i)} = j\}x^{(i)}}{\sum_{i=1}^{m} 1\{c^{(i)} = j\}} \tag{2}$$

where $c^{(i)}$ is the class in which sample i is nearest in k classes, whose value ranges from 1 to K, which is divided into two categories (K = 2). The centroid μ_j may be the center point of the samples, which belongs to the same category. First, we choose the point at which K samples concentrate as K centroids. The first step is to calculate the distance of every mean point to the centroid, and then select the nearest of the categories as $c^{(i)}$, so that after the first step each mean point has its own category. The second step is re-calculating its centroid μ_j of every mean point. Repeat the first step and the second step until the center of mass is the same or the change is very small.

Figure 2 shows the clustering results of historical measured daily mean radiation data. The boxes in the figure represent sunny day samples and circles represent cloudy samples; by data clustering, we can distinguish these two types of samples.

3.2 Classification algorithm for sunny and sunny cloudy days

The main difference between sunny and sunny cloudy direct radiation is radiation change tendency. Therefore, select the radiation data between the daily sunrise and sunset time to find the difference between the data using formula 3, and then calculate the mean data of the daily difference. Then, by the K-means clustering method, we obtain two categories: sunny and sunny cloudy. Finally, we can obtain the classification extreme data, and if the mean is higher than the value, then it is a sunny cloudy day. Otherwise, it is a sunny day:

$$Y_j = \frac{1}{n-1} \sum_{i=2}^{n} (X_{i+1} - X_i) \tag{3}$$

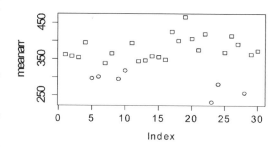

Figure 2. Classification of sunny and cloudy samples.

54

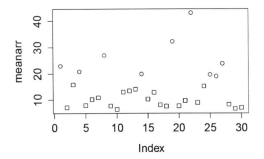

Figure 3. Classification of sunny and sunny cloudy samples.

Table 1. Weather type of a monitoring station in Northwest China in March 2015.

Date	Weather type	Date	Weather type
2015/3/1	Sunny	2015/3/17	Sunny cloudy
2015/3/2	Sunny cloudy	2015/3/18	Sunny cloudy
2015/3/3	Cloudy	2015/3/19	Cloudy
2015/3/4	Sunny	2015/3/20	Cloudy
2015/3/5	Sunny	2015/3/21	Sunny cloudy
2015/3/6	Sunny cloudy	2015/3/22	Sunny cloudy
2015/3/7	Sunny	2015/3/23	Sunny cloudy
2015/3/8	Sunny	2015/3/24	Sunny cloudy
2015/3/9	Sunny	2015/3/25	Cloudy
2015/3/10	Sunny cloudy	2015/3/26	Sunny cloudy
2015/3/11	Sunny cloudy	2015/3/27	Sunny cloudy
2015/3/12	Sunny cloudy	2015/3/28	Cloudy
2015/3/13	Sunny	2015/3/29	Cloudy
2015/3/14	Cloudy	2015/3/30	Sunny cloudy
2015/3/15	Sunny cloudy	2015/3/31	Sunny
2015/3/16	Sunny		

where X_{i+1} represents the direct radiation value of the i+1 time of day i, n is the number of daily direct radiation samples, and Y_j represents the difference mean of day j. Figure 3 shows the clustering results of historical measured daily difference mean radiation data, where the boxes represent sunny day samples and circles represent sunny cloudy samples.

The classification algorithm is used to classify the data of a monitoring station in the northwest of China, and the following results are obtained.

4 DIRECT RADIATION SIMULATION

According to extreme value of the sunny cloudy sample classification, we can divide the history samples into sunny, cloudy, and sunny cloudy. For each type of sample, we can use loess regression to model radiation curve.

4.1 Loess regression

Loess regression is a type of local polynomial regression, which is similar to moving average technique. That is, within the specified window, we can get each point value by weighted regression on data nearest window. The method includes the following steps:

1. Calculated initial weight of each direct radiation data points between daily sunrise and sunset time, the weight function expression is cubic function of the Euclidean distance ratio on radiation data;
2. Using initial weight to perform regression estimation, we can get an estimate formula. Get weight function using residuals of estimate formula and recalculate new weights;
3. Using new weight, repeat step (2), keep modifying weight function after the nth step is converged. We can get any smooth value based on polynomial and weight.
4. Finally, we can get predict direct radiation curve for the next day.

4.2 Case analysis

For the thermal power station in the northwest region, we select data of monitoring station near the thermal power station to model. First, according to the weather type, select the most recent direct radiation data near the forecast date for modeling.

If the next-day regular weather forecast is sunny, select the fitting curve as the next-day forecast curve and obtain the next 0–24 h direct radiation prediction results. The direct radiation history data near sunny days are selected, and loess method is used to fit the solar direct radiation curve. Using a similar method, we can forecast sunny cloudy and cloudy days, and the results are shown in Figures 5 and 6.

As can be seen from Figure 4, after inputting conventional weather forecast data into regression model, the predicted curve is similar to the measured curve.

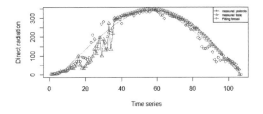

Figure 4. Comparison of direct radiation measurement and prediction in sunny day.

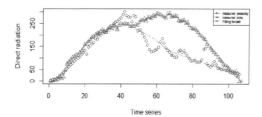

Figure 5. Comparison of direct radiation measurement and prediction in sunny cloudy days.

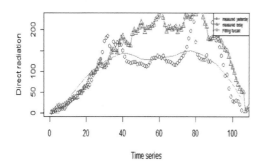

Figure 6. Comparison of direct radiation measurement and prediction in cloudy day.

Table 2. Error contrast.

Weather type	Days	Mean absolute error (W/m²)
Sunny	9	12.572
Sunny cloudy	15	40.05
Cloudy	7	46.12

4.3 *Error contrast*

Using the above method, the radiation prediction of the station was carried out in March 2015, and the error is shown in Table 2.

The error obtained by comparison is shown in Table 2. In sunny days, the radiation is more stable and the prediction is more accurate, that is, in March 2015, it is 12.572 W/m², and its tendency is very similar. However, in cloudy days, because of the random movement of clouds, error is large and the cloudy and cloudy days are similar.

5 POWER GENERATION SIMULATION

To simulate power generation capacity in thermal power generation station, we choose SAM software. We can calculate power output each hour, and get an array of 8760 items of power generation capacity, which is the output in 1 year. Using these data from different areas, we can provide location of the station[10]. These 8760 items can choose

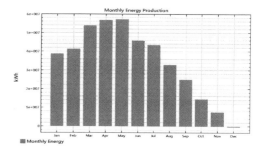

Figure 7. Simulation of power generation capacity in 2015.

conventional weather forecast in 1 year, and then we choose three types of line curve as the forecast line and obtain the prediction results of 2015.

We model a 100M trough solar thermal power plant in the northwest using SAM software. By inputting the radiation and temperature data, we can simulate power generation capacity, as shown in Figure 7.

6 CONCLUIONS

Using prediction method through conventional weather forecasting direct radiation, we can predict the radiation directly. This method effectively solves the problem of missing direct radiation data of power station. Through the simulation of direct radiation data, we can further simulate data generation in power stations, to provide reference for the operation of power stations. As sunny cloudy fitting curve is more complex, the prediction results have certain deviation, and further experiments are needed to select the better regression model to forecast.

REFERENCES

Luo Zhi-hui, Long Xin-feng. (2006). State and Trend of Solar Parabolic Trough Power Generation Technology., Electrical Equipment, 7(11), 29–32.

Zhao Ming-zhi, Song Shi-jin, Zhang Xiao-ming. (2013). A selection method of trough solar thermal power station siting, 31(03):18–22.

Chen Runze, Sun Hongbin, Li Zhengshuo, Liu Yibing. (2014). Grid Dispatch Model and Interconnection Benefit Analysis of Concentrating Solar Power Plants with Thermal Storage[J]. Automation of Electric Power Systems, 19, 1–7.

Qu Hang, Yu Xiao, Zhao Jun. (2009). Study and Simulation of Siting Factor of Solar Radiation for the Parabolic Trough Solar Thermal Power Plant, Ludong University Journal (Natural Science Edition), 25(1), 78–84.

Liao Shike, Wang Xiangyan, Zhu Lingzhi, Yang Wenhua. (2013). Research of water consumption in concentrated solar power system, 1, 35–40.

NREL. (2010). Parabolic trough power plant system technology[EB/OL]. http://www.nrel.gov/csp/trough-net /power_plant_systems.html,January 28.

A study on the external power demand and scale of a UHV synchronous power grid

Mingxin Xu & Long Zhao
Inner Mongolia Eastern Electric Power Co. Ltd., Economic and Technical Research Institute, Inner Mongolia, China

Yuan Meng, Jianhua Zhang, Yang Liu, Hao Hu, Tao Tan & Rishang Long
State Key Laboratory of New Energy Power System, North China Electric Power University, Beijing, China

ABSTRACT: Because of the different distribution of energy distribution and power demand in China, the study of demand and scale of the external power receiving is particularly important. In this paper, we use the convolution method based on probabilistic criteria of calculation to obtain the reliability index, calculate the external power demand of the regional power grid by using the reliability theory, and then through the static security and stability of power grid, plan the capability of transportation scale. Finally, an example is analyzed to obtain the external demand and scale. The method is proved to be correct.

1 INTRODUCTION

Because of the inconsistency in China's energy and electricity space and the increase of its economic development and power demand, in the future, China will form large-scale "West to East power transmission" and "Nortel to South power transmission". For reasonable planning tie line scale, structure, placement, and other issues into a grid, we must first clear external power demand. On this basis, it should be speculated that calculation is performed if the power grid meets the demand, margin, and the maximum transmission capacity.

The study on the effect of the external power on the regional power grid has been carried out elsewhere, which is related to the research work. References Yu (1995); Cai (2016); Cheng (2015); and Hu (2006) pointed out that the cross-transmission electrical advantages realize the optimization of resource allocation in large interval, and calculated the external power consumptive ability. References Chen (2015) and Chen (2015), with the background of AC/DC external power, analyzed Shanghai Grid Static reactive power voltage control, capacitive reactive power balance, the normal way and small DC mode system inductive reactive power balance, peaking resources allocation, development strategies. The static security and stability analysis method of large power grid has been studied elsewhere (Hou, 2004; Wu, 2007; Su, 2014; Qi, 2006). An analytical calculation model for the reliability evaluation of large power grid was proposed in Zhao (2013); Zhao (2013); Zeng (2013); Dai (2014); and Liu (2004).

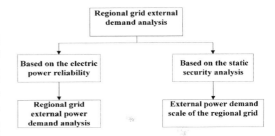

Figure 1. Regional grid external demand analysis process.

The main research ideas of this paper are shown in Fig. 1. First, the external power demand of the regional power grid is calculated based on power generation reliability theory, and then the external demand scale of the power grid can be accounted for based on the static security analysis method.

2 EXTERNAL DEMAND ANALYSIS OF REGIONAL POWER GRID

2.1 External demand analysis method based on the electric power reliability

In this paper, the convolution method based on probability criterion establishes the discrete probability distribution of the generating capacity and a discrete probability distribution of the load level, and then convolutes two distributions to calculate the corresponding reliability index. By gradually increasing the external power demand size and iterative calculation, external power demand with certain reliability can be obtained. The analysis-specific calculation flow chart is shown in Figure 2.

2.2 External demand analysis model based on the electric power reliability

2.2.1 Generation capacity reliability model
1. Outage model of generating set
Defined equivalent forced outage rates are as follows:

EFOR

$$= \frac{\text{Fored outage hours+ Equivalent outage hours}}{\text{Operating hours+Forced outage hours}} \tag{1}$$

In the equivalent process, the unit derating operating hours is equivalent to the outage hours, such as a 1000MW unit outputs 800MW for 10 h, then its equivalent outage time is 2 h.
2. Discrete probability distribution model of generating capacity
This method commonly stipulates reasonable rate of power-generating capacity. The capacity grade into the rate is rounded to the closest to a prescribed level. Therefore, we can obtain the corresponding generating capacity and probability.

2.2.2 Discrete probability distribution model of load level
Load level of a discrete probability distribution model is usually sort size load in descending order during the period of statistics. Then, the load distribution curve is drawn, and the multi-load level model is obtained through the analysis of the original load curve. If the load uncertainty is included, a generalized load level classification table is established, using the normal distribution

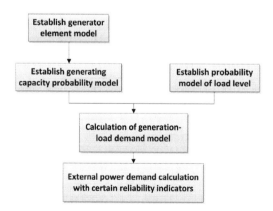

Figure 2. External demand analysis process based on the electric power reliability.

Table 1. Generating capacity and probability.

Generation capacity	Probability
G1	P1
G2	P2
…	…
Gk	Pk
…	…
Gn	Pn

Table 2. Generalized load level classification and probability.

Segment	Probability
Lk(1–3δ)	Pk1 = 0.0062Tk1/T
Lk(1–2δ)	Pk2 = 0.0606Tk2/T
Lk (1–δ)	Pk3 = 0.2417Tk3/T
Lk	Pk4 = 0.3830Tk4/T
Lk(1+δ)	Pk5 = 0.2417Tk5/T
Lk (1+2δ)	Pk6 = 0.0606Tk6/T
Lk (1+3δ)	Pk7 = 0.0062Tk7/T

characteristics and uncertainty of the load. Taking seven segments as example, each segment takes the center point data, and normal distribution table is established as follows:

2.3 External power demand analysis reliability index

Reliability index of power generation system generally used for reliability assessment via outage time and insufficient power expectation:
1. *LOLE* (Loss Of Load Expectation)

$$LOLE = \sum_{j=1}^{Ng} \sum_{i=1}^{Nl} P_i P_j I_{ij} * T \tag{2}$$

$$I_{ij} = \begin{cases} 0 & L_i \le G_j \\ 1 & L_i > G_j \end{cases} \tag{3}$$

where L_i is the ith load level; P_i is the probability of the ith load level; N_l is the grading number of load level in load level probability table; G_j is the jth generation capacity; P_j is the probability of the jth class generation capacity; N_g is the power generation capacity grading number in grading table; and t is total load duration.
2. *LOEE* (Loss Of Energy Expectation)

$$LOEE = \sum_{j=1}^{Ng} \sum_{i=1}^{Nl} P_i P_j * \max(0, L_i - G_j) * T \tag{4}$$

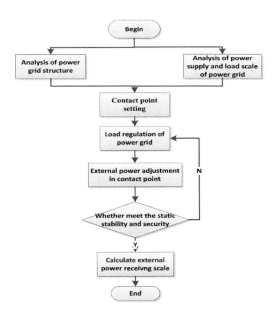

Figure 3. Scale of the external power receiving analysis process.

3 ANALYSIS OF EXTERNAL POWER RECEIVING SCALE

3.1 *Method for the analysis of external power receiving scale*

The prerequisites of the analysis of the external power receiving scale of the regional power grid are that the static security level is guaranteed and the method to calculate the maximum value of the external connection power supply is accepted by the grid. The analysis process is shown in Figure 3:

4 CASE STUDY

This chapter mainly selects the BPA data of a synchronous power grid in 2020 and the power load of the power grid. The total number of nodes is 24927. The active load is 7.68 million KW, the regional active output is 5.03 million KW, and the vacancy of active power output is 2.65 million KW. The analysis of regional active load and scale distribution of 13 regions is shown in Fig. 4. This paper mainly selects the BPA data of a synchronous power grid in 2020 and the power load of the power grid. The total number of nodes is 24927. The active load is 7.68 million KW, the regional active output is 5.03 million KW, and the vacancy of active power output is 2.65 million KW. The analysis of regional active load and scale distribution of 13 regions is shown in Fig. 4.

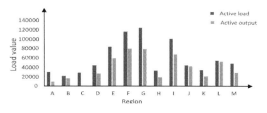

Figure 4. Analysis of regional active load and scale distribution of 13 regions.

The foregoing analysis showed a positive correlation with the load of each region. Active load, output, and power gap of regional G is the largest.

The external power demand analysis of regional power grid can use the above analysis results to establish generating system reliability model, select reliability index, setting index level, calculate the external power demand, and then appropriately adjust to the active power output or load in order to satisfy the external power limit under the static security level.

4.1 *External demand analysis result*

To further analyze system variation relationship of reliability and power demand, the case calculated the external power demand under the condition of different LOLE by changing LOLE to change index convergence condition and establish a relationship between LOLE and external power demand as shown in Fig. 5.

According to the requirements of reliability guidelines, in order to guarantee the reliability of the system level, the system power supply availability rate is not less than 99.9%, i.e. LOLE is less than or equal to 8.76 h/year, The region's external power demand should be 201043MW at least.

4.2 *Result of external power receiving scale analysis*

1. The result of external power receiving scale analysis

The area external to the electric contact point has 35 contact points, including 14 DC placement and 21 AC placement.

2. The result of load adjustment in each region

The case increases the load level of the area slowly for each region with 0.5% growth interval, adjusts the power supply of the external power supply at the same time, and then obtains the power supply scale by the system static security check.

3. Calculation results of the power transmission scale of the contact point and results of the external power receiving scale.

The results of a region is shown in the table below

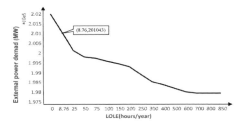

Figure 5. External power demand under different LOLE level.

Table 3. The result of load adjustment in each region.

Region	2020 load /MW	External power receiving/MW	Load increase /MW	Increment percentage
A	30559	33007	2448	8.01
B	21674	23845	2171	10.02
C	29357	35801	6443	21.95
D	124659	139634	14975	12.01
...	...	....	...	...
Total	767980	844665	76686	9.99%

Table 4. Calculation results of the power transmission scale of the contact point and the external power receiving scale.

Line	Load in high-flow heavy operation mode in 2020/MW	Load in external power receiving/MW	Additional output of contact point/MW	External power receiving scale/MW
DC line	8000	8000	4442	12442
	...	...	...	...
	10000	10000	6190	16190
	4000	4000	600	4600
AC line	2386	356	5553	5909
	...	...	...	...
	16560	16930	9	16939
Total	227686	227793	79084	306877

It is evident from the tables that, in high-flow heavy operation mode in 2020, external power receiving scale is 307 million KW, in which original power load is 228 million KW. We can know the power grid mainly in the UHV DC transmission line access. In addition, the maximum power supply scale of DC high-voltage transmission line of each contact point is 16190 MW and the largest AC transmission scale is 16939 MW. Therefore, the synchronous power grid in case study meets the level of reliability requirements, and the external power change interval is [2.01 3.07] million kW. The results not only satisfy a higher level of reliability of the regional power grid, but also continue to increase the external power receiving about 79 million KW, that is, the growth margin is 25.73%.

5 CONCLUSIONS

In this paper, we studied external power demand and scale of UHV power grid. First, an analytical method of external power demand based on power generation reliability is proposed, which is based on the convolution method, and power demand is

obtained by the reliability calculation. In addition, in the case of the static security level, the maximum value of the external contact transmission power is used to obtain the external power supply scale. In the case study, the method is proved to be reliable and can increase the margin. Finally, the feasibility of the method is verified.

REFERENCES

Cai Hui, Meng Fanjun, GE Yi, Huang Junhui, Luo Jinshan, Xie Zhenjian, QI Wanchun. (2016) Accommodation Capacity of Jiangsu Power Grid External Electricity Based on UHV AC / DC Hybrid. J. Electric Power Construction. 37(2), 100–106.

Chen Bo, Xu, Yiqing Huang, Yichao, Tao Yanfeng. (2015) Research on Peak Regulation and Resources Sharing Mechanisms of Shanghai Power Grid. J. Power System and Clean Energy. 31(7), 37–43.

Chen Bo, Yin Ting, Huang Yichao, Cao Minmin, Zhang Kai. (2015) Reactive Power Issues of Shanghai Power Grid under the Background of AC-DC Electric Power from External Regions. J. Power System and Clean Energy. 31(4), 66–71.

Cheng Xiong, Cheng Chuntian, Shen Jianjian, Li Gang, LU Jianyu. (2015) Hierarchically Structure and Model Driven Automatic Generation of Distriction Network

Graphs. J. Automation of Electric Power System. 39(1), 151–158+232.

Dai Hongyang. (2014) Research on the Reliability Evaluation Method for AC/DC Power Grid. D. North China Electric Power University. 15–58.

Hou Ru-feng, Cai Ze-xiang, Yin Liang, Wang Chang-zhao, MA Jie-ran. (2004) Security Analysis Of District Power Network Based On The Maximal Load Forecasting. J. Power System Teehnology. 28(23), 38–42.

Hu Wei, Liu Jin-guan, Liu Hua-wei. (2006) Influence of power from outside on Jiangsu Power Grid and the countermeasures. J. East China Electric Power. 34(1): 52–56.

Liu Hai-tao, Cheng Lin, Sun Yuan-zhang, Zheng Wang-qi. (2004) Reliability Evaluation of Hybrid AC/DC Power System. J. Power System Technology. 28(23): 27–31.

Qi Xu, Zeng De-wen, Shi Da-jun, Fang Xiao-song, LI Lan, Su Hong-tian, Wu Wei. (2006) Study on Impacts of UHVDC Transmission on Power System Stability. Power System Technology. 30(2):1–6.

Su Weibo, Feng Yanwei, Yang Yude. (2014) A Method of Static Security Analysis and Correction Based on BPA Data. J. Power System and Clean Energy. 30(5), 13–17.

Wu Zhifeng. (2007) Static Security Analysis Considering Uncertainty in Power System. D. Tianjin University. 12–30

Yu Zhen, Zhao Kezheng, Yu Dunyao. (1995) The Necessity Of International Interconnected Power System In Viewing Of Developing East China Power System. J. Power System Teehnology. 19(8):24–28+33.

Zeng Qingyu. (2013) Analysis on Reliability of UHVAC and UHVDC Transmission Systems. J. Power System Technology. 37(10), 2681–2688.

Zhao Yuan, Guo Yin, Xie Kaigui. (2013) Research on Probability Distribution Characteristics of Bulk Power System Reliability Considering Parameter Uncertainty. J. Power System Technology. 37(8):2165–2172.

Zhao Yuan, Zhang Jin, Lu Jingjing, Xie Kaigui. (2013) A Convolution Model for Bulk Power Grid Reliability Evaluation. J. Power System Technology. 09, 2466–2473.

Research on a predictive maintenance system of coal washing equipment

D. Li & M.Y. Ren

School of Mechanical Electronic and Information Engineering, China University of Mining and Technology (Beijing), Beijing, China

ABSTRACT: The washing process of coal is of great significance to the efficient use of energy and environmental protection before burning. Vibration screen is one of the important equipment in the process of coal washing. However, faults cannot be avoided due to various reasons. 4M system is a new type of mining equipment health diagnosis system, which includes four combined functional modules. This system can change the periodical maintenance or break maintenance to predictive maintenance, minimize potential safety problems, and prolong equipment lifecycle. Thus, it can ensure the safe and efficient operation of the equipment.

1 INTRODUCTION

1.1 Coal washing

Coal is the main source of energy of China. The prospective reserves of coal are more than 5 trillion tons and the proven reserves are 1.2 trillion tons, which is the third most after Russia and the United States. The annual demand of coal in China is about 4 billion tons, which is increasing every year. With the continuous development of new energy technology, China adjusts the energy structure constantly. Even so, coal will also be the basic energy source quite a long time in the future.

As the demand of coal increases, coal mining gradually achieves mechanization and the quality of raw coal differs in different geological conditions. The use of raw coal, which has high ash content, high sulfur content, and high gangue content, will cause incomplete combustion leading to environment pollution and wastage of resource. Therefore, the raw coal needs to be washed and cleaned before use, in order to reduce the impurity.

According to the statistical data of China Coal Industry Association, the quantity of raw coal washed is 2.214 billion tons (59.8% of the total coal used). This proportion should reach 70%–95% and even 100%. According to the goal of *Atmospheric Pollution Prevention Action Plan* published by the state council in 2013, the ratio of raw coal washing should be over 70% in 2017. Thus, it can be seen that raw coal washing and cleaning will play increasingly important roles in coal production. Coal washery is an important part in the process of coal washing and processing. Its main functions are:

1. To remove the impurities and lower the ash content and sulfur content;
2. To classify the coal by different quality and specifications to meet different needs;
3. To remove the gangue and reduce invalid transportation.

1.2 Vibration screen and common faults

Modern coal washery is highly automatic and mechanized. The vibration screen (shown in Figure 1), which is one of the most important washing equipment in coal washery, is mainly used to achieve raw coal classification, medium recovery, etc. And it is related to the safety of coal production.

With the increasing coal production requirements, the production efficiency of equipment is

Figure 1. Vibration screen.

Table 1. Common faults of vibration screen.

Faults	Cause analysis
Poor effect of sieving	(1) Screen blinding
	(2) High moisture content
	(3) Feeding is not uniform
	(4) Feeding too thick
	(5) Screen is not tight
Low rotation speed	Conveyor is not tight
Bearing heating	(1) Bearing is oil-less
	(2) Bearing is dirty
	(3) Too much oil
	(4) Bearing wears
Wake vibration	Incorrect installation
Strong vibration	Eccentricity is different
The spindle is not turned	Seal sleeve plugs
Abnormal voice	(1) Bearing failure
	(2) Screen is not tight
	(3) Bolt loosens
	(4) Spring failure
	(5) Gear failure

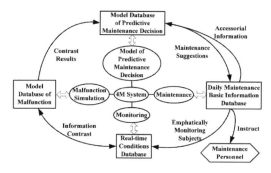

Figure 2. Structure and function diagram of 4M system.

significantly improved, and the vibration screen has become larger with high processing capacity and higher efficiency. Continuous running under poor working condition leads to the vibration being operated upon by not only the huge impact force, but also the exciting force. If the vibration could not get the maintenance timely, some faults would appear like screen box crack, beam fracture, and bearing failure. The common faults of vibration are shown in Table 1.

2 "4M SYSTEM"

At present, most mining enterprises have adopted the periodical maintenance or break maintenance to the key equipment. For example, Chengzhuang coalmine's coal washery of Jincheng Coal Group, which is located in Shanxi Province, has adopted the periodical maintenance system. The workers check all the equipment in coal washery from 9 a.m. to 12 a.m. every day. This approach is not purposeful and all the equipment need to be stopped. In addition, the most important problem is that some faults like cracks would be left unnoticed sometimes because of the dusky environment. Modern coal washery adopts the assembly line work. Once damage occurs to certain equipment, it would stop the whole production line.

Therefore, if we change periodical maintenance or break maintenance to predictive maintenance, we would know the equipment running state and fault occurrence regularity, diagnose the equip-

ment detect in the early stage, and predict the remaining life. Safe production in coalmine is of great significance for the coal washery.

Mining Equipment Health Diagnostic System (4M system) is a new type of quaternity mining equipment management mode. The system mainly includes daily maintenance information management, on-line monitoring, malfunction simulation, and predictive maintenance. Four databases respectively correspond to four functions. The structure of 4M system and its function are shown in Figure 2.

The basic operation procedure of 4M system is as follows: original design information and historical maintenance information are stored in basic information database of daily maintenance. They can be used for the prescient maintenance decision-making. Real-time monitoring system can collect the information data of working equipment, which is stored in real-time database. We can obtain the faults characteristic parameters through the malfunction simulation, which can then be compared with the real-time monitoring data. The results of the comparison can be used to judge the condition of equipment and give the maintenance suggestion, which can also be stored in basic information database and guide the further maintenance work.

3 DESIGN OF "4M SYSTEM"

3.1 Daily maintenance basic information database

The daily maintenance basic information database includes two parts: basic information of equipment and routine maintenance account. Basic information of equipment includes equipment name, model, design parameters, design diagram, production units, and enable time information. Daily maintenance account includes historical fault information, history and detailed notes of maintenance, and the

Table 2. Historical maintenance information.

Num	Projects	Time	Location	Details	Result
1	Replace 2 beams	2002.2.6	4th floor	2 beams cracked, removed and replaced them	Operating well
2	Replace 2 beams	2002.3.26	4th floor	Removed 4 beams and replaced them all	Operating well
3	Replace 2 beams	2002.3.25	4th floor	Abnormity in vibrator, moved and replaced it	Operating well
4	Replace 2 beams	2002.7.27	4th floor	2 beams cracked, removed and replaced them	Operating well
5	Replace 2 beams	2002.9.6	4th floor	Abnormity in vibrator, moved and replaced it	Operating well

Table 3. Basic information of vibration screen.

		Quantity
Vibration screen (315)	2ZKX2448 Separation size 20 mm Sieving area 12m^2 Inclination angle 6 Mesh size 20 × 20 Feed size 0–80 mm Capacity 80–125 t/h	
Motor	Y180L-4 22KW 42.5A 1470 rpm	1
Vibrator	ZKX120	1
Bearing	4E3012324Q	8
Gear	ZKX120-12	4
–	–	–

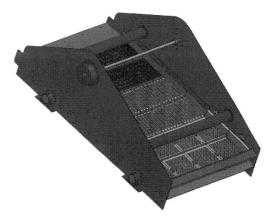

Figure 4. 3D model of vibration screen.

Figure 3. Monitoring of vibration screen.

predicting maintenance decision system that put forward in the next stage of maintenance content. The examples are shown in Tables 2 and 3.

3.2 *Monitring*

This module can obtain the information of related parameters such as vibration, temperature, or noise signals in the working condition of equipment through on-line monitoring system platform (shown in Figure 3).

3.3 *Malfunction simulation*

The model database of malfunction includes the typical malfunction model and historical malfunction model. The historical malfunction model needs the monitoring data of equipment for a long time, which is not easy to achieve. Therefore, we can obtain the typical malfunction model using computer simulation. We can build a three-dimensional (3D) model (shown in Figure 4) and add the typical malfunction to it, in order to obtain the typical fault characteristics.

3.4 *Model of predictive maintenance decision*

To establish the predicting maintenance decision model, we should refer to the existing theory of predicting maintenance decisions, for example, projection pursuit. The common methods are summarized as follows: data analysis method,

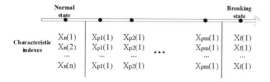

Figure 5. Development process of equipment status.

qualitative analysis method, time series analysis method, casual relationship analysis method, mathematical modeling method, and combined method.

The future state of equipment can be reflected by the historical and current states, so we can study the relationship between them and predict the future state. The development process of equipment status is shown in Figure 5.

4 CONCLUSION

Coal washery is an important part in the process of the coal washing and processing. Reliable operation of coal preparation equipment is of great significance for the coal washing process. Therefore, it is necessary to establish a set of predicting maintenance system for coal washery equipment, to replace the existing artificial periodic maintenance way. 4M system, Mining Equipment Health Diagnostic System, is a type of new equipment management mode, which can obtain equipment running state, find out the early faults, forecast the malfunction development trend, and evaluate residual life. The establishment of the system is of great significance for the safe and reliable operation of the coal washery and even the whole coalmine.

REFERENCES

Loparo, K.A. Falah, A.H. & Adams, M.L. 2003. Model-Based Fault Detection and Diagnosis in Rotating Machinery. *Proceedings of the Tenth International Congress on Sound and Vibration*, 2003:1299–1306.

McArthur, S.D.J. Booth, C.D. McDonald, J.R. McFadyen, I.T. 2004. An agent-based anomaly detection architecture for condition monitoring. *IEEE Transactions on Power Systems*, 20(4):1675–1682.

Li, W. Meng, G.Y. Cheng, X.H & Yang, J. 2011. Design on Precognition Maintenance System of Mine Equipment. *Coal Engineering*, 2011(5):125–126.

Cheng, X.H. Meng, G.Y. Li, W & He, X. 2011. Design of Mining Equipment Health Diagnostic System (4M system). *Computer Science and Automation Engineering (CSAE), 2011 IEEE International Conference on IEEE*, 2011(1):273–276.

Zorriassatine, F. Ashraf, & B.Notini, L. 2005. Smarter maintenance through Internet-based condition monitoring with indirect sensing, novelty detection, and XML. *Journal of Systems and Control Engineering*, 219 (I4):283–293.

Advances in Energy, Environment and Materials Science – Wang & Zhou (Eds)
© 2017 Taylor & Francis Group, London, ISBN 978-1-138-03600-0

Study on the method to realize the automatic compensation of power factor in situ using equivalent variable capacitor

Shufan Mao
Department of Engineering Training Center, Tianjin University of Technology, Tianjin, China

Xianglong Li
Department of Computer and Information Engineering, Henan University, Kaifeng, Henan Province, China

Xv Han & Zhuang Chen
Tianjin University of Technology, Tianjin, China

ABSTRACT: This paper presents an automatic compensation method based on the principle of capacitor charge and discharge voltage. The switch circuit is composed of two self-turn off devices which are respectively connected in reverse parallel with the self-turn off device. By logic driving circuit, pulse width modulation circuit, zero crossing detection circuit and power factor, the detection circuit control switch and the self-turn off device of the circuit are switched off. In order to control the charging and discharging voltage of the capacitor, the compensation current of the capacitor is controlled, and the automatic control of the power factor compensation of the load is realized. The principle and working process of the circuit and the selection of the main parameters of the circuit are analyzed and described. The automatic power factor compensation control device of self-turn off device is made by using the method, which has the advantages of simple structure, small volume, high efficiency, and high automatic power factor compensation of the power factor of the electric load.

1 INTRODUCTION

It is an effective technique to use the capacitor to compensate the power factor of AC power, which is widely used at present. And the field (random) compensation is the most effective compensation method (Li, 2009; Ma, 2010). Field random compensation requires small size, high efficiency, can automatically control the power factor compensation. The circuit used in the current power factor automatic compensation device can be broadly divided into two categories (Su, 2007): Firstly, SVC, is the abbreviation of "Static Var Compensator" (Liu, 2015; Yang, 2000). The main control elements of the device are thyristor, capacitor and reactor. The current of the thyristor controlled reactor is "static", but its function is the dynamic reactive power compensation. The method is adopted to control the conduction angle of the thyristor, and control the current of the saturated reactance to adjust the reactive power of the compensation. The compensation equipment made by this kind of method not only has large volume of saturated reactor, but also has great loss of thyristor switch. So it cannot meet the requirements of random field. Second, SVG, also known as ASVG

or STATCOM, is the abbreviation of "Var Generator Static" (Hu, 2006; Huang, 2007). The basic principle of the circuit is that the self-commutated bridge type circuit is connected to the power grid through a reactor or a direct parallel connection. The method is suitable for the AC side output voltage of the bridge circuit, the phase and the amplitude of the side output voltage, or directly control the AC side current of the device. In this way, the circuit can be absorbed or sent to meet the requirements of reactive current (inductive or capacitive), to achieve the purpose of automatic control of power factor. Although this kind of method is compared with the one kind of method to remove the large loss of saturation reactance, the efficiency is improved. But the compensation device is made up of large reactance volume, self-turn off device switching losses, and is not easy to make the field random compensation equipment.

In this paper, an automatic compensation method based on automatic control capacitor charging and discharging voltage is presented. The method is based on the self-turn off device power factor and the automatic compensation control circuit. By using the logic driving circuit, the pulse width modulation circuit, the zero point detecting circuit and the power factor detecting circuit, the two self-shut down devices in the switching circuit are respectively controlled to pass through and break off at the zero crossing of the AC current. In order to control the charging and discharging voltage of the capacitor, the compensation current of the capacitor is realized, and the control goal of the automatic compensation of the power factor of the capacitor is achieved. It can also be said that the circuit is the capacitor voltage automatic regulation and compensation circuit. The circuit not only overcomes the disadvantages of the existing SVC and SVG losses, but also has the advantages of small volume, high efficiency and automatic control compensation power factor. The circuit can meet the requirements of the market in the field of urgent demand for random compensation. This kind of circuit has not been reported so far.

2 THE BASIC PRINCIPLES OF THE METHOD

The basic principle of the compensation circuit is shown in Figure 1. 50 Hz electricity by input A and B access equipment power supply power supply Fz. Because the electrical equipment Fz is inductive load, capacitive impedance in Figure 1 of the reactor L1 and capacitor C compared to 1%. So it can be ignored. Fig. 2 is an equivalent circuit diagram of Figure 1. In Figure 2, the Fz is equivalent to a resistance RL, and the inductor is connected in series with the compensation capacitor C. Self-turn off device T1 and T2 are equivalent switches.

2.1 Capacitors in AC circuit

The energy stored by the capacitor in the DC (Yuan, 2006):

$$w = \int_9^t uidt = \int_0^t cudu = \frac{1}{2}cu^2 \qquad \text{(2–1 type)}$$

As can be seen from the 2–1 type, the energy stored in the capacitor is proportional to the square of the voltage at both ends. To change the terminal voltage of the capacitor, the energy stored in the capacitor can be changed.

$$P = \frac{1}{T}\int_0^t Pdt = \frac{1}{T}\int_0^t -\frac{U^2}{2C}\sin 2\omega t\, dt = 0 \qquad \text{(2–2 type)}$$

It can be seen from the 2–2 type that the charge current is equal to the discharge current when the capacitor is operating in an AC circuit. So the average power is equal to 0 (see 2–2 type).

2.2 No non sinusoidal reactive power and active power

According to the definition, in the circuit containing non sinusoidal, active power P, apparent power S, and reactive power Q have a relationship is:

$$S = \sqrt{P^2 + Q^2} \qquad \text{(2–3 type)}$$

Due to the voltage waveform of the capacitor charging and discharging, the waveform of the voltage is only the sine wave which is the boundary of the switch, while the central part is not a sine wave. So the reactive power of the capacitor compensation is:

$$Q = UI\sin\varphi \qquad \text{(2–4 type)}$$

This type can be seen as the maximum value when UI sin90°. This circuit switch controls the charging voltage on π/2 coinciding with the maximum compensation. In the actual circuit, taking into account the loss and compensation, allowance can be taken 1.2 UI. Apparent power S is not in compliance with the law of conservation of energy, S is only possession, and P is the use of the law of conservation of energy.

2.3 Time domain control of charge and discharge voltage to control reactive power compensation

By the 2–5 type can be seen in the cycle of alternating current, the realization of self-turn off device T1 and T2 on the capacitor C charging and discharging voltage level control, you can control the AC load power factor compensation of the current size. About the ic in Fig. 2:

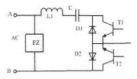

Figure 1.   Basic principle of the compensation circuit.

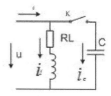

Figure 2.   Equivalent circuit diagram of Figure 1.

$$i\sin\varphi = i_c - i_l\sin\varphi$$
$$i\cos\varphi = i_l\cos\varphi$$
(2–5 type)

$$P = \frac{1}{T}\int_0^T P\,dt = \frac{1}{T}\int_0^T -\frac{U^2}{2C}\sin 2\omega t\,dt = 0$$
(2–6 type)

Active power P in the circuit is:

$$\begin{aligned}
p &= \frac{1}{2}\int_0^{2\pi} uid(\omega t)\\
&= UI\cos\omega\\
&= \frac{1}{2}\int_0^{2\pi} (UI\cos\varphi - UI\cos\varphi\cos 2\omega t)d(\omega t)\\
&\quad - \frac{1}{2}\int_0^{2\pi}(-UI\sin\omega\sin 2\omega t)d(\omega t)\\
&= UI\cos\omega
\end{aligned}$$
(2–7 type)

By the 2–6 type can be seen in the work of the cycle of alternating current, capacitor charging power, can achieve self-turn off device on the capacitor voltage U, and then control the size of the capacitor current. Because of the change of voltage U, the instantaneous active power P is changed to achieve the purpose of automatic compensation of the power factor of the AC load (see 2–7 type).

3 IMPLEMENTATION OF THE METHOD

The circuit can be opened and closed controlled by the driving circuit, the capacitor C charging and discharging (see Figure 3). The black area in Fig. 4 is the charge voltage. AC L through the current limiting inductor AC, flow to the compensation capacitor C. The switching circuit is composed of two self-turn off devices T1, T2 and two diodes D1, D2. In the positive half cycle of the AC voltage, T1, in the interval (0, π /2), controlled capacitor D2 through time adjustable charging mode. Control waveforms are shown in Figure 5. In (π/2, π) time, the r end of driving the T2 control the

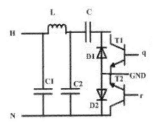

Figure 3. Driving circuit.

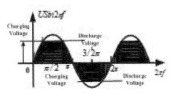

Figure 4. Both ends of the capacitor charge and discharge voltage waveform.

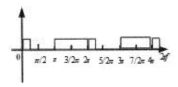

Figure 5. Point q drive waveform.

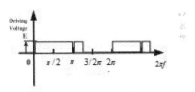

Figure 6. Point q drive waveform.

discharging of capacitor D1. In the negative half cycle of AC voltage, T2, in the interval (π, 3π/2), controlled capacitor D1 through time adjustable charging mode. Control waveforms are shown in Figure 6. In the period of (3π/2, 2π), the T1 control the discharging of capacitor D2. The capacitor charging is realized in the periodic operation of alternating current.

4 METHOD OF APPLICATION

4.1 Case (Li, 2011)

Figure 7 is the circuit diagram, using the above principle design, introduces a self-turn off device capacitor power factor compensation control device. When the input side of the N and the common ground terminal H access AC, the output M and N connection load. In the filtering circuit 6, the current is filtered by the filter capacitor C1, the inductor L and the capacitor C2, and the circuit is composed of a capacitor 7 and a switch circuit 1. At work, the power factor detects the current of circuit 5, mutual inductor DL will pass the current waveform to the phase comparator yx. Compared with the voltage waveform detected by the

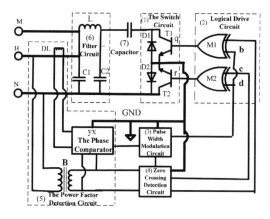

Figure 7. Compensation control device.

transformer B in the circuit 5, it will pass to pulse width modulation circuit in the form of voltage 3. At the same time, the zero detection circuit 4 is based on the voltage waveform detected by the secondary coil of the transformer B in the circuit 5. A zero sync pulse voltage is sent to the circuit 3. The circuit 3 outputs the corresponding pulse width voltage waveform according to the received zero synchronous pulse voltage and the modulation voltage. Output pulse width in alternating positive and negative half cycles are respectively varied at zero and π point as a starting point in $(0 - \pi/2)$ and $(\pi - \pi/2)$ in the periodic.

In the positive half cycle of alternating current, the control end of the self-turn off device q is controlled by T1, and the capacitor 7 is charged by the diode D2 in the circuit 1. Negative half periodic square wave output to the point A of the circuit M1, by the Q of the T1 control, the diode D2 in the switch circuit 1 discharges the capacitor 7. In the work of alternating current, the zero terminal voltage of the self-turn off device in the circuit are started capacitor charging and zero current discharging.

4.2 The main parameters of the device selection

When the reactor is opened at the zero point potential, the current limiting inductance L can be greatly reduced. In the design of the self-turn off device to choose the opening of the tube voltage drop, coupled with the diode voltage drop, the voltage value is calculated by 2V. Current limiting reactance value XL = 2V/Working current. When the working current is 10, n = 2/220 ≈ 0.009Ω. Consider the margin by 0.01, so the reactance value of XL = 0.01Xc = 0.01 × (1/2π • f C). Xc can be used for the electrical capacitance C's compensation. Now select XL = 0.01, which can be seen, the reac-

tance of this circuit is 1/10 of the reactance of the current compensation circuit, which greatly reduces the volume of the current limiting inductance.

In order to further reduce the volume of the whole machine, a new type of film capacitor is used. The volume is 1/3 of the average capacitor, and the resistance ESR only needs MΩ unit level, which can further reduce the loss, and the service life is more than 100 thousand hours. In the 380V AC working voltage, we will choose the voltage 800VDC. And in the 220V alternating current, we will choose the pressure 450VDC.

5 THE TEST RESULTS

By four prototypes of 220V, 4.2A and two prototypes of three-phase 380V, 2.5A, the experimental results show that, the power factor compensation control device of self-turn off device is made by using the method, which has the advantages of small loss, small volume and low noise. Device can automatically control the power factor in the 0.97–98 range, fully capable of the requirements of random field compensation. Through the oscilloscope observation, after compensation, the voltage waveform has no obvious distortion, and the compensation of the power supply voltage is not increased. In particular, the three-phase AC power factor automatic compensation control device is composed of three circuits. The power factor compensation is carried out separately for each phase power source in the three phase operation. And the device may have a three-phase unbalanced power. The feasibility of the automatic compensation method of the capacitor charging and discharging voltage is proved by the running of the prototype.

In the laboratory, we made a three-phase 380V, 3A self-turn off devices of power factor automatic compensation device. And the 5 KW three-phase air condition obtained the actual data as follows:

	Working voltage (V)	Operating current (A)	Power factor
Before compensation	386	9	0.6
After compensation	386	6.5	0.985

Before compensation, the input current has a dynamic change in 8.7–9A, and after compensation, the current changes in 6.2–6.5A. There is no higher harmonic exists here. As a result, the power grid can improve the utilization rate of 27–29%. The power saving rate can reach 5–7%.

6 CONCLUSION

This paper introduces a method of automatic compensation of power factor of AC power factor by controlling the charge and discharge voltage of capacitor. This method has been proved in theory. By utilizing the self-turn off characteristic of the device, we component switch circuit, and power factor compensation control circuit of capacitor. The method is verified by means of an example. The zero current charge and discharge circuit is adopted to reduce the switching loss, and the volume of the current limiting reactance is reduced, and the working efficiency is improved. The soft switching operation of self-turn off device has the advantages of low loss, high efficiency, small size, simple structure, low cost, convenient use and maintenance, etc. It provides a blue print to compensate the power factor of capacitor.

REFERENCES

Cheng Yang, "Analysis of main circuit connection type ± 200MVA chain type static compensator STATCOM," in press.

Heping Su, Guiying Liu, "Static wattless power compensation technology," China Power Press.

Ming Hu, Yan Du, Ling Gan, "Application of SVG in Xuzhou Mining Group 6 kV power supply system in the governance of the harmonic".

Shun Yuan, Shui Han, "Distribution network reactive power optimization and no wattless compensation device," China Power Press, May 2006.

Xianglong Li, "Self turn off device of power factor compensation control circuit," CN patent, 2011, 20489240.5.

Xiaojun Li, "The calculation of wattless power compensation macro asynchronous motor," Power capacitor and wattless power compensation.

Xixia Liu, "Application of SVC static type dynamic wattless power compensation system".

Yuqun Huang, "SVG application of non-reactive power compensation in distribution network," Mechanical and electrical information.

Zhaoxiang Ma, "Research of oilfield distribution network wattless optimization compensation," Northeast Electric Power Technology, April 2010.

Advances in Energy, Environment and Materials Science – Wang & Zhou (Eds)
© 2017 Taylor & Francis Group, London, ISBN 978-1-138-03600-0

Responses due to hot fluid injection into a cross-anisotropic thermoelastic full space

J.C.-C. Lu
Department of Civil Engineering, Chung Hua University, Hsinchu, Taiwan, R.O.C.

F.-T. Lin
Department of Naval Architecture and Ocean Engineering, National Kaohsiung Marine University, Kaohsiung, Taiwan, R.O.C.

ABSTRACT: An analysis method based on analytic technique is presented to determine the responses of hot fluid injection into a poroelastic full space. A mathematical model is developed for the distribution of land deformation, excess pore fluid pressure and temperature changes of the full space aquifer. Closed-form solutions are derived through the application of Hankel transform with respect to the radial coordinate. The results can provide better understanding of the hot fluid injection induced responses of the cross-anisotropic strata.

1 INTRODUCTION

Thermal mechanical responses of strata due to the hot fluid injection into deep underground are important petroleum engineering issues. For the impact on engineering safety, many studies were conducted to understand the mechanical, thermal and hydraulic behavior of deep hot fluid injection. Excessive hydraulic and thermal disturbance usually result in a volumetric change of fluid and solid skeleton. This change can increase excess pore fluid pressure and lead to decrease in effective stress, which can result in a hydraulic or thermal failure in the strata due to loss of shear resistance of solid skeleton. The simulation of these physical features is a complex task, and its validation is a major concern for the safety improvement of the hot fluid injection.

In recovering heavy oil from heterogeneous reservoirs through fine-mesh numerical simulations, Alajmi *et al.* (2009) investigated the performance of hot water flooding compared to conventional water flooding. Bear and Corapcioglu (1981) developed a mathematical model for the areal distribution of land subsidence, horizontal displacements, fluid pressure and temperature due to hot water injection into thermoelastic confined and leaky aquifers. Rosenbrand *et al.* (2014) addressed permeability change in sandstone due to heating from 20°C to 70~200°C. In the study of Sasaki *et al.* (2009), a system of gas production from methane hydrate layers involving hot water injection using dual horizontal wells was investigated.

A three-dimensional middle-size reactor was used by Yang *et al.* (2010) to simulate gas production from methane hydrate-bearing sand by hot-water cyclic injection. Lu and Lin (2006) displayed transient ground surface displacement produced by a point heat source or sink through analog quantities between poroelasticity and thermoelasticity. Based on three-dimensional thermoelastic theory of homogeneous isotropic media, the golden ratio shown in the maximum ground surface horizontal displacement and corresponding vertical displacement of a half space subjected to a circular plane heat source was presented by Lin and Lu (2010).

In general, soils are deposited through a geologic process of sedimentation over a long period of time. Under the accumulative overburden pressure, strata display significant anisotropy on mechanical, seepage and thermal properties. Both stratified soil and rock masses show the phenomenon of anisotropy. For this reason, theoretical and numerical models should be able to simulate the layered soils and rocks as cross-anisotropic media (Amadei *et al.* 1988, Lee & Yang 1998, Sheorey 1994, Tarn & Lu 1991, Wang & Tzeng 2009).

The present investigation is focused on the closed-form solutions of a cross-anisotropic thermoelastic medium due to a point of deep hot fluid injection which still have not been derived in previous studies. In this paper, the soil or rock mass is modelled as a linearly elastic medium with cross-anisotropic properties. The hydraulic fluid flow, thermal conductivities and mechanical properties are assumed to be cross-anisotropic. By using

Hankel transform with respect to the radial coordinate, closed-form solutions of long-term displacements, excess pore fluid pressure and temperature changes of the strata due to a point of hot fluid injection at large depth are obtained. The results can provide better understanding of the hot fluid injection induced responses of the strata.

2 MATHEMATICAL MODEL

2.1 Basic equations

Figure 1 displays a point of hot fluid injection into a poroelastic full space. The layer of cross-anisotropic soil or rock is modeled as a homogeneous full space. For simplicity, the plane of symmetry of the strata is in the horizontal direction. Let (r, θ, z) be a cylindrical coordinate system for this layer of strata where the plane of isotropy coincides with the $r - \theta$ horizontal plane. The constitutive law for an elastic medium with linear axisymmetric deformation can thus be expressed by

$$\sigma_{rr} = A\frac{\partial u_r}{\partial r} + (A - 2N)\frac{u_r}{r} + F\frac{\partial u_z}{\partial z} - \beta_r \vartheta - p, \quad (1a)$$

$$\sigma_{\theta\theta} = (A - 2N)\frac{\partial u_r}{\partial r} + A\frac{u_r}{r} + F\frac{\partial u_z}{\partial z} - \beta_r \vartheta - p, \quad (1b)$$

$$\sigma_{zz} = F\frac{\partial u_r}{\partial r} + F\frac{u_r}{r} + C\frac{\partial u_z}{\partial z} - \beta_z \vartheta - p, \quad (1c)$$

$$\sigma_{rz} = L\left(\frac{\partial u_r}{\partial z} + \frac{\partial u_z}{\partial r}\right), \quad (1d)$$

where u_r and u_z are the displacements of the elastic medium in the radial and vertical directions, respectively. The symbol σ_{ij} is the total stress components of the full space, ϑ is the temperature change of

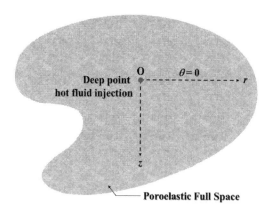

Deep point
hot fluid injection

O $\theta = 0$

r

z

Poroelastic Full Space

Figure 1. Deep point hot fluid injection into a poroelastic full space.

the thermoelastic medium, and p is the excess pore fluid pressure of the strata. The constants A, C, F, L, N of the cross-anisotropic strata are defined by Love (1944). For axially symmetric problem, it can be noted that the shear stresses $\sigma_{r\theta}$, $\sigma_{\theta z}$, and circumferential displacement u_θ would vanish as the vertical z-axis is located through the hot fluid injection point. In equations (1a) to (1c), β_r and β_z represent the thermal expansion factors along and normal to the symmetric plane, respectively. The expression for the thermal expansion factors are:

$$\beta_r = 2(A - N)\alpha_{sr} + F\alpha_{sz}, \quad (2a)$$

$$\beta_z = 2F\alpha_{sr} + C\alpha_{sz}, \quad (2b)$$

where α_{sr} and α_{sz} are the linear thermal expansion coefficients of the strata in the horizontal and vertical directions, respectively.

The constants A, C, F, L, N employed in equations (1a) to (1d) are related through the following equations:

$$A = \frac{E_r(1 - \nu_{rz}\nu_{zr})}{(1 + \nu_{r\theta})(1 - \nu_{r\theta} - 2\nu_{rz}\nu_{zr})}, \quad (3a)$$

$$C = \frac{E_z(1 - \nu_{r\theta})}{1 - \nu_{r\theta} - 2\nu_{rz}\nu_{zr}}, \quad (3b)$$

$$F = \frac{E_z\nu_{rz}}{1 - \nu_{r\theta} - 2\nu_{rz}\nu_{zr}} = \frac{E_r\nu_{zr}}{1 - \nu_{r\theta} - 2\nu_{rz}\nu_{zr}}, \quad (3c)$$

$$L = G_{rz}, \quad (3d)$$

$$N = \frac{E_r}{2(1 + \nu_{r\theta})}. \quad (3e)$$

Here, the symbols E_r and E_z are defined as Young's moduli with respect to directions lying in the plane of isotropy and perpendicular to it, respectively; $\nu_{r\theta}$ is the Poisson's ratio for strain in the horizontal direction due to a horizontal direct stress; ν_{rz} is the Poisson's ratio for strain in the vertical direction due to a horizontal direct stress; ν_{zr} is the Poisson's ratio for strain in the horizontal direction due to a vertical direct stress; and G_{rz} is shear modulus for planes normal to the plane of isotropy.

For the cases of isotropic poroelastic medium, the constants A, C, F, L, N and thermal expansion factors $\beta_i(i = r, z)$ can be denoted as

$$A = C = \lambda + G, \quad (4a)$$

$$F = \lambda, \quad (4b)$$

$$L = N = G, \quad (4c)$$

$$\beta_r = \beta_z = (2G + 3\lambda)\alpha_s. \quad (4d)$$

Here, λ and G are the Lame moduli of the isotropic poroelastic medium, and α_s is the linear thermal expansion coefficient of the isotropic solid skeleton.

For a general problem, these axially symmetric total stresses σ_{ij} must satisfy the equilibrium equations:

$$\frac{\partial \sigma_{rr}}{\partial r} + \frac{\sigma_{rr} - \sigma_{\theta\theta}}{r} + \frac{\partial \sigma_{rz}}{\partial z} + f_r = 0, \tag{5a}$$

$$\frac{\partial \sigma_{rz}}{\partial r} + \frac{\sigma_{rz}}{r} + \frac{\partial \sigma_{zz}}{\partial z} + f_z = 0, \tag{5b}$$

where f_r and f_z denote the body force components. For axisymmetric problems with body forces neglected, the equilibrium equations can be expressed in terms of displacements, excess pore fluid pressure, and temperature change of the medium as follows:

$$A\left(\frac{\partial^2 u_r}{\partial r^2} + \frac{1}{r}\frac{\partial u_r}{\partial r} - \frac{u_r}{r^2}\right) + L\frac{\partial^2 u_r}{\partial z^2} + (F+L)\frac{\partial^2 u_z}{\partial r\partial z}$$
$$- \beta_r\frac{\partial \vartheta}{\partial r} - \frac{\partial p}{\partial r} = 0, \tag{6a}$$

$$(F+L)\left(\frac{\partial^2 u_r}{\partial r\partial z} + \frac{1}{r}\frac{\partial u_r}{\partial z}\right) + L\left(\frac{\partial^2 u_z}{\partial r^2} + \frac{1}{r}\frac{\partial u_z}{\partial r}\right) + C\frac{\partial^2 u_z}{\partial z^2}$$
$$- \beta_z\frac{\partial \vartheta}{\partial z} - \frac{\partial p}{\partial z} = 0. \tag{6b}$$

The hot fluid injection point is considered at the location of great depth point $(0, 0)$. Using the laws of mass balance and energy conservation, the continuity equation and heat conduction equation can be obtained as below:

$$\frac{k_r}{\gamma_f}\left(\frac{\partial^2 p}{\partial r^2} + \frac{1}{r}\frac{\partial p}{\partial r}\right) + \frac{k_z}{\gamma_f}\frac{\partial^2 p}{\partial z^2} + \frac{Q_f}{2\pi r}\delta(r)\delta(z) = 0, \tag{7a}$$

$$\lambda_{tr}\left(\frac{\partial^2 \vartheta}{\partial r^2} + \frac{1}{r}\frac{\partial \vartheta}{\partial r}\right) + \lambda_{tz}\frac{\partial^2 \vartheta}{\partial z^2} + \frac{Q_t}{2\pi r}\delta(r)\delta(z) = 0, \tag{7b}$$

in which γ_f is the unit weight of injected hot fluid, and the constants λ_{tr} and λ_{tz} denote the horizontal thermal conductivity of heat flow in the plane of isotropy and the corresponding vertical thermal conductivity in the plane perpendicular to the isotropic plane, respectively. The permeability of the full space aquifer in the horizontal and vertical directions are expressed as k_r and k_z, respectively. The symbols $\delta(r)$ and $\delta(z)$ are the Dirac delta functions. The injected hot fluid is considered as a constant thermal strength Q_t corresponding with fluid volume Q_f per unit time.

For a linearly thermoelastic medium with cross-anisotropic properties, the differential equations expressed by equations (6a), (6b), (7a) and (7b) govern the steady state responses of the porous

medium subjected to axisymmetric disturbance of a point hot fluid injection.

2.2 Boundary conditions

The point hot fluid injection at great depth is assumed no impact on the ground surface. This implies that the ground surface can be treated as a remote boundary, and the strata can be modeled as a full space. Thus, the effect of the deep thermally disturbance vanishes at the remote boundaries $z \to \pm\infty$. In other words, the displacements, excess pore fluid pressure and the temperature change of the strata at the remote boundaries should be vanished. Therefore, the remote boundary conditions can be expressed by

$$\lim_{z\to\pm\infty}\begin{Bmatrix} u_r(r,z) \\ u_z(r,z) \\ p(r,z) \\ \vartheta(r,z) \end{Bmatrix} = \begin{Bmatrix} 0 \\ 0 \\ 0 \\ 0 \end{Bmatrix}. \tag{8}$$

The responses can be derived from the differential equations (6a), (6b), (7a) and (7b) corresponding with the remote boundary conditions (8).

3 CLOSED-FORM SOLUTIONS

The closed-form solutions of poroelastic deformation, excess pore fluid pressure, and temperature increments of the strata due to a point hot fluid injection into in a cross-anisotropic poroelastic full space can be obtained by using Hankel transform with respect to the radial coordinate (Erdelyi et al. 1954, Gradshteyn & Ryzhik 1980, Sneddon 1951) as:

$$u_r(r,z) = \frac{Q_f\gamma_f}{4\pi k_z}\left(a_{1f}\frac{r}{R_{1f}^*} + a_{2f}\frac{r}{R_{2f}^*} + a_{3f}\frac{r}{R_{3f}^*}\right)$$
$$+ \frac{Q_t}{4\pi\lambda_{tz}}\left(a_{1t}\frac{r}{R_{1t}^*} + a_{2t}\frac{r}{R_{2t}^*} + a_{3t}\frac{r}{R_{3t}^*}\right), \tag{9a}$$

$$u_z(r,z) = \frac{Q_f\gamma_f}{4\pi k_z}\left(b_{1f}\sinh^{-1}\frac{\mu_{1f}z}{r} + b_{2f}\sinh^{-1}\frac{\mu_{2f}z}{r}\right.$$
$$\left. + b_{3f}\sinh^{-1}\frac{\mu_{3f}z}{r}\right) + \frac{Q_t}{4\pi\lambda_{tz}}\left(b_{1t}\sinh^{-1}\frac{\mu_{1t}z}{r}\right.$$
$$\left. + b_{2t}\sinh^{-1}\frac{\mu_{2t}z}{r} + b_{3t}\sinh^{-1}\frac{\mu_{3t}z}{r}\right), \tag{9b}$$

$$p(r,z) = \frac{Q_f\gamma_f}{4\pi k_z}\frac{1}{\mu_{3f}R_{3f}^*}, \tag{9c}$$

$$\vartheta(r,z) = \frac{Q_t}{4\pi\lambda_{tz}}\frac{1}{\mu_{3t}R_{3t}^*}. \tag{9d}$$

Here, the hydraulic constants $a_{if}(i = 1, 2, 3)$, $b_{if}(i = 1, 2, 3)$, thermal constants $a_{it}(i = 1, 2, 3)$ and $b_{it}(i = 1, 2, 3)$ are defined by

$$a_{1f} = \frac{L + \mu_{1f}^2 (F + L - C)}{CL\mu_{1f} \left(\mu_{1f}^2 - \mu_{2f}^2\right)\left(\mu_{1f}^2 - \mu_{3f}^2\right)}, \tag{10a}$$

$$a_{2f} = \frac{L + \mu_{2f}^2 (F + L - C)}{CL\mu_{2f} \left(\mu_{2f}^2 - \mu_{1f}^2\right)\left(\mu_{2f}^2 - \mu_{3f}^2\right)}, \tag{10b}$$

$$a_{3f} = \frac{L + \mu_{3f}^2 (F + L - C)}{CL\mu_{3f} \left(\mu_{3f}^2 - \mu_{1f}^2\right)\left(\mu_{3f}^2 - \mu_{2f}^2\right)}, \tag{10c}$$

$$a_{1t} = \frac{L\beta_r + \left[(F + L)\beta_z - C\beta_r\right]\mu_{1t}^2}{CL\mu_{1t} \left(\mu_{1t}^2 - \mu_{2t}^2\right)\left(\mu_{1t}^2 - \mu_{3t}^2\right)}, \tag{11a}$$

$$a_{2t} = \frac{L\beta_r + \left[(F + L)\beta_z - C\beta_r\right]\mu_{2t}^2}{CL\mu_{2t} \left(\mu_{2t}^2 - \mu_{1t}^2\right)\left(\mu_{2t}^2 - \mu_{3t}^2\right)}, \tag{11b}$$

$$a_{3t} = \frac{L\beta_r + \left[(F + L)\beta_z - C\beta_r\right]\mu_{3t}^2}{CL\mu_{3t} \left(\mu_{3t}^2 - \mu_{1t}^2\right)\left(\mu_{3t}^2 - \mu_{2t}^2\right)}, \tag{11c}$$

$$b_{1f} = \frac{L\mu_{1f}^2 + (F + L - A)}{CL\left(\mu_{1f}^2 - \mu_{2f}^2\right)\left(\mu_{1f}^2 - \mu_{3f}^2\right)}, \tag{12a}$$

$$b_{2f} = \frac{L\mu_{2f}^2 + (F + L - A)}{CL\left(\mu_{2f}^2 - \mu_{1f}^2\right)\left(\mu_{2f}^2 - \mu_{3f}^2\right)}, \tag{12b}$$

$$b_{3f} = \frac{L\mu_{3f}^2 + (F + L - A)}{CL\left(\mu_{3f}^2 - \mu_{1f}^2\right)\left(\mu_{3f}^2 - \mu_{2f}^2\right)}, \tag{12c}$$

$$b_{1t} = \frac{L\beta_z\mu_{1t}^2 + (F + L)\beta_r - A\beta_z}{CL\left(\mu_{1t}^2 - \mu_{2t}^2\right)\left(\mu_{1t}^2 - \mu_{3t}^2\right)}, \tag{13a}$$

$$b_{2t} = \frac{L\beta_z\mu_{2t}^2 + (F + L)\beta_r - A\beta_z}{CL\left(\mu_{2t}^2 - \mu_{1t}^2\right)\left(\mu_{2t}^2 - \mu_{3t}^2\right)}, \tag{13b}$$

$$b_{3t} = \frac{L\beta_z\mu_{3t}^2 + (F + L)\beta_r - A\beta_z}{CL\left(\mu_{3t}^2 - \mu_{1t}^2\right)\left(\mu_{3t}^2 - \mu_{2t}^2\right)}. \tag{13c}$$

In addition, the characteristic roots $\mu_{1f} = \mu_{1t}$, $\mu_{2f} = \mu_{2t}$, and $\mu_{ij}(i = 1, 2; j = f, t)$ must satisfy the following characteristic equation:

$$CL\mu_{ij}^4 - \left[AC - F(F + 2L)\right]\mu_{ij}^2 + AL = 0. \tag{14}$$

The fluid and thermal characteristic roots μ_{3f} and μ_{3t} are defined by

$$\mu_{3f} = \sqrt{k_r/k_z}, \tag{15a}$$

$$\mu_{3t} = \sqrt{\lambda_{tr}/\lambda_{tz}}. \tag{15b}$$

In equations (9a), (9c) and (9d), the distance symbols are shown below:

$$R_{ij}^* = \sqrt{r^2 + \mu_{ij}^2 z^2} + \mu_{ij}|z|, (i = 1, 2; j = f, t), \tag{16a}$$

$$R_{3i} = \sqrt{r^2 + \mu_{3i}^2 z^2}, (i = f, t), \tag{16b}$$

$$R_{3i}^* = R_{3i} + \mu_{3i}|z|, (i = f, t). \tag{16c}$$

The derived closed-form solutions, equations (9a) to (9d), illustrated that all field quantities are functions of the distance from the hot fluid injection source. Those quantities are inversely proportional to the hydraulic permeability or thermal conductivity. Besides, the mechanical moduli do not have influence on excess pore fluid and temperature increments of the strata. Examination of the closed-form solutions are completed by letting the radial and axial coordinates approach infinity.

4 CONCLUSIONS

Based on the theory of thermo-poroelasticity, the closed-form solutions of the homogeneous cross-anisotropic elastic full space for axially symmetric deformation of the horizontal displacement, vertical displacement, excess pore fluid pressure, and temperature increments subjected to a point hot fluid injection are presented by equations (9a) to (9d). Those quantities are inversely proportional to the hydraulic permeability or thermal conductivity. Besides, the mechanical moduli do not have influence on excess pore fluid and temperature increments of the strata. The results can provide better understanding of the hot fluid injection induced responses of the cross-anisotropic strata.

ACKNOWLEDGEMENTS

This work is supported by the Ministry of Science and Technology of Taiwan, R.O.C., through grants NSC102-2221-E-216-022 and NSC87-2211-E-216-009.

NOTATION OF SYMBOLS

$a_{if}(i = 1, 2, 3)$	Hydraulic constants defined in equations (10a) to (10c) (Pa^{-1})
$a_{it}(i = 1, 2, 3)$	Thermal constants defined in equations (11a) to (11c) $(°C^{-1})$
A, C, F, L, N	Material constants of the strata defined by Love (Pa)
$b_{if}(i = 1, 2, 3)$	Hydraulic constants defined in equations (12a) to (12c) (Pa^{-1})
$b_{it}(i = 1, 2, 3)$	Thermal constants defined in equations (13a) to (13c) $(°C^{-1})$

E_r, E_z — Young's modulus in horizontal/vertical direction (Pa)

$f_i (i = r, z)$ — Body forces of the strata (N/m^3)

G — Shear modulus of the isotropic strata (Pa)

G_{rz} — Shear modulus for planes to the plane of isotropy of the normal isotropic strata (Pa)

k_r, k_z — Permeability of the strata in the horizontal/vertical direction (m/s)

p — Excess pore fluid pressure of the strata (Pa)

Q_f — Fluid volume of the injected hot fluid per unit time (m^3/s)

Q_t — Thermal strength of the injected hot fluid (J/s)

(r, θ, z) — Cylindrical coordinates system ($m, radian, m$)

$R_{3i} (i = f, t)$ — Distance parameter defined in equation (16b) (m)

$R_{3i}^* (i = f, t)$ — Distance parameter defined in equation (16c) (m)

$R_{ij}^* (i = 1, 2; j = f, t)$ — Distance parameter defined in equation (16a) (m)

$u_i (i = r, z)$ — Displacement components of the strata (m)

α_s — Linear thermal expansion coefficient of the solid skeleton of the isotropic strata ($^\circ C^{-1}$)

α_{sr}, α_{sz} — Linear thermal expansion coefficient of the cross-anisotropic strata in horizontal/vertical direction ($^\circ C^{-1}$)

β_r, β_z — Thermal expansion factors of the cross-anisotropic strata ($Pa/^\circ C$)

γ_f — Unit weight of injected hot fluid (N/m^3)

$\delta(x)$ — Dirac delta function (m^{-1})

ϑ — Temperature changes of the strata ($^\circ C$)

λ — Lame constant of the isotropic strata (Pa)

λ_t — Thermal conductivity of the isotropic thermoelastic strata ($J/sm^\circ C$)

$\lambda_{tr}, \lambda_{tz}$ — Thermal conductivity of the cross-anisotropic thermoelastic strata in the horizontal/vertical direction ($J/sm^\circ C$)

$\mu_{3i} (i = f, t)$ — Characteristic root defined in equations (15a) and (15b) (Dimensionless)

$\mu_{ij} (i = 1, 2; j = f, t)$ — Characteristic roots of the characteristic equation (14) (Dimensionless)

ν — Poisson's ratio of the isotropic strata (Dimensionless)

ν_{rz} — Poisson's ratio for strain in the vertical direction due to a horizontal direct stress (Dimensionless)

$\nu_{r\theta}$ — Poisson's ratio for strain in the horizontal direction due to a horizontal direct stress (Dimensionless)

ν_{zr} — Poisson's ratio for strain in the horizontal direction due to a vertical direct stress (Dimensionless)

$\sigma_{ij} (i, j = r, \theta, z)$ — Total stress components of the strata (Pa)

REFERENCES

Alajmi, A.F., Gharbi, R. & Algharaib, M. 2009. Investigating the performance of hot water injection in geostatistically generated permeable media. *Journal of Petroleum Science and Engineering* 66(3–4): 143–155.

Amadei, B., Swolfs, H.S. & Savage, W.Z. 1988. Gravity-induced stresses in stratified rock masses. *Rock Mechanics and Rock Engineering* 21(1): 1–20.

Bear, J. & Corapcioglu, M.Y. 1981. A mathematical model for consolidation in a thermoelastic aquifer due to hot water injection or pumping. *Water Resources Research* 17(3): 723–736.

Erdelyi, A., Magnus, W., Oberhettinger, F. & Tricomi, F.G. 1954. *Tables of Integral Transforms*, New York: McGraw-Hill.

Gradshteyn, I.S. & Ryzhik, I.M. 1980. *Table of Integrals, Series, and Products*, New York: Academic Press.

Lee, S.L. & Yang, J.H. 1998. Modeling of effective thermal conductivity for a nonhomogeneous anisotropic porous medium. *International Journal of Heat and Mass Transfer* 41(6–7): 931–937.

Love, A.E.H. 1944. A Treatise on the Mathematical Theory of Elasticity, New York: Dover Publications.

Lu, J.C.-C. & Lin, F.-T. 2006. The transient ground surface displacements due to a point sink/heat source in an elastic half-space, *Geotechnical Special Publication No. 148*: 210–218.

Lu, J.C.-C., Lin, W.-C. & Lin, F.-T. 2010. Closed-form solutions of the homogeneous isotropic elastic half space subjected to a circular plane heat source. *Geotechnical Special Publication No. 204*: 79–86.

Rosenbrand, E., Haugwitz, C., Jacobsen, P.S.M., Kjoller, C. & Fabricius, I.L. 2014. The effect of hot water injection on sandstone permeability. *Geothermics* 50: 155–166.

Sasaki, K., Ono, S., Sugai, Y., Ebinuma, T., Narita, H. & Yamaguchi, T. 2009. Gas production system from methane hydrate layers by hot water injection using dual horizontal wells. *Journal of Canadian Petroleum Technology* 48(10): 21–26.

Sheorey, P.R. 1994. A theory for in situ stresses in isotropic and transversely isotropic rock. *International Journal of Rock Mechanics and Mining Sciences and Geomechanics Abstracts* 31(1): 23–34.

Sneddon, I.N. 1951. *Fourier Transforms*, New York: McGraw-Hill.

Tarn, J.-Q. & Lu, C.-C. 1991. Analysis of subsidence due to a point sink in an anisotropic porous elastic half space. *International Journal for Numerical and Analytical Methods in Geomechanics* 15(8): 573–592.

Wang, C.D. & Tzeng, C.S. 2009. Displacements and stresses due to nonuniform circular loadings in an inhomogeneous cross-anisotropic material. *Mechanics Research Communications* 36: 921–932.

Yang, X., Sun, C.-Y., Yuan, Q., Ma, P.-C. & Chen, G.-J. 2010. Experimental study on gas production from methanehydrate-bearing sand by hot-water cyclic injection. *Energy Fuels* 24(11): 5912–5920.

Advances in Energy, Environment and Materials Science – Wang & Zhou (Eds)
© 2017 Taylor & Francis Group, London, ISBN 978-1-138-03600-0

Identification of oil spill by gas chromatography-isotopic ratio mass spectrometry

Haixia Wang
Navigation College, Dalian Maritime University, Liaoning, P.R. China

Yu Liu & Wenjing Chen
College of Environmental Sciences and Engineering, Dalian Maritime University, Liaoning, P.R. China

Jixiang Xu
College of Environmental Sciences and Engineering, Dalian Maritime University, Liaoning, China
Maritime Safety Administration, P.R. China

ABSTRACT: Gas Chromatography-Isotopic Ratio Mass Spectrometry (GC-IRMS) was used as the analytical method to determine the stable carbon isotope composition characteristics of 380# fuel oil and Brazil crude oil in this study. A short-term simplified weathering experiment was performed under natural condition to evaluate the effects of natural weathering process on the isotopic compositions of individual n-alkanes. Results demonstrate the following: a. Short-term weathering has no significant effect on both $\delta^{13}C$ value and profile of individual-alkanes (C_{11}–C_{32}), which is useful for weathering oil spill identification; The samples being analyzed under repeatability conditions show that the absolute difference of $\delta^{13}C$ values of n-alkanes is less than the repeatability limit r_{95}%; c. Through stable carbon isotope composition of n-alkanes can distinguish between the 380# crude oil and Brazil crude oil.

1 INTRODUCTION

Recent statistical data shows that the average number of incidents involving large oil spills from tankers has reduced progressively for the last three and a half decades and since 2010 has stood at an average of 1.8 large oil spills per year, but the total recorded amount of oil lost to the environment in 2015 was approximately 7,000 tons (ITOPF 2015). Particularly, with the industrial demand for petroleum increasing, marine oil spills by wrecked tanker or natural seepage has been increasingly frequent (Han et al. 2015). Therefore, to unambiguously characterize the types and sources of oil spill is extremely important for making an appropriate spill response, understanding the fate and environmental impact of the oil spills, as well as to investigate the responsibility and legal liability of oil spill accidents (Wang et al. 2004; Hayworth et al. 2015).

Compound-Specific Isotope Analysis (CSIA) of n-alkanes (Hayes et al. 1987) is a highly specialized analytical method, improving the accuracy and reliability of source characterization for oil. It has been successfully applied to petroleum-related studies, focusing on organic geochemistry and environmental forensics, which is associated with spills investigations (Wang et al. 2006; Yu et al. 2011), for example, to establish oil–oil and oil–source rock correlations (Bjorøy et al. 1992; Li et al. 1997; Odden et al. 2002; Inaba and Suzuki 2003; Li and Guo 2010), to assess the fate and environmental damage of the oil spill (Lemkau et al. 2010; Betti et al. 2011), to infer organic matter source and depositional paleo environment (Ruble et al. 1994; Al-Areeqand Maky 2015; Wang et al. 2015). Based on the stable carbon isotope profile of n-alkanes, Li et al. (2009) confirmed that GC–IRMS can be a powerful tool fortracing oil spill sources. In certain cases, particularly for weathered oils or those with relatively low concentration biomarkers, GC-IRMS can provide valuable correlation data in the absence of biomarker data. Mansuy et al. (1997) conducted three independent weathering experiments, which also proved that the $\delta^{13}C$ values of n-alkanes are minimally affected by evaporation, water-washing or biodegradation.

This paper aims to accurately distinguish the crude oil from fuel oil by analyzing stable carbon isotope composition. It is expected to provide the technical basis for oil spill identification. This paper carries out the study from the following aspects: a. Stable carbon isotopic ratios of n-alkanes ($\delta^{13}C$) from Brazil crude oil and

380# fuel oils samples were analyzed by a Gas Chromatography-Isotopic Ratio Mass Spectrometry (GC-IRM) and provided their stable carbon isotope composition characteristics; b. A short-term simplified weathering experiment under natural condition was performed to evaluate the effects of weathering process on the isotopic composition of individual n-alkanes and to reveal the effect of weathering on the chemical composition of spilled oil at the molecular level.

2 MATERIALS AND METHODS

2.1 Weathering simulation experiment

A simplified natural weathering simulation experiment was performed on two fuels oil and three crude oil (Table 1). Oil samples were weighed (about 8 g each) and then added to a series of beakers containing 1600 ml of seawater. The beakers were then placed on a lab building rooftop for 28 days. For each single-source oil, the weathering time was 0, 1, 3, 7, 14, 21, and 28 days respectively.

2.2 n-Alkane isolation

The procedure isolated the aliphatic hydrocarbons by traditional column chromatography on activated alumina followed by separating the aliphatic hydrocarbons into n-alkanes and branched hydrocarbon fractions by urea adduction (Harvey et al. 2012). Firstly, about 0.2 g of the oil sample was dissolved in 10 ml hexane HPLC, Tedia, USAand then was centrifuged at 3000 r/min speed for 10 minutes. The fractionation of saturates was conducted in achromatographic column (0.47 cm i.d × 12 cm). The column was dry-packed with 10 g of activated alumina (activated 4 h at 200°C, 100–200 mesh, 100–200 mesh, AR) and was topped with about 1 cm of anhydrous sodium sulphate (activated 4 h at 350°C, AR). The column was conditioned with hexane and the eluent was discarded, just prior to exposure of the sodium sulphate layer to air, about 200 ul concentrated extracted were quantitatively transferred. The aliphatic hydrocarbons were eluted with 15.0 mL of hexane. N-Alkanes purified by urea adduction were determined using Gas Chromatography–Isotopic Ratio Mass Spectrometry (GC–IRMS).

Table 1. Samples backgrounds.

Samples no.	Locations	Types
380	China	Fueloil
BX	Brazil	Crude oil

2.3 Gas Chromatography-Isotopic Ratio Mass Spectrometry (GC-IRMS)

The n-alkanes obtained from oil samples were analyzed by a Gas Chromatography-Isotopic Ratio Mass Spectrometry (GC-IRMS) instrument (TRACE GC ULTRA, Thermo Fisher Scientific, USA). The GC was equipped with a DB-5MS capillary column (60 m × 0.25 mm × 0.25 um) with helium as the carrier gas. The GC oven temperature was started at 60°C, and ramped to 100°C at 20°C/min, held for 2 min, then to 280°C at a rate of 6°C/min and held for 35 min isothermally. In splitless mode, 1 mL of samples was injected. The injector, interface and ion source were kept at 280, 250, and 200°C, respectively. Electron Impact (EI) mass spectra was taken at 70 eV acquired from 50 to 550 amu at a scan rate of 0.2 s scan^{-1}.

Carbon isotope ratios for individual alkanes were calculated using CO_2 as a reference gas that was automatically introduced into the IRMS at the beginning and end of each analysis and the stable isotope date are presented as delta (δ) values representing the deviation in parts per thousand (‰, per mil, or ppt) from an accepted standard. The isotope ratio for carbon was expressed as:

$$\delta^{13}C = \left[\left(R_{samples}/R_{standard}\right)-1\right]\times 10^3 \qquad (1)$$

where $R_{samples}$ and $R_{standard}$ are the isotope ratio ($^{13}C/^{12}C$) of the samples and recognized international standards Vienna Pee Dee Belemnite (VPDB) for $\delta^{13}C$ values, respectively. The analytical precision was < 0.02‰ (1σ) for ^{13}C.

A standard mixture of n-alkanes (n-C_{10}–n-C_{32}) from was purchased from Sigma (St. Louis, MO, USA).

3 RESULTS AND DISCUSSION

3.1 Effect of weathering on the carbon isotopic composition of the individual n-alkanes

Distribution and isotopic profiles of individual n-alkanes from 380# fuel oil and Brazil crude oils are shown in Fig. 1(a–b).

The 380# fuel oil (Fig. 1(a)) has relative depleted $\delta^{13}C$ values, ranging from −30‰ to −24‰ and the values are almost constant or become slightly heavier with increasing carbon number. In addition, the distribution of n-alkanes in Brazil crude oil (Fig. 1(b)) has a narrow range (from C_{10} to C_{21}) and the Brazil crude oil has relatively heavy $\delta^{13}C$ values (from −28‰ to −22‰), which are significantly different from that of the 380# fuel oil.

3.2 Repeatability limit (r_{95} %)

r_{95}% is that the Value less than or equal to the difference between two test results obtained under

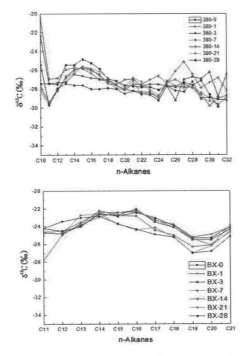

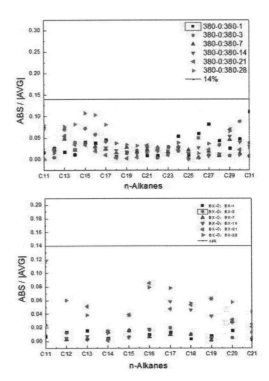

Figure 1. Distribution and isotopic profiles of individual n-alkanes: (a) 380# fuel oils (b) Brazil crud oils.

Figure 2. Repeatability limit test based on $\delta^{13}C$ values of individual n-alkanes: (a) 380# fuel oils (b) Brazil crud oils.

repeatability condition which may be expected to be within a probability of 0.95 (ISODIN. 5725-22002).

The isotope ratio for carbon is expressed as:

$$r_{95\%} = 2\sqrt{2}S_r = 2.8S_r \qquad (2)$$

Generally, the RSD = 5% limit can be seen as a quality criterion. Therefore, The repeatability limit $r_{95}\%$ is calculated by multiplying the fixed RSD (Sr) with a factor 2.8 (CEN/TR 15522-22012) by using the following equation:

$$r_{95\%} = 2.8 \times 5\% = 14\% \qquad (3)$$

This means when the samples are analyzed under repeatability condition, each ratio with an Sr of 5% may not differ more than 14% relative.

In this study, the repeatability limit parameter was used for the identification of weathered oil samples. With repeatability limit analysis, the experimental results is shown in Fig. 2(a–b). The results show that the absolute difference of $\delta^{13}C$ values of n-alkanes from the unweathered samples and their corresponding weathered samples were less than the repeatability limit 95%.

With repeatability limit analysis, the experimental results show that $\delta^{13}C$ values of n-alkanes from fuel oil and crude oil are suitable as the

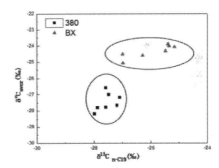

Figure 3. The cross plot of $\delta^{13}C$avervalues versus $\delta^{13}C$of C_{19} n-alkane.

characteristic ratios for oil identification in a shorter weathering process.

3.3 Classification diagrams based on individual n-alkane $\delta^{13}C$ ratios

A plot of the average $\delta^{13}C$ values (from C_{11} to C_{32}) n-alkanes' versus $\delta^{13}C$ of C_{19} n-alkane from three crude oil and two fuel oil is given in Fig. 3. The Brazil crude oil is characterized by less negative values of $\delta^{13}C$ (isotopically heavy). The 380# fuel oil with more

^{13}C depleted values which are far away from Brazil crude oils samples. As can be seen, the Brazil crude oil can be distinguished from the 380# fuel oils through the cross-plot diagram (Fig. 3).

4 CONCLUSIONS

The weathering simulation experiment conducted in this paper shows short-term weathering has no significant effect on $\delta^{13}C$ values of individual n-alkanes (C_{12}–C_{33}), suggesting the isotopic compositions of individual n-alkanes can be a useful tool for tracing the source of an oil spill.

The 380# fuel oil has relative depleted $\delta^{13}C$ values, ranging from −30‰ to −24‰; however, the Brazil crude oil has relatively heavy $\delta^{13}C$ values (from −28‰ to −22‰), which are significantly different with the 380# fuel oil.

ACKNOWLEDGEMENTS

This study was supported by The National Key Technologies R&D Program (2015 BAD17B05) and the Fundamental Research Funds of the Central Universities (01760129, 01760515).

REFERENCES

Al-Areeq N M, Maky A F (2015). Organic geochemical characteristics of crude oils and oil-source rock correlation in the Sunah oilfield, Masila Region, Eastern Yemen. Marine and Petroleum Geology 63: 17–27.

Betti M, Boisson F, Eriksson M et al. (2011). Isotope analysis for marine environmental studies. International Journal of Mass Spectrometry 307(1): 192–199.

Bjorøy M, Hall P B, Hustad E et al. (1992). Variation in stable carbon isotope ratios of individual hydrocarbons as a function of artificial maturity. Organic Geochemistry 19(1): 89–105.

CEN/TR 15522-2 (2012). Oil spill identification. Waterborne petroleum and petroleum products. Part 2: Analytical methodology and interpretation of results based on GC-FID and GC–MS low resolution analyses. CEN, Technical Report.

Denity Meter Ruble T E, Bakel A J, Philp R P (1994). Compound specific isotopic variability in Uinta Basin native bitumens: paleoenvironmental implications. Organic Geochemistry 21: 661–671.

Han E, Park H J, Bergamino L et al. (2015). Stable isotope analysis of a newly established macrofaunal food web 1.5 years after the Hebei Spirit oil spill [J]. Marine pollution bulletin 90(1): 167–180.

Hayes J M, Takigiku R, Ocampo R et al. (1987). Isotopic compositions and probable origins of organic molecules in the Eocene Messel shale. doi:10.1038/329048a0.

Harvey S D, Jarman K H, Moran J J et al. (2012). Characterization of diesel fuel by chemical separation combined with capillary gas chromatography (GC) isotope ratio mass spectrometry (IRMS). Talanta 99: 262–269.

Hayworth J S, Clement T P, John G F et al. (2015). Fate of Deepwater Horizon oil in Alabama's beach system: Understanding physical evolution processes based on observational data. Marine pollution bulletin 90(1): 95–105.

Inaba T, Suzuki N (2003). Gel permeation chromatography for fractionation and Isotope ratio analysis of steranes and Triterpanes in oils. Organic Geochemistry 34: 635–641.

ISODIN (2002). 5725-2: Accuracy (trueness and precision) of measurement methods and results–Part 2: Basic method for the determination of repeatability and reproducibility of a standard measurement method (12.02). Beuth, Berlin.

ITOPF (International Tanker Owners Pollution Federation) (2015). Oil Tanker Spill Statistics. http://www.itopf.com/knowledge-resources/data-statistics/statistics/. Accessed 26 December 2015.

Lemkau K L, Peacock E E, Nelson R K et al. (2010). The M/V CoscoBusan Spill: Source Identification and Short-term Fate, Marine Pollution Bulletin 60: 2123–2129.

Li M, Riediger C L, Fowler M G, Snowdon L R (1997). Unusual polycyclic aromatic hydrocarbons in the Lower Cretaceous Ostracode Zone sedimentary and relatednoils of the western Canada Sedimentary Basin. Organic Geochemistry 27:439–448.

Li S M, Guo D (2010). Characteristics and application of compound specific isotope in oil-source identification for oils in Dongying Depression, Bohai Bay Basin. Geoscience 24:252–258 (in Chinese).

Li Y, Xiong Y, Yang W et al. (2009). Compound-specific stable carbon isotopic composition of petroleum hydrocarbons as a tool for tracing the source of oil spills[J]. Marine Pollution Bulletin 58(1): 114–117.

Mansuy L, Philp R P, Allen J (1997). Source identification of oil spills based on the isotopic composition of individual compo nents in weathered oil samples. Environ Sci Technol 31(12):3417–3425.

Odden W, Barth T, Talbot M R (2002). Compound-specific carbon isotope analysis of natural and artificially generated hydrocarbons in source rocks and petroleum fluids from offshore Mid-Norway. Organic Geochemistry 33: 47–65.

Wang L, Song Z, Cao X et al. (2015). Compound-specific carbon isotope study on the hydrocarbon biomarkers in lacustrine source rocks from Songliao Basin. Organic Geochemistry 87: 68–77.

Wang Z, Fingas M, Lambert P et al. (2004). Characterization and identification of the Detroit River mystery oil spill (2002). Journal of Chromatography A 1038(1): 201–214.

Wang Z, Stout S A, Fingas M (2006). Forensic fingerprinting of biomarkers for oil spill characterization and source identification. Environmental Forensics 7(2): 105–146.

Yu S, Pan C, Wang J et al. (2011). Molecular correlation of crude oils and oil components from reservoir rocks in the Tazhong and Tabei uplift of the Tarim Basin, China. Organic Geochemistry 42: 1241–1262.

Advances in Energy, Environment and Materials Science – Wang & Zhou (Eds)
© 2017 Taylor & Francis Group, London, ISBN 978-1-138-03600-0

Research on the simulation and application of a spray quenching system of 1 million kW nuclear power rotor

Zhanjun Wang
Beijing Research Institute of Mechanical and Electrical Technology, Beijing, China

Xiaoyu Zhang
Department of Materials, University of Science and Technology Beijing, Beijing, China

Xianjun Li
Beijing Research Institute of Mechanical and Electrical Technology, Beijing, China

ABSTRACT: The cooling process of 1 million kW nuclear power rotor was studied using numerical simulation, and a cooling curve was obtained. A water density of 10 L/m²·s is determined to meet the requirement in the treating technology of the rotor through combining with the CCT curve of rotor material. The average cooling rate and water consumption with different water densities were compared. Thus, the optimal amount of water for the rotor is determined to be 3000 m³/h. The effective stress field of rotor was analyzed. It is noted that, in the cooling process of the rotor, the stress at the site with abrupt change in diameter is too large. The cooling technology was optimized with the layered cooling form to reduce the effective stress. The analysis of the actually produced rotor reveals that the corresponding mechanical properties conform to the product requirements. The development of the equipment conforms to the concept of intellectualization and greenization in heat treatment. The cooling system can be widely used for the cooling process of heavy castings and forgings.

1 INTRODUCTION

The greenization and intellectualization is the next major development trend of heat treatment industry (Jin, 2015). The heat treatment is the key technique for such heavy workpieces as the rotor to exhibit excellent performances. The determination of technological parameters for the heat treatment of nuclear power rotor depends mainly on experience and error trial, and the function and control level of heat treatment equipment is low, causing the poor performance of rotor after heat treatment (Xu, 2005). Liu Chuncheng (2015) studied the heat treatment processes of 6MW rotor fabricated with 26Cr2Ni4MoV steel, and compared various cooling schemes after heating program. Wang Xiaoyan (2011) studied both water quenching technology and combined air–water quenching technology for the low-pressure rotor. Chen Ruikai (2012 & 2013) and Yu Lina (2014) et al. investigated the microstructural transformation products of nuclear power rotor in the cooling process. However, studies on the quenching system of rotor are very few. Previously, quenching of nuclear power rotors adopted either the immersing form in quenching tank or the water–air alternate form, which could not achieve controlled cooling at sites with abrupt

diameter change, and cracks occur easily on the rotor. On the contrary, quenching of oil will generate more smoke, which causes pollution.

Numerical simulation, as the main research tool for the quenching process (Li, 2003; Yuan, 2005), can intuitively reflect the temperature and stress fields of the workpiece and medium (Hamouda, 2001; Zhou, 2003; Lee, 1999) as well as the surface heat transfer coefficient of the workpiece in the cooling process, and thus can provide the important basis for the design of quenching equipment and optimization of quenching technology. In this paper, the automatic jet quenching machine tool and intelligent cooling system, which can achieve layered cooling for 1 million kW nuclear power rotor, whose size and weight are given in Table 1, are developed. In addition, numerical simulation for the cooling process of the rotor is conducted to guide the precise control of the cooling technology of the rotor.

Table 1. Material, size and weight for one million kilowatts nuclear power rotor.

Material	Size (mm)	Weight (t)
30Cr2Ni4MoV	$\Phi 2826 \times 5196$	291.7

2 FINITE ELEMENT EQUILIBRIUM EQUATIONS AND MODELING FOR QUENCHING PROCESS

2.1 *Finite element equations*

The temperature field is the main factor of the quenching process. The thermal partial differential equations can be expressed as follows:

$$\rho c \frac{\partial T}{\partial t} = \frac{\partial}{\partial x}\left(\lambda \frac{\partial T}{\partial x}\right) + \frac{\partial}{\partial y}\left(\lambda \frac{\partial T}{\partial y}\right) + \frac{\partial}{\partial z}\left(\lambda \frac{\partial T}{\partial y}\right) \quad (1)$$

where ρ = density (kg/m³), c = the specific heat capacity (J/kg · °C), and λ = thermal conductivity (W/m · °C). By solving with the third type of boundary condition, the convective heat transfer coefficient and temperature of fluid medium can be obtained:

$$-\lambda \left[\frac{\partial T}{\partial h}\right]_s = h\left(T_w - T_q\right) \quad (2)$$

The convective heat transfer coefficient can be expressed as follows (R. D. Morals, 1990; Hodgon, 1991):

$$h = 3.15 \times 10^6 W_f^{0.616} \left[700 + \frac{T_s - 700}{e^{\left(\frac{T_s - 700}{10}\right)} + 1} \right]^{-2.455}$$

$$\times \left[1 - \frac{1}{e^{\left(\frac{T_s - 250}{40}\right)} + 1} \right] \quad (3)$$

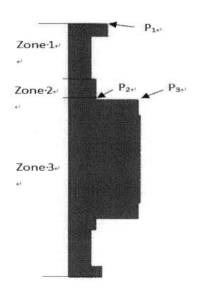

Figure 1.   Meshing of nuclear power rotor.

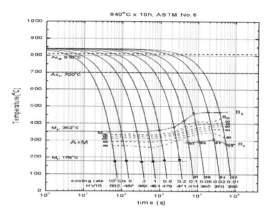

Figure 2.   CCT diagram of 30Cr2Ni4MoV steel[4].

Table 2.   Cooling process of 1 million kW nuclear power rotor.

No.	Cooling process	
1	Air cooling	20min
2	Spraying water in 1 to 3 zones	Stop spraying water in 1 zone at temperature below 170°C, spraying wind
3	Spraying water in 2 to 3 zones	Stop spraying water in 2 zones at temperature below 170°C, spraying wind
4	Spraying water in 3 zones	Stop spraying water in 3 zones at temperature below 170°C, spraying wind
5	Spraying wind in 3 zones	Stop spraying wind in 3 zones at temperature below 50°C

where T_s = surface temperature of workpiece (°C) and W_f = spraying water density (L/m²s).

2.2 *Numercal modeling process*

In the modeling process for the rotor, the axial symmetric model is adopted, the unit structure with four nodes is used, and meshing is performed with finite element software. Meshing near the surface of the workpiece is refined to increase the calculation accuracy, as shown in Figure 1. The whole quenching process involves both water and wind spraying technologies, as illustrated in Figure 2.

3 NUMERICAL SIMULATION RESULTS AND OPTIMIZATION

The quenching process at the spraying water density of 10 L/m²s is simulated, the temperature field

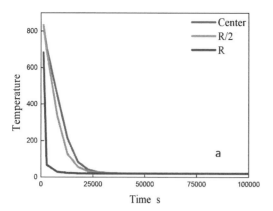

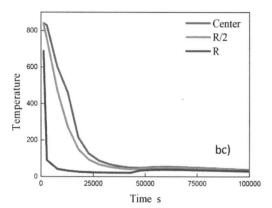

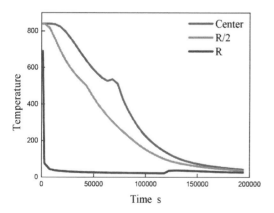

Figure 3. Cooling curves for (a) part 1, (b) part 2, and (c) part 3.

and stress field are obtained, and the corresponding results are shown in Figure 3.

It can be seen from Figure 3 that the temperature changes obviously with time. The surface temperature decreases rapidly in each zone, but the temperature in the central region decreases slowly because the inner surface of the rotor cannot directly contact water, and thus cooling depends only on the heat conduction of the material. The thickness of workpiece in each zone is different, and the temperature decrease is the highest in zone 1 and lowest in zone 3. The temperature is hindered because of phase transformation in the cooling process. The hindering effect is not obvious in both zones 1 and 2, whose thickness is small. In zone 3 (Figure 3c), the quenching process is hindered obviously by the latent heat of phase transformation.

According to the process requirement, the microstructure of rotor after quenching is composed of martensite and bainite. Figure 4 shows the relationship between the cooling rate and spraying water density in the temperature range of 300–500°C. The cooling rates are 0.0096, 0.011, 0.016, 0.0123, and 0.013°C/s at spraying water densities of 5, 10, 15, 40, and 100L/m²s. According to the CCT curves in Figure 2, the minimum cooling rate to obtain bainite is 0.01°C/s. Therefore, the spraying water density of 5 L/m²s is not sufficiently large. It can be seen from Figure 4 that the cooling rate increases slowly with an increase in the spraying water density. The water consumption values are 1265, 2530, 3795, 10,120, and 25,300 m³/h at the aforementioned spraying water densities, respectively. According to the process requirement and environmental protection concept, the water consumption is determined to be 3000 m³/h.

Thermal stress is generated and the microstructural transformation occurs in the quenching process. The excessive quenching stress may cause cracking or plastic deformation of the rotor. Figure 5 shows the varying curves in the equivalent stress of rotor with time at three diameter—varying points. It can be seen from Figure 5a that the

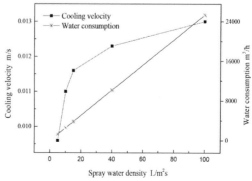

Figure 4. Cooling curves in the temperature range of 300–500°C.

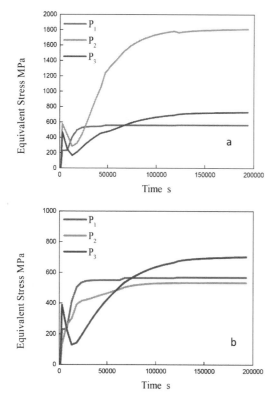

Figure 5. Curves for equivalent stress (a) uniform cooling and (b) layered cooling.

equivalent stress first increases gradually, and then becomes stable. Under the same quenching condition in each cooling layer, the equivalent stress in the site P_2 is too large, and cracking occurs easily. Therefore, it is necessary to optimize cooling in that site and control the quenching process with the layered cooling form.

When the spraying water density in the cooling layer corresponding to the P_2 site is reduced to 3 L/m²s, the corresponding optimized results are illustrated in Figure 5b. It can be seen that the equivalent stress at that site decreases by 50%.

4 EXPERIMENTAL RESULTS

The microstructures and mechanics performances of rotor after quenching and tempering are shown in Figure 6 and Table 3, respectively. It can be seen that the microstructures and mechanical properties are complied with the requirements. It means that the proposed heat treatment processes and designed equipment are suitable for the practical production of 1 million kW nuclear power rotor.

Figure 6. Microstructures of rotor (a) 100 mm from surface, (b) 300 mm from surface, and (c) 600 mm from surface.

Table 3. Mechanical properties.

Sample number	$\sigma_{0.2}$/MPa	σ_b/MPa	A/%	ψ/%
1	630	740	25.0	76.5
2	635	750	25.5	73.0
3	635	745	24.0	70.5
4	640	750	26.0	71.5
5	635	745	23.0	69.0

5 CONCLUSIONS

1. The temperature in the central zone of 1 million kW nuclear power rotor decreases slowly. The quenching process is hindered by the latent heat of microstructural transformation.
2. The optimal water consumption to satisfy the requirement is 3000 m³/h.
3. The equivalent stress at the site P_2 site is too large if the spraying water density in each cooling layer is the same. Reduction of the spraying water density to 3 L/m²s in that site by the layered cooling form can decrease the equivalent stress by 50%.
4. The microstructures and mechanical properties of the rotor can satisfy the product requirements.

REFERENCES

Chen, Ruikai. Gu, Jianfeng. Pan, Jiansheng. 2012. Kinetics of isothermal pearlite transformation of 30Cr2Ni4MoV steel for low pressure rotors. *HEAT Treatment of Metals*, 37(5): 2–5.

Chen, Ruikai. Gu, Jianfeng. Han, Lizhan. et al. 2013. Austenitization kinetics of 30Cr2Ni4MoV. steel *Transactions of Materials and Heat Treatment*. (1): 173–177.

Hamouda, A.M.S. Sulaiman, S. 2001. Lau C K. Finite element analysis on the effect of workpiece geometry on the quenching of ST50 steel. *Journal of Materials processing Technology*. 119: 354–360.

Hodgon, P.D. Browne, K.M. et al. 1991. A Mathematical Model to Simulate the Thermomechanical Processing of Steel. *In 3rd International Seminar on Quenching and Carburising. Melbourne Australia*: 139–159.

Jin, Jiayu, Zheng, Jianbo, Zhao, Xichun. 2015. Cleaning heat treatment of heavy forgings and energy saving. *heat treatment of metals*. 40(12): 139–142.

Liu, Chuncheng. Xu, Xuejun. Liu, Zhuang. et al. 2015. Explore the Heat treatment Process of Rotors. *heavy casting and forging*. 40(9): 1–15.

Li, Qiang. Wang, Ge. Chen, Nailu. et al. 2003. Computational Simulation of flow fields for quenching medium. *Chinese Science Bulletin*. 48(8):739–742.

Lee, D.Y. Vafai, K. 1999. Comparative Analysis of Jet Impingentment and Microchannel Cooling for High Heat Flux Application. *Int Heat Mass Transfer*. 42(9): 1555–1568.

Morals R.D., A.G. Lopez, I.M. Olivares. Heat transfer analysis during water spray cooling of steel rods. ISIJ Int. 1990 30(1): 48–57.

Wang, Xiaoyan. 2011. Study on the Computer Simulation of LP Rotor's Quenching Process. *Shanghai Jiao Tong University*.

Xu, Yueming. Li, Qiao. progress in heat treatment. *heat treatment of metals*. 40(9): 1–15.

Yu, Lina. Liu, Shaokun. 2014. Dynamic Analysis of Austeniting of Steel 30Cr2Ni4MoV Based on Computer Simulation. *Foundry Technology*. 35(10): 2184–2186.

Yuan, Jian. Zhang, Weimin. Liu, Zhancang. et al. 2005. Research on the cooling characteristic and heat transfer coefficient of dynamic quenchant. *Transactions of Materials and Heat Treatment*. 26(4): 115–119.

Zhou, Z.Q. Thomson, P. F. Lam, Y. C. et al. 2003. Numerical analysis of residual stress in hot-rolled steel strip on the run-out table. *Journal of Materials Processing Technology*. 132(1/2/3): 184–197.

Advances in Energy, Environment and Materials Science – Wang & Zhou (Eds)
© 2017 Taylor & Francis Group, London, ISBN 978-1-138-03600-0

An overview of the mechanism and analysis method for power oscillation of large-scale wind power transmission

Yanhong Gao & Huajun Wang
School of Control Science and Engineering, Hebei University of Technology, Tianjin, China

Linlin Wu
Jibei Electric Power Research Institute, Beijing, China

Jiaan Zhang
School of Control Science and Engineering, Hebei University of Technology, Tianjin, China

ABSTRACT: With the rapid development of wind power in China, the stability of the large-scale wind power transmission system is becoming more prominent. In this paper, a detailed classification of power oscillation induced by large-scale wind power generation is carried out, and then the generation mechanism is analyzed and summarized. Then, the analysis method appropriate for sub-synchronous oscillation in large-scale wind power transmission system is summarized. Finally, the research prospects of this project are discussed. Owing to the lack of research on the power oscillation of wind turbines and the complexity of wind power integration, the research on this topic is of great challenge and innovation.

1 INTRODUCTION

In recent years, the wind power has developed rapidly in China and the scale is expanding gradually. However, wind power has power fluctuation characteristics due to the randomness and intermittence of wind speed, and large-scale wind power integration has certain impact on the grid stability. This large-scale wind power grid connected to the power system may lead to power oscillation, which can be roughly divided into two types: low-frequency oscillation and sub-synchronous oscillation. Low-frequency oscillation of power system refers to oscillation phenomenon whose frequency is 0.2 ~ 2.5 Hz caused by the relative swing of the parallel synchronous generator rotors when the power system is disturbed. Sub-synchronous oscillation refers to a significant energy exchange phenomenon state between the electrical systems and generating set's shafts at one or several sub-synchronous frequency. With reference to the IEEE working group for the definition of sub-synchronous oscillation, it does not include rigid oscillation mode of the rotor shaft. Wind power and other new energies connecting to the grid make the system power oscillation phenomenon occur frequently, which may also lead to cascading failures and split by out-of-step operation between units and contribute to great harm to the operation of power system. The more reasonable method can be used to analyze the specific mode by more clear classification of the power oscillation, and then more reasonable measures can be taken to suppress the mode of oscillation.

2 MECHANISM OF LOW-FREQUENCY OSCILLATION

The explanation mechanism of the low-frequency oscillation mainly includes the negative damping mechanism, the forced oscillation mechanism, bifurcation theory, and the chaotic mechanism.

2.1 *Negative damping mechanism*

F. P. Demellon and C. Concordia were the first to put forward the positive and negative damping torque method for analysis of low-frequency oscillation phenomenon. The sufficient research and analysis is in reference by Demello F in 1969. The increasing oscillation is due to lack of damping torque in the single machine infinite system. Besides, it points out that with high amplification multiple of excitation system application, it increases the synchronizing torque of the system. At the same time, it also increases the negative damping system due to the existence of excitation system inertia, which makes the system positive damping offset. If the system is disturbed at this time, it will cause the rotor oscillations. In 2013, Yang Yue takes a power system containing a complete dynamic model of DFIG as an example and then analyzes the damping characteristics of the electrical system containing DFIG. The results indicate that the damping ratio is not monotonic with the increase in wind power integration capacity, but there is an optimum integration capacity of wind power. The installation of high-response-rate power electronic devices—inverters in power system containing DFIG—is good for small signal stability of the system. The damping characteristic of the wind farm is closely related to the electric distance between the access position of the wind farm and the fault point.

2.2 Forced resonance mechanism

The cyclical fluctuations in the individual node load power of the power system make the constant and periodic disturbance source appear. When the perturbation frequency is close to inherent frequency, local periodic small disturbance may cause the obvious oscillation phenomenon in the system, and this oscillation phenomenon is called forced oscillation. This mechanism is different from the negative damping mechanism, which describes the system by non-homogeneous differential equations. Forced oscillation is a system resonance caused by periodic disturbance, and hence, its suppression is worked by eliminating the disturbance source and reducing the amplitude. This mechanism is obviously different from the negative damping mechanism, and it has the characteristics of oscillation. Based on the model of single machine infinite bus system of conventional generator, the basic theory of forced power oscillation in power system is expounded in a paper by TANG Yong in 2006. In a paper by FAN Wei in 2009 and a paper by ZHAO Shuqiang in 2009, the mechanism of forced power oscillation of a common asynchronous wind generator is analyzed.

2.3 Mode resonance mechanism

Mode resonance mechanism is that where the small changes of system parameters may cause resonance of two or more modes, whose characteristic value is close in the system; the resonance may lead to one pair of characteristic value crossing the imaginary axis rapidly, causing system oscillation instability. With interconnection of the large-scale power grid, the orders of system and the number of electromechanical mode are increasing, and the corresponding characteristic value distribution is highly dense in the complex plane. The change in system parameters may cause characteristic roots' movement and encounter when the system operating conditions change, and thus, the resonance

appears. In the course of southern power grid DC modulator design, mode resonance phenomenon which is caused by HVDC modulation is found, and this mechanism is verified in the actual power system. In a paper by Dobson I in 2001, oscillation modes that exist simultaneously in multi-machine system are analyzed and researched, and the relations between the characteristic roots corresponding to each mode are studied.

2.4 Nonlinear theory mechanism

Nonlinear theory mechanism explains that the power system is a complex nonlinear system, and its dynamic behavior contains complex nonlinear oscillations of the mechanical and electrical systems. It can be studied by bifurcation theory, chaos, nonlinear methods, etc. Nonlinear anomalies may cause low-frequency oscillations.

3 MECHANISM OF SUB-SYNCHRONOUS OSCILLATION IN WIND POWER GENERATION SYSTEM

Sub-synchronous oscillation accidents occur in the grid sometimes, especially when the large-scale wind farm power system accesses the grid, and some of the specific characteristics of wind power may increase the risk of sub-synchronous oscillation accident. The sub-synchronous oscillation has become one of the key problems that threat the safe and stable operation of the power system and restrict the power grid transmission capacity. It is necessary to fully acquaint the sub-synchronous oscillations of power systems. Therefore, this section describes the mechanism of the sub-synchronous oscillations when the wind power generation system accesses the grid by analyzing the mechanism of sub-synchronous oscillation in the wind power system, and provides the basis for selecting appropriate analysis method and analyzing its factors.

3.1 Induction Generator Effect (IGE)

Induction generator effect refers to that "self-excitation" phenomenon may occur when the generator is connected with capacitive load or is connected to the power grid through series compensation circuit. Under the resonant frequency of the stator circuit inductance and capacitance, synchronous generator runs as an asynchronous generator to provide energy for oscillations, so this self-excited manner is also known as "induction generator effect". Figure 1 is a schematic generator set connected to the infinite system by series capacitor compensation, where R and L are the transmission line lumped parameter resistance and

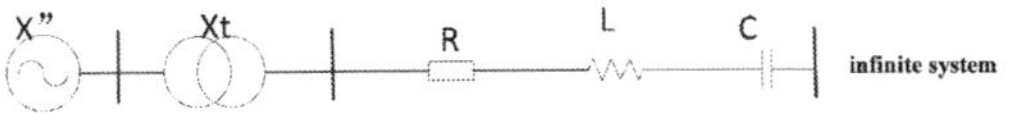

Figure 1. Generating set with compensation circuit.

inductance, respectively, C is the series capacitor compensation, X'' is the sub-transient reactance of generator, and X_t is the equivalent reactance of transformer.

There is a resonance circuit in the external transmission system shown in Figure 1. The natural resonance frequency is:

$$\omega_{er} = \omega_0 \sqrt{\frac{X_C}{X_\Sigma}} < \omega_0 \qquad (1)$$

where ω_0 is the synchronous rotation frequency. When the small disturbance of the frequency occurs in the grid, in addition to the generator stator power frequency current, there is sub-synchronous frequency current component ω_{er}, which is equivalent to a rotor of an asynchronous machine. As the rotation speed of the rotor is higher than that of the secondary synchronous rotating magnetic field generated by the stator sub-synchronous current component, the equivalent resistance of the rotor to the sub-synchronous current is negative from the stator. When this apparent negative resistance is greater than the equivalent resistance of the stator plus the transmission system at the electric resonance frequency, the electrical self-excitation will be generated, which is the effect of the induction generator. The induction generator effect is due to the apparent negative resistance characteristic of the rotor of the synchronous generator, which is lower than the system power frequency. The induction generator effect belongs to the self excitation phenomenon which only considers the dynamic behavior of the electric system, and has nothing to do with the generator shaft. Therefore, the simple induction generator effect will not lead to the occurrence of the torsional vibration of the shaft.

3.2 Sub-Synchronous Control Interaction (SSCI)

Sub-synchronous control interaction is the interaction between wind power, photovoltaic, and other new energy turbine converter with series compensated lines. The problem of sub-synchronous oscillation caused by the controller of the wind turbine is a new sub-synchronous oscillation phenomenon, which is produced with the rapid development of wind power generation. Unlike this, for the sub-synchronous oscillation caused by the electronics devices, there is no contact between the sub-synchronous oscillation caused by the controller of the wind turbine and the mechanical

system. The frequency and damping rate of such sub-synchronous oscillations are determined by the wind power controller parameters and transmission system parameters, and the natural mode frequency of the shaft is completely independent. In addition, the oscillation divergence speed of the voltage and current is much faster than the conventional power oscillation regardless of the mechanical system. The research shows that the sub-synchronous control interaction problem in doubly fed induction generator is the most serious problem in the four types of wind turbines. Furthermore, double-fed induction wind turbine and permanent magnet synchronous wind turbine exhibit converter control inside, and if the converter controller parameters are unreasonable, it will exhibit negative damping characteristics in the second synchronous zone and may also become a wind turbine sub-synchronous oscillation excitation source.

3.3 Sub-Synchronous Torque Interaction (SSTI)

Sub-synchronous torque interaction's occurrence is due to the unreasonable controller parameter setting of High-Voltage Direct Current transmission (HVDC) and FACTS or other power electronic devices. The power and current rapid control or response of power electronic controller in the sub-synchronous frequency will affect the generator electromagnetic torque and phase difference of the rotational speed, and if the electromagnetic torque and phase difference of rotational speed is more than 90 degrees, these devices will bring negative damping into the generator, triggering the power oscillation of the generator shaft. Similar to the study of sub-synchronous resonance in turbines, a more precise shaft model must be established in the research of wind turbine and power grid torsional interaction. It is generally believed that the shaft system of a two mass model can meet the requirements of most of the simulation accuracy. Compared with the large turbine generator, the stiffness of wind turbine shaft is smaller, shaft natural torsional vibration has lower frequencies (less than 10 Hz), and the required sub-synchronous torsional interaction has a high degree of series compensation, and therefore, sub-synchronous torque interaction is not the main power oscillation problem of wind turbines.

4 POWER OSCILLATION PROBLEM ANALYSIS METHOD

At present, there are many methods to analyze the power oscillation; each has its own application scope, advantages, and disadvantages. The main analysis methods are explained as follows.

4.1 Eigenvalue analysis method

Eigenvalue analysis method is also known as mode analysis method, which determines the stability of the system by solving the system of the coefficient matrix of the characteristic values after linearization of the power system model under small perturbation condition. Its theoretical basis is Lyapunov's second theorem of stability. The basic steps of eigenvalue analysis are as follows: first, the dynamic modeling of each component (including the wind turbine shaft) and the power network is carried out in the power system, and the unified differential algebraic equations are formed. Then, differential algebraic equations which describe the dynamic process of generator mechanical part and electromagnetic transient process of network are linearized on the system operating point and transformed into uniform rotation, and thus, coefficient matrix of the linear state equation of the whole system is obtained. By analyzing the structure characteristics of the coefficient matrix, characteristic values, and corresponding eigenvectors, correlation factors can be calculated. According to these results, the shafting torsional vibration mode stability, damping characteristics and the shafting quality block torsional vibration amplitude and phase relationship can be analyzed, and the torsional vibration mode which is strongly related to the quality of the block can be found in order to perform monitoring. In addition, the sensitivity analysis on the torsional vibration mode, especially the mode which has the risk of synchronous oscillation, can be performed in order to take effective prevention and control measures.

4.2 Frequency scanning method

The frequency scanning method is an approximate linear method proposed by Kilgore. It can be used to screen out the operating conditions of the system with potential sub-synchronous frequency threat, and it can be used to determine the partial system or operating conditions which have no effect or little effect on the sub-synchronous oscillation. The three-phase symmetrical unit current is injected into the generator, which has potential SSO threat, and the voltage on the bus is recorded. Change the injected current frequency continuously in the sub-synchronous oscillation frequency domain. The frequency of the abrupt change of impedance is the electrical resonance frequency of the power grid. If the frequency and a mechanical resonance frequency (that is a shafting mode) of the generator are equal to the power frequency, unstable sub-synchronous resonance may occur in this structure or operation mode.

4.3 Complex torque coefficients method

The complex torque coefficient method is proposed by the famous BBC Company. It judges whether the system will have a sub-synchronous resonance by comparing the generator's electrical complex torque coefficient and the mechanical torque coefficient. In a sense, it can be considered as a combination of the frequency scanning method and the eigenvalue analysis method. Similar to the eigenvalue analysis method, it is also a method based on the system linearization principle, and it also needs to get a linear model of the whole system. However, the eigenvalue analysis method may encounter the problem of "dimension disaster", and the complex torque coefficients method does produce to overcome this problem. The specific method of the complex torque coefficient method is: exert mandatory small mechanical oscillation $\Delta\dot{\delta}$, whose frequency is h $(h < f_0)$ on a generator rotor's relative angle of the system, and the change in the generator mechanical system and the electrical system with respect to $\Delta\dot{\delta}$ can be obtained by the calculation, that is mechanical complex torque increment $\Delta\dot{T}m$ and electric complex torque increment $\Delta\dot{T}e$. Moreover, we define the mechanical complex torque coefficient Km and the electric complex torque coefficient Ke on this account:

$$Km(jh) = \Delta\dot{Tm}\Big/\Delta\dot{\delta} = Km(h) + jhDm \qquad (2)$$

$$Ke(jh) = \Delta\dot{Te}\Big/\Delta\dot{\delta} = Ke(h) + jhDe \qquad (3)$$

4.4 Time domain simulation method

The so-called time-domain simulation method is to solve the differential equations of the whole system by the numerical integration method. The mathematical model used in this method can be linear or nonlinear. The network components can adopt lumped parameter model or distributed parameter model. The spring mass of the generator set's shafts can be divided in a very detailed manner, and even the distributed parameter model can be used. This method can simulate the generator, the system controller, the system fault, the switch operation, etc., elaborately. The method is the main method for studying the transient torque amplification effect, which can simulate system controller, a generator, a system failure, and network operating, elaborately. The more important point is that the transient process of a variety of nonlinear devices can be taken into account in this method. The electromagnetic transient simulation software commonly used include DigSilent/powerfactory, ATP, PSCAD/EMTDC, PSASP, BPA, etc. The advantage of the time domain simulation method is that the curves in which variables change with the time can be obtained considering all kinds of nonlinear factors. It can be used not only for large disturbance analysis of sub-synchronous oscillation, but also for small disturbance sub-synchronous oscillation study. However, it is difficult to identify the modes of system torsional oscillation and damping characteristics, which do not provide information of influence factors on sub-synchronous oscillation, generation mechanism, and restraining measures.

5 CONCLUSION

When compared with the traditional thermal power generation unit, there are essential differences in the wind power generation system structure mode of grid-connection. Wind farms/group is composed of various types of wind turbines, which make power oscillation of large-scale wind power transmission very complicated. The mature power oscillation mechanism analysis results of thermal power generating units and the transmission system cannot be directly applied to the wind turbine, which needs further research. With the construction of large-scale wind power base in China, the mechanism of analysis, modeling, and suppression of power oscillation are the urgent problems to be solved. In this paper, the mechanism and analysis method of power oscillation caused by large-scale wind power integration are summarized preliminarily, which lays a foundation for further research.

REFERENCES

Bi Tianshu, Kong Yongle & Xiao Shiwu (2012). Large scale wind power transmission in the times synchronous oscillation. *J. Power science and Technology Journal of.* 27 (1), 10–15.

Chen X, Zhang W & Zhang W (2005). Chaotic and sub-harmonic oscillations of a nonlinear power system. *J. Circuits and Systems II: Express Briefs, IEEE Transactions on.* 52(12), 811–815.

Demello F (1969). Concepts of synchronous machine stability as affected by excitation control. *J. IEEE Trans on Power Apparatus and Systems.* 889(4), 316–329.

Deng Jixiang & Zhao Lili (2005). Study on the two order nonlinear correlation of dominant low frequency oscillation modes. *J. proceedings of the Chinese Society for electrical engineering.* 25 (7), 75–80.

Dobson I, Zhang J & Greene S (2001). Is strong modal resonance a precursor to power system oscillation. *J. IEEE Transactions on Circuits and Systems.* 48(3), 340–349.

Fan Wei & Zhao Shuqiang (2009). Analysis of power fluctuations in asynchronous wind turbines with linear dynamic model. *J. Electric Power Science and Engineering.* 25(30), 1–5.

Ghasemi H, Gharehpetian G B & Nabavi-Niaki S A (2013). Overview of subsynchronous resonance analysis and control in wind turbines. *J. Renewable & Sustainable Energy Reviews.* 27(6), 234–243.

Irwin G D, Jindal A K & Isaacs A L (2011). Sub synchronous control interaction between type 3 wind turbines and series compensated AC transmission systems. *In IEEE PES General Meeting, Detroit MI, USA,* pp. 1–6.

Jafarian M & Ranjbar A M (2013). The impact of wind farms with doubly fed induction generators on power system electromechanical oscillations. *J. Renewable Energy.* 50(3):780–785.

Li Jifang (2004). Complex torque coefficients method extened to the rectification of multi-machine power systems. *J. Relay.* 32 (8), 5–7.

Parimi A M, Elamvazuthi I & Saad N (2008). Interline Power Flow Controller (IPFC) Based Damping Controllers for Damping Low Frequency Oscillations in a Power System. *In IEEE International Conference on Sustainable Energy Technologies, Univ. of Nottingham, Nottingham, UK.* pp. 334–339.

Tang Yong (1995). The analysis of forced power oscillation in power system. J. Power System Technology. 19(12), 6–10.

Tang Yong (2006). Fundamental theory of forced power oscillation in power system. *J. Power System Technology.* 30(10), 29–33.

Varma R K, Auddy S & Semsedini Y (2008). Mitigation of Subsynchronous Resonance in a Series-Compensated Wind Farm Using FACTS Controllers. *J. IEEE Transactions on Power Delivery.* 23(3), 1645–1654.

Wang L, Xie X & Jiang Q (2014). Investigation of SSR in Practical DFIG-Based Wind Farms Connected to a Series-Compensated Power System. *J. IEEE Transactions on Power Systems.* 30(5), 1–8.

Xue Yusheng, Zhou Haiqiang & Gu Xiaorong (2002). Review of research on Bifurcation and chaos in power systems. *J. automation of electric powersystems.* 26(16), 9–15.

Yang Yue, Li Guoqing & Li Jiang (2013). Damping characteristics of low frequency oscillation in power system of doubly fed induction generator. *J. Journal of Shenyang University of Technology.* 1, 25–30.

Zhao Shuqiang & FAN Wei (2009). Power system forced power oscillation induced by wind speed disturbance in wind farms. *J. East China Electric Power.* 37(1), 98–102.

Advances in Energy, Environment and Materials Science – Wang & Zhou (Eds)
© 2017 Taylor & Francis Group, London, ISBN 978-1-138-03600-0

Effect of carbon sources on the denitrifying phosphorus removal electrogenesis system

Licheng Zhang
Architecture Design and Research Institute, Shenyang Jianzhu University, Shengyang, Liaoning, China

Jiale Sun
School of Municipal and Environmental Engineering, Shenyang Jianzhu University, Shengyang, Liaoning, China

ABSTRACT: Sodium acetate and glucose were successively used as the carbon source for the device of denitrifying phosphorus removal electrogenesis to investigate the impact of carbon source on the denitrifying phosphorus removal electrogenesis system. The device had been successfully activated and ran stably. This experiment shows that the system had good capacity of resisting the impact of changes in carbon source. When glucose was used instead of sodium acetate as carbon source, the efficiency on denitrification and phosphorus removal and power production efficiency turned bad. The removal rate of phosphorus in the system was becoming lower with the system operation. However, the efficiency of electricity production was increasing with the running of the system.

1 INTRODUCTION

Denitrifying phosphorus removal technology is a new efficient sewage treatment technology, which breaks through the traditional biological phosphorus removal process of anaerobic phosphorus release and aerobic phosphorus uptake mechanism, using denitrifying phosphorus removal bacteria under anoxic conditions with nitrate or nitrite instead of oxygen as an electron acceptor, in the realization of excess phosphorus uptake and denitrification. Owing to the large amount of energy present in municipal sewage, it can alleviate the problem of energy shortage to a certain extent if it is used. Microbial Fuel Cell (MFC) is a kind of sewage treatment technology which uses the microorganism as a catalyst, and the chemical energy of organic matter in sewage can be converted into electric energy. When conducting in-depth researches, experts generally combine microbial fuel cells with wastewater treatment technology in order to degrade the pollutants in water and recover electrical energy at the same time. This new technique is coupled with the denitrification and phosphorus removal process and microbial fuel cell and establishes a system for denitrifying phosphorus removal electrogenesis that can remove nitrogen, phosphorus, and organic matter, and also produce electricity.

Both of the two biochemical processes of denitrifying phosphorus removal and microbial fuel cell need carbon source, and Volatile Fatty Acid (VFA) is the key factor of the two kinds of biochemical reactions. Obviously, carbon source is also an important influencing factor of the denitrifying phosphorus removal system. In recent years, some studies have demonstrated that acetic acid and propionate are the two most important kinds of volatile fatty acids in domestic wastewater (60%–70% and 20%–30%, respectively, accounting for the total of VFA), which can be used as high-quality carbon source[1]. In addition to acetic acid and propionic acid, glucose was another widely used carbon source. Some studies had run the denitrifying phosphorus removal system successfully by using glucose as the sole carbon source. However, there were also studies showing that the advantage of the nitrification and phosphorus removal bacteria would gradually disappear with glucose as the sole carbon source, then concentrate large quantities of GAOs and the phosphorus removal deteriorate[2]. Sun Xiaoying believed that using sodium acetate instead of propionate as the sole carbon source could help improve the performance of the electricity generation. Therefore, this study, respectively, chose sodium acetate and glucose as the sole carbon sources to study the effect of the denitrifying phosphorus removal electrogenesis system.

2 MATERIALS AND METHODS

2.1 Experimental equipment

The process of denitrifying phosphorus removal electrogenesis was based on the double sludge process, and one anaerobic tank and two settling tanks were used as anode chamber and cathode chamber of microbial fuel cell, respectively. The anode chamber and the cathode chamber were separated by an ion exchange membrane. The experimental technological process is shown in Figure 1.

2.2 Experimental methodology

The device of denitrifying phosphorus removal electrogenesis had been started successfully, and the carbon source used was sodium acetate. Then, we changed the type of carbon source from sodium acetate to glucose. The temperature of the system was controlled at 22–25 degree Celsius, and the pH was controlled at 7.5. The dissolved oxygen concentrations in the anaerobic tank, aerobic tank, and anoxic tank were 0.2 mg/L, 1.0–2.0 mg/L, and 0.5 mg/L, respectively. MLSS was controlled at 4000 mg/L. We detected the pollutant indicators continuously until the denitrifying phosphorus removal electrogenesis system entered a long-term stable state, so that the effects of different kinds of carbon sources on the denitrifying phosphorus removal electrogenesis system can be studied.

2.3 Analytical items and methods

COD was evaluated by the potassium dichromate resolution method. The phosphate concentration $(PO_4^{3-}-P)$ was determined by molybdenum-antimony-ascorbic acid spectrophotometry. TN was determined by the ultraviolet spectrophotometric method using potassium sulfate. NH_4^+-N was estimated with Nessler's reagent spectrophotometer. NO_2^--N was detected by N-(1-naphthyl)-ethylenediamine spectrophotometry. NO_3^--N was determined by the ultraviolet spectrophotometric method. The voltage was measured using a voltmeter.

2.4 The raw water quality

Table 1. The raw water quality.

Index	The unit of mass concentration: mg/L				
	COD	NH_4^+-N	TP	Alkalinity	pH
Range	130–160	35–45	7–10	300–400	7–8
Average	152.34	39.69	7.85	347	7.63

3 RESULTS AND DISCUSSION

Although the device of denitrifying phosphorus removal electrogenesis had started successfully, and the reaction system also generated fluctuations when we changed the type of carbon source, there really existed different effects on the removal of nutrients in domestic sewage after the system was stable again.

3.1 Efficiency of COD removal

Figure 2 shows the influence on the removal effect of COD when sodium acetate and glucose were respectively used as influent carbon source on the system of the denitrifying phosphorus removal electrogenesis. We could see that when we used sodium acetate as the carbon source, the removal of total organic matter fluctuated between 50% and 60%. The influence on the system was relatively large by using glucose as the carbon source, and fluctuations in the system last for a long time. After about 15 days, the system would tend to be stable, and the organic matter removal rate has remained at around 60%. We could know from Figure 2 that the removal effect of the system on COD was not quite different, and the anaerobic stage played an important role in the removal effect of COD. A large proportion of COD are removed in the anaerobic stage. This was because not only de- nitrifying phosphorus accumulating bacteria need to absorb large amounts of organic matter to finish phosphorus release, but electrogenic bacteria also need to decompose the organic matter to generate electricity.

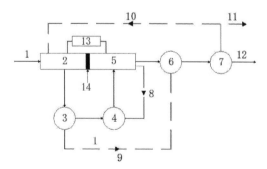

Figure 1. Flow diagram of denitrifying phosphorus removal electrogenesis process. 1, Inlet water; 2, Anaerobic tank (Anode chamber); 3, Primary sedimentation tank; 4, Aerobic tank; 5, Secondary sedimentation tank; 6, Anoxic tank; 7, Final sedimentation tank; 8, Recycled sludge; 9, Beyond sludge; 10, Recycled sludge; 11, Sludge discharge; 12, Effluent water; 13, Resistance box; 14, Ion exchange membrane.

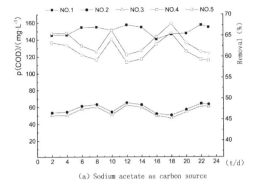

(a) Sodium acetate as carbon source

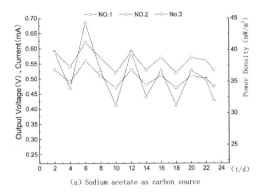

(a) Sodium acetate as carbon source

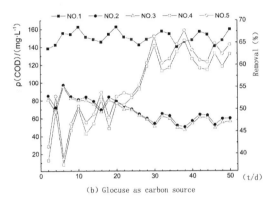

(b) Glocuse as carbon source

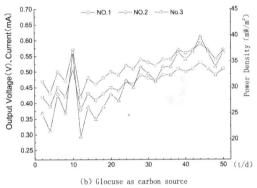

(b) Glocuse as carbon source

Figure 2. Effect of different carbon sources on the efficiency of COD removal No. 1, Inlet COD; No. 2, Anaerobic effluent COD; No. 3, Anoxic effluent COD; No. 4, COD removal rate in anaerobic stage; No. 5, COD removal rate.

Figure 3. Effect of different carbon sources on the efficiency of electricity production. No. 1, Output voltage; No. 2, Current; No. 3, Power density.

3.2 Efficiency of electricity production

Figure 3 shows the effect on the property of electric power when sodium acetate and glucose were respectively used as influent carbon sources on the system of the denitrifying phosphorus removal electrogenesis. When sodium acetate was used as the carbon source, the output voltage was stable which fluctuated between 0.45V and 0.60V, and the average voltage was 0.5V, and the maximum output power could reach 44.28 mW/m². When glucose was used as the sole carbon source, the system was stable after 15 days. The effect of producing electricity was worse than the system with sodium acetate as the sole carbon source at first, but with the increasing reaction, the output voltage was between 0.4V and 0.53V, and the average voltage was 0.49V, and the maximum output power could reach 39.66 mW/m². We analyzed the reason for this result that the microbial population was more complex in the system of activated sludge. The dominant bacteria in the anaerobic pond

were the denitrification phosphorus-accumulating bacteria and the electrogenesis bacteria or the denitrification phosphorus-accumulating bacteria capable of electrogenesis. While glucose was used as the carbon source, it would promote the growth of GAOs, so that the advantage of the nitrification and phosphorus removal bacteria would be threatened. However, we knew that electrogenesis bacteria also need the carbon source, which meant that GAOs and the electrogenesis bacteria would compete for organic matter, and the electrogenesis bacteria would hinder the growth of GAOs.

3.3 Efficiency of denitrification removal

Figure 4 shows the effect on denitrification when sodium acetate and glucose were respectively used as influent carbon sources on system of the denitrifying phosphorus removal electrogenesis. The influent of the two carbon source system would have a small amount of nitrate nitrogen that might be due to the final sedimentation tank sludge

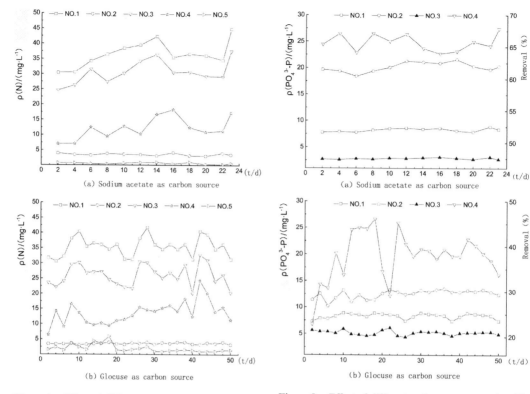

Figure 4. Effect of different carbon sources on the efficiency of denitrification removal. No. 1, Inlet ammonia nitrogen; No. 2, Inlet nitrate nitrogen; No. 3, Effluent nitrate nitrogen at nitrification stage; No. 4, Effluent nitrate nitrogen; No. 5, Effluent nitrite nitrogen.

Figure 5. Effect of different carbon sources on the efficiency of phosphorus removal. No. 1, Inlet phosphorus; No. 2, Anaerobic effluent phosphorus; No. 3, Effluent phosphorus; No. 4, phosphorus removal rate.

reflux. Based on the experimental data, the denitrification capacity was bad when glucose was used as the carbon source, and the mass concentration of nitrate nitrogen removal has changed between 10 and 15 mg/L, which accounted for 60% of the average mass concentration of nitrate removal when sodium acetate was used as the carbon source. This may be due to the change in the carbon source, which influenced the phosphorus release in anaerobic phase and then affected denitrification.

3.4 Efficiency of phosphorus removal

According to Figure 5, when glucose was used as the carbon source, the phosphorus removal rate was much lower than that of the system with sodium acetate. As it is shown in Figure 5 (b), phosphorus release had been maintained at 4.3–6.5 mg/L, when the system became stable again, and the phosphorus removal rate gradually decreased to 37%. When sodium acetate and propionate were respectively used as sole carbon sources, the typical biological phosphorus removal process was carried out in the

systems of denitrifying phosphorus removal electrogenesis. This meant using denitrifying phosphorus removal bacteria to adsorb the organic matter in the anaerobic stage to finish the phosphorus release, and under anoxic conditions with nitrate or nitrite as an electron acceptor, in the realization of excess phosphorus uptake and denitrification. While using glucose as the carbon source, the concentration of organic matter was decreased and the decrement was similar to that in the system with sodium acetate, so that the release of phosphorus was poor, which was a typical metabolic process of GAOs. We could believe that the microbial environment of original system has changed when glucose was used as the carbon source. Mino's study suggested that glucose in the raw water could replace glycogen to provide energy and reducing power for the synthesis of PHA, and then the dependence of DPAOs on the hydrolysis of Poly-P was decreased, which led to the decrease in phosphorus release and the accumulation of GAOs in the system[3-4]. We analyzed the causes of the phenomenon, which indicates that glucose might provide energy instead of glycogen for the synthesis of PHA. Then, this promoted the accumulation of GAOs, so that this might depress the capability of anaerobic phosphorus release, which would influence the effect of denitrification and phosphorus uptake in anoxic stage.

According to the changes in the concentration of the substances under the conditions of two kinds of carbon sources, the carbon source would affect the system. However, it would soon be back to normal, which meant that the system had a good adaptability to the impact of the carbon source transformation, so that the change in the carbon source will affect the system.

Glucose is a carbon source leading to the gradual loss of the dominant position of the phosphorus accumulating bacteria, resulting in a decrease in the rate of phosphorus removal. The average removal rate of phosphorus was only down to approximately 40% in the fifty days experiment. The results were different from those obtained in the study by Bao Linlin[5] on the effects of different types of carbon sources on the denitrification and phosphorus removal. Binding figure 3, it shows that when glucose was used as the carbon source, the performance of the electricity generation had been improved, and the maximum output voltage of the latter can reach 0.53V. While glucose was used as the carbon source, it would promote the growth of GAOs, and we could see from section 3.2 that the electrogenesis bacteria would hinder the growth of GAOs in this system. To sum up, the process of denitrifying phosphorus removal electrogenesis was more stable than the conventional denitrification and phosphorus removal process.

4 CONCLUSIONS

1. The stable operation of the denitrification phosphorus removal electrogenesis system had good

ability to bear the shock caused by the change in the carbon source. It could also assure the better effluent equalities and electricity generation.
2. It would reduce the effect of denitrification and phosphorus removal, when glucose was used as the carbon source, but the effect of electricity generation would be improved.

ACKNOWLEDGMENTS

This project was sponsored by Liaoning BaiQian-Wan Talents Program.

REFERENCES

Bao Linlin, Li Xiangkun, Zhang Jie. 2011. Effects of carbon source types on the system of denitrification and phosphorus removal [J]. Techniques and Equipment for Environmental Pollution Control. 5(7): 1567–1571.

Canizares P., De Lucas A., Rodriguez L., et al. 2002. Anaerobic uptake of different organic substrates by an enhanced biological phosphorus removal sludge [J]. Environ. Technol. 21(4):397–405.

Mino T., Van Loosdrecht, M C M., Heijnen, JJ. 1998. Microbiology and biochemistry of the enhanced biological phosphate removal process [J]. Wat Res. 32(11): 3193–3207.

Thomas M, Wright P., Blackall L. 2003. Optimization of Noosa BNR plant to improve performance and reduce operating costs [J]. Wat. Sci. Technol. 47(12):141–148.

Wang Y., Jiang, F., Zhang, Z. 2010. The long-term effect of carbon source on the competition between poly-phosphorus-accumulating organisms and glycogen accumulating organism in a continuous plug-flow anaerobic/aerobic (A/O) process [J]. Biores Tech. 101(1): 98–104.

Advances in Energy, Environment and Materials Science – Wang & Zhou (Eds)
© *2017 Taylor & Francis Group, London, ISBN 978-1-138-03600-0*

Transient analysis of the cooling system for coal-fired plants

Dong Wang
He Nan Electrical Survey and Design Institute, Zheng Zhou, China

ABSTRACT: When cooling water pumps for power plant restart and stop, it will influence the flow rate of the pipe network system, and trigger water hammer and cavitation. Based on the theory of unsteady flow for pressure pipeline, using the PIPENET software, the calculation model for the cooling water system was established to carry out transient calculation, to study the relation between pump trip and exhaust pressure of a condenser. With these analysis and calculation, the design and operation mode of the cooling water system can be optimized for power plants.

1 INTRODUCTION

The cooling water system for thermal power plants is mainly used to cool the exhaust steam in turbine, achieving to maintain the normal operation of generator units and improve power generation efficiency. It plays a very important role in safe and economic operation for generators.

Using the PIPENET software, the cooling water system calculation model was established, and the relationship between water hammer and pump trip was analyzed. Besides, the influence of closing style of pump outlet valve was studied, and exhaust pressure and flow rate variation curve was obtained when the pump stopped operation.

2 CALCULATION METHOD

2.1 Basic equations

The basic method of calculation of hydraulic transient process is based on the pressure pipelines unsteady flow (also called transient flow) calculation theory, and the elastic water hammer calculation uses the method of characteristics. Basic equations of transient calculation are the equation of motion and the continuity equation:

Equations of motion:

$$\frac{\partial V}{\partial t} + V\frac{\partial V}{\partial x} + g\frac{\partial H}{\partial x} + \frac{fV|V|}{2D} = 0 \tag{1}$$

Continuity equation:

$$\frac{\partial H}{\partial t} + V\frac{\partial H}{\partial x} - V\sin\alpha + \frac{a^2}{g}\frac{\partial V}{\partial x} = 0 \tag{2}$$

where H is the pressure head, V is the velocity, x is the distance from the left end of the pipe section, g is the acceleration of gravity, f is the frictional head loss, D is the diameter, a is the water hammer wave velocity, and t is the time.

With the equation of motion and continuity equation, universal application characteristics equation can be obtained:

$$C+: H_P = C_P - B \cdot Q_P \tag{3}$$

$$C-: H_P = C_M + B \cdot Q_P \tag{4}$$

$$C_P = H_A + B \cdot Q_A - R \cdot |Q_A| \cdot Q_A \tag{5}$$

$$C_M = H_B - B \cdot Q_B + R \cdot |Q_B| \cdot Q_B \tag{6}$$

$$B = \frac{a}{gA}$$

$$R = f \cdot \Delta x \cdot \frac{1}{(2gDA^2)}$$

where H_A and Q_A are the number i-1 node pressure head and flow rate, respectively, at t-Δt time, H_B and Q_B are the number $i + 1$ node pressure head and flow at t-Δt time, Δx is the distance between two adjacent nodes, R is the resistance coefficient, and C_P and C_M are, respectively, related to pressure head and flow rate at t-Δt time, and both are known at t time.

2.2 Boundary condition

Cooling water pump, valve, inlet, outlet and pipe fitting are the same boundary equation with the conventional cooling system. Siphon well and condenser utilize the receiving vessel and the pipe bundle from the PIPENET software.

3 COOLING WATER SYSTEM TRANSIENTS CALCULATION

3.1 Cooling water system

The cooling system of the power plant is the sea-water once-through cooling system. Two generators configurate five cooling pumps (including one standby pump in common). Single pump rated flow rate is 15.47 m³/s, and hydraulic head is 16.5 m. Circulating water pipes material is glass reinforced plastic, and the main pipe diameter is DN3800. Branch pipes linked to the condenser are welding carbon steel pipe with cathodic protection, and the pipe diameter is DN2600.

Each generator is equipped with two cooling pumps, and the performance of these pumps is as follows:

flow rate: Q = 15.47 m³/s

head: H = 16.5 m

power capacity: N = 3000 kW

When the seawater temperature is 30.2°C in the summer, the steam turbine exhaust pressure is 7.76 kPa. When the average seawater temperature is 23.1°C, the condenser exhaust pressure is 5.41 kPa.

3.2 Transient calculation for cooling system

Case 1: Pumps stop and outlet valves do not operate

Two cooling pumps stopped at 10th second, and the outlet valves do not operate. Figure 1 shows that the main pipe of cooling pump outlet appears to have reverse flow at 55th second, causing the cooling pumps to operate reversely, and the maximum reverse flow is −7.5 m³/s; the maximum value

exceeds the allowable value. Maximum pressure head is 25.4 m, and the minimum pressure head is −9.77 m. Both appear at the position of the condenser top. Figure 2 shows that vaporization occurs at the position of the condenser top, and the maximum volume of vaporization is about 4.5 m³. This is due to the position of the condenser located at the relatively high point. When the pump stops, this position is less than the minimum vaporized pressure, resulting in vaporization. Therefore, it is necessary to take measures to prevent water hammer and cavitation.

Case 2: Outlet valves close in two stages (fast close and slow close)

Two cooling pumps stop running at 10th second, and the outlet valve closes in two stages. The first stage is closing the valve to 75% at a fast speed in 8 seconds, and the other stage is closing the valve completely slowly in 35 seconds. The outlet valve start operation time is 10 seconds, 30 seconds, and 55 seconds, respectively. The results are given in Table 1.

Table 1 shows that the outlet valve starts operation in three modes, cooling pumps appear to have reverse flow, and, when the operation time is too long, the reverse flow rate is much greater. Taking two stages to close the outlet valve can decrease the reverse flow to ensure the value within the allowable range. Therefore, in the actual project, when the cooling pumps stop, the outlet valve operation should respond as soon as possible in order

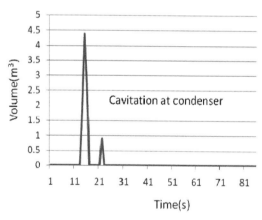

Figure 2. Volume of vaporization in the condenser under case 1.

Table 1. Outlet valve operation time and reverse flow rate relation.

Start time (s)	50s	30s	10s
Reverse flux (m³/s)	−5.6	−3.7	−1.1
Minimum pressure (m)	−9.77	−9.77	−9.77

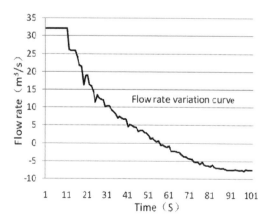

Figure 1. Outlet flow rate variation in the cooling pump under case 1.

to decrease reverse flow. In addition, under three operating mode conditions, even if the outlet valve is closed in two stages (fast close and slow close), the minimum pressure in the cooling water system is still less than the vaporized pressure at the downstream of the condenser.

Case 3: The outlet valve closes in two stages with the venting valve

Two cooling pumps stop at 10th second, and meanwhile, the outlet valve closes in two stages. The first stage is closing the valve to 75% at a fast speed, and the second stage is completely closing the valve in 35 seconds with DN200 safety venting valves at the downstream of the condenser. The results indicate that the maximum pressure head is 15.7 m, and minimum pressure head is −5.69 m. The negative pressure is generated, but it is greater than the vaporized pressure. Therefore, with the venting valves, it is effective to avoid the vaporization.

3.3 Transient calculation and analysis of condenser exhaust pressure

The cooling system for power plant function is mainly to provide sufficient cooling water for the steam turbine condenser, cool exhaust steam after the completion of energy conversion, and maintain thermal equilibrium. The cooling water system will directly affect the value of the condenser exhaust pressure, and it can influence the efficiency and output capacity of the steam turbine. Therefore, the study of relation between cooling pump and condenser exhaust pressure is of great significance.

Working conditions: each turbine is equipped with two cooling pumps. One is in normal operation, and the other stops at 10th second. As shown in Figure 3, when the two pumps run normally, under the conditions of seawater temperature of 30.2°C in the summer, the condenser exhaust pressure is

7.76 kpa, and the value does not exceed 11.8 kpa, which is the highest exhaust pressure for the 100% power output capacity; therefore, the cooling water system is safe and stable. When a pump stops running and the other operates, the condenser exhaust pressure will increase. The maximum exhaust pressure is 11.96 kpa, and it is over 11.8 kpa, causing the steam turbine not to run at full capacity and even stop running. This will seriously affect the normal operation of the steam turbine.

4 CONCLUSION

1. When two pumps stop, the outlet valve can be closed in two stages (fast and slow). This can avoid the reverse operation of cooling water pumps and decrease water hammer.
2. When the cooling pump stops running, the transient positive pressure is small, and it is easy to have a greater negative pressure and vaporization; therefore, it is necessary to add venting valves to protect the cooling water system.
3. With the calculation data and analysis, a correlation curve between flow rate variation and exhaust pressure can be obtained. When a cooling water pump stops operation, due to the decrease in flow rate, the exhaust pressure will increase immediately. In the summer, under the condition of seawater temperature of 30.2°C, when one cooling pump stops by accident, the exhaust pressure will exceed the safety value. Therefore, it is necessary to decrease heat load, operate spare pump, and other measures to ensure the generator operate stably and safely.

REFERENCES

Chen fushan, Liu deyou, Study the transient process of water power plant cooling water system[J] Energy Research and Utilization. 2005. 3: 14–17.
Jia Faxian, Transition process calculation for Wang tan power plant Circulating Water System [D] Beijing: Beijing University of Technology, 2006.
Jinzhui, Jiang naichang, Wang xinghua, Guan xingwang, Water hammer and protection measure of stopping pump [M] Beijing: China Building Industry Press, 2015.
Li jiaba. 2 × 660 MW thermal power plant circulating water system operation optimization [D]. Northeast Electric Power University, 2013.
Wangdinglei, Wang Lin, Zhang Zhiyuan. 660 MW supercritical units optimization analysis on circulating water system[J] Shenyang Institute of Engineering (Natural Science Edition), 2010, 03: 215–217.
Yang kailin, power plant and pump station hydraulic transient calculation and regulation [M] Beijing: China Water Power Press, 1999.

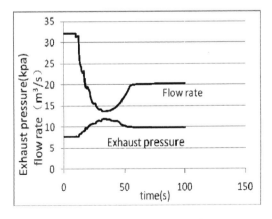

Figure 3.   Cooling water flow rate and exhaust pressure relation curve.

Advances in Energy, Environment and Materials Science – Wang & Zhou (Eds)
© *2017 Taylor & Francis Group, London, ISBN 978-1-138-03600-0*

Research on the key technologies of a biomass briquette-fueled curing barn heating system

Yali Guo, XianHong Zhao & Feng Wang
Guizhou Provincial Tobacco Company Qianxinan Branch, Qianxinan, Guizhou, China

Dexiang Hu, Dabin Zhang & Yang Cao
College of Mechanical Engineering of Guizhou University, Guiyang, Guizhou, China
Guizhou Mechanical and Electrical Equipment Engineering Technology Research Center Co. Ltd., Guiyang,
Guizhou, China

ABSTRACT: Environmental pollution problems involve traditional coal-fired curing barn and its high curing costs. This paper develops an intelligent biomass briquette-fired curing barn heating system, with biomass energy replacing traditional coal, by transforming bulk curing barn heating system. It also solves the key technical problems arising during system development. The problems include deepening biomass briquetting technology, replacing coal-fired stoves with biomass stoves, and performing structural modification for heat exchangers. Automatic feeding system and intelligent control baking system are adopted for regulating and controlling the whole process. Moreover, air supply and combustion system have been proposed based on tertiary inlet air technology with innovation. In addition, by comparing the baking tests of biomass briquette-fired curing barn and traditional coal-fired curing barn, the results show that in terms of economic benefit, the former is 1 yuan lower than the latter as for 1 kg dry tobacco leaves, in which energy cost accounts for 30% and labor cost makes up 70%. For the quality and average price of cured tobacco leaves, the rate of first-class tobacco leaves of the former increases by 6% and the overall average price of dry tobacco leaves is 0.95 yuan higher than that of the latter. As for combustion emissions, such as CO_2, SO_2, and NO_x, the former is 0.3 kg, 0.001 kg, and 0.001 kg lower than the latter for 1 kg dry tobacco leaves successively. The above data show that the application of biomass briquette-fired curing barn can further promote the concept of energy saving and emission reduction, efficiency-improving as well as work simplification, which has great practical significance for tobacco industry.

1 INTRODUCTION

Traditional bulk curing barn mainly adopts coal as the heating energy. However, in recent years, with the exhaustive exploitation and consumption of coal and other non-renewable resources, these resources are facing to dry up. In addition, large quantities of waste, like CO_2, dust, and SO_2, generated during combustion process have triggered a series of global environmental problems. Therefore, looking for new sources of energy is the major development trend of bulk curing barn, which is also a new and important issue for the world (Chen & Tao, 2004). Biomass energy is a kind of chemical energy stored in plant, transformed from solar energy through photosynthesis, having the advantages of renewability, extensive sources, low cost, no pollution, etc. Thus, biomass energy is an ideal substitute of coal and other fossil energies.

The most important feature of biomass is its wide variety of sources. Waste wood and crop straw can be used as the raw material. According to the statistics, the annual tobacco planting area in our country reaches over 15 million mu, and the annual output amounts to about 3.6 million tons. At present, tobacco stems, tobacco waste, and waste tobacco stalks are mainly treated through burning, which not only pollutes the environment, but also needs a great deal of costs. Therefore, processing tobacco stems and stalks into biomass briquette, to replace coal as fuel for baking tobacco leaves that turn waste into wealth, not only can reduce environmental protection pressure and save a large amount of fuel costs, but also can prevent tobacco leaves from the corrosion of the smoke from burning coal during first baking, which can greatly enhance the quality of tobacco leaves. At the same time, the collection of tobacco straws can reduce tobacco virus, pests, and disease damage. What is more important, it lessens the burden of tobacco growers, and improves their enthusiasm of planting tobacco, which completely complies with the great trend of modern tobacco agriculture development. Besides, it is also an important measure to benefit the nation and the people.

2 BIOMASS BRIQUETTING

2.1 Key technologies of biomass briquetting

In order to enlarge the application range of biomass energy and improve its utilization efficiency, it is required to convert original biomass into other forms of energy. In implementing the project, biomass curing molding technology is adopted (Ayhan 2004, Yao & Tian 2005, Shi & Hua 2006), which is also the most commonly used in the current industrial production. Biomass curing molding technology is a processing technology (Zhang 1999, Wang & Cai 2008, Angelo 2007), to crumble original biomass into certain particles by using a crusher, which are then extruded into solid fuel with high density and regular shape under certain pressure and temperature. Its main purpose is to transform low-density biomass energy into high-density biomass energy.

For biomass curing molding technology, the compressing process is the key. In general (Liu, Niu & Zhang et al 2005), biomass curing molding technology is classified into carbonization briquetting, wet briquetting, as well as the hot briquetting, adopted in this project. After original biomass is dried and crushed, the lignin contained in raw material is a monomer with aromatic structure, which is non-crystal, with no melting point, but there is softening point. During processing, when the temperature reaches up to 70~100°C, the adhesive force of lignin starts to increase. For 200~300°C, softening degree increases, causing it to liquefy. At this time, certain pressure to make it tightly bond with cellulose, and cement with adjacent particles, can be applied. After cooling, particles, rod-like or bulk materials, can be made (E. Granada, L.M. López Gonzá lez, J.L. Mguez, et al 2002, Tian 2009, Chen, Wu & Tian et al 2010).

2.2 Technical requirements and conditions for launching project

The technical requirement for molding in this project is as follows: after original biomass is crushed, its length should be lower than 5 mm, and water content is controlled within 14~18%. Then, the material is delivered into a feeding throat by a feeding conveyer. With spindle rotation driving roller and compression roller rotation, the bulk material is extruded from the model hole; the briquettes fuel is a kind of solid particle fuel, with diameter between 8 and 18 mm, length between 30 and 40 mm, and density between 10 and 14 g/mm3, which can be fueled directly. After dropping from the discharge hole and cooling (water content should be over 14%), the material is packed.

Biomass material is mainly from various crop straws (like tobacco straws), industrial waste (such as vinasse and sugar residue), forestry waste materials (like deadwood), etc. According to investigation, the forest land in Qianxinan prefecture covers an area of 7.5638 million mu, with a great amount of forestry waste materials. Moreover, there is large-scale wine-making industry in Guizhou province, which produces large amounts of vinasse annually. There are sufficient agricultural straw resources, with about 1.18 million tons all year round. In addition, there are abundant sugarcane bagasse resources, with over 0.5 million tons annually, used as biomass fuel resources. In order to further reduce production cost, the government can plan regions in Qianxinan and choose the right location to build 2~3 biomass briquette manufacturing plants, covering an area of about 6000 m², equipped with production line with complete briquetting equipment according to the construction standard.

3 BIOMASS ENERGY CURING BARN HEATING SYSTEM

3.1 Stove shell material

The original coal-fired curing barn heating system (coal-fired stove) is built with a brick-concrete structure, with an outer width of wall of 1.4 meters, height of 2.8 meters, and total area of wall of 11.76 square meters. The inner layer can absorb heat and the outer layer can release heat. Based on thermal exchange principle, the outer wall exposed in air will lose a large amount of heat. In order to solve this problem, the stove shell size and material are redesigned as follows: the outer width of stove on four sides is changed into 1.1 meters, the height of 2.5 meters, and the total area of shell of 11 square meters. The Q235 steel with a wall thickness of 1.5 mm is wrapped with a 40 mm-thick heat-insulating material and its sectional drawing is shown in Fig. 1.

Under the environment temperature of 20°C, the heat dispersion of two types of wall is inspected. The temperature of outer wall during high - and low-temperature baking is measured. The results are shown in Table 1.

According to the data in Table 1, it can be found that under the environment of 20°C, during high - and low-temperature baking stages, the temperature of the outer wall of biomass stove is 38°C lower and 16°C higher than that of original

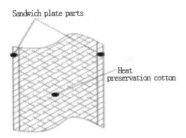

Figure 1.   Heat insulation layer sectional drawing.

Table 1.   Statistical table of the temperature of coal-fired stove outer wall and biomass stove shell (°C).

Name	Average temperature of coal-fired stove outer wall	Average temperature of biomass stove shell	Temperature difference of biomass stove and coal-fired stove
High temperature (48°C–68°C) stage	70	32	38
Low temperature (35°C–42°C) stage	40	24	16

coal-fired stove, respectively, namely a thermal insulation material is adopted, which can reduce the surface area of wall, decreasing the energy consumption to a great extent.

3.2   Automatic feeding system

During the use of bulk curing barn stove, the quantity of coal is added according to the temperature in curing barn and the baker's experience. Too much coal will lead to inadequacy burning and insufficient coal cannot satisfy baking requirements. In addition, the curing barn temperature and the quality of tobacco leaves cannot be controlled well. According to this phenomenon, this project designs the automatic feeding system in accordance with the biomass briquette size, as shown in Fig. 2.

This system is mainly made up of hopper, frequency control servo motor, screw conveyer, and other parts. Automatic feeding system is controlled by controller with higher intelligent degree in accordance with the temperature in curing barn. When the temperature is lower than 0.1°C, the objective temperature is set by intelligent controller, the automatic feeding system will start, and variable frequency motor will drive screw conveyer.

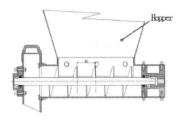

Figure 2.   Automatic feeding device for biomass stove.

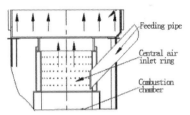

Figure 3.   Feeding pipe for automatic feeding system of biomass stove.

Thus, the biomass fuel in hopper will be fed into the furnace of curing barn stove through the feeding device of coal-fired stove. At the same time, the fire in furnace will increase and the temperature in the curing barn will rise. When the temperature is higher than the objective temperature range set by intelligent controller, 0.1~0.3°C, the automatic feeding system will stop.

In order to prevent the flame in furnace igniting the fuel in the feeding system, an air inlet can be added in the feeding pipe to make wind blow into the furnace downward. The wind is called wind shield. The wind shield is always in working state from the beginning of stove burning to the end of stove burning, as shown in Fig. 3.

According to the experiment, biomass stove has high degree of automation, which can save labor and time. Compared with coal-fired stove, it can save 70–80% labor, it can provide automatic feeding in accordance with the requirements for temperature in curing barn (precision: ±0.1°C), in order to reduce unreasonable fuel burning. The time of constant temperature in curing barn is prolonged by the wind shield flame retardant system of the automatic feeding system, which guarantees the baking quality of tobacco leaves and enhances the operating performance and safety performance of the automatic feeding system.

3.3   Air-supplying combustion system for biomass stove

For the problems, such as inadequate burning of new biomass stove, this project conducts a research on the combustion technologies of biomass stove and develops an air-supplying combustion system, as shown in Fig. 4.

The air-supplying combustion system for biomass baking stove is composed of a tertiary air intake system and a wind shield. In addition, the tertiary air intake system is made up of an upper air inlet pipe, a central air inlet pipe, a main fan interface, a lower air inlet pipe, and a lower air inlet ring. The wind shield comprises an interface of fan for wind shield, a solenoid valve, and a wind shield. The upper air inlet pipe and central inlet system of tertiary air intake system can lead to better adequacy burning of biomass fuel. When the main fan is started, the air in wind shield will enter the feeding pipe through the wind shield to form a negative pressure with the main fan, to ensure the normal operation of feeding device. When the indoor dry-bulb temperature is higher than the set dry-bulb temperature, the main fan will stop and the solenoid valve will be opened at the same time. Moreover, partial air in wind shield will enter into the lower air inlet system in the tertiary air supply system to form a secondary air supply system with the wind in the wind shield, which realizes low fire burning after the main fan stops running, facilitating the burning of biomass fuel and reducing combustible gas emissions after it stops running. Based on multiple tests, a reasonable air supply rate of the tertiary air supply system and wind shield is obtained, as shown in Table 2.

Tertiary air supply system and wind shield are the innovative points of this project, which realizes three-stage combustion and solves secondary and tertiary burning problems. At the same time, with heat storage function, it facilitates sufficient burning of biomass fuel, reducing combustible gas emissions. Burning heat value reaches up to 4000–4500 large calories/kg and the overall thermal efficiency is up to 80%, which ensures the heat required for tobacco baking.

3.4 Heat exchange device

Heat exchanger is a kind of device to convert the chemical energy in biomass stove into thermal energy, which is mainly made up of loop heat exchanging cabinet and straight plate heat exchanging cabinet, as shown in Figure 5.

The heat-exchange efficiency is proportional to heat dissipating surface and the path length of hot air. Therefore, to realize higher conversion efficiency, the project group conducts an experimental research on heat exchanger area and the path length of hot air and compares the heater exchanger for biomass stove with the heat exchanger for coal-fired stove. The experimental data are given in Table 3.

From Table 3, it can be observed that the total effective heat-dissipation area of the heat

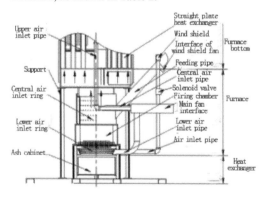

Figure 4. Air-supplying combustion system for biomass stove.

Table 2. Reasonable supply air rate of the devices for the air-supplying combustion system of biomass stove (m³/h).

	Main fan	Upper air inlet pipe	Central air inlet pipe	Lower air inlet pipe	Wind shield fan	Wind shield	Low fire air inlet pipe
Air volume	2258	293.5	667.4	1287	6.5	3.05	3.45

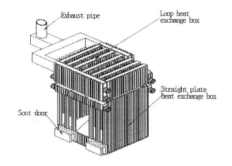

Figure 5. Heat exchanger of biomass stove.

Table 3. Comparison table of the efficiency of heat exchanger for intelligent biomass energy curing barn and traditional bulk curing barn.

	Heat exchanger for biomass stove	Heat exchanger for coal-fired stove	Ratio
Heat-dissipation area (m²)	14.3	9	1.59
Path length of hot air (mm)	3760	2790	1.35

exchanger for biomass stove is 1.59 times that of the heat exchanger for coal-fired stove and the path length of hot air is 1.35 times that of the coal-fired stove, that is to say, the total heat utilization rate increases by 2.15 times, reducing the heat loss and fuel cost.

4 INTELLIGENT BIOMASS STOVE CONTROL SYSTEM I

In order to better implement the concept "work simplification and cost reduction" and achieve the intelligent control of the above various devices, the project group develops an intelligent electrical apparatus control system for biomass stove, namely, main control instrument. Its control loop and picture are shown in Figure 6.

The working principle of this intelligent control system: pressing SB1 to start air blower (normally closed during stove operation); pressing SB2 to start speed governor manually for hand feeding; pressing SB3 to start self-priming pump; when the dry-bulb temperature of controller for bulk curing barn is lower than the subjective temperature of dry bulb, KM1 pulls in, M1starts, and speed governor starts for feeding and solenoid valve closes at the same time; when the dry-bulb temperature of controller for bulk curing barn is higher than the subjective temperature of dry bulb, KM1 closes, M1stops, speed governor starts for feeding, and solenoid valve starts at the same time.

During tobacco baking, when the dry-bulb temperature in curing barn is higher than 0.1°C, the set subjective dry-bulb temperature, the main fan, and the feeding motor will be disconnected by the main control instrument and main fire will stop burning. At this time, the solenoid valve will start and the wind shield maintains low fire burning. When the dry-bulb temperature in curing barn is lower than 0.1°C, the set subjective dry-bulb temperature, the main fan, and the feeding motor will be connected

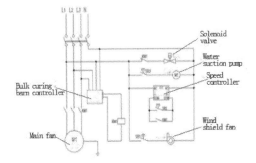

Figure 6. Intelligent control loop of biomass stove and picture.

by the main control instrument, the solenoid valve will be closed, and main fire starts to burn. When the wet-bulb temperature in curing barn is higher than 0.1°C, the set subjective wet-bulb temperature and the cold air valve will be open for moisture removal. Otherwise, cold air valve will be closed to stop removing moisture. The experiment shows that the baking curve error is between 0.1 and 0.2°C, with few bias-temperature phenomena, in order to control the temperature and humidity in curing barn effectively and reduce the tobacco baking risk significantly.

5 PREFERENCES, SYMBOLS, AND UNITS

In order to better understand the baking benefit of intelligent biomass stove, the project group conducted the same baking efficiency test for traditional bulk curing barn while developing this project in Anlong County in 2015.

5.1 Baking economic benefit

From Table 4, it can be known that, for baking 1 kg dry tobacco leaves, the baking cost of traditional coal-fired curing barn is 4.55 yuan, while the baking cost of intelligent biomass curing barn is 1.58 yuan, which is about 1 yuan lower than the former. For the difference between the two, the fuel cost accounts for about 30%, and the labor cost for about 70%. Thus, the requirement for energy saving and labor reduction is achieved, which greatly increases economic benefits.

5.2 Quality of backed tobacco leaves

From the data in Table 5, it can be observed that the ratio of first-class tobacco from intelligent biomass curing barn is 6% higher than that from traditional coal-fired curing barn, while the ratio of low-grade tobacco from the former is 2% lower than that from the latter. Combining the average price of various tobacco ratios, it can be known that the average price of tobacco leaves baked by intelligent biomass curing barn is 34.45 yuan per kg and the average price of tobacco leaves baked by traditional coal-fired curing barn is 33.5 yuan per kg. The former is 1 yuan higher than the latter per kilogram. Therefore, the intelligent biomass curing barn can increase the efficiency.

5.3 Waste emissions

From Table 6, we can observe that, for baking 1 kg dry tobacco leaves by burning biomass energy, there will be 4.3 kg CO_2, 0.014 kg SO_2, and 0.012 kg NO_X emissions, which is 0.3 kg, 0.001 kg, and 0.001 kg

Table 4. Comparison table of the main economic benefits of intelligent biomass curing barn and traditional bulk curing barn.

Types of curing barn	Baking dry soot	Fuel quantity	Fuel cost	Baking time	Workshop for feeding and soot cleaning	Labor cost of feeding and soot cleaning	Unit dry soot cost
	kg	kg	yuan	h	h	yuan	yuan
Traditional coal-fired curing barn	500	880	775	220	10	500	2.55
Intelligent biomass curing barn	500	820	640	200	3	150	1.58

Note: the coal cost is 880 yuan per ton, and the biomass fuel cost is 780 yuan per ton. Owing to non-continuity of labor, the labor cost is 50 yuan per hour.

Table 5. Comparison table of the quality of tobacco leaves baked by intelligent biomass curing barn and traditional bulk curing barn.

Types of curing barn	First-class tobacco ratio	Medium tobacco ratio	Low-grade tobacco ratio	Average price	Average price improvement
	%	%	%	yuan/kg	yuan/kg
Traditional coal-fired curing barn	63.54	33.17	3.29	33.50	–
Intelligent biomass curing barn	69.69	28.88	1.43	34.45	0.95

Note: according to the statistic in 2015, the average price of first-class tobacco was 38 yuan per kg, medium tobacco, 27 yuan per kg, and low-grade tobacco, 12 yuan per kg.

Table 6. Comparison table of waste emissions of intelligent biomass curing barn and traditional bulk curing barn (kg).

	CO_2	*SO_2	*NO_x
Traditional coal-fired curing barn	4.6	0.015	0.012
Intelligent biomass curing barn	4.3	0.014	0.011
Emissions	0.3	0.001	0.001

Note: (Streets, DG, Waldhoff, ST 1998): for burning 1 ton coal, 2620 kg CO_2, 8.50 kg SO_2, and 7.00 kg NO_x will be produced. For 1 ton crop straws, there will be 791.3 kg CO_2, 0.53 kg SO_2, and 1.29 kg NO_x.

lower than that of the burning coal. It can be known that adoption of biomass energy as fuel can obtain an excellent effect of emission reduction. Besides, the research shows that the solid waste from burning briquette fuel is reduced significantly. The ash contained in the waste generated by burning biofuel accounts for 5%, which is plant ash and can be used for planting tobacco, fruit trees, and flowers to achieve resources recycling.

6 CONCLUSION

The following main conclusions can be drawn based on the research of these project topics:

1. Proposing to replace coal with biomass briquette for tobacco baking and deepening biomass briquette technology.
2. The reselection of the material for the outer wall of firing furnace and the redesigning of heat exchanger structure can improve the energy utilization rate.
3. During the air supply and combustion of biomass stove, tertiary air intake, wind shield, and other combustion modes have been proposed to ensure complete burning of biomass in the firing furnace.
4. A complete set of intelligent biomass heat supply control systems is designed to remove the labor required in the whole baking process. In addition, human' experience is not required for controlling the baking process, which can save time and simplify work.
5. Compared with the test data of traditional coal-fired bulk curing barn, intelligent biomass curing barn has greater advantages in baking

economic benefits, quality of baked tobacco leaves, and waste emissions, which indicates that intelligent biomass curing barn is a big trend of future development.

ACKNOWLEDGMENTS

This work was funded by Guizhou Provincial Tobacco Company Qianxinan Branch S&T project "The Popularization and Application of Intelligent Biomass Barn and Briquette Fuels".

REFERENCES

Bainiang, Z. (1999). Rural Energy Engineering *M. Beijing: China agriculture press*, 201–211.

Demirbas, A. (2004). Combustion characteristics of different biomass fuels *J. Progress in Energy and Combustion Science*. 30(2), 219–230.

Granada, E. & López Gonzá lezJ. L. Mguez, L.M. et al (2002). Fuel lignocel lulosic briquettes die design and products Study *J. Renewable Energy*. 27(4), 561–573.

Jianxiang, W. & Hongzhen, C. (2008). Physical Quality of Biomass Briquettes and Molding Technology *J. Agricultural Mechanization Research*. (1), 203–205.

Jun, C. & Zhanniang, T. (2004). Energy Chemistry *M.* Beijing: chemical industry press . (3),216–217.

Mazz, A. (2007). Study design and prototyping of an animal traction cambased press for biomass Densification *J. Mechanism and Machine Theory*. 42 (6), 652–667.

Ronghou, L. Weisheng, N. & Dalei, Z. et al (2005). Biomass Thermochemical Conversion Technology *M.Chemical Industry Press*. 105–109.

Streets, DG, Waldhoff, ST (1998). Biofuel use in Asia and acidifying emissions *J. Energy*. 23(12), 1029–10.

Xiangjun, Y. & Yishui, T. (2005). Biomass Energy Resources Clean Transformation and Utilization Technology *M.* Beijing: *chemical industry press*. 108–114.

Yanhong, C. pei, W. & Xueyan, T. et al (2010). The Current Situation of Manufacturing Technology of Biomass Briquettes *J. Agricultural Mechanization Research*. 1(1), 206–211.

Yishui, T. (2009). Development Status and Outlook of Biomass Briquettes Industry *J. Agricultural Engineering Technology New Energy Industry*. (3). 20–26.

Zhongping, S. & Yaozhe, H. (2006). Japanese Energy Institute. Biomass and Biomass Energy Manual *K.* Beijing: *chemical industry press*. 123–12.

Advances in Energy, Environment and Materials Science – Wang & Zhou (Eds)
© 2017 Taylor & Francis Group, London, ISBN 978-1-138-03600-0

A study on the thermodynamic characteristics of coal in temperature variation desorption in a closed system

Tingting Cai, Dong Zhao & Zengchao Feng
Taiyuan University of Technology, Taiyuan, Shanxi, China

ABSTRACT: Methane adsorption or desorption in coal depends on pressure and temperature, and desorption volume increases with the raised temperature and decreased pressure. In a closed system, once temperature is raised, desorption volume grows correspondingly, and the pressure of the system increases to inhibit further desorption. Energy changes in mutual transformation of methane between free state and adsorbed state. Based on real gas state equation, Boltzmann energy distribution theory, and two-energy-state model, mathematical expression of adsorption heat was obtained, and verified by physical experiments in this work. The experiment results showed when temperature was raised in a closed system, desorption promotion by temperature and adsorption inhibition by pressure interacted, but desorption promotion was much more significant, so more methane get desorbed in this process. The two-energy-state model can be used to describe the changing relationship among adsorption heat, pressure, and temperature very well. Besides, adsorption heat was the function of temperature and pressure, and its value was concerned with the initial equilibrium condition. The larger the initial equilibrium pressure, the smaller the adsorption heat, and the shorter the time to get dynamitic equilibrium.

1 INTRODUCTION

The capacity of coal to adsorb methane and other gases depends on the surrounding temperature and pressure, decreasing as the temperature rises but increasing as the pressure rises. In a closed system, especially for large-scale coal samples, coal is heated and some adsorbed gas becomes free gradually due to the rising temperature, and such free gas makes the pressure in the system larger than before to promote adsorption. Therefore, when the temperature rises, the interaction inside coal is coupled by two opposite processes, on the one hand, the rising temperature promotes desorption and, on the other hand, the increasing pressure accelerates adsorption, and the two processes work together to make the system get equilibrium. The relationship between adsorption capacity and gas pressure at constant temperature is described by the Langmuir equation. At constant temperature, the adsorption capacity of coal is a constant value, whereas the adsorption volume increases with pressure. Once the temperature of the adsorption system varies, the Langmuir equation is no longer valid. The effect of temperature on the parameters in the Langmuir equation was studied in the literature (Li D.Y. et al. 2010, Zhao D. et al. 2012), but their study failed to establish a precise equation to describe the effect of temperature or pressure variation on adsorption volume, because it is the variation in both

temperature and pressure that makes the Langmuir equation cannot match with test data very well. The relationship between adsorption capacity and system temperature or adsorption pressure is also described by the Dubinin-Astakhov (D-A) equation (Amankwah & Schwarz. 1995), and indeed, for temperature or pressure changes, it is still valid. However, this equation is based on saturated vapor pressure (P_s) to calculate the adsorption potential. For some super-critical substances, such as methane and hydrogen, the saturated vapor pressure does not exist, thus, an approximate saturated vapor pressure (f_s) is considered instead of P_s (Li M. et al. 2003, Huan X. et al. 2015). In this method, the involved parameter is usually determined by trial calculation, and it fails to describe adsorption phenomena precisely (Jiang W. et al. 2011, Su X.B. et al. 2008). Although modified equation can be used to conform to experimental results, the physical significance is made unclear.

From the potential theory proposed by Polany, the adsorption potential is a function of both temperature and pressure, and it reflects the potential energy difference of gas molecular switching between free state and adsorbed state. This theory can be used to describe the adsorption characteristics of coal when the temperature or pressure varies quite well.

Although the adsorption volume is an important parameter to evaluate the characteristics of coal adsorption, many scholars started from energy to discuss and analyze the adsorption ability of coal and thermodynamic characteristics in the system. Surface free energy variation (Lu S.Q. et al. 2014, Liu S.S. et al. 2015) and isosteric heat of adsorption (Bai J.P. et al. 2014, Zhou L. et al. 2011) before and after adsorption are often used.

In view of this, based on the adsorption potential theory, a series of temperature variation desorption experiments in a closed system were conducted to study the relationship among temperature, pressure, and adsorption heat in a closed adsorption system, and their effect on adsorption/desorption was also studied in this work.

2 ADSORPTION HEAT

The adsorption heat in methane adsorption of coal reflects the energy change on the surface of coal in the process of adsorption. It is a macro behavior of the interaction of methane and coal surface.

The methane adsorption of coal is exothermic, and it can be calculated as follows:

molecule (free) $\rightleftharpoons$ molecule (adsorbed);

$\Delta H = \varepsilon < 0$

where H is the thermodynamic enthalpy and ε is the adsorption heat of each methane molecule.

Heat is exhausted when adsorption occurs, while on the contrary, desorption is endothermic. It is quite difficult to measure and determine adsorption heat directly. Indirect calculation shows that it ranges at 0–30 kJ/mol (Jiang W.P. et al. 2006, Liu Z.X. et al. 2012). By the method of molecular simulation, Jiang W.P. (2007) calculated adsorption heat, and it was 4–9 kJ/mol.

According to the real gas-state equation, when equilibrium is reached in a closed system, the thermodynamic state of free gas inside the instrument can be obtained as follows:

$$PV = nRTZ \qquad (1)$$

where P is the equilibrium pressure, MPa; V is the volume of free gas, L; n is the quantity of substance, mol; R is the gas constant, 8.314 J/(mol·K); T is the temperature in a closed system, K; and Z is the gas compressible coefficient.

In a closed adsorption system, methane molecules exist in free state or adsorbed state. Assuming that the minimum energy required for a molecule to escape from coal surface is ε_0, molecules with energy larger than ε_0 are in the free state, and the molecules with energy less than ε_0 are in the adsorbed state. Based on the Boltzmann energy distribution law, once a system reaches equilibrium, all of the molecules in this system exist over a well-defined energy distribution (Yan J.M. et al. 1979, Liu Z.X. et al. 2012, Feng Z.C. et al. 2016). Thus, based on the two-state-energy model, the number of free-state molecules is calculated as follows:

$$N = A\exp(-\varepsilon_0 / kT) \qquad (2)$$

where A is the proportionality constant and K is the Boltzmann constant, 1.38×10–23 J/K.

Substituting equation (2) into equation (1) yields:

$$P = BT \exp(-\varepsilon/kT) \qquad (3)$$

where $B = ARZ / VN_a$.

The above equation can be transformed into an adsorption heat expression and rearranged to yield equation (4):

$$\varepsilon = -kT \ln(P/BT) \qquad (4)$$

From the above thermodynamic derivation process, we can see that adsorption heat is a function of pressure and temperature, and both temperature and pressure affect adsorption heat.

3 EXPERIMENTAL METHODOLOGY

3.1 *Experimental coal samples and instruments*

The lump coal materials were taken from Gucheng mine and Gaohe mine in Qinshui coalfield in North China. After field sampling, the coal materials were wax sealed, and then be machined into cylindrical specimens with the size of ⌀100 mm × 100 mm by a core-taking drilling machine. The two specimens were numbered as 1# and 2#, respectively. Both specimens are lean coal ($R_{o,max1}$ = 2.26, $R_{o,max2}$ = 2.02). Owing to the core-taking direction, which is along the joint in coal, both specimens have a smooth surface without large fractures. In this work, the coals with identical degree of metamorphism were chosen for mutual verification.

The main experimental equipment includes adsorption instrument, high-temperature adsorption platform, and GW-1200 A temperature controller. The temperature controller has a high sensitivity in temperature rising and controlling with error range less than 1°C, and such accuracy can meet the experimental requirement. There are also some auxiliary experimental devices, such as the precise digital pressure gauge, vacuum pump, thermocouple thermometer, methane gas cylinder, and water drainage device. The accuracy of the precise digital pressure gauge is 0.001 MPa, and it can record and store the real-time gas pressure value inside the adsorption instrument in the whole experimental process. As for the water drainage device, in order to make sure that the exhaust methane volume was equivalent each time, the graduate with a measurement range of 1 L and the least calibration of 10 ml was used to drain water for methane collection. The experimental system is shown in Figure 1.

3.2 Temperature variation desorption experiments

After the experimental system was assembled and fixed well, a series of temperature variation desorption experiments were conducted, and these experimental steps can be divided into three stages. The first stage dealt with system airtightness by the method of high-pressure helium. After accomplishment, the vacuum pump was taken on to degas the free air in adsorption instrument. After 24 hours, it was taken off with the vacuum degree lower than 70 Pa, and then the methane gas cylinder was taken on to inject methane to a certain volume. In the second stage, the temperature variation desorption experiments were conducted. In these experiments, eight temperature spots were set, ranging from 20 to 90°C at the interval of 10°C. In the whole process, the precise digital pressure gauge was used to record the real-time pressure of free methane inside the adsorption instrument. At the time when the pressure was quite stable and the reading change was no more than 0.002 MPa in 20 minutes, it was considered that the adsorption process had reached equilibrium, and then the final equilibrium pressure was recorded. It took about

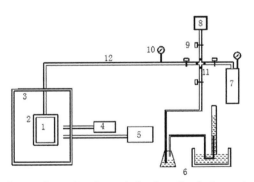

1. experimental coal sample 2. adsorption instrument
3. high temperature adsorption platform
4. thermocouple thermometer 5. temperature controller
6. water drainage device 7. methane gas cylinder
8. vacuum pump 9. valve 10. digital pressure gauge
11. four directions device 12. pipeline

Figure 1. Experimental system.

6 hours for each desorption equilibrium. After the equilibrium, the temperature controller was set to the next temperature spot to continue desorption. Last but not least, after the equilibrium of 90°C, the adsorption instrument was cooled naturally to room temperature, and then the device, which was used to drain water for methane collection, was open to exhaust some free methane by certain volume to make the free methane inside the adsorption instrument get a new pressure state. In the third stage, the temperature variation desorption experiments at the temperature of 20–90°C in the second stage were repeated under the new gas pressure. The above operational steps were repeated 6 times and ensured that the exhaust free methane gas volume was all the same each time. The concrete test scheme is given in Table 1.

4 RESULTS

4.1 Temperature and pressure in a closed system

Both specimens were subjected to temperature variation desorption experiments at 6 initial pressures and 8 temperature spots. According to the equilibrium pressure, the precise digital pressure gauge recorded, desorption equilibrium pressure versus temperature curves in a closed system are obtained, shown in Figure 2.

In Figure 2, the equilibrium pressure at different initial pressures of both specimens increases with the temperature. This means that, when temperature rises in a closed system, desorption promotion from increasing temperature and desorption inhibition from increasing pressure both occur until an equilibrium is reached, in this process, promotion from temperature is larger than inhibition from pressure, so the desorption volume is becoming larger and larger, which makes the pressure of free methane inside the instrument increase gradually.

Besides, for each single curve in certain initial pressure, the equilibrium pressure increases nonlinearly in the same temperature gradient, concretely, and it increases smoothly in low-temperature range, but it presents a quick increase in high tem-

Table 1. Test scheme.

No.	Pressure /MPa	Temperature/°C
1[#]	P_1	20,30,40,...,90
2[#]	exhaust ΔP, $P_2 = P_1 - \Delta P$	20,30,40,...,90
	exhaust ΔP, $P_3 = P_2 - \Delta P$	20,30,40,...,90
	exhaust ΔP, $P_4 = P_3 - \Delta P$	20,30,40,...,90
	exhaust ΔP, $P_5 = P_4 - \Delta P$	20,30,40,...,90
	exhaust ΔP, $P_6 = P_5 - \Delta P$	20,30,40,...,90

perature range. When curves in different initial pressures are compared, such increase trend in low initial pressure is not as significant as that in high initial pressure. As more adsorbed gas get desorbed to make pressure in instrument increase when the temperature rises, the promotion of pressure on adsorption varies in the same gradient. In detail, it is promoted much in low pressure range, but little in high pressure range (Zhang Q.L. et al. 2004, Zhang Q. et al. 2008); thus, the larger the pressure, the smaller the promotion on adsorption. It is the reason for larger pressure increase in high temperature and high pressure range.

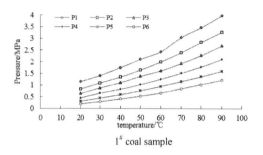

1# coal sample

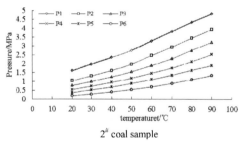

2# coal sample

Figure 2. Equilibrium desorption pressure versus temperature curves.

4.2 *Initial pressure and adsorption heat*

As shown in equation (4), adsorption heat is a function of both pressure and temperature. In desorption process, temperature and pressure act together to create adsorption heat. Curves in Figure 2 show that, there is an obvious regularity between equilibrium pressure and temperature. Equation (3) was used to simulate the curves, and the results are shown in Table 2.

In Table 2, the correlation coefficients of simulated results are nearly 1, which means that equation (3) derived from two-state-energy model can be used to describe the relationship between desorption equilibrium pressure and temperature very well. Besides, the adsorption heat involves initial equilibrium conditions, and the higher initial equilibrium pressure, the smaller adsorption heat, that is to say, in a closed system with higher initial pressure, the adsorption heat for coal to free more methane is smaller. Compared with a system with lower equilibrium pressure, heat for coal to adsorb to get final desorption equilibrium is smaller. In the same temperature gradient, heat offered by system is equal, so one can infer that, for same temperature rise, the system with higher initial pressure is prone to reach equilibrium more quickly.

5 DESORPTION RATE AND DESORPTION TIME

As analyzed before, when the temperature was raised from 20 to 90°C, the whole process in this closed system presented desorption. The free gas pressure and temperature were measured all the time, and the real gas-state equation was combined to calculate the free gas volume in standard condition until an equilibrium was reached, then desorption curves were obtained. Curves of desorption in different initial pressures from 20 to 90°C were almost the same. Restricted by the length of paper,

Table 2. Simulated results.

Coal sample	Initial pressure (MPa)	Simulated results	ε	B	R^2
1#	1.145	$P = 1.1664\,T\exp(-1580.2/T)$	1580.2	1.1664	0.9980
	0.821	$P = 1.1533\,T\exp(-1754.9/T)$	1754.9	1.1533	0.9983
	0.630	$P = 1.0970\,T\exp(-1808.3/T)$	1808.3	1.0970	0.9930
	0.445	$P = 1.3371\,T\exp(-1963.8/T)$	1963.8	1.3371	0.9919
	0.305	$P = 1.8109\,T\exp(-2165.1/T)$	2165.1	1.8109	0.9918
	0.190	$P = 3.1230\,T\exp(-2460.3/T)$	2460.3	3.1230	0.9922
2#	1.595	$P = 1.6928\,T\exp(-1367.4/T)$	1367.4	1.6928	0.9974
	1.032	$P = 1.3753\,T\exp(-1747.6/T)$	1747.6	1.3753	0.9986
	0.765	$P = 1.5371\,T\exp(-1866.0/T)$	1866.0	1.5371	0.9997
	0.539	$P = 1.7566\,T\exp(-2000.9/T)$	2000.9	1.7566	0.9984
	0.362	$P = 2.2102\,T\exp(-2175.3/T)$	2175.3	2.2102	0.9942
	0.189	$P = 5.2735\,T\exp(-2618.2/T)$	2618.2	5.2735	0.9944

the desorption curve of 1# with an initial pressure of 0.445 MPa is shown in Figure 3. Seven desorption phases were sectioned by temperatures, and all phases went with common regularity.

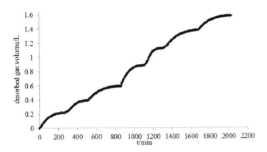

Figure 3. Curves of desorption.

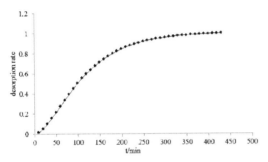

Figure 4. Curves of desorption rate at temperatures ranging from 20°C to 30°C.

In each desorption phase, at the beginning, heat was adsorbed by coal to desorb gas, and the instantaneous desorbed gas volume was quite large. As time went, the instantaneous desorbed gas volume decreased and the cumulative desorbed gas volume increased until equilibrium was reached. 1# coal sample with an initial pressure of 0.445 MPa at the temperature from 20 to 30 °C was taken as an example. Curve of desorption rate changing with time is shown in Figure 4. In Figure 4, the desorption rate was large in the early, but it slowed down with time, and the curve became smooth until equilibrium was reached.

Desorption rate and desorption time are two significant dynamic parameters to evaluate the speed of desorption process. Desorption time is usually defined as the cumulative time when the desorption rate reaches 65% in practical engineering. Desorption curves with different initial pressures in different temperature ranges were simulated, and each desorption time was calculated. The results are shown in Tables 3 and 4.

The results show that, in one initial pressure, the desorption time calculated varies even in the same temperature range. Generally, the desorption time increases with the rising temperature. On the one hand, at high temperature, the adsorbed methane is less than that at lower temperature, and hence, it is harder for methane to desorb at high temperature. On the other hand, the adsorption promotion from pressure in a closed system to inhibit desorption still exists, and the larger the pressure, the smaller the adsorption promotion; thus, it takes more time to get equilibrium.

Table 3. Simulated results of desorption rate and desorption time of 1# coal sample.

Pressure (MPa)	Temperature (°C)	Simulated results	r^2	Desorption time (min)	Pressure (MPa)	Temperature (°C)	Simulated results	(r^2)	Desorption time (min)
1.145	20–30	$y = 0.356\ln x - 0.895$	0.9790	76.7	0.445	20–30	$y = 0.322\ln x - 0.781$	0.9534	85.1
	30–40	$y = 0.312\ln x - 0.743$	0.9648	86.9		30–40	$y = 0.347\ln x - 0.920$	0.9786	92.2
	40–50	$y = 0.376\ln x - 1.057$	0.9891	93.7		40–50	$y = 0.406\ln x - 1.205$	0.9780	96.4
	50–60	$y = 0.338\ln x - 0.886$	0.9537	94.1		50–60	$y = 0.373\ln x - 1.083$	0.9645	104.2
	60–70	$y = 0.347\ln x - 0.954$	0.9528	101.7		60–70	$y = 0.329\ln x - 0.922$	0.9890	118.8
	70–80	$y = 0.335\ln x - 0.935$	0.9601	113.4		70–80	$y = 0.355\ln x - 1.030$	0.9783	113.6
	80–90	$y = 0.366\ln x - 1.104$	0.9587	120.6		80–90	$y = 0.364\ln x - 1.124$	0.9792	130.8
0.821	20–30	$y = 0.314\ln x - 0.735$	0.9742	82.3	0.305	20–30	$y = 0.323\ln x - 0.805$	0.9671	90.4
	30–40	$y = 0.401\ln x - 1.135$	0.9864	85.7		30–40	$y = 0.376\ln x - 1.063$	0.9742	95.2
	40–50	$y = 0.376\ln x - 1.063$	0.9723	95.2		40–50	$y = 0.396\ln x - 1.176$	0.9848	100.6
	50–60	$y = 0.353\ln x - 0.970$	0.9668	98.4		50–60	$y = 0.342\ln x - 0.965$	0.9508	112.4
	60–70	$y = 0.347\ln x - 0.973$	0.9675	107.5		60–70	$y = 0.366\ln x - 1.068$	0.9743	109.3
	70–80	$y = 0.355\ln x - 1.059$	0.9872	123.3		70–80	$y = 0.388\ln x - 1.219$	0.9614	123.6
	80–90	$y = 0.362\ln x - 1.080$	0.9639	119.1		80–90	$y = 0.392\ln x - 1.270$	0.9828	134.0
0.630	20–30	$y = 0.405\ln x - 1.149$	0.9737	84.9	0.190	20–30	$y = 0.343\ln x - 0.921$	0.9879	97.5
	30–40	$y = 0.363\ln x - 0.985$	0.9038	90.4		30–40	$y = 0.329\ln x - 0.883$	0.9905	105.6
	40–50	$y = 0.371\ln x - 1.052$	0.9785	98.3		40–50	$y = 0.354\ln x - 1.001$	0.9682	106.0
	50–60	$y = 0.352\ln x - 0.958$	0.9457	96.4		50–60	$y = 0.378\ln x - 1.120$	0.9661	108.0
	60–70	$y = 0.328\ln x - 0.919$	0.9863	119.6		60–70	$y = 0.358\ln x - 1.053$	0.9597	116.4
	70–80	$y = 0.334\ln x - 0.928$	0.9646	112.8		70–80	$y = 0.344\ln x - 1.021$	0.9403	128.7
	80–90	$y = 0.339\ln x - 0.997$	0.9812	128.8		80–90	$y = 0.328\ln x - 0.968$	0.9738	138.8

Table 4. Simulated results of desorption rate and desorption time of 2# coal sample.

Pressure (MPa)	Temperature (°C)	Simulated results	(r²)	Desorption time (min)	Pressure (MPa)	Temperature (°C)	Simulated results	(r²)	Desorption time (min)
1.595	20–30	y = 0.384lnx–1.063	0.9679	86.6	0.539	20–30	y = 0.347lnx–0.940	0.9620	97.7
	30–40	y = 0.332lnx–0.852	0.9508	92.2		30–40	y = 0.328lnx–0.871	0.9577	103.3
	40–50	y = 0.329lnx–0.846	0.9668	94.4		40–50	y = 0.377lnx–1.118	0.9666	108.8
	50–60	y = 0.367lnx–1.058	0.9753	105.0		50–60	y = 0.367lnx–1.099	0.9946	117.4
	60–70	y = 0.409lnx–1.299	0.9489	117.4		60–70	y = 0.396lnx–1.251	0.9487	121.6
	70–80	y = 0.371lnx–1.138	0.9424	123.9		70–80	y = 0.323lnx–0.927	0.9778	131.9
	80–90	y = 0.358lnx–1.086	0.9766	127.6		80–90	y = 0.335lnx–1.007	0.9820	140.7
1.032	20–30	y = 0.346lnx–0.930	0.9892	96.2	0.362	20–30	y = 0.367lnx–1.057	0.9815	104.7
	30–40	y = 0.353lnx–0.948	0.9335	92.5		30–40	y = 0.346lnx–0.971	0.9398	108.3
	40–50	y = 0.341lnx–0.931	0.9527	103.2		40–50	y = 0.391lnx–1.201	0.9849	113.8
	50–60	y = 0.359lnx–1.071	0.9809	120.8		50–60	y = 0.388lnx–1.217	0.9426	123.0
	60–70	y = 0.320lnx–0.856	0.9893	110.5		60–70	y = 0.343lnx–1.020	0.9788	130.2
	70–80	y = 0.396lnx–1.255	0.9924	122.8		70–80	y = 0.329lnx–0.957	0.9726	132.2
	80–90	y = 0.334lnx–0.969	0.9778	127.4		80–90	y = 0.337lnx–1.005	0.9834	135.8
0.765	20–30	y = 0.352lnx–0.956	0.9625	95.8	0.189	20–30	y = 0.344lnx-0.952	0.9823	105.3
	30–40	y = 0.323lnx–0.839	0.9588	100.5		30–40	y = 0.371lnx-1.106	0.9536	113.7
	40–50	y = 0.358lnx–1.030	0.9411	109.2		40–50	y = 0.354lnx-1.043	0.9811	119.4
	50–60	y = 0.353lnx–1.017	0.9515	112.4		50–60	y = 0.347lnx-1.025	0.9912	124.9
	60–70	y = 0.339lnx–0.982	0.9782	123.2		60–70	y = 0.326lnx-0.942	0.9862	132.2
	70–80	y = 0.379lnx–1.188	0.9878	127.7		70–80	y = 0.355lnx-1.083	0.9764	131.7
	80–90	y = 0.322lnx–0.926	0.9862	133.5		80–90	y = 0.362lnx-1.140	0.9840	140.4

When comparing curves with different initial pressures, one can see that desorption times and initial pressures are in close negative correlation irrespective of the low temperature range (20–30°C) or high temperature range (80–90°C). Concretely, the larger the initial pressure, the less the time taken for desorption. In the same interval of temperature, less heat needs to be adsorbed by coal in high initial pressure, so the time for equilibrium is shorter than that in low initial pressure; correspondingly, desorption time in each desorption phase is short, too. The results go inherently with former analysis.

6 CONCLUSIONS

The thermodynamic characteristics of coal in temperature variation desorption in a closed system are studied in this work. When the temperature rises, desorption promotion from temperature and desorption inhibition from pressure both exist to make the system get equilibrium, and the whole process presents desorption. Parameters, such as temperature, pressure, adsorption heat, desorption time, and their relationship, were analyzed. The conclusions are as follows:

1. When temperature rises, desorption and adsorption both occur until equilibrium in a closed system, and the relationship between temperature and pressure can be expressed in a well-defined function when the temperature varies. Con-

cretely, the two-state-energy model can be used to describe the relationship among adsorption heat, pressure, and temperature quite well. Adsorption heat involves initial pressure, and the larger the initial pressure, the smaller the adsorption heat.

2. In desorption process in certain initial pressure, the desorption time in each desorption phase increases gradually in the same gradient of temperature when the temperature rises. For different initial pressures, the larger the initial pressure, the shorter the desorption time for coal to get equilibrium.

ACKNOWLEDGMENTS

The authors sincerely acknowledge the funding supports from the National Natural Science Foundation of China (51304142, 21373146) and the Basic Research Program of Shanxi Province (2013021029-3).

REFERENCES

Amankwah K. and Schwarz J. 1995, A modified approach for estimating pseudo-vapor pressures in the application of the Dubinin-Astakhov equation. Carbon, 33(9):1313–1319.

Bai J.P., Zhang D.K., Yang J.Q., et al. 2014. Thermodynamic characteristics of adsorption-desorption of methane in coal seam 3 at Sihe coal mine [J]. Journal of China Coal Society. 39(9), 1812–1819.

Feng Z.C., Zhao D., Zhao Y.S., et al. 2016. Effects of temperature and pressure on gas adsorption in coal in an enclosed system: a theoretical and experimental study [J]. International Journal of Oil, Gas and Coal Technology, 11(2): 193–203.

Huan X., Zhang X.B., Wei H.W. 2015. Research on parameters of adsorption potential via methane adsorption of different types of coal [J]. Journal of China Coal Society, 40(8): 1859–1864.

Jiang W., Wu C.F., Jiang W. et al. 2011.Application of adsorption potential theory to study on adsorption-desorption of coal-bed methane [J]. Coal Science and Technology, 39(5):102–104.

Jiang W.P., Cui Y.J., Zhang Q., et al. 2006. The quantum chemical study on the coal surface interacting with CH_4 and CO_2[J]. Journal of China Coal Society. 31(2), 237–240.

Jiang W.P., Cui Y.J., Zhang Q., et al. 2007. The quantum chemical study on different rank coals surface interacting with methane [J]. Journal of China Coal Society, 32(3), 292–295.

Li D.Y., Liu Q.F., Philipp Weniger, et al. 2010. High-pressure sorption isotherms and sorption kinetics of CH4 and CO2 on coals [J]. Fuel, (89):569–580.

Li M., Gu A.Z., Lu X.S., et al. 2003. Study on methane adsorption above critical temperature by adsorption potential theory [J]. Chemical Engineering of Gas, (5):28–31.

Liu S.S., Meng Z.P. 2015. Study on energy variation of different coal-body structure coals in the process of isothermal adsorption [J]. Journal of China Coal Society. 40(6), 1422–1427.

Liu Z.X., Feng Z.C. 2012. Theoretical study on adsorption heat of methane in coal [J]. Journal of China Coal Society, 37(4), 647–653.

Lu S.Q., Wang L., Qin L.M. 2014. Analysis on adsorption capacity and adsorption thermodynamic characteristics of different metamorphic degree coals [J]. Coal Science and Technology, 42(6), 130–150.

Su X.B., Chen R., Lin X.Y., et al. 2008. Application of adsorption potential theory in the fractionation of coal-bed methane during the process of adsorption/desorption [J]. Acta Geologica Sinica. 82(10), 1382–1389.

Yan J.M., Zhang Q.Y. 1979. Adsorption and Cohesion-Solid Surface and Pore. Science Press, Beijing.

Zhang Q., Cui Y.J., Zhong L.W., et al. 2008. Temperature pressure comprehensive adsorption model for coal adsorption of methane [J]. Journal of China Coal Society, 33(11), 1272–1278.

Zhang Q.L., Cui Y.J., Cao L.G. 2004. Influence of pressure on adsorption ability of coal with different deterio-ratio level [J]. Natural Gas Industry. 24(1):98–100.

Zhao D., Zhao Y.S., Feng Z.C., et al. 2012. Experiments of methane adsorption on raw coal at 30–270°C [J]. Energy sources, 34(4): 324–331.

Zhou L., Feng Q.Y., Qin Y. 2011. Thermodynamic analysis of competitive adsorption of CO2 and CH4 on coal matrix [J]. Journal of China Coal Society, 36(8), 1307–1311.

Advances in Energy, Environment and Materials Science – Wang & Zhou (Eds)
© *2017 Taylor & Francis Group, London, ISBN 978-1-138-03600-0*

A comparative experiment on the curing effect of different heating energy sources of bulk tobacco-curing barn

Shubin Lan
Guizhou Provincial Tobacco Company Qianxinan Branch, Xingyi, China

Ze Lu
School of Mechanical Engineering, Guizhou University, Guiyang, China

Ying Ma, Weilin Chen & Feng Wang
Guizhou Provincial Tobacco Company Qianxinan Branch, Xingyi, China

Dabin Zhang & Yang Cao
School of Mechanical Engineering, Guizhou University, Guiyang, China
Guizhou Mechanical and Electrical Equipment Engineering Technology Research Center Co. Ltd., Guiyang,
Guizhou, China

ABSTRACT: To improve the curing quality of tobacco, the greenhouse gas emission from fossil energy, energy consumption, and labor cost of curing were reduced. In this paper, an experiment was conducted to compare the economic characters and appearance quality of Yunyan 97 cured in bulk curing barn with the combination of coal, electricity, and diesel with that cured in bulk curing barn with the combination of biomass, electricity, and diesel. The results indicate that comparison between the curing barn with the combination of biomass, electricity, and diesel and the curing barn with the combination of coal, electricity, and diesel shows that the cadmium orange tobacco rate of the former is higher than that of the latter by 16.73%; the main chemical components of the former reach the range of excellent quality; the curing cost of the former is lower than that of the latter by 0.8 yuan/kg; and the economic character of the former is higher than that of the latter by 2.84%.

1 INTRODUCTION

Many researches on heating energy source of tobacco-curing barn have been carried out domestically. Heating energy sources of tobacco-curing barn mainly include coal, diesel, electrical heating, biomass energy, and solar energy (Tie 2009, Wang 2013, Gong 2005, Sun 2013, Zhang 2010). Since the cost of coal energy is low, coal has been used as the main heating energy source from natural ventilation airflow curing barn to mechanical ventilation circulation curing barn. The burning of coal, however, is of retardance and hysteresis (Liu 2014). Therefore, it is difficult to fast heat up by burning coal. The control accuracy of coal combustion is low. It needs to frequently replenish coal and the labor intensity is high. The combustion efficiency of coal is low, and it causes fuel waste. Concerning diesel, the combustion efficiency is high, fast heating up can be realized, the effect of tobacco curing is good, but the cost is high. Hence, it is not recommended to use diesel throughout the curing

process. Electrical heating refers to electrical heating element emitting heat and transferring heat to curing barn. Electrical heating element is of high temperature resistance and high humidity. The air heated by electrical heating element is of no objectionable odor. The electrical heating element can be directly installed in curing barn without casing, which can guarantee curing quality and improve thermal efficiency. However, the cost is high. Thus, it is not recommended to use the electrical heating element throughout the curing process (Xu 2008). Biomass energy is a renewable energy source, helping reduce greenhouse gas emission of fossil energy and development circulatory ecological tobacco agriculture. Solar energy is a renewable energy source, easy to transport, and causes no pollution. The curing quality with solar energy is high. In Southwest China where it is always cloudy and drizzly for days on end, however, solar energy is not widely promoted. Through analyzing the characteristics of coal, diesel, and electrical heating based on the actual condition of heating energy in Guizhou, the following conclusions were drawn: with diesel, it can be realized to fast heat up, but the cost of diesel is high; concerning electrical heating, accurate adjustment can be realized, but sharply heating up cannot be realized. The combination of coal, electricity, and diesel is less than satisfactory. Biomass energy is a clean energy source easy to transport and store. It is necessary to compare the combination of biomass, electricity, and diesel with the combination of coal, electricity, and diesel via experiment to observe the cost and heating effect of different energy sources, so as to make high-efficiency energy-saving strategy, make the actual temperature value be more closely to the set value, and realize high curing effect.

In this paper, different heating energy sources were combined for comparison based on their heating characteristics, to analyze the energy cost and curing quality of tobacco: as to tobacco cured in barn with biomass energy, electricity, and diesel as heating energy, the curing quality is high, and the cost is low, of good application prospect.

2 DESIGN OF MULTI-ENERGY BULK CURING BARN

2.1 Design of heating system

Multi-energy heating system design refers to adopting coal, electricity, biomass, electrical heating, and diesel as heating energy for curing, respectively. Gallery burner is stretched into the furnace chamber of coal and biomass, air is blown with a fan into the combustion chamber to be heated via the fins, and the electrical heater is heated, heat flow is guided into the curing barn, and cold air is discharged for heating. A schematic diagram of the heating system is shown in Fig. 1.

2.2 Design of control strategy

According to Fig. 2, the combustion of coal (or biomass) runs through the whole curing process. After the temperature sensor feeds back the collected temperature to the control system, the system calculates the set temperature and the actual temperature. $\Delta T = |$set temperature—actual temperature$|$.

If $\Delta T > 4°C$, the system needs to be heated up rapidly to reach the set temperature, and the diesel heater is started; if $1°C \leq \Delta T \leq 4°C$, the temperature difference is small, the diesel heater is turned off, and the electrical heater is started to make the actual temperature be in line with the set temperature; if $\Delta T \leq 1°C$, electrical heater is turned off, and only coal is used for heating.

3 MATERIALS AND METHODS

3.1 Material and equipment

The comparative experiment was conducted in Anlong County of Qianxinan prefecture of Guizhou province. For this experiment, middle leaves of Yunyan 97 growing in the same field and ripening at the same time were taken, and curing barns of polyurethane plank frame with combined heating energy sources of biomass, electricity, and diesel (barn A) and with combined heating energy sources of coal, electricity and diesel (barn B) were adopted (of which the panels are 50 mm thick, the internal size of the tobacco holding chamber is 8.0 m × 2.7 m × 3.5 m, and three sheds were provided). For both barns, air blowers and circulating fans of the same model and same power were employed, but different heating equipment were employed (biomass fuel burner for barn B). Fig. 3 shows the scene of tobacco feeding.

Conventional control instrument was employed for barn A. Self-developed single neuron PID controller was employed for barn B, as shown in Fig. 4. The controller weight coefficient w(k) was determined and updated according to the Delta learning rule. The Delta learning rule is a supervised

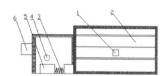

1 Electric control system; 2 Tobacco feeding chamber; 3 Electrical heater; 4 Fuel oil burner (biomass combustion

Figure 1. Schematic diagram of the hybrid energy heating system of curing barn.

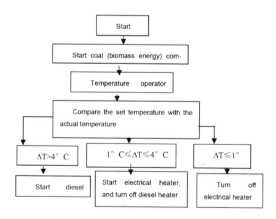

Figure 2. Schematic diagram of the control strategy.

122

Figure 3. Tobacco feeding scene.

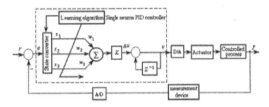

Figure 4. Single neuron PID controller.

learning algorithm based on the steepest descent method, which acts to minimize the introduced performance index so as to correct the weight coefficient (Ren 2008).

The control algorithm is as follows:

$$u(k) = u(k-1) + K\sum_{i=1}^{3} wi(k)xi(k),$$

i = 1,2,3

  $\Delta u(k) = u((k)) - u(k-1)$
  $W1(k) = w1(k-1) + \eta I\ (k)u(k)[2e(k) - e(k-1)]$
  $W2(k) = w2(k-1) + \eta P\ (k)u(k)[2e(k) - e(k-1)]$
  $W3(k) = w3(k-1) + \eta D\ (k)u(k)[2e(k) - e(k-1)]$
  ηI, ηP and ηD are learning rate.
  where u(k) refers to the control signal of curing barn system;

  Wi(k) refers to the weighting coefficient of x(k);

  ηP refers to the learning rate of the proportionality factor;

  ηI refers to the learning rate of the integration coefficient;

  ηD refers to the learning rate of the differential coefficient;

  K refers to the proportionality factor of neuron, k > 0;

  X1(K) refers to the deviation value of temperature and humidity of curing barn;

  X2(K) refers to the order difference;

X3(K) refers to the accumulative error of the system.

The curing barn system performed data sampling by means of MODBUS—RTU communication to obtain the process set value P of curing barn and the actual temperature and humidity value Y; x1(K), x2(K), and x3(K) stand for the integral term, proportional term, and differential term of conventional PID adjuster, respectively.

3.2 Experimental treatment process

For purpose of the experiment, two treatment processes were adopted. A: curing barn with combined heating energy sources of biomass, electricity, and diesel; and B: curing barn with combined heating energy sources of coal, electricity, and diesel. For both curing barns, a three-step curing technology was adopted.

3.3 Observation record

1. The tobaccos cured in the two barns were subject to appearance quality evaluation according to GB2635-92, and the ratio of cadmium orange tobacco to slightly green tobacco to smooth tobacco to varicolored tobacco to bluish yellow tobacco to black tobacco was calculated.
2. Samples were taken from tobaccos cured in both barns and sent to Institute of Tobacco Sciences of Guizhou Province for chemical component analysis and sensory quality evaluation.
3. The cured tobaccos were graded according to GB2635-92, and the proportion of each grade and average purchasing price were calculated.
4. Energy cost of each barn was calculated.

4 EXPERIMENTAL RESULTS

4.1 Field control effect of single neuron PID

Based on the pre-input curing technology parameters, single neuron PID is effective to pre-judge the changing trends of temperature and humidity, so that the weight coefficients of the two data can be adjusted timely, the opening of dehumidifying windows and the start-up and shut down of air blower can be well controlled, and the differences between the actual temperature and humidity and the process temperature and humidity become smaller and smaller, to improve the control accuracy, and reduce green tobacco rate and black tobacco rate (Zhao 2006). Based on the data of field curing experiment stored in the controller, a curve was drawn, as shown in Fig. 5. According to the figure, the controller controlled the heating up speed well in the section from 46°C to 50°C, reducing browning reaction. At the stem-drying phase, the heating up rate was high from 54°C to 68°C, when the required heating up rate could not be achieved in sole reliance upon coal, and therefore, the fuel heating equipment and electrical heating equipment were started up by the single neuron PID controller to heat up fast and reduce the time of curing. The single neuron PID controller can fit the set curve well, to realize accurate control with a temperature and humidity error within ±0.5°C.

4.2 Appearance quality of cured tobacco

The appearance qualities of tobaccos cured in the two curing barns are given in Table 1. According to Table 1, the cadmium orange tobacco rate of barn A is higher than that of barn B by 16.73%, and the varicolored tobacco rate is lower by 12.12%; the slightly green tobacco rate of barn B is higher; comprehensive analysis shows that the appearance quality of tobaccos cured in the barn with combined energy sources of biomass, electricity, and diesel is superior.

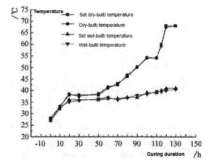

Figure 5.   Single neuron PID control curve.

4.3 Main chemical components of cured tobacco

The temperature at which tobacco turns yellow is 38°C. If this temperature is exceeded, tobacco will not thoroughly turn yellow, and it is possible to produce green tobacco, and tobacco becomes liable to decay due to high temperature and high humidity. The temperature at which tobacco turns yellow directly influences the activity and action time of main enzymes contained in tobacco, to influence the degradation of macromolecular substances. Accurate temperature control contributes to full degradation of chlorophyll and accumulation of aroma components in tobacco. At a relatively low temperature, the activity of amylase is high, and the enzymatic activity lasts for a long time, helpful to further degrade starch, and reduce the starch content and nicotine content in cured tobacco. In the yellowing stage, the heat produced by biomass (coal) and electrical heating tube is sufficient to meet the temperature requirement, when the control of the controller over the electrical heating tube is especially important. Based on the pre-judgment of changing trend of temperature, the starting-up and shut down of electrical heating tube are controlled to accurately set the temperature for yellowing tobacco, and make up the defect of coal that the temperature cannot be accurately controlled, so as to guarantee the activity and action time of enzymes and the rationality of chemical components in tobacco. Table 2 shows that, according to the range as specified in the Guizhou the Standard for Quality Evaluation of Tobacco (2005 Trial), the chemical components remain within the range of excellent quality.

4.4 Economic character of cured tobacco

Dry tobaccos were sent to local tobacco station for grading. The purchasing prices of dry tobac-

Table 1.   Comparison of appearance quality of tobaccos cured in different barns.

Barn type	Cadmium Orange	Slightly Green	Smooth	Varicolored	Bluish Yellow	Black
Barn A	75.52	15.71	5.14	3.9	0	0
Barn B	58.52	18.35	7.10	16.02	0	0

Table 2.   Chemical component contents in tobaccos cured in different barns.

Barn type	Total sugar	Reducing sugar	Total nitrogen	Potassium	Chlorine	Glycoprotein	Nicotine
Excellent quality range	19.81–26.92	18.55–22.25	1.50–2.46	1.97–3.12	0.16–0.54	7.32–7.59	1.47–2.94
Barn A	21.67	19.74	1.98	2.15	0.45	6.65	1.79
Barn B	27.29	21.34	1.76	2.22	0.84	6.63	1.80

Table 3. Comparison of incomes from tobaccos cured in different barns.

Barn type	Total sugar	Reducing sugar	Total nitrogen	Potassium	Chlorine	Glycoprotein	Nicotine
Excellent Quality Range	19.81–26.92	18.55–22.25	1.50–2.46	1.97–3.12	0.16–0.54	7.32–7.59	1.47–2.94
Barn A	21.67	19.74	1.98	2.15	0.45	6.65	1.79
Barn B	27.29	21.34	1.76	2.22	0.84	6.63	1.80

Table 4. Influence of different heating energy combinations on curing cost.

	Dry tobacco weight (kg/shed)	Coal (gas) consumption (kg/shed) or (m³/shed)	Coal and gas consumption (yuan/shed)	Electricity consumption (yuan/shed)	Diesel cost (yuan/shed)	Dry tobacco energy consumption cost (yuan/kg)
A barn	692.4	10011.7	851	321	351.28	2.2
B barn	685.5	2515.5	1132	285	639.5	3

Note: Electricity price: 0.45 yuan/KWH, coal price: 450 yuan/T; unit price of straw gasification: 0.085 yuan/m³.

cos offered by local tobacco station are 26 yuan/kg (first-grade), 22 yuan/kg (middle-grade), and 18 yuan/kg (low-grade). Based on the prices, the incomes from selling tobaccos were calculated.

The average price of dry tobacco of B barn is 21.49 yuan/kg, and that of A barn is 21.92 yuan/kg. The latter is higher than the former by 0.43 yuan. The proportion of first-grade and middle-grade tobaccos of B barn is 74.55%, and that of A barn is 81.46%. The curing quality of A is better than that of B. Adopting A barn can improve the quality of tobacco curing and economic income of tobacco growers.

4.5 Influence of different heating energy combinations on curing cost

According to Table 4, the average cost of A barn is lower than that of B by 0.8 yuan/kg. The high electricity consumption of A barn is because biomass energy gasification needs electricity. From the aspect of energy, the combination of biomass, electricity, and diesel is more suitable to be promoted.

5 DISCUSSIONS

An experiment was conducted to compare the economic character and appearance quality of tobaccos cured in barns with combined heating energy sources of coal, electricity, and diesel, with that combined heating energy sources of biomass energy, electricity, and diesel. The results indicate that comparison between the curing barn with the combination of biomass, electricity, and diesel and the curing barn with the combination of coal, electricity, and diesel shows that the cadmium orange tobacco rate of the former is higher than that of the latter by 16.73%; the main chemical components of the former reach the range of excellent quality; the curing cost of the former is lower than that of the latter by 0.8 yuan/kg; the economic character of the former is higher than that of the latter by 2.84%. However, gasification of straw is required before utilization of biomass energy, and gasification equipment is needed to supply gas for curing, which should be tackled at the time of building straw gasification curing workshop. First, supporting tobacco-curing workshops are needed. The heat demands of 18 curing barns can be supplied simultaneously. Second, high cost of straw gasification equipment is a difficulty in promoting the technology. Once the equipment cost decreases, this technology can be widely promoted and used. Thus, it is urgent to solve problems relating to building of gasification and intensive energy utilization.

ACKNOWLEDGMENTS

This work was supported by Science and technology project of Guizhou Provincial Tobacco Company Qianxinan Branch: Development and application of intelligent biomass baking equipment and molding fuel.

REFERENCES

Gong, C.R & Pan, J.B. (2005). Evolution and Research Progress of Tobacco-curing Equipment in China [J]. *Tobacco Science & Technology*, 11:35–38.

Liu, J.Y & Kan, H.W. (2014). Effect of Tobacco-curing with Hanging Basket of Barn Integrating Gathering, Transporting and Curing [J]. *Guizhou Agricultural Sciences*, 11:79–82.

Ren, W. & W., X. (2008). Design and Simulation of Linear Quadratic Type-based Single Neuron PID Optimal Controller [J]. *Computer Application and Software*, 05:123–124+154.

Sun, G.W & Chen, Z.G. (2013). Current Situation and Development Direction of Energy Utilization of Bulk Curing Barn [J]. *Journal of Anhui Agricultural Sciences*, 20:8691–8693.

Tie, Y, & He, Z.J. (2009). Luo Huilong. Current Situation and Outlook of Application of Bulk Tobacco-curing Barn [J]. *Chinese Agricultural Science Bulletin*, 13:260–262.

Wang, F, & Jing, J.L, (2013). A Study on Influence of Optimization, Upgrading and Transformation of Bulk Curing Barn on Cured Tobacco Quality [J]. *Chinese Agricultural Science Bulletin*, 08:135–139.

Xu, L. (2006). Intelligent Control and Intelligent System [M]. Beijing: *China Machine Press*.

Xu, X.H & Sun, F.S. (2008). Discussions on Current Situation and Development Direction of Research and Application of Bulk Curing Barn in China [J]. *Chinese Tobacco Science*, 04:54–56+61.

Zhang, L.L. & Wang, X.Y. (2010). Drying characteristics and colorchanges of infrared drying eggplant [J]. *Transactions ofthe Chinese Society of Agricultural Engineering*, (28):291-296

Zhao, M.Q & Wang, B.X. (2006). Analysis of Enzymatic Activity and Related Chemical Component of Tobacco at Different Ageing Periods [J]. *Journal of China Agricultural University*, 04:7–10.

Advances in Energy, Environment and Materials Science – Wang & Zhou (Eds)
© 2017 Taylor & Francis Group, London, ISBN 978-1-138-03600-0

The design and study of a composite box-type heat exchanger used in biomass stoves

Shubin Lan, Feng Wang, Ying Ma & Weilin Chen
Guizhou Provincial Tobacco Company Qianxinan Branch, Xingyi, China

Dabin Zhang, Yang Cao & Hao Hu
School of Mechanical Engineering, Guizhou University, Guiyang, China
Guizhou Mechanical and Electrical Equipment Engineering Technology Research Center Co. Ltd., Guiyang, Guizhou, China

ABSTRACT: In this paper, a composite box-type heat exchanger used in biomass stoves is designed and studied. Firstly, a biomass stove structure which employs biomass as fuel is designed. Then the heat exchanger in the stove is designed and studied. The composite box-type, consisting of the loop heat exchanger and the straight plate heat exchanger, is proposed. A mathematical model of the heat exchanger is established and then performance of this heat exchanger and the traditional one is compared. Finally such biomass stove is applied to conduct the tobacco curing experiment. The result shows that fuel cost has decreased greatly after employing the biomass stove of the composite box-type heat exchanger to cure a load of lower leaves, middle leaves and upper leaves. And the study shows that the composite box-type heat exchanger can effectively realize energy efficiency in tobacco curing process, which has good practical significance.

1 INTRODUCTION

Nowadays, with the growing awareness of environmental protection, people have paid more and more attention to the environmental pollution of atmosphere caused by coal combustion, especially exhaust gases like CO_2, SO_2, and NOX given off through the combustion of non-renewable coal resources, having great impact on the environment and causing hazy weather in many cities in China, which has already reached the level of governance. [1–3] Under such a background, many scholars have focused on the study of and the society has paid great attention to the bio-energy instead of coal technology, which is also in line with the trend of development of the industry. [4–7] And the promotion of bio-energy mainly depends on the stove technology suitable for bio-energy burning. At present, many scholars have studied on the stove technology [8–9] but done little research on the stove technology suitable for bio-energy burning. Among them, Yao Zonglu and his fellow colleagues have done researches on Anti-slagging Biomass Pellet Fuel Burner, and Tan Wenying and his fellow colleagues have designed the multi-functional biomass stove and tested its performance. [9–10] It is easy to notice that there are only few researches having been done on bio-energy curing equipment, yet there is no breakthroughs in such aspects as the structural optimization of bio-energy curing equipment, the improvement of key technologies, and the optimization of heat exchange of stoves.

According to above-mentioned problems, a piece of biomass curing equipment based on the composite box-type heat exchanger is proposed in this paper. The equipment employs the new-type heat exchanger to make the stove more suitable for the combustion and the heat exchange of biomass fuels, thus applying biomass fuels instead of coals during tobacco curing process to reduce the emission of polluting gases, to meet the standards of dust emission, sulfur dioxide emission and ringelman emittance, and to improve the heat change rate and heating performance during the curing process and reduce energy loss.

2 BIOMASS STOVE

The biomass stove designed in the paper is shown in Figure 1. The equipment mainly consists of an ash box, props, a lower air-inlet ring, a combustor, a middle air-inlet ring, feeding pipes, a heat exchanger, and so on. And the stove is also divided into three parts—the bed of the stove, the firepot and the heat exchanger. The bed of the stove consists of an ash box, props, and so on to pack up ashes produced by combustion. The firepot

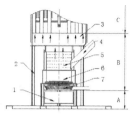

1.Ash box 2. Props 3. Heat Exchanger 4. Feeding Pipes 5. Middle Air-inlet Ring 6. Combustor 7. Lower Air-inlet Ring A.Bed   B. Firepot   C. Heat Exchanger

Figure 1.   Firepot of the biomass stove.

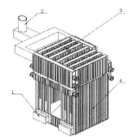

1. Soot Door 2. Exhaust Pipe
3. Loop Heat Exchanger
4. Straight Plate Heat Exchanger

Figure 2.   The composite box-type heat exchanger.

consists of props, the lower air-inlet ring, the combustor, the middle air-inlet ring, feeding pipes, and so on, used for the biomass combustion. The heat exchanger is a part for exchanging and utilizing the stove heat, whose performance will directly influence the efficiency and fuel utilization of the stove. Thus, one of the key technologies of the biomass stove design is to design rationally the heat exchanger of the biomass stove.

3   BIOMASS STOVE

3.1   *The design of the heat exchanger*

Since tow key parameters—the unit heat dissipation area and the path length of heat flow—will influence the heat exchanger, in this paper, a composite box-type heat exchanger, based on the study of traditional heat exchanger of the coal-fired stove, is proposed, as shown in Figure 2:

The composite box-type heat exchanger consists of the loop heat exchanger and the straight plate heat exchanger. The heat flow generated in the combustor firstly flows through the straight plate heat exchanger to conduct heat exchange once, then into the loop heat exchanger to conduct secondary heat exchange. Through the design of such heat

Figure 3.   The heat-flow path of the traditional heat exchanger.

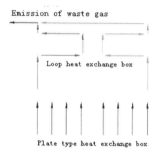

Figure 4.   The heat-flow path of the composite box-type heat exchanger.

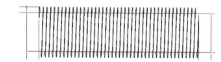

Figure 5.   The finned-tube structure.

exchanger, the length of the heat flow is enhanced, enabling the heat energy to complete heat exchange twice, which improves the exchange effect. The paths of the heat flow of the heat exchangers of the traditional coal-fired stove and the composite box type are respectively shown in Figures 3 and 4. As is easy to notice that the path of the heat flow of the latter one is longer than that of the former one.

Meanwhile, in order to increase the unit heat exchange area of the heat sink, the finned-tube structure is applied, as shown in Figure 5.

3.2   *Mathematical modeling*

Based on the heat transfer theory, the effect of the heat exchanger can be expressed with the following equation (1)

$$Q = kS\left(\frac{T_1 - T_2}{2} - T_0\right) \tag{1}$$

In the equation: Q means the heat dissipation index, k means heat transfer coefficient, S the cooling area, T1 inlet-water temperature, T2 outlet-water temperature, and T0 the room temperature.

In order to increase the heat dissipation area, the composite box-type heat exchanger designed in this paper is finned-tube structured. So the equation of the heat dissipation area can be expressed as the following one:

$$S = \left[\pi\left(d_1^2 - d_0^2\right) + \pi d_1 t_1 \right] l_0 n + \pi d_0 \left(1 - n t_1\right) l_0 \qquad (2)$$

In the quotation, $d_1 = d_0 + 2l_1$, in which d_1 means the total diameter of the finned tube, l_1 means the fin height and d_0 the tube diameter. The parameter t_1 means the fin thickness, l_0 means the tube length. $n = \frac{1}{t_1 + t_0}$, in which n means the number of fins, t_0 means fin spacing.

Then the quotation (2) is put substituted into the quotation (1), obtaining the following quotation:

$$Q = k \cdot \left[\left[\pi\left(d_1^2 - d_0^2\right) + \pi d_1 t_1 \right] \cdot l_0 \cdot n + \pi d_0 \left(1 - n t_1\right) \cdot l_0 \right]$$
$$\cdot \left(\frac{T_1 - T_2}{2} - T_0 \right) \qquad (3)$$

The above-mentioned quotation (3) is exactly the mathematical model of heat dissipation of the finned heat sink. According to the model, it is easy to notice that the heat dissipation depends mainly on the the length of the heat flow (or tube length: l_0) as well as the unit cooling area $S_0 = [\pi(d_1^2 - d_0^2) + \pi d_1 t_1] \cdot n + \pi d_0 (1 - n t_1)$. The effect of the heat exchanger is proportional to the length of the heat flow and the unit cooling area. The larger the unit cooling area of the heat exchanger is, the better the effect will be. The longer the heat flow path, the better the effect. Therefore, in the design both the heat flow and the cooling area of the heat exchanger shall be increased in order to enhance the exchange effect.

3.3 Comparison and analysis

With the above mathematical model of the heat sink, the effective cooling area and the length of the heat flow of heat exchangers of the traditional coal-fired stove and the composite box type are calculated, as are shown in Table 1:

From the Table 1, the total effective cooling area of the composite box-type heat exchanger is 1.59 times larger than that of the traditional coal-fired heat exchanger, i.e., the thermal efficiency is improved 1.59 times. The length of the heat flow of the biomass heat exchanger is 1.35 times longer than that of the traditional coal-fired heat exchanger, i.e., the thermal efficiency is improved 1.35 times. Therefore, the composite box-type heat exchanger reduces heat loss and fuel cost.

Table 1. Statistics of the cooling area and the length of the heat flow of coal-fired stove and biomass stove.

Name	Total effective cooling area (m2)	Length of the heat flow
Biomass stove	14.3	3760
Coal-fired stove	9	2790
ratio		
(Biomass stove/ coal-fired stove)	1.59	1.35

Table 2. Fuel cost comparison of lower leaf curing.

Type	Traditional stove	Biomass stove
Fuel consumption (kg)	745	712
Fuel price (yuan/ton)	830	780
Fuel cost (yuan/time)	618.35	555.36
Reduction of fuel cost (yuan/load)	–	9.00

4 EXPERIMENT RESULT AND ANALYSIS

On the basis of the foregoing results, the steel of 16 Mn, δ2.0 is employed to manufacture the heat exchanger, the loop heat exchange box, the straight plate heat exchange box, and the finned tube. The composite box-type heat exchanger and the traditional coal-fired heat exchanger are respectively employed to conduct the tobacco curing tests. And the fuel consumption during the processes of lower leaf curing, middle leaf curing and upper leaf curing is compared. The actual curing amount is 360 bars/pile. The amount of fuel consumed is the average of actual amount of curing coals (the stove of the traditional coal-fired heat exchanger) and the amount of biomass fuel (the stove of the composite box-type heat exchanger). According to the professional curing data of the experimental site, coal fuel costs 830 yuan/ton, and biomass fuel costs 780 yuan/ton. When the experiment of lower leaf curing is carried out, the average of coal consumption of the stove of the traditional coal-fired heat exchanger is 745 kg/time, while the average of biomass fuel consumption of the stove of the composite box-type heat exchanger is 712 kg/time. The fuel costs of the stove of the traditional coal-fired heat exchanger is 618.35 yuan/time, while that of the the stove of the composite box-type heat exchanger is 555.36 yuan/time, as shown in Table 2.

When the experiment of middle leaf curing is carried out, the average of coal consumption of the stove of the traditional coal-fired heat exchanger is 815 kg/time, while the average of biomass fuel consumption of the stove of the composite box-type

Table 3. Fuel cost comparison of middle leaf curing.

Type	Traditional stove	Biomass stove
Fuel consumption (kg)	815	774
Fuel price (yuan/ton)	830	780
Fuel cost (yuan/time)	676.45	603.72
Reduction of fuel cost (yuan/load)	–	8.08

Table 4. Fuel cost comparison of upper leaf curing.

Type	Traditional stove	Biomass stove
Fuel consumption (kg)	885	827
Fuel price (yuan/ton)	830	780
Fuel cost (yuan/time)	734.55	645.06
Reduction of fuel cost (yuan/load)	–	8.14

heat exchanger is 774 kg/time. The fuel costs of the stove of the traditional coal-fired heat exchanger is 676.45 yuan/time, while that of the the stove of the composite box-type heat exchanger is 603.72 yuan/time, as shown in Table 3.

When the experiment of middle leaf curing is carried out, the average of coal consumption of the stove of the traditional coal-fired heat exchanger is 885 kg/time, while the average of biomass fuel consumption of the stove of the composite box-type heat exchanger is 827 kg/time. The fuel costs of the stove of the traditional coal-fired heat exchanger is 734.55 yuan/time, while that of the the stove of the composite box-type heat exchanger is 645.06 yuan/time, as shown in Table 4.

From Table 2, it is easy to notice that applying the biomass stove of the composite box-type heat exchanger can reduce the fuel amount used for lower leaf curing and the curing cost, i.e., the fuel cost reduces 9.00 yuan when a load of lower leaf is under curing. From Table 3, it is easy to notice that applying the biomass stove of the composite box-type heat exchanger can reduce the fuel amount used for middle leaf curing and the curing cost, i.e., the fuel cost reduces 8.08 yuan when a load of middle leaf is under curing. From Table 4, it is easy to notice that applying the biomass stove of the composite box-type heat exchanger can reduce the fuel amount used for upper leaf curing and the curing cost, i.e., the fuel cost reduces 8.14 yuan when a load of upper leaf is under curing. Therefore, the composite box-type heat exchanger designed in this paper can effectively realize energy efficiency in tobacco curing process, which has good practical significance.

5 CONCLUSION

In this paper, the biomass stove is firstly designed, which consists of an ash box, props, a lower air-inlet ring, a combustor, a middle air-inlet ring, feeding pipes, a heat exchanger, and so on. Then the heat exchanger in the stove is studied specifically, which, designed as a composite box type, consists of the loop heat exchanger and the straight plate heat exchanger. The cooling pipes are of finned-tube structure, thus increasing the unit cooling area and the length of the heat flow. When compared with the traditional coal-fired heat exchanger, the effective cooling area increases by 1.59 times and the length of the heat flow increases by 1.35 times. Besides, the mathematical model of the finned-type heat exchanger is established to provide a theoretical basis for the calculation of the amount of the heat sink. Finally, tobacco curing tests are carried out by employing the stove of the composite box-type heat exchanger and the stove of the traditional coal-fired heat exchanger. The results show that by employing the stove of the composite box-type heat exchanger, the fuel costs for curing a load of lower leaf, middle leaf, and upper leaf respectively reduce 9.00 yuan, 8.08 yuan, and 8.14 yuan. Therefore, the composite box-type heat exchanger designed in this paper can effectively realize energy efficiency in tobacco curing process, which has good practical significance.

ACKNOWLEDGMENTS

Science and technology project of Guizhou Provincial Tobacco Company Qianxinan Branch: Development and application of intelligent biomass baking equipment and molding fuel.

Corresponding Author: *Wang Feng, Male, Professor, mainly engaged in tobacco production technology, Email: yancaowangfeng@163.com.*

REFERENCES

De Best, C.J.J. M & van Kemenade, H.P. 2008. Particulate emission reduction in small-scale biomass combustion plants by a condensing heat exchanger. *Energy and Fuels*, (22): 587–597.

Gilbe, C & Öhman, M. 2008. Slagging characteristics during residential combustion of biomass pellets. *Energy & Fuels.*, 22(5): 3536–3543.

Liu, Y.L & Sun, L. 2011. Hazards of inhalableparticulates PM2.5 on human healt. *International Journal of Pharmaceutical Research*, 38(6): 428–431.

Song, S.J & Wu, Y. 2012. Chemical characteristics of size-resolved PM2.5 at a roadside environment in Beijing, China. *Environmental Pollution*, (161): 215–221.

Yao, Z.L & Zhao, L.X. 2010. Ronnback M, et al. Comparison on characterization effect of biomass pellet fuel on combustion behavior. *Transactions of the Chinese Society Agricultural Machinery*. 41(10): 97–102.

Advances in Energy, Environment and Materials Science – Wang & Zhou (Eds)
© 2017 Taylor & Francis Group, London, ISBN 978-1-138-03600-0

Experimental research on the denitration of the simulated flue gas by strong ionization dielectric barrier discharge coordinating with ozone

Ping Li, Chengwu Yi, Jing Wang, Jue Li, Ying Zhu & Shu Wang
School of the Environment and Safety Engineering, Jiangsu University, Zhenjiang, Jiangsu, China

ABSTRACT: Strong ionization dielectric barrier discharge method, coordinating with ozone, was used to oxidize and remove nitrogen oxides in the simulated flue gas. The main factors of denitration efficiency, including water content, oxygen content and discharge voltage were researched and analyzed. Experimental results showed that the denitration efficiency of the simulated flue gas was higher than 90% under the situation shown as the follow: the total flow was 4 L/min; water and oxygen content were 2.2% (v/v) and 20% (v/v) respectively; the concentration of ozone and NO_x were 186 ppm and 300 ppm respectively; discharge voltage was at 2.4 kV; the carrier gas was N_2. And the product of denitration was nitric acid rather than nitrous acid.

1 INTRODUCTION

Nitrogen oxide (NO_x) is one of the atmospheric pollutants, which has posed a great threat to the environment and human health (Yi et al. 2009, Li et al. 2016). The newly promulgated "Emission Standard of Air Pollutants for Thermal Power Plants (GB-13223-2011)" stipulated that NO_x emission must not exceed 100 mg/m³. Therefore, research on theory and technology is crucially important.

The common flue gas denitration technologies mainly include Selective Catalytic Reduction (SCR) and Selective Non-Catalytic Reduction (SNCR). But there are some disadvantages, such as ammonia leakage, secondary pollution and catalyst poisoning deactivation and high investment and operating cost, etc. Nowadays, the low temperature plasma technologies, applied to the flue gas control fields, gradually attract the researchers' attention. These technologies has many advantages, including a simpler system structure, higher denitration efficiency, less investment, less land occupation and lower operating cost, etc. Therefore, it has wide prospects for application (Du et al. 2011, Yin et al. 2010). Low temperature plasma technologies mainly include electron beam radiation, pulsed corona discharge, Dielectric Barrier Discharge (DBD) and strong ionization dielectric barrier discharge. Even though the average electron energy of electron beam radiation method was up to 33 eV and the efficiency of the simulation results in the coal-fired power plant were as high as 85%~95%, Chang et al. (1992) pointed out that X-ray radiation and high cost hindered the

application of the electron beam method in the air pollution control field. The industrial application of the pulsed corona discharge method was also limited by the small ionization region, the low efficiency and the high energy consumption (Yi et al. 2006, Basfar et al. 2008). Stamate et al. (2013) used plasma to generate Ozone (O_3) to remove NO_x, without the existence of H_2O. The denitration efficiency was less than 50%. Saavedra et al. (2007) removed NO_x by DBD and the removal efficiency was as high as 98%. But the final product nitrous acid accounted for 7.8%~34.8%.

In order to solve the problem of low denitration efficiency and low average electron energy in the process of NO_x removal, the strong ionization dielectric barrier discharge method was used to remove NO and the average electron energy is higher than 13 eV (Yi et al. 2006). Mok et al. (1998) found that O_3 is the most important active substance for NO oxidation and the effect of O and ·OH on the oxidation of NO is relatively small, compared with O_3. Therefore, advanced oxidant O_3 is added in this study, coordinating with strong ionization, to remove NO_x. The O_3 was mixed with the simulated flue gas. So that the denitration efficiency will be improved and HNO_3 can be generated. Furthermore, it can promote the industrial application.

2 EXPERIMENT

2.1 *Experimental system*

The experimental flow chart of the denitration by strong ionization dielectric barrier discharge coordinating with O_3 is shown in Figure 1.

2.2 Experimental apparatus

The output voltage and frequency range of the low temperature plasma power supply (CTP-2000 K/P, Nanjing Suman Electronics Co., Ltd., China) are 0~30 kV and 3~100 kHz respectively. The internal structure of the plasma reactor is shown in Figure 2. The discharge gap of the plasma reactor is about one millimeter wide. The dielectric layer α-Al_2O_3 is rather thin, which is covered on the discharge electrode and the earth electrode by the method of plasma smelting. There is a double discharge channel. The effective volume of the reaction chamber is 100 mm × 190 mm × 2 mm. The excitation voltage is 0~3.5 kV, the frequency is 4.5 kHz~6.5 kHz, the average electron energy is higher than 10 eV and the average electron density is above $10^{15}/m^3$.

2.3 Analysis and measure instruments

The main analysis and measure instruments used in the experiment are shown in Table 1.

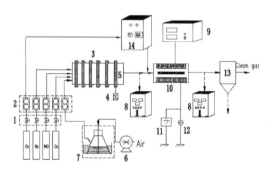

Figure 1. The experimental flow chart of the denitration by strong ionization discharge coordinating with ozone. 1. Pressure-releasing valve; 2. Glass flow meter; 3. Electric heating belt; 4. Temperature controller; 5. Gas mixing chamber; 6. Atmospheric pump; 7. Electric-heated thermostatic water bath; 8. Flue gas analyzer; 9. Plasma power supply; 10. Plasma generator; 11. Q3-V Voltmeter; 12. Oscilloscope; 13. Product collect device; 14. Ozone generator.

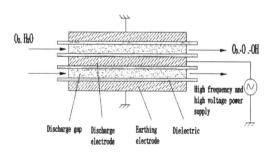

Figure 2. The structure of strong ionization discharge.

Table 1. The list of the main analysis and measure instruments.

Name	Model	Parameters
Portable integrated flue gas analyzer	NOVA5003-S	NO & SO_2: 0~5000 ppm, NO_2: 0~800 ppm
Handled T & RH instrument	CENTER314	T: ± 0.7°C; RH: ± 2.5%
Temperature controller	KZ810	T: –200~400°C;
Ion chromatograph	Metrohm Compact 861	Resolution: 0.028 ns/cm; Conductivity: 0~5000 µs/cm

3 RESULTS AND DISCUSSIONS

3.1 Effects of H_2O content on NO removal efficiency

Figure 3 shows the NO removal efficiency at 5.15 kHz, 30°C, 4 L/m^3 of total flow, 20.9% (v/v) of O_2 added. The relationship between H_2O content and denitration efficiency (with 430 ppm of NO) can be concluded from Figure 3. When the water content was lower than 1.6% (v/v), the denitration efficiency decreased with the increase of water content, without the existence of O_3. In that condition, the reactor produced a lot of high-energy electrons to dissociate and ionize N_2 molecules into large amounts of N and other reductive particles. The reduction reaction was mainly in the process of denitration (Zhang et al. 2014). When the water content was higher than 1.6% (v/v), the efficiency was improved rapidly with the increase of O_3 content and H_2O content. This is because the H_2O in flue gas is the main reactants in the process that O, O_3 and other reactive oxygen species convert into ·OH. O_3 can be decomposed into O (Eliasson et al. 2000) and then reacted with H_2O to generate ·OH. Under this circumstance, the process of denitration was mainly realized by oxidation (Mätzing et al. 2007). Therefore, good removal efficiency can be obtained with the water content between 2.0% (V/V)~2.4% (V/V).

3.2 Effects of O_2 content on NO removal efficiency

Figure 4 shows the NO removal efficiency for three different voltages: 2.4 kV, 2.8 kV and 3.2 kV. It can be seen from Figure 4 that as the discharge voltage increased, the variation trend of the denitration rate with the increase of the oxygen content is identical. Due to the presence of O_3, discharge still

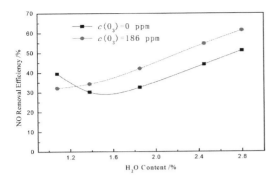

Figure 3. Effects of O_2 content on NO removal efficiency.

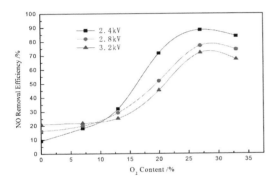

Figure 4. Effects of O_2 content on NO removal efficiency.

produced a certain amount of reactive oxygen species even without oxygen. In the reaction system, the oxidative removal was still dominant. The NO was oxidized to NO_2 and NO_3. And then the NO_2 and NO_3 were eventually transformed into HNO_3 (Wang et al. 2007 and Bai et al. 2014). But with the constant voltage and water content, the amount of ·OH was constant. So when the ·OH was consumed, it led to the decrease of denitration efficiency.

From the above analysis, it can be concluded that the efficiency will be enhanced by appropriately increasing the oxygen content in the flue gas. When the oxygen content is about 27% and the voltage is 2.4 kV, the efficiency can reach 88%.

3.3 Effects of discharge voltage on NO removal efficiency

Figure 5 shows the NO removal efficiency for nine different discharge voltages at 5.15 kHz, 30°C, 4 L/m³ of total flow rate, 20.9% (v/v) and 1.8% (v/v) of O_2 and H_2O added respectively and 186 ppm of

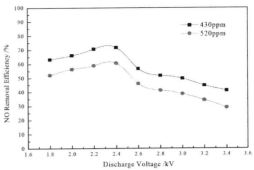

Figure 5. The effects of discharge voltage on NO removal efficiency.

O_3 added. Note from Figure 5 that under different initial concentrations, the variation trend of the denitration rate with the discharge voltage is identical. As the voltage increased, the rate increased and then decreased. Due to the required excitation energy of gas molecules N_2, O_2 and H_2O decreased in turn. When the voltage is lower than 2.4 kV, the electron energy obtained from the strong ionization discharge electric field isn't high enough to dissociate large number of N_2 molecules. With the increased voltage, the concentration of high energy electrons increased. Then O_2 and H_2O molecules are dissociated to generate ·OH, O and other oxygen active particles to aggravate the oxidative removal of NO and improve the efficiency. When the voltage increased beyond 2.4 kV, more and more N_2 molecules were dissociated and ionized to produce N, N(^{2}D), N_2(A), etc (Sathiamoorthy et al. 1999). However, there quickly generated extra NO_X by a large number of active particles reacting with N, N(^{2}D), N_2(A). Not only extra NO_X were produced, but also ·OH, O and other oxygen active particles were consumed. So the removal efficiency decreased.

4 PRODUCT ANALYSIS

The product acid liquid of denitration, under the initial conditions shown in 3.1 section, was collected to analyse the concentration and composition of the anions. Metrohm Compact 861 ion chromatography was used in this study. The chromatogram was shown in Figure 6. In addition, standard solution of NO_3^- was prepared to contrast with the product solution. Figure 7 shows the chromatogram of standard solution. By contrast, the peak time of the two chromatographic was the same and there was only one peak. Therefore, it can be concluded that the product of denitration was HNO_3 rather than HNO_2.

133

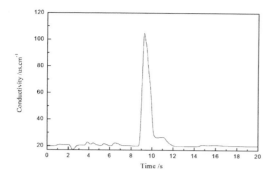

Figure 6.   Ion chromatogram of denitration acid liquid.

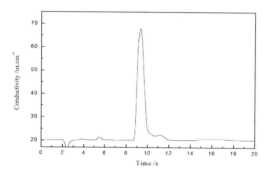

Figure 7.   Ion Chromatogram of NO_3^- in standard solution.

5   CONCLUSIONS

1. Experimental results showed that the denitration efficiency of simulated flue gas was higher than 90% under the situation shown as the follow: the total flow was 4 L/min; water and oxygen content are respectively around 2.2% and 20%; the O_3 concentration was 186 ppm and NO_x concentration was 300 ppm; discharge voltage was at 2.4 kV. And the product of denitration was HNO_3 rather than HNO_2.
2. The strong ionization dielectric barrier discharge method can produce high concentration of active particles and the process of removal was mainly oxidation reaction, which greatly improved the NO_x removal efficiency. Therefore, it can effectively solve the problems of low electron energy, high investment cost and low efficiency of the conventional plasma discharge methods. Furthermore, this study is of great significance for the industrialization of low temperature plasma technology.

ACKNOWLEDGMENT

This work was supported financially by National Natural Science Foundation of China (Project 51278229) and Six Talent Peaks Funded Project of Jiangsu Province, China (Project JNHB—018). *Corresponding Author: Chengwu Yi, E-mail: yichengwu0943@163.com.

REFERENCES

Basfar, A. A. et al. 2008. Electron Beam Flue Gas Treatment (EBFGT) technology for simulated removal of SO_2 and NO_x, from combustion of liquid fuels, *Fuel* 87, 1446–1452.

Bai, M. D. et al. 2014. Simulated desulfurization and denitration by O_2^+, O_3, *Chin Envron Sci.* 34, 324–330.

Chang, M. B. et al. 1992. Removal of SO_2 and the simulated removal of SO_2 and NO from simulated flue gas streams using dielectric barrier discharge plasmas, *Plasma Chem. Plasma Process.* 12, 565–580.

Du, X, et al. 2011. Influence of ethylene on NO oxidation in a dielectric barrier discharge reactor, *J. Chin. Soci. Power Engin.* 31, 882–886.

Eliasson, B. et al. 2000. Ozone synthesis from oxygen in dielectric barrier discharges, *J. Phys. D Appl. Phys.* 20, 1421–1437.

Li, Y. et al. 2016. Experimental research on the sterilization of escherichia coli and bacillus subtilis in drinking water by dielectric barrier discharge, *Plasma Sci. Technol.* 18, 173–178.

Mätzing, H. 2007. Chemical kinetics of flue gas cleaning by irradiation with electrons, *Adv. Chem. Phys.* 80, 315–402.

Mok, Y. S. et al. 1998. Mathematical analysis of positive pulsed corona discharge process employed for removal of nitrogen oxides, *IEEE Trans. Plasma Sci.* 26, 1566–1574.

Sathiamoorthy, G. et al. 1999. Chemical reaction kinetics and reactor modeling of NO_x removal in a pulsed streamer corona discharge reactor, *Ind. Eng. Chem. Res.* 38, 1844–1855.

Saavedra, H. M. et al. 2007. Modeling and experimental study on nitric oxide treatment using dielectric barrier discharge, *IEEE Trans. Plasma Sci.* 35, 1533–1540.

Stamate, E. et al. 2013. Investigation of NO_x reduction by low temperature oxidation using ozone produced by dielectric barrier discharge, *Jpn. J. Appl. Phys.* 52, 79–83.

Wang, Z et al. 2007. Simulated removal of NO_x, SO_2 and Hg in nitrogen flow in a narrow reactor by ozone injection: experimental results, *Fuel Process. Technol.* 88, 817–823.

Yi, C. W. et al. 2006. Study on plasma technology of simulating flue gas desulfurization and resource utilization, *High Voltage Engin.* 32, 81–83.

Yi, C. W. et al. 2006. Analysis of operation parameters of plasma desulfurization, *Environ Engin.* 24, 34–37.

Yi, C. W. et al. 2009. Harmfulness and treatment technology of organic waste gas process. *J. Anhui Agri. Sci.,* 37, 351–352.

Yin, S. E. et al. 2010. Effect of relative humidity on removal of NO and SO_2 from flue gas in dielectric barrier discharge reactor, *J. Chin. Soci. Power Engin.* 30, 41–46.

Zhang, L. et al. 2014. Removal dynamics of Nitric Oxide (NO) pollutant gas by pulse-discharged plasma technique, *Sci. World J.* 2014, 171–192.

Advances in Energy, Environment and Materials Science – Wang & Zhou (Eds)
© *2017 Taylor & Francis Group, London, ISBN 978-1-138-03600-0*

Measurement of data centers' energy efficiency: A data envelopment analysis approach

Changgeng Yu & Liping Lai
School of Mechanical and Electronic Engineering, Hezhou University, Hezhou, China

ABSTRACT: With the high costs of power, energy efficiency has rapidly become a critical consideration issue when evaluating data centers. Three of the most important data center metrics, PUE, DCEP, and CUE, are examined closely. Data Envelopment Analysis (DEA) is a fractional linear programming-based technique that has gained wide acceptance in centers years due to its effectiveness in comparing energy efficiency of data centers. First, expounds feasibility of the data centers' energy efficiency assessment with DEA is introduced. Second, the application method and model properties are discussed. Finally, based on the DEA method for evaluation of energy efficiency, the results are concluded. The results indicate that the proposed method can overcome the limitations of single index parameters (PUE), without subjective analysis of the input element weights. Some proposals are given for improving the conservation planning and construction in data center energy efficiency evaluation.

1 INTRODUCTION

Energy consumption in data centers is increasing year by year with the growth in data centers market. In order to mitigate global warming, it is an urgent task to reduce power consumption in data centers. Generally, ICT equipment (e.g. servers, routers, switches, and storage units) and facility equipment (e.g. power delivery components, Heating Ventilation, and Air Conditioning (HVAC) system components) account for a large percentage of energy consumption in data centers.

To strengthen the energy efficiency evaluation of enterprises, improving the efficiency of data center energy has become the priority of data center monitoring system. Global industry and academia widely focus on problems of energy efficiency evaluation, for example, the U.S. Department of Energy (DOE) was built in an organization, in order to improve the energy efficiency, as the primary goal of Industrial Assessment Centers, IAC (J. Koomey, 2005). The center, relying on 29 colleges, universities, and industrial sectors in the United States, has set up a file in more than 15000 U.S. companies to implement suggestions on the industrial energy audit and evaluation of projects, such as Green Grid organization (Green Grid) founded in 2007 (Belady C., 2008), the combined company, government, industry, and research and tries to provide the best data center energy saving methods, assessment methods, and techniques. Founded in the same year, Green500 group, the group in June and December each year, respectively, issued to MegaFlops/W (that is, per watt power can complete one million floating point calculations per second) as an index,

efficiency of 500 sets of the world's fastest supercomputer, sorting; EPA and the Department of Energy, DOE launched the national data center Energy efficiency information project, ASHRAE, The Green Grid, such as eight professional association to data centers in Washington, hopes to make join data center energy efficiency evaluation standard, in order to improve data center energy efficiency and reduce energy consumption.

2 LITERATURE REVIEW

Data Envelopment Analysis (DEA), Originally developed by Charnes et al. (1978), is a non-parametric technique analysis method used to evaluate the relative effectiveness of Decision-Making Units (DMUs) with a sample (Samoilenko, 2008). The DEA method, as a new achievement in the field of operations research and management science, is a new system evaluation method in the study to improve energy efficiency and environmental research (Forsund, 2002 & 2005). Data Envelopment Analysis (DEA) model is a non-deterministic efficiency measurement model based on the concept of relative efficiency used to the single index input or single index output efficiency of the project concept development to multi-index input or multi-index output of the DMUs relatively effective evaluation method. DMU can be with types of departments or units, input through a number of factors of production produced a number of *products*, because the output is DMU *decision* results. The DMU is the DEA evaluation model object.

In the DEA model, the relative efficiency is defined as the weighted average of the output and input ratios, the value distribution is [0,1], weighted statistics can be calculated directly from the observable "input and output", the DEA model does not require output between input of the functional relationship, by mathematical programming is constructed out of all input/output possible combinations of external effective production segment Frontier (that is, envelope), and make the constraints of the model of all input/output observation points located within the envelope. Only DEA efficiency (or weak) of decision-making units corresponding points in the envelope, and located in the envelope surface of DUMs for "effective element", its relative efficiency is 1; and decision-making units outside the envelope for unit are invalid, and the relative efficiency is less than 1. The invalid unit can find the corresponding effective unit through the projection mode, and the distance from the invalid unit to the front face shows that the relative efficiency of the decision-making unit is high and low. The difference between the DEA method and parameter method of regression analysis is shown in Figure 1.

There are eight examples given for the DMU of single input and output. The horizontal and vertical coordinates are represented by the input and the output, and the relative efficiency between the output and input were expressed by point P_1–P_8. The parametric method of linear regression line is represented by the dotted line, describing the P_1–P_8 data point distribution trend. The parameter method, where the DMU in the linear regression line is above or below the regression line, is considered to be the effective decision-making unit (P_2, P_3, and P_4). The DEA is an efficient decision-making unit P_2, P_3, P_4, and P_1 to form a piecewise envelope. P_5 is not in the envelope, for the invalid decision-making unit, its relative efficiency is less than 1. The invalid unit P_5, hypothesis in the corresponding cell of the frontier *projection* just as the decision-making units P_1, decision-making units P_5 use too many

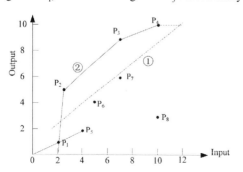

Figure 1. Difference between DEA method and regression analysis.

resources and its relative efficiency is $OP_1/OP_5 < 1$, which needs to be improved. Based on this idea, through the decision-making unit input/output observation data, the mathematical programming is used to find out the relative efficiency of each decision-making unit and evaluation methods.

3 METHODOLOGY

Energy assessment technology mainly includes parameter and non-parameter methods. The parameter method is to hypothesize precise function expressions between input and output. By a group of observed data of input and output, complying with the constrained data, it can use the mathematical programming or regression analysis method to determine the parameters of the function expression. Parameter methods need to know more about the function expression and distribution error. It is also very difficult to describe the statistical properties of the estimated values. The estimated result is not stable. The DEA method is the parameter method.

3.1 Model formulation

Suppose that there are K DMUs represented by DMU_k ($k = 1, 2, ..., K$). For each DMU_k ($k = 1, 2, ..., K$) have N inputs indexes and M output index. v_i ($i = 1, 2, ..., N$) and w_j ($j = 1, 2, ..., M$) are the input and output weights, respectively. The efficiency of entity evaluated is obtained as a ratio of the weighted output to weighted input subject to the condition that the ration for every entity. Mathematically, it is described as follows:

$$h_k = \frac{\sum_{j=1}^{s} w_j y_{jk}}{\sum_{i=1}^{m} v_i x_{ik}}, k = 1, 2, \cdots, n \quad (1)$$

Consider the relative efficiency of k DMUs, where k use inputs x_{ik} ($i = 1, 2, ..., m, k = 1, 2, ..., K$) to produce s outputs y_{jk} ($j = 1, 2, ..., s, k = 1, 2, ..., K$). The well-known CCR model for measuring the relative efficiency scores of DMUs is formulated as the following linear program problem:

$$\max \sum_{j=1}^{s} w_j y_{jk_0}$$

$$s.t. \begin{cases} \sum_{j=1}^{s} w_j y_{jk} - \sum_{i=1}^{m} v_i x_{ik} \le 0, k = 1, 2, \cdots, n \\ \sum_{i=1}^{m} v_i x_{ik_0} = 1 \\ v_i, w_j \ge 0, i = 1, 2, \cdots, m; j = 1, 2, \cdots, s \end{cases} \quad (2)$$

Where v_i $(i=1,2,\ldots,m)$ and w_j $(j=1,2,\ldots,s)$ are the input and output weights assigned to the ith input and sth output, respectively, and DMU_0 refers to the DMU under evaluation.

To resolve the uncertainty of classification by model (2), when dual-role factors are excluded from consideration, the CCR efficiency of DMU_0 can be calculated by model (3).

$$\min \; \theta$$
$$s.t. \begin{cases} \sum_{i=1}^{m} x_{ik}\lambda_k \leq \theta x_{ik_0}, i=1,2,\cdots,m \\ \sum_{j=1}^{s} y_{jk}\lambda_k \geq y_{jk_0}, j=1,2,\cdots,s \\ \lambda_k \geq 0, \; k=1,2,\cdots,n \end{cases} \quad (3)$$

where λ_k denotes the weight of input and output indicators, and θ denotes the coefficients of the input relative to the output.

3.2 Decision variables

V_i – Weight of input i (unitless)
W_j – Weight of output j (unitless)
λ_k – Weight of input and output indicators
θ – The coefficients of the input relative to the output

4 CASE STUDY

4.1 The selection of input measures and output measures

Based on DEA, the mathematical programming method is used to construct the evaluation stand-ard of the envelope surface. According to the basic principle of the DEA model, envelope surface and decision-making unit of input and output meas-ures were selected to measure the efficiency for data centers.

We collect both input and output measures that reflect the evaluation objective and concerned index by the managers, and input and output measures must be positive, and for input measures smaller is better, whereas for output measures bigger is better, in order to increase the relative efficiency of numerical accu-racy. Decision units were selected without unit, but all decision units are similar evaluation indexes.

According to the basic principle, when using the DEA model for data center energy efficiency, per-formance evaluation should be correctly selected for related input and output measures. The input measures are energy consumption for equipment, data centers area, and operation of the effective load, whereas the output measures are PUE and DCeP. For actual using, according to the require-ments of the DEA model, the input and output measures were processed.

4.2 Input and output measures collection and calculation

This paper were selected for four different data center regions as an example of evaluation, the com-munications room using the same power supply, its cue index value is not considered, collect data from room construction area S, server numbers N_1, server consumption power P_s, switch consumption P_w, infrastructure consumption power P_c, the amount of data transmitted L_d, etc. The data on these meas-ures are given in Table 1. Tables 1 and 2, respectively,

Table 1. Data for the centers.

Center	Construction area /m³	Server numbers (U)	Server consumption (degree)	Switch consumption (degree)	Infrastructure consumption (degree)	Data transmitted (T)
Center A	1920	1600	185.9047	13.9336	311.9558	7.4706
Center B	24000	10354	795.5744	38.4	1388.7946	103.9487
Center C	21600	1000	1425.2250	38.4	2561.1230	65
Center D	1344	1300	188.4764	13.9336	338.5540	10.16

Table 2. DEA model of the input and output measures and result.

Center	Input M_1	Output M_2	M_3	Result DCeP	PUE	Efficiency θ
Center A	0.2666	0.1162	7.4706	0.5362	2.5610	0.9693
Center B	0.0926	0.0768	103.948	2.7070	2.6652	1
Center C	0.1863	0.1425	65	1.6927	2.7498	0.7664
Center D	0.4025	0.1450	10.16	0.7292	2.6726	0.8781

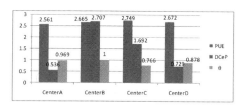

Figure 2.　Result for each data center.

provide the evaluation of data center energy efficiency of the collection data, and the DEA model of the input and output measures. According to the requirement of the DEA model to the data, P_S, P_W, P_C, etc. and S_1, N_1 is related to the transformation of quantities related to "efficiency factor", thus forming with DEA model input and output measures, total electricity consumption $P_{total} = P_s + P_w + P_c$; electricity $M_1 = P_{total}/(kW \cdot h / m^2)$ consumption of per unit area; low power consumption per unit of server $M_2 = Ps /N_1$ $(kW \cdot h / U)$; the unit electricity consumption transmission $DCeP = L_d /P_w$ $(T/ kW \cdot h)$. Input and output measures substitute DEA model, data calculation using MATLAB, calculate the relative efficiency η. The calculation results are shown Table 2. The data center efficiency is obtained by comparing the relative efficiency of η, the higher the value, the better the plant efficiency.

5 RESULTS

The results for the four dater centers are illustrated in Figure 2. These results show that Center-B is kept constant at 1 for the highest energy efficiency, Center-C is kept constant at 0.7664 for the minimum, four centers energy efficiency from high to low ranking: Center-B, Center-A, Center-D, Center-C, Do not consider the transmission of data, the evaluation is not comprehensive if simple to *PUE* index to evaluate the energy efficiency of data center, example: communication data centers.

6 CONCLUSION

In this paper, we have a DEA model for evaluating the data centers efficiency values and ranking of DMUs with input and output that may not necessarily be symmetrical. We may draw the following conclusions.

First of all, the DEA method has a good scientific nature. The sample data were established in the production frontier projection. The decision-making unit is compared with the data production frontier. At the same time, the relative efficiency start sorting, and data center energy efficiency is obtained. Energy efficiency in the process of input and output weights is produced by the data ele-

ments, and they are not given weight coefficient, which is not affected by subjective factors.

Second, the DEA model was proposed by data centers energy efficiency evaluation. Data centers energy efficiency index mainly includes the data centers PUE/DCiE, Carbon energy Utilization Efficiency (CUE) and energy efficiency (DCeP), which overcome the limitations that single index parameters' data center PUE.

Finally, from the results of the examples, it can be concluded that the proposed method performs better than the other multi factor performance testing and evaluation methods.

ACKNOWLEDGMENTS

This research was supported by the National Natural Science Foundation of China (Grant No. 61540055), the doctor's scientific research foundation of Hezhou University (No. HZUBS201506), and the project of Guangxi University of Science and Technology Research (No. 2013YB242).

REFERENCES

Belady C. Green Grid Data Center Power Efficiency Metrics: PUE and DCiE [R/OL]. http://www.THE-GREE-NGRID.org, 2008.

Charnes A, Cooper W W, Rhodes E. Measuring the efficiency of decision making units[J]. European Journal of Operation Research, 1978, 2(6):429–444.

Data center industry leaders reach agreement on guiding principles for energy efficiency metrics [EB/OL]. (2010-02-01) [2013-07-01]. http:www.energystar.gov/ia/parners/prod_develoment/downloads/DataCenters_AgreementGuidingPrinciples.pdf.

Forsund F R, Sarafoglou N. On the origins of data envelopment analysis[J]. Journal of Productivity Analysis, 2002 (17):207–224.

Forsund F R, Sarafoglou N. The tale of two research communities: The diffusion of research on productive efficiency[J]. International Journal of Production Economics, 2005 (98): 17–40.

Greenpeace "likes" facebook's New Datacenter, But Wants a Greener Friendship [Online]. Available: http://www.greenpeace.org/internation/en/press/releases/Greenpease-likes-Face-books-new-datacentre-but wants-a-greener-friendship.

J.Koomey. Growth in Data Center Electricity Use: 2005 to 2010[Online]. Available: http://www.analyticspress.com/datacenters.html.

Samoilenko S, Osei-Bryson K M., Increasing the discriminatory power of DEA in the presence of the sample heterogeneity with cluster analysis and decision tress [J]. Expert Systems with Applications, 2008, 34(2):1568–1581.

U.S. Energy information admin. annual energy review 2011 [R]. DOE/EIA-0384 (2011). Washington D. C., 2011: 3–4.

The temperature control of planar blackbody radiation source based on nonlinear PID

D.X. Huang
Division of Thermophysics and Process Measurements, National Institute of Metrology, Beijing, China
School of Electrical Engineering, Tianjin University of Technology, Tianjin, China

J.H. Wang
Division of Thermophysics and Process Measurements, National Institute of Metrology, Beijing, China

Z.G. Wang
School of Electrical Engineering, Tianjin University of Technology, Tianjin, China

ABSTRACT: This paper designs a planar blackbody temperature control based on Nonlinear PID (NPID). The nonlinear Tracking-Differentiator (TD) is used to arrange the transition process and extract the differential signal, which is combined with the nonlinear error feedback to transform the conventional PID control. Temperature control and data acquisition is achieved by the control system software graphical programming, which is realized in the LabVIEW software, combined with Compact RIO real-time control system. The experimental results show that the temperature control effect of the NPID control algorithm can effectively improve planar blackbody radiation source, the quality of temperature control reached within 20 mK/10 min, and it has a stronger performance of disturbance rejection compared with the conventional PID control.

1 INTRODUCTION

Planar blackbody radiation source is the main reference source of airborne infrared imaging system calibration and field infrared calibration. We need to ensure that the radiation source has high enough temperature control accuracy and temperature stability, because the infrared remote sensing instrument needs real-time high-precision calibration source (Yang et al. 2011).

As one of the earliest control strategies, PID control is widely used in industrial process control because of its simple algorithm, good robustness and high reliability (Calmette & De Caussin 2006, Lv et al. 2015, Wu et al. 2010, Silva et al. 2002). The PID controller is still mainly used in the current process control in the world, if including all kinds of improved PID, then its share is more than 90% (Wang et al. 2000). Nevertheless, the PID controller still exists many flaws that need to be improved, such as the contradiction between the dynamic and static state of system, tracking set point and eliminate perturbations, and robustness and control performance and so on (Su & Duan 2003). The temperature control process of the planar blackbody radiation source has the characteristics of inertia, time delay and parameter variation, and the conventional PID cannot be used to produce a satisfactory control effect. Nonlinear correction of the conventional PID con-

trol, which makes it a NPID control, can effectively improve the control performance (Wang & Lin 2015, Naso et al. 2012, Kianfar et al. 2011, Hu 2006).

Tracking differentiator is proposed by the researcher J.Q. Han, compared to classical differentiator prone to noise amplification effect, it can reasonably track discontinuous input signal and extract the approximate differential signals, thus effectively solving the practical engineering problem that reasonable extraction differential signal from the signal of noise pollution (Han & Wang 1994, Wu et al. 2004). NPID is modified by nonlinear tracking differentiator, and the temperature control of planar blackbody radiation source is carried out on the Compact RIO real-time control platform. The performance and quality of temperature control of planar blackbody radiation source can be improved effectively by combining the advantages of NPID and Compact RIO platform.

2 CONTROL SYSTEM DESIGN

The structure of the temperature control system of the planar blackbody radiation source is shown in Fig. 1.

In the figure, K1 is an intermediate relay; the event module triggers K1 to disconnect the power supply of the heater when the temperature of the

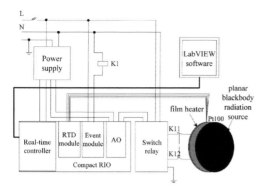

Figure 1.   Temperature control system of planar black-body radiation source.

black body exceeds the normal value, thus ensuring the system security. The NPID program is compiled in LabVIEW software, and then it is imported to the compact RIO real-time controller. The output of the control signal is 4–20 mA current, which is converted to the PWM wave to supply power to the film heater. Moreover, the temperature of the planar blackbody radiation source is collected by the RTD module and is fed back to the real-time controller and the LabVIEW host computer. In addition, the real-time temperature curve and control volume curve can also be observed in the LabVIEW platform, so as to better adjust the controller parameters.

2.1 Temperature control requirement and characteristic analysis

As a real-time thermal reference source, the planar blackbody radiation source needs to be controlled in a very small range, and can be kept stable for a long time. In addition, the controller also needs to have faster response speed and disturbance rejection ability, because the external environment temperature can change greatly when the field calibration is carried out. According to the actual situation, the quality requirement of the temperature control is within 50 mK/10 min.

According to Fig. 1, the material of the planar type of radiation source is copper; the film heater is attached to the back of the blackbody for uniform heating. The following transfer function model of the temperature control can be regarded as a first-order plus time delay process. It can be expressed as following:

$$G(s) = \frac{K}{Ts+1} e^{-\tau s} \qquad (1)$$

where K is the system gain, T is the inertia time constant, τ is the time delay.

By analyzing the model of the controlled system, it is concluded that the temperature deviation is get-

ting bigger when the temperature is starting to heat up, and the large control input is needed to speed up the heating up. However, measuring temperature cannot reflect the thermal state of the system in real time and accurately, and the temperature control process of the controlled system has high nonlinear characteristics, because of inertia, heat transfer delay and parameter time-varying of the system. When the temperature reaches the set value, even if the control input is reduced to zero, the temperature will be further increased, which leads to the overshoot of the temperature, and increases the time of the temperature control. On the other hand, in the stability of temperature control, the control strategy should be adjusted accordingly, as a result of the different dynamic characteristics that the temperature rising and cooling process has shown.

2.2 Control strategy

Tracking differentiator can arrange transient process reasonably and extract differential signal effectively, so that the actual temperature fast track set value changes and the control signal is given in advance. In this way, it can achieve rapid temperature control process, and reduce or avoid overshoot. For the structure of tracking differentiator as shown in Fig. 2, the output signal v_1 can quickly track the input signal v; v_2 is the differential signal of v_1. In this paper, we use the most speed discrete tracking differentiator (Han 1994), it is expressed as following:

$$\begin{cases} fh = fhan(v_1(k) - v(k), v_2(k), r, h_0) \\ v_1(k+1) = v_1(k) + h \cdot v_2(k) \\ v_2(k+1) = v_2(k) + h \cdot fh \end{cases} \qquad (2)$$

where $fhan$ is the most speed control synthesis function, its call formula is $u = fhan(x_1, x_2, r, h)$, and the specific algorithm formula is as follows:

$$\begin{cases} d = r \cdot h_0^2 \\ a_0 = h_0 \cdot x_2 \\ y = x_1 + a_0 \\ a_1 = \sqrt{d(d+8|y|)} \\ a_2 = a_0 + sign(y)(a_1 - d)/2 \\ fsg(x,d) = (sign(x+d) - sign(x-d))/2 \\ a = (a_0 + y) fsg(y,d) + a_2(1 - fsg(y,d)) \\ fhan = -r(\frac{a}{d}) fsg(a,d) - r \cdot sign(a)(1 - fsg(a,d)) \end{cases}$$

$$(3)$$

Figure 2.   The structure of tracking differentiator.

where h_0 is a filter factor, h is the integral step size, r is the velocity factor, and they determine the filtering effect, tracking accuracy and tracking speed respectively. In addition, the speed of the transient process can be achieved by adjusting the velocity factor r.

Fig. 3 shows the schematic diagram of the NPID controller.

The dashed frame is a NPID control which is realized by two tracking differentiator. v is the system set value; u, y represent the control input and output of the system respectively. Tracking differentiator TD_1 is used to arrange the transient process v_1 of v and give its differential signal v_2; tracking differentiator TD_2 is mainly used to fast track y and gives its approximate differential signal, z_1, z_2 represent the tracking value of system output y and its differential signal respectively. $e_0 = \int e$, $e_1 = e$ and $e_2 = \dot{e}$ each represent the integral of error between the transient process v_1 and the tracking value z_1 of output signal, the error between the v_1 and z_1, and the error between v_2 and z_2. Nonlinear State Error Feedback Control Law (NLSEF) is generated by the nonlinear combination of these three variables to determine the control input of the system u, so as to alleviate effectively the contradiction between overshoot and fast response produced by linear combination, and to improve the control effect of the controller. In this paper, the power function is used to construct the nonlinear error feedback control law, the power function, i.e., the saturation function is a nonlinear function, and it is expressed as following:

$$fal(e, \alpha, \delta) = \begin{cases} \dfrac{e}{\delta^{1-\alpha}}, |e| \leq \delta \\ |e|^{\alpha} sign(e), |e| \geq \delta \end{cases} \quad (4)$$

where δ is the interval size of the linear segment of the function, α is the degree of nonlinearity, and determines the nonlinear shape of the function. When $0 < \alpha < 1$, it is actually an empirical knowledge of control engineering circles: mathematical fitting of "error is large, small gain; error is

small, large gains". The essence of the methods of fuzzy control, intelligent control, and variable gain PID control and so on is based on this experience knowledge, and the function *fal* describes the experience knowledge with a simple nonlinear structure merely. Moreover, for the differential link, the differential gain is also small when the differential error is small; on the contrary, the differential gain is large when the differential error is large. So taking $\alpha > 1$, the differential effect will be reduced when the system output is close to steady state, and it is beneficial to improve the performance of the control system (Han 1944, Han 2008, Huang et al. 2005). In addition, joining the integral function is more beneficial to improve the precision and stability of temperature control. Ultimately, the resulting nonlinear error feedback control law is expressed as following:

$$u = \beta_0 e_0 + \beta_1 fal(e_1, \alpha_1, \delta) + \beta_2 fal(e_2, \alpha_2, \delta) \quad (5)$$

where β_0, β_1 and β_2 represent the feedback gain of each error respectively, parameter α_1, α_2 and δ can be determined in advance, here $\alpha_1 = 0.75$, $\alpha_2 = 1.5$, $\delta = 2h$, and h is the sampling time. In this way, it is similar to the conventional PID, adjusting the three parameters of β_0, β_1 and β_2 to generate the appropriate control signal u.

3 EXPERIMENTAL PROCEDURE AND RESULTS

According to Figs. 1 and 3, the system experimental platform was built and the control system software was compiled in LabVIEW. Due to the limitation of the input power of the film heater, the analog output of Compact RIO is limited to 4–8 mA. Its range is extended evenly to the 0–100 so as to improve the control precision, the range of the control input signal u is limited to 0–100 correspondingly, and the system output is the actual measured temperature value.

Sampling time of control system is $h = 0.1 \ s$, the parameters of conventional PID controller are obtained by tuning, proportional gain $P = 50$, integral time constant $T_i = 20s$, differential time constant $T_d = 2s$. In the NPID controller, the parameters of the tracking differentiator TD_1 are $h_1 = h$, $r_1 = 60$, the parameters of the tracking differentiator TD_2 are $h_2 = h$, $r_2 = 50$, and the feedback gains of the error are $\beta_0 = 2$, $\beta_1 = 60$, $\beta_2 = 100$.

The heating process of planar blackbody radiation source is shown in Fig. 4. The experimental temperature is from ambient temperature to the set value of 50°C. By comparison of conventional PID and NPID curve, it can be seen that the

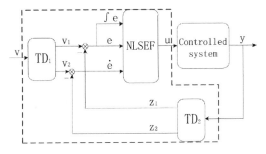

Figure 3. Principle block diagram of the NPID controller.

system response is faster in the NPID. The transition process time of NPID is 150 s, and decreased by 43 s compared to the conventional PID control. The heating process is without overshoot, and the overall control effect is significantly better than the conventional PID control.

Fig. 5 shows the response of the system when the set value is steeply increased from 50°C to 55°C. The control effect is similar to Fig. 4. Compared with the conventional PID control, the system can track the change of the set value faster in the NPID control, and the whole process is not overshoot. Moreover, it also shows that the NPID algorithm has good performance in different temperature rising process, and it has a certain effect on preventing the super temperature of the blackbody.

Fig. 6 shows the response of the system when the set value is steeply reduced from 50°C to 55°C. Because it is a natural cooling by the heat exchange with the environment, the system has the fastest cooling rate when the control input is zero. However, when the system output is close to the set value, it is easy to cause overshoot due to the system characteristics. It can be seen from the figure that, NPID has obviously faster response speed than the conventional PID control and can effectively reduce the overshoot. This is because the tracking differentiator can reasonably extract the differential signal, give advance regulation, and be coupled with nonlinear error feedback control law to further improve the control effect.

Fig. 7 shows the response of the system when the radiation source is equipped with a fan mode, which is similar to the external environment. At $t = 0s$, the fan power supply is switched on and continued for 60 s. Comparison between the two control curves can be seen that in the conventional PID control system, output fluctuations are larger

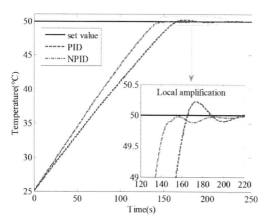

Figure 4. Control effect of heating from 25°C to the set value 50°C.

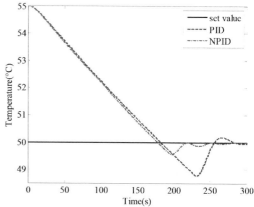

Figure 6. The control effect of the setting value from 55°C to 50°C.

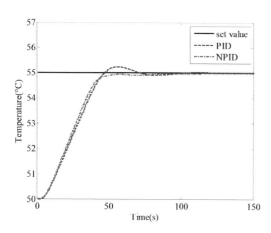

Figure 5. The control effect of the setting value from 50°C to 55°C.

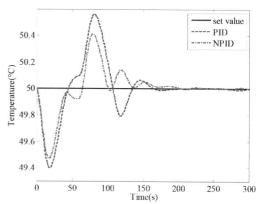

Figure 7. Control effect when external disturbance is given.

142

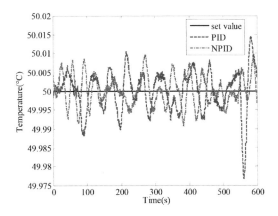

Figure 8.   Control effect of temperature stability.

than the NPID. This shows that the NPID has better anti-interference ability in the same situation.

Fig. 8 shows the temperature stability curves of two kinds of control modes. In 10 minutes, the temperature stability of the conventional PID control is 37.8 mK, and the temperature stability of the NPID control is 18.9 mK, and its temperature control quality is better than that of the conventional PID control.

4   CONCLUSIONS

In this paper, the NPID is used to control the temperature of planar blackbody radiation source. The conventional PID control is transformed by two nonlinear tracking differentiator and nonlinear error combination, which makes it a NPID control. It can be seen that the NPID control effect is obviously superior to the conventional PID control in practical application, the system has a faster response speed, non-overshoot warming in the heating process, stronger anti-jamming ability and better quality of the temperature control, and the expected temperature stability requirements are achieved. On the other hand, although the parameters of NPID that need to be tuned are more than the PID, the tuned parameters can also have excellent adaptability when the condition of the controlled system characteristics changes greatly. That is to say, good control result can also be obtained and its parameter tuning is relatively easier than conventional PID when controlling other.

REFERENCES

Calmette, W. De Caussin, G. (2006). "Move with modified PID loops," Motion System Design.

Han, J.Q. & Wang, W. (1944). "Nonlinear tracking differential," Systems science and Mathematics, vol. 4, pp. 177–183.

Han, J.Q. (1994). "Nonlinear PID controller," Acta Automatica Sinica, vol. 4, pp. 487–490.

Han, J.Q. (2008). "Active Disturbance Rejection Control Technique—the Technique for Estimating and Compensating the Uncertainties," Beijing, National Defense Industry Press.

Hu, B.G. (2006). "A Study on Nonlinear PID Controllers—Proportional Component Approach," American Journal of Dermatopathology, vol. 35, pp. 541–554.

Huang, H.P. Wu, L.Q. Gao, F. & Wei, W.C. (2005). "Main steam temperature control of thermal power plant based on active disturbance rejection control," Journal of System Simulation, vol. 17, pp. 241–244.

Kianfar, K. Amiri, R. & Bozorgmehr, A. (2011). "Designing the Non-Linear PID Controller for a Missile," First International Conference on Informatics and Computational Intelligence. IEEE, pp. 171–175.

Lv, L. Chang, C.Y. Zhou, Z.Q. & Yuan, Y.B. (2015). "An FPGA-Based Modified Adaptive PID Controller for DC/DC Buck Converters," Journal of Power Electronics, vol. 15, pp. 346–355.

Naso, D. Cupertino, F. & Turchiano, B. (2012). "NPID and Adaptive Approximation Control of Motion Systems With Friction," IEEE Transactions on Control Systems Technology, vol. 20, pp. 214–222.

Silva, G.J. Datta, A. & Bhattacharyya, S.P. (2002). "New results on the synthesis of PID controllers," IEEE Transactions on Automatic Control, vol. 47, pp. 241–252.

Su, Y.X. & Duan, B.Y, (2003). "A new class of nonlinear PID controller," Control and Decision, vol. 18, pp. 126–128.

Wang, D.C. & Lin, H. (2015). "A new class of dual support vector machine NPID controller used for predictive control," Ieej Transactions on Electrical & Electronic Engineering, vol. 10, pp. 453–457.

Wang, W. Zhang, J.T. & Chai, T.Y. (2000). "A survey of advanced PID parameter tuning methods," Acta Automtica Sinica, vol. 26, pp. 347–355.

Wu, J. Huang, J. Wang, Y. G. & Xing, K. X. (2010). "RLS-ESN based PID control for rehabilitation robotic arms driven by PMTS actuators," Modelling, Identification and Control (ICMIC), The 2010 International Conference on. IEEE, pp. 511–516.

Wu, L.Q. Lin, H. & Han, J.Q. (2004). "Study of tracking differentiator on filtering," Journal of System Simulation, vol. 16, pp. 651–653.

Yang, J.B. Zhang, W.R. Bai, S. & Liu, Y.K. (2011). "Temperature Analysis for Infrared Radiation Calibration Plane Blackbody," Vacuum & Cryogenics, vol. 17, pp. 23–27.

Advances in Energy, Environment and Materials Science – Wang & Zhou (Eds)
© 2017 Taylor & Francis Group, London, ISBN 978-1-138-03600-0

Closed-form solutions of a transversely isotropic half space subjected to hot fluid injection

J.C.-C. Lu
Department of Civil Engineering, Chung Hua University, Hsinchu, Taiwan, R.O.C.

F.-T. Lin
Department of Naval Architecture and Ocean Engineering, National Kaohsiung Marine University, Kaohsiung, Taiwan, R.O.C.

ABSTRACT: The analytic methods based on integral transforms are used to derive the long-term responses of hot fluid injection into a homogeneous transversely isotropic thermally poroelastic half space. The study developed a mathematical model for the distribution of land deformation, excess pore fluid pressure, and temperature changes of the half space aquifer. Analytic solutions are derived through the application of Hankel transform and Fourier transform with respect to the radial coordinate and axial coordinate, respectively. The results can provide better understanding of the hot fluid injection induced half space responses of the transversely isotropic porous strata.

1 INTRODUCTION

Responses of strata due to the hot fluid injection into a formation below the surface of the ground are an important petroleum engineering issues. Many studies were concentrated on understanding the mechanical, hydraulic, and thermal behavior of hot fluid injection for the impact on engineering safety. Hydraulic and thermal disturbance usually result in a volumetric change of fluid and solid skeleton. This change can increase excess pore fluid pressure and lead to decrease in effective stress, which can result in a hydraulic or thermal failure in the strata due to loss of shear resistance of solid skeleton. The simulation of these physical features is a complex task, and its validation is a major concern for the safety improvement of the hot fluid injection.

A mathematical model for the areal distribution of land subsidence, horizontal displacements, fluid pressure and temperature due to hot water injection into thermoelastic confined and leaky aquifers was developed by Bear and Corapcioglu (1981). In recovering heavy oil from heterogeneous reservoirs through fine-mesh numerical simulations, Alajmi *et al.* (2009) investigated the performance of hot water flooding compared to conventional water flooding. In the study of Sasaki *et al.* (2009), a system of gas production from methane hydrate layers involving hot water injection using dual horizontal wells was investigated. Rosenbrand *et al.* (2014) addressed permeability change in sandstone due to

heating from 20°C to 70~200°C. To simulate gas production from methane hydrate-bearing sand by hot-water cyclic injection, a three-dimensional middle-size reactor was used by Yang *et al.* (2010). Lin and Lu (2010) presented the golden ratio shown in the maximum ground surface horizontal displacement and corresponding vertical displacement of a half space subjected to a circular plane heat source on the basis of three-dimensional thermoelastic theory of homogeneous isotropic media. Transient ground surface displacements produced by a point heat source or sink through analog quantities between poroelasticity and thermoelasticity were displayed by Lu and Lin (2006).

The strata are deposited through a geologic process of sedimentation over a long period of time in general. Strata display significant anisotropy on mechanical, seepage and thermal properties under the accumulative overburden pressure. Both stratified soil and rock masses show the phenomenon of anisotropy. For this reason, theoretical and numerical models shall be able to simulate the layered soils and rocks as transversely isotropic media (Amadei *et al.* 1988, Lee & Yang 1998, Sheorey 1994, Tarn & Lu 1991, Wang & Tzeng 2009).

The present investigation is focused on the closed-form solutions of a transversely isotropic half space due to a point hot fluid injection which still have not been derived in previous studies. The soil or rock mass is modelled as linearly elastic medium with transversely isotropic properties in this paper. The hydraulic fluid flow, thermal conductivities and mechanical properties are assumed to be transversely isotropic. Hankel transform and Fourier transform are used with respect to the radial coordinate and axial coordinate, respectively. Closed-form solutions of the long-term displacements, excess pore fluid pressure, and temperature changes of the strata due to a point of hot fluid injection into a half space are obtained. The results can provide better understanding of the hot fluid injection induced responses of the half space.

2 MATHEMATICAL MODEL

2.1 Basic equations

Figure 1 displays a point hot fluid injection into a poroelastic half space. The transversely isotropic soil or rock is modeled as a homogeneous elastic half space. For simplicity, the plane of symmetry of the stratum is positioned in the horizontal direction. Let (r,θ,z) be a cylindrical coordinate system for this layer of stratum where the plane of isotropy coincides with the r-θ horizontal plane. The constitutive law for an elastic medium with linear axially symmetric deformation can thus be expressed by

$$\sigma_{rr} = A\varepsilon_{rr} + (A-2N)\varepsilon_{\theta\theta} + F\varepsilon_{zz} - \beta_r\vartheta - p, \qquad (1a)$$

$$\sigma_{\theta\theta} = (A-2N)\varepsilon_{rr} + A\varepsilon_{\theta\theta} + F\varepsilon_{zz} - \beta_r\vartheta - p, \qquad (1b)$$

$$\sigma_{zz} = F\varepsilon_{rr} + F\varepsilon_{\theta\theta} + C\varepsilon_{zz} - \beta_z\vartheta - p, \qquad (1c)$$

$$\sigma_{rz} = 2L\varepsilon_{rz}. \qquad (1d)$$

The symbol σ_{ij} is the total stress components of the half space aquifer; ϑ is the temperature changes of the stratum, and p is the excess pore fluid pressure of the half space. The material constants A, C, F, L, N of a transversely isotropic stratum are defined by Love (1944). In equations (1a) to (1c), β_r and β_z represent the thermal expansion factors of the half space aquifer along and normal to the symmetric plane, respectively. Here, the relations between strain ε_{ij} and displacement u_i of the stratum is expressed by the linear law:

$$\varepsilon_{rr} = \frac{\partial u_r}{\partial r}, \qquad (2a)$$

$$\varepsilon_{\theta\theta} = \frac{u_r}{r}, \qquad (2b)$$

$$\varepsilon_{zz} = \frac{\partial u_z}{\partial z}, \qquad (2c)$$

$$\varepsilon_{rz} = \frac{1}{2}\left(\frac{\partial u_r}{\partial z} + \frac{\partial u_z}{\partial r}\right), \qquad (2d)$$

where u_r and u_z are the displacements of the stratum in radial and vertical directions, respectively. For axially symmetric problem, it is noted that the shear stresses $\sigma_{r\theta}, \sigma_{\theta z}$, and circumferential displacement u_θ would vanish as the vertical z-axis is located through the hot fluid injection point. The expression for the thermal expansion factors β_r in radial direction and β_z in vertical direction are:

$$\beta_r = 2(A-N)\alpha_{sr} + F\alpha_{sz}, \qquad (3a)$$

$$\beta_z = 2F\alpha_{sr} + C\alpha_{sz}, \qquad (3b)$$

where α_{sr} and α_{sz} are the linear thermal expansion coefficients of the stratum in the horizontal and vertical directions, respectively.

The material constants A, C, F, L, N employed in equations (1a) to (1d) are related through the following equations:

$$A = \frac{E_r(1-\nu_{rz}\nu_{zr})}{(1+\nu_{r\theta})(1-\nu_{r\theta}-2\nu_{rz}\nu_{zr})}, \qquad (4a)$$

$$C = \frac{E_z(1-\nu_{r\theta})}{1-\nu_{r\theta}-2\nu_{rz}\nu_{zr}}, \qquad (4b)$$

$$F = \frac{E_z\nu_{rz}}{1-\nu_{r\theta}-2\nu_{rz}\nu_{zr}} = \frac{E_r\nu_{zr}}{1-\nu_{r\theta}-2\nu_{rz}\nu_{zr}}, \qquad (4c)$$

$$L = G_{rz}, \qquad (4d)$$

$$N = \frac{E_r}{2(1+\nu_{r\theta})}. \qquad (4e)$$

Here, the symbols E_r and E_z are defined as Young's moduli with respect to directions lying in and perpendicular to the plane of isotropy,

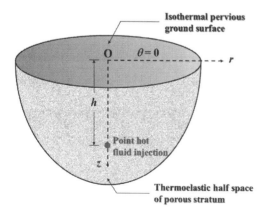

Figure 1. Point hot fluid injection into a poroelastic half space.

respectively; $v_{r\theta}$ is the Poisson's ratio for strain in the angular direction due to a horizontal direct stress; v_{rz} is the Poisson's ratio for strain in the vertical direction due to a horizontal direct stress; v_{zr} is the Poisson's ratio for strain in the horizontal direction due to a vertical direct stress; and G_{rz} is shear modulus for planes normal to the plane of isotropy.

For the cases of isotropic poroelastic medium, the material constants A, C, F, L, N can be denoted as

$$A = C = \lambda + G, \tag{5a}$$

$$F = \lambda, \tag{5b}$$

$$L = N = G, \tag{5c}$$

$$\beta_r = \beta_z = (2G + 3\lambda)\alpha_s. \tag{5d}$$

Here, λ and G are the Lame moduli of the homogeneous isotropic poroelastic half space, and α_s is the linear thermal expansion coefficient of the isotropic solid skeleton.

In general, these axially symmetric total stresses σ_{ij} must satisfy the equilibrium equations:

$$\frac{\partial \sigma_{rr}}{\partial r} + \frac{\sigma_{rr} - \sigma_{\theta\theta}}{r} + \frac{\partial \sigma_{rz}}{\partial z} + f_r = 0, \tag{6a}$$

$$\frac{\partial \sigma_{rz}}{\partial r} + \frac{\sigma_{rz}}{r} + \frac{\partial \sigma_{zz}}{\partial z} + f_z = 0, \tag{6b}$$

where f_r and f_z denote the body force components. For axisymmetric problems with effect of body forces neglected, the equilibrium equations can be expressed in terms of displacements, excess pore fluid pressure, and temperature change of the porous medium as follows:

$$A\left(\frac{\partial^2 u_r}{\partial r^2} + \frac{1}{r}\frac{\partial u_r}{\partial r} - \frac{u_r}{r^2}\right) + L\frac{\partial^2 u_r}{\partial z^2} + (F + L)\frac{\partial^2 u_z}{\partial r \partial z} \\ - \beta_r \frac{\partial \vartheta}{\partial r} - \frac{\partial p}{\partial r} = 0, \tag{7a}$$

$$(F + L)\left(\frac{\partial^2 u_r}{\partial r \partial z} + \frac{1}{r}\frac{\partial u_r}{\partial z}\right) + L\left(\frac{\partial^2 u_z}{\partial r^2} + \frac{1}{r}\frac{\partial u_z}{\partial r}\right) \\ + C\frac{\partial^2 u_z}{\partial z^2} - \beta_z\frac{\partial \vartheta}{\partial z} - \frac{\partial p}{\partial z} = 0. \tag{7b}$$

The hot fluid injection point is considered at a depth h of point $(0,h)$ as shown in Figure 1. Using the laws of mass balance and energy conservation, the continuity equation and heat conduction equation can be obtained as below:

$$\frac{k_r}{\gamma_f}\left(\frac{\partial^2 p}{\partial r^2} + \frac{1}{r}\frac{\partial p}{\partial r}\right) + \frac{k_z}{\gamma_f}\frac{\partial^2 p}{\partial z^2} + \frac{Q_f}{2\pi r} \\ \delta(r)\delta(z - h) = 0, \tag{8a}$$

$$\lambda_{tr}\left(\frac{\partial^2 \vartheta}{\partial r^2} + \frac{1}{r}\frac{\partial \vartheta}{\partial r}\right) + \lambda_{tz}\frac{\partial^2 \vartheta}{\partial z^2} + \frac{Q_t}{2\pi r} \\ \delta(r)\delta(z - h) = 0, \tag{8b}$$

in which γ_f is the unit weight of injected fluid, and the constants λ_{tr} and λ_{tz} denote the horizontal thermal conductivity of heat flow in the plane of isotropy and the corresponding vertical thermal conductivity in the plane perpendicular to the isotropic plane, respectively. The permeability of the half space aquifer in the horizontal and vertical directions are expressed as k_r and k_z, respectively. The symbols $\delta(r)$ and $\delta(z)$ are the Dirac delta functions. The injected hot fluid is considered as a constant thermal strength Q_t corresponding with fluid volume Q_f per unit time.

For a linearly thermoelastic medium with transversely isotropic properties, the differential equations expressed by equations (7a), (7b), (8a) and (8b) govern the steady state responses of the porous medium subjected to axisymmetric disturbance of a point hot fluid injection.

2.2 Boundary conditions

The half space ground surface at $z = 0$ is treated as a traction-free, pervious and isothermal boundary for all times $t \geq 0$. Therefore, its mathematical statements of the traction-free mechanical boundary conditions are:

$$\sigma_{rz}(r,0) = 0, \tag{9a}$$

$$\sigma_{zz}(r,0) = 0. \tag{9b}$$

The mathematical statements of the pervious isothermal condition at the boundary $z = 0$ are given as the following equations (10a) and (10b), respectively:

$$p(r,0) = 0, \tag{10a}$$

$$\vartheta(r,0) = 0. \tag{10b}$$

The point hot water injection source is assumed no effect at the remote boundary of $z \to \infty$ for all times $t \geq 0$. Hence

$$\lim_{z \to \infty} \begin{Bmatrix} u_r(r,z) \\ u_z(r,z) \\ p(r,z) \\ \vartheta(r,z) \end{Bmatrix} = \begin{Bmatrix} 0 \\ 0 \\ 0 \\ 0 \end{Bmatrix}. \tag{11}$$

The responses can be derived from the differential equations (7a), (7b), (8a) and (8b) corresponding with the half space boundary conditions (10a)-(10b) and remote boundary conditions (11).

147

3 CLOSED-FORM SOLUTIONS

The closed-form solutions of poroelastic deformation, excess pore fluid pressure, and temperature increment of the strata due to a point hot fluid injection into in a transversely isotropic poroelastic half space can be obtained by using Hankel transform and Fourier transform (Erdelyi *et al.* 1954, Gradshteyn & Ryzhik 1980, Sneddon 1951) with respect to the radial coordinate and axial coordinate, respectively, as below:

$$
u_r(r,z) = \frac{Q_f \gamma_f}{4\pi k_z} \left(a_{1f} \frac{r}{R_{1f}} + a_{2f} \frac{r}{R_{2f}} + a_{3f} \frac{r}{R_{3f}} \right.
$$

$$
+ a_{4f} \frac{r}{R_{4f}} + a_{5f} \frac{r}{R_{5f}} + a_{6f} \frac{r}{R_{6f}}
$$

$$
+ a_{7f} \frac{r}{R_{7f}} + a_{8f} \frac{r}{R_{8f}} + a_{9f} \frac{r}{R_{9f}}
$$

$$
\left. + a_{10f} \frac{r}{R_{10f}} \right)
$$

$$
+ \frac{Q_t}{4\pi\lambda_{tz}} \left(a_{1t} \frac{r}{R_{1t}} + a_{2t} \frac{r}{R_{2t}} + a_{3t} \frac{r}{R_{3t}} \right.
$$

$$
+ a_{4t} \frac{r}{R_{4t}} + a_{5t} \frac{r}{R_{5t}} + a_{6t} \frac{r}{R_{6t}}
$$

$$
+ a_{7t} \frac{r}{R_{7t}} + a_{8t} \frac{r}{R_{8t}} + a_{9t} \frac{r}{R_{9t}}
$$

$$
\left. + a_{10t} \frac{r}{R_{10t}} \right),
$$

(12a)

$$
u_z(r,z) = \frac{Q_f \gamma_f}{4\pi k_z} \left[b_{1f} \sinh^{-1} \frac{\mu_{1f}(z-h)}{r} \right.
$$

$$
+ b_{2f} \sinh^{-1} \frac{\mu_{2f}(z-h)}{r}
$$

$$
+ b_{3f} \sinh^{-1} \frac{\mu_{3f}(z-h)}{r}
$$

$$
+ b_{4f} \sinh^{-1} \frac{\mu_{1f}(z+h)}{r}
$$

$$
+ b_{5f} \sinh^{-1} \frac{\mu_{2f}(z+h)}{r}
$$

$$
+ b_{6f} \sinh^{-1} \frac{\mu_{3f}(z+h)}{r}
$$

$$
+ b_{7f} \sinh^{-1} \frac{\mu_{1f}z + \mu_{2f}h}{r}
$$

$$
+ b_{8f} \sinh^{-1} \frac{\mu_{1f}z + \mu_{3f}h}{r}
$$

$$
+ b_{9f} \sinh^{-1} \frac{\mu_{2f}z + \mu_{1f}h}{r}
$$

$$
\left. + b_{10f} \sinh^{-1} \frac{\mu_{2f}z + \mu_{3f}h}{r} \right]
$$

$$
+ \frac{Q_t}{4\pi\lambda_{tz}} \left[b_{1t} \sinh^{-1} \frac{\mu_{1t}(z-h)}{r} \right.
$$

$$
+ b_{2t} \sinh^{-1} \frac{\mu_{2t}(z-h)}{r}
$$

$$
+ b_{3t} \sinh^{-1} \frac{\mu_{3t}(z-h)}{r}
$$

(12b)

$$
+ b_{4t} \sinh^{-1} \frac{\mu_{1t}(z+h)}{r}
$$

$$
+ b_{5t} \sinh^{-1} \frac{\mu_{2t}(z+h)}{r}
$$

$$
+ b_{6t} \sinh^{-1} \frac{\mu_{3t}(z+h)}{r}
$$

$$
+ b_{7t} \sinh^{-1} \frac{\mu_{1t}z + \mu_{2t}h}{r}
$$

$$
+ b_{8t} \sinh^{-1} \frac{\mu_{1t}z + \mu_{3t}h}{r}
$$

$$
+ b_{9t} \sinh^{-1} \frac{\mu_{2t}z + \mu_{1t}h}{r}
$$

$$
\left. + b_{10t} \sinh^{-1} \frac{\mu_{2t}z + \mu_{3t}h}{r} \right],
$$

$$
p(r,z) = \frac{Q_f \gamma_f}{4\pi k_z \mu_{3f}} \left[\frac{1}{\sqrt{r^2 + \mu_{3f}^2 (z-h)^2}} \right.
$$

$$
\left. - \frac{1}{\sqrt{r^2 + \mu_{3f}^2 (z+h)^2}} \right],
$$

(12c)

$$
\vartheta(r,z) = \frac{Q_t}{4\pi\lambda_{tz}\mu_{3t}} \left[\frac{1}{\sqrt{r^2 + \mu_{3t}^2 (z-h)^2}} \right.
$$

$$
\left. - \frac{1}{\sqrt{r^2 + \mu_{3t}^2 (z+h)^2}} \right].
$$

(12d)

Here, the hydraulic constants a_{if} ($i = 1,\dots, 10$), b_{if} ($i = 1,\dots, 10$), thermal constants a_{it} ($i = 1,\dots, 10$) and b_{it} ($i = 1,\dots, 10$) are defined by

$$
a_{1f} = \frac{L + (F + L - C)\mu_{1f}^2}{CL\mu_{1f}\left(\mu_{1f}^2 - \mu_{2f}^2\right)\left(\mu_{1f}^2 - \mu_{3f}^2\right)},
$$

(13a)

$$
a_{2f} = \frac{L + (F + L - C)\mu_{2f}^2}{CL\mu_{2f}\left(\mu_{2f}^2 - \mu_{1f}^2\right)\left(\mu_{2f}^2 - \mu_{3f}^2\right)},
$$

(13b)

$$
a_{3f} = \frac{L + (F + L - C)\mu_{3f}^2}{CL\mu_{3f}\left(\mu_{3f}^2 - \mu_{1f}^2\right)\left(\mu_{3f}^2 - \mu_{2f}^2\right)},
$$

(13c)

$$
a_{4f} = \frac{\mu_{1f} + \mu_{2f}}{\mu_{1f} - \mu_{2f}} a_{1f},
$$

(13d)

$$
a_{5f} = \frac{\mu_{2f} + \mu_{1f}}{\mu_{2f} - \mu_{1f}} a_{2f},
$$

(13e)

$$
a_{6f} = -a_{3f},
$$

(13f)

$$a_{7f} = \frac{2\mu_{2f}}{\mu_{1f} - \mu_{2f}} \frac{m_{2f}}{m_{1f}} a_{2f}, \tag{13g}$$

$$a_{8f} = \frac{2}{\mu_{1f} - \mu_{2f}} \frac{\mu_{3f} - S_{3f}}{m_{1f}} a_{3f}, \tag{13h}$$

$$a_{9f} = \frac{2\mu_{1f}}{\mu_{2f} - \mu_{1f}} \frac{m_{1f}}{m_{2f}} a_{1f}, \tag{13i}$$

$$a_{10f} = \frac{2}{\mu_{2f} - \mu_{1f}} \frac{\mu_{3f} - S_{3f}}{m_{2f}} a_{3f}, \tag{13j}$$

$$a_{1t} = \frac{L\beta_r + \left[(F+L)\beta_z - C\beta_r\right]\mu_{1t}^2}{CL\mu_{1t}\left(\mu_{1t}^2 - \mu_{2t}^2\right)\left(\mu_{1t}^2 - \mu_{3t}^2\right)}, \tag{14a}$$

$$a_{2t} = \frac{L\beta_r + \left[(F+L)\beta_z - C\beta_r\right]\mu_{2t}^2}{CL\mu_{2t}\left(\mu_{2t}^2 - \mu_{1t}^2\right)\left(\mu_{2t}^2 - \mu_{3t}^2\right)}, \tag{14b}$$

$$a_{3t} = \frac{L\beta_r + \left[(F+L)\beta_z - C\beta_r\right]\mu_{3t}^2}{CL\mu_{3t}\left(\mu_{3t}^2 - \mu_{1t}^2\right)\left(\mu_{3t}^2 - \mu_{2t}^2\right)}, \tag{14c}$$

$$a_{4t} = \frac{L\left(\mu_{1t}a_{1t}^* - b_{1t}^*\right) - \mu_{2t}\left(Fa_{1t}^* + C\mu_{1t}b_{1t}^*\right)}{CL^2 m_{1t}\left(\mu_{1t} - \mu_{2t}\right)\left(\mu_{1t}^2 - \mu_{2t}^2\right)\left(\mu_{1t}^2 - \mu_{3t}^2\right)}, \tag{14d}$$

$$a_{5t} = \frac{L\left(\mu_{2t}a_{2t}^* - b_{2t}^*\right) - \mu_{1t}\left(Fa_{2t}^* + C\mu_{2t}b_{2t}^*\right)}{CL^2 m_{2t}\left(\mu_{2t} - \mu_{1t}\right)\left(\mu_{2t}^2 - \mu_{1t}^2\right)\left(\mu_{2t}^2 - \mu_{3t}^2\right)}, \tag{14e}$$

$$a_{6t} = \frac{\left[AC\alpha_{sr} - C(F+L)\alpha_{sz}\right]\mu_{3t}^2 - AL\alpha_{sr}}{\mu_{3t}\left\{CL\mu_{3t}^4 + \left[F(F+2L) - AC\right]\mu_{3t}^2 + AL\right\}}, \tag{14f}$$

$$a_{7t} = \frac{\mu_{2t}\left(Fa_{2t}^* + C\mu_{2t}b_{2t}^*\right) - L\left(\mu_{2t}a_{2t}^* - b_{2t}^*\right)}{CL^2 m_{1t}\left(\mu_{2t} - \mu_{1t}\right)\left(\mu_{2t}^2 - \mu_{1t}^2\right)\left(\mu_{2t}^2 - \mu_{3t}^2\right)}, \tag{14g}$$

$$a_{8t} = \frac{1}{Lm_{1t}S_{4t}\left(\mu_{2t} - \mu_{1t}\right)}\left[C\mu_{2t}S_{3t} + L\frac{S_{3t}}{\mu_{3t}} + \frac{C\mu_{2t}\mu_{3t}b_{3t}^* - L\left(2\mu_{3t}a_{3t}^* - b_{3t}^*\right)}{\mu_{3t}a_3}\right], \tag{14h}$$

$$a_{9t} = \frac{L\left(\mu_{1t}a_{1t}^* - b_{1t}^*\right) - \mu_{1t}\left(Fa_{1t}^* + C\mu_{1t}b_{1t}^*\right)}{CL^2 m_{2t}\left(\mu_{1t} - \mu_{2t}\right)\left(\mu_{1t}^2 - \mu_{2t}^2\right)\left(\mu_{1t}^2 - \mu_{3t}^2\right)}, \tag{14i}$$

$$a_{10t} = \frac{1}{Lm_{2t}S_{4t}\left(\mu_{1t} - \mu_{2t}\right)}\left[C\mu_{1t}S_{3t} + L\frac{S_{3t}}{\mu_{3t}} + \frac{C\mu_{1t}\mu_{3t}b_{3t}^* - L\left(2\mu_{3t}a_{3t}^* - b_{3t}^*\right)}{\mu_{3t}a_{3t}^*}\right], \tag{14j}$$

$$b_{1f} = \frac{L\mu_{1f}^2 + (F+L-A)}{CL\left(\mu_{1f}^2 - \mu_{2f}^2\right)\left(\mu_{1f}^2 - \mu_{3f}^2\right)}, \tag{15a}$$

$$b_{2f} = \frac{L\mu_{2f}^2 + (F+L-A)}{CL\left(\mu_{2f}^2 - \mu_{1f}^2\right)\left(\mu_{2f}^2 - \mu_{3f}^2\right)}, \tag{15b}$$

$$b_{3f} = \frac{L\mu_{3f}^2 + (F+L-A)}{CL\left(\mu_{3f}^2 - \mu_{1f}^2\right)\left(\mu_{3f}^2 - \mu_{2f}^2\right)}, \tag{15c}$$

$$b_{4f} = \frac{\mu_{1f} + \mu_{2f}}{\mu_{1f} - \mu_{2f}} S_{1f} a_{1f}, \tag{15d}$$

$$b_{5f} = \frac{\mu_{2f} + \mu_{1f}}{\mu_{2f} - \mu_{1f}} S_{2f} a_{2f}, \tag{15e}$$

$$b_{6f} = -b_{3f}, \tag{15f}$$

$$b_{7f} = \frac{2\mu_{2f}S_{1f}}{\mu_{1f} - \mu_{2f}} \frac{m_{2f}}{m_{1f}} a_{2f}, \tag{15g}$$

$$b_{8f} = \frac{2S_{1f}}{\mu_{1f} - \mu_{2f}} \frac{\mu_{3f} - S_{3f}}{m_{1f}} a_{3f}, \tag{15h}$$

$$b_{9f} = \frac{2\mu_{1f}S_{2f}}{\mu_{2f} - \mu_{1f}} \frac{m_{1f}}{m_{2f}} a_{1f}, \tag{15i}$$

$$b_{10f} = \frac{2S_{2f}}{\mu_{2f} - \mu_{1f}} \frac{\mu_{3f} - S_{3f}}{m_{2f}} a_{3f}, \tag{15j}$$

$$b_{1t} = \frac{L\beta_z\mu_{1t}^2 + (F+L)\beta_r - A\beta_z}{CL\left(\mu_{1t}^2 - \mu_{2t}^2\right)\left(\mu_{1t}^2 - \mu_{3t}^2\right)}, \tag{16a}$$

$$b_{2t} = \frac{L\beta_z\mu_{2t}^2 + (F+L)\beta_r - A\beta_z}{CL\left(\mu_{2t}^2 - \mu_{1t}^2\right)\left(\mu_{2t}^2 - \mu_{3t}^2\right)}, \tag{16b}$$

$$b_{3t} = \frac{L\beta_z\mu_{3t}^2 + (F+L)\beta_r - A\beta_z}{CL\left(\mu_{3t}^2 - \mu_{1t}^2\right)\left(\mu_{3t}^2 - \mu_{2t}^2\right)}, \tag{16c}$$

$$b_{4t} = S_{1t}a_{4t}, \tag{16d}$$

$$b_{5t} = S_{2t}a_{5t}, \tag{16e}$$

$$b_{6t} = S_{3t}a_{6t}, \tag{16f}$$

$$b_{7t} = S_{1t}a_{7t}, \tag{16g}$$

$$b_{8t} = S_{1t}a_{8t}, \tag{16h}$$

$$b_{9t} = S_{2t}a_{9t}, \tag{16i}$$

$$b_{10t} = S_{2t}a_{10t}. \tag{16j}$$

In addition, the characteristic roots $\mu_{1f} = \mu_{1t}$, $\mu_{2f} = \mu_{2t}$, and μ_{ij} $(i = 1, 2; j = f, t)$ must satisfy the following characteristic equation:

$$CL\mu_{ij}^4 - \left[AC - F(F+2L)\right]\mu_{ij}^2 + AL = 0. \tag{17}$$

The fluid and thermal characteristic roots μ_{3f} and μ_{3t} are defined by

$$\mu_{3f} = \sqrt{k_r/k_z}, \tag{18a}$$

$$\mu_{3t} = \sqrt{\lambda_{tr}/\lambda_{tz}}. \tag{18b}$$

149

In equation (12a), the distance symbols are shown below:

$$R_{1f} = \sqrt{r^2 + \mu_{1f}^2(z-h)^2} + \mu_{1f}|z-h|, \qquad (19a)$$

$$R_{2f} = \sqrt{r^2 + \mu_{2f}^2(z-h)^2} + \mu_{2f}|z-h|, \qquad (19b)$$

$$R_{3f} = \sqrt{r^2 + \mu_{3f}^2(z-h)^2} + \mu_{3f}|z-h|, \qquad (19c)$$

$$R_{4f} = \sqrt{r^2 + \mu_{1f}^2(z+h)^2} + \mu_{1f}(z+h), \qquad (19d)$$

$$R_{5f} = \sqrt{r^2 + \mu_{2f}^2(z+h)^2} + \mu_{2f}(z+h), \qquad (19e)$$

$$R_{6f} = \sqrt{r^2 + \mu_{3f}^2(z+h)^2} + \mu_{3f}(z+h), \qquad (19f)$$

$$R_{7f} = \sqrt{r^2 + (\mu_{1f}z + \mu_{2f}h)^2} + \mu_{1f}z + \mu_{2f}h, \qquad (19g)$$

$$R_{8f} = \sqrt{r^2 + (\mu_{1f}z + \mu_{3f}h)^2} + \mu_{1f}z + \mu_{3f}h, \qquad (19h)$$

$$R_{9f} = \sqrt{r^2 + (\mu_{2f}z + \mu_{1f}h)^2} + \mu_{2f}z + \mu_{1f}h, \qquad (19i)$$

$$R_{10f} = \sqrt{r^2 + (\mu_{2f}z + \mu_{3f}h)^2} + \mu_{2f}z + \mu_{3f}h, \qquad (19j)$$

$$R_{1t} = \sqrt{r^2 + \mu_{1t}^2(z-h)^2} + \mu_{1t}|z-h|, \qquad (20a)$$

$$R_{2t} = \sqrt{r^2 + \mu_{2t}^2(z-h)^2} + \mu_{2t}|z-h|, \qquad (20b)$$

$$R_{3t} = \sqrt{r^2 + \mu_{3t}^2(z-h)^2} + \mu_{3t}|z-h|, \qquad (20c)$$

$$R_{4t} = \sqrt{r^2 + \mu_{1t}^2(z+h)^2} + \mu_{1t}(z+h), \qquad (20d)$$

$$R_{5t} = \sqrt{r^2 + \mu_{2t}^2(z+h)^2} + \mu_{2t}(z+h), \qquad (20e)$$

$$R_{6t} = \sqrt{r^2 + \mu_{3t}^2(z+h)^2} + \mu_{3t}(z+h), \qquad (20f)$$

$$R_{7t} = \sqrt{r^2 + (\mu_{1t}z + \mu_{2t}h)^2} + \mu_{1t}z + \mu_{2t}h, \qquad (20g)$$

$$R_{8t} = \sqrt{r^2 + (\mu_{1t}z + \mu_{3t}h)^2} + \mu_{1t}z + \mu_{3t}h, \qquad (20h)$$

$$R_{9t} = \sqrt{r^2 + (\mu_{2t}z + \mu_{1t}h)^2} + \mu_{2t}z + \mu_{1t}h, \qquad (20i)$$

$$R_{10t} = \sqrt{r^2 + (\mu_{2t}z + \mu_{3t}h)^2} + \mu_{2t}z + \mu_{3t}h. \qquad (20j)$$

Besides, the symbols $a_{it}^*(i=1,2,3)$, $b_{it}^*(i=1,2,3)$, $m_{if}(i=1,2)$, $m_{it}(i=1,2)$, $S_{if}(i=1,\cdots,4)$ and $S_{it}(i=1,\cdots,4)$ in equations (13) to (16) are defined in equations (21) to (26) as below:

$$a_{1t}^* = \frac{1}{\mu_{1t}}\{L\beta_r + [(F+L)\beta_z - C\beta_r]\mu_{1t}^2\}, \qquad (21a)$$

$$a_{2t}^* = \frac{1}{\mu_{2t}}\{L\beta_r + [(F+L)\beta_z - C\beta_r]\mu_{2t}^2\}, \qquad (21b)$$

$$a_{3t}^* = \frac{1}{\mu_{3t}}\{L\beta_r + [(F+L)\beta_z - C\beta_r]\mu_{3t}^2\}, \qquad (21c)$$

$$b_{1t}^* = L\beta_z\mu_{1t}^2 + [(F+L)\beta_r - A\beta_z], \qquad (22a)$$

$$b_{2t}^* = L\beta_z\mu_{2t}^2 + [(F+L)\beta_r - A\beta_z], \qquad (22b)$$

$$b_{3t}^* = L\beta_z\mu_{3t}^2 + [(F+L)\beta_r - A\beta_z], \qquad (22c)$$

$$m_{1f} = \frac{C\mu_{1f}^2 + F}{C\mu_{1f}^2 - L}, \qquad (23a)$$

$$m_{2f} = \frac{C\mu_{2f}^2 + F}{C\mu_{2f}^2 - L}, \qquad (23b)$$

$$m_{1t} = \frac{C\mu_{1t}^2 + F}{C\mu_{1t}^2 - L}, \qquad (24a)$$

$$m_{2t} = \frac{C\mu_{2t}^2 + F}{C\mu_{2t}^2 - L}, \qquad (24b)$$

$$S_{1f} = \frac{L\mu_{1f}^2 - A}{(F+L)\mu_{1f}} = \frac{(F+L)\mu_{1f}}{L - C\mu_{1f}^2}, \qquad (25a)$$

$$S_{2f} = \frac{L\mu_{2f}^2 - A}{(F+L)\mu_{2f}} = \frac{(F+L)\mu_{2f}}{L - C\mu_{2f}^2}, \qquad (25b)$$

$$S_{3f} = \frac{L\mu_{3f}^2 + F + L - A}{L + (F+L-C)\mu_{3f}^2}, \qquad (25c)$$

$$S_{4f} = \frac{CL\mu_{3f}^4 + [F(F+2L) - AC]\mu_{3f}^2 + AL}{L + (F+L-C)\mu_{3f}^2}, \qquad (25d)$$

$$S_{1t} = \frac{L\mu_{1t}^2 - A}{(F+L)\mu_{1t}} = \frac{(F+L)\mu_{1t}}{L - C\mu_{1t}^2}, \qquad (26a)$$

$$S_{2t} = \frac{L\mu_{2t}^2 - A}{(F+L)\mu_{2t}} = \frac{(F+L)\mu_{2t}}{L - C\mu_{2t}^2}, \qquad (26b)$$

$$S_{3t} = \frac{\mu_{3t}\{L\beta_z\mu_{3t}^2 + [(F+L)\beta_r - A\beta_z]\}}{L\beta_r + [(F+L)\beta_z - C\beta_r]\mu_{3t}^2}, \qquad (26c)$$

$$S_{4t} = \frac{CL\mu_{3t}^4 + [F(F+2L) - AC]\mu_{3t}^2 + AL}{L\beta_r + [(F+L)\beta_z - C\beta_r]\mu_{3t}^2}. \qquad (26d)$$

The displacement expressions of the closed-form solutions (12a) and (12b) contains two terms, one for the constant thermal strength Q_t and another describing the fluid volume Q_f effect per unit time under the injection of hot fluid into a poroelastic half space. All of the derived field quantities are functions of the distance from the hot fluid injection source. Those quantities are inversely proportional to the hydraulic permeability or thermal conductivity. Besides, the mechanical moduli do not have influence on the long-term excess pore fluid and temperature increment of the strata.

150

4 CONCLUSIONS

Based on the theory of thermally poroelasticity, the long-term closed-form solutions of a homogeneous transversely isotropic elastic half space for axially symmetric deformations, excess pore fluid pressure, and temperature increment subjected to a point injection of hot fluid are presented by equations (12a) to (12d). The displacement expressions of the closed-form solutions contains two terms, one for the constant thermal strength Q_t and another describing the fluid volume Q_f effect per unit time under the injection of hot fluid into the poroelastic half space. The results can provide better understanding of the hot fluid injection induced responses of a transversely isotropic poroelastic half space.

ACKNOWLEDGEMENTS

This work is supported by the National Science Council of Republic of China through grants NSC102-2221-E-216-022 and NSC88-2218-E-216-004.

REFERENCES

Alajmi, A.F., Gharbi, R. & Algharaib, M. 2009. Investigating the performance of hot water injection in geostatistically generated permeable media. *Journal of Petroleum Science and Engineering* 66(3–4): 143–155.

Amadei, B., Swolfs, H.S. & Savage, W.Z. 1988. Gravity-induced stresses in stratified rock masses. *Rock Mechanics and Rock Engineering* 21(1): 1–20.

Bear, J. & Corapcioglu, M.Y. 1981. A mathematical model for consolidation in a thermoelastic aquifer due to hot water injection or pumping. *Water Resources Research* 17(3): 723–736.

Erdelyi, A., Magnus, W., Oberhettinger, F. & Tricomi, F.G. 1954. *Tables of Integral Transforms*, New York: McGraw-Hill.

Gradshteyn, I.S. & Ryzhik, I.M. 1980. *Table of Integrals, Series, and Products*, New York: Academic Press.

Lee, S.L. & Yang, J.H. 1998. Modeling of effective thermal conductivity for a nonhomogeneous anisotropic porous medium. *International Journal of Heat and Mass Transfer* 41(6–7): 931–937.

Love, A.E.H. 1944. *A Treatise on the Mathematical Theory of Elasticity*, New York: Dover Publications.

Lu, J. C.-C. & Lin, F.-T. 2006. The transient ground surface displacements due to a point sink/heat source in an elastic half-space, *Geotechnical Special Publication No. 148*: 210–218.

Lu, J. C.-C., Lin, W.-C. & Lin, F.-T. 2010. Closed-form solutions of the homogeneous isotropic elastic half space subjected to a circular plane heat source. *Geotechnical Special Publication No. 204*: 79–86.

Rosenbrand, E., Haugwitz, C., Jacobsen, P.S.M., Kjoller, C. & Fabricius, I.L. 2014. The effect of hot water injection on sandstone permeability. *Geothermics* 50: 155–166.

Sasaki, K., Ono, S., Sugai, Y., Ebinuma, T., Narita, H. & Yamaguchi, T. 2009. Gas production system from methane hydrate layers by hot water injection using dual horizontal wells. *Journal of Canadian Petroleum Technology* 48(10): 21–26.

Sheorey, P.R. 1994. A theory for in situ stresses in isotropic and transversely isotropic rock. *International Journal of Rock Mechanics and Mining Sciences and Geomechanics Abstracts* 31(1): 23–34.

Sneddon, I.N. 1951. *Fourier Transforms*, New York: McGraw-Hill.

Tarn, J.-Q. & Lu, C.-C. 1991. Analysis of subsidence due to a point sink in an anisotropic porous elastic half space. *International Journal for Numerical and Analytical Methods in Geomechanics* 15(8): 573–592.

Wang, C.D. & Tzeng, C.S. 2009. Displacements and stresses due to nonuniform circular loadings in an inhomogeneous cross-anisotropic material. *Mechanics Research Communications* 36: 921–932.

Yang, X., Sun, C.-Y., Yuan, Q., Ma, P.-C. & Chen, G.-J. 2010. Experimental study on gas production from methanehydrate-bearing sand by hot-water cyclic injection. *Energy Fuels* 24(11): 5912–5920.

NOTATION OF SYMBOLS

$a_{if}(i=1,\cdots,10)$	Hydraulic constants defined in equations (13a) to (13j) (Pa^{-1})
$a_{it}(i=1,\cdots,10)$	Thermal constants defined in equations (14a) to (14j) $(^\circ C^{-1})$
$a_{it}^*(i=1,2,3)$	Thermal constants defined in equations (21a) to (21c) $(Pa^2/^\circ C)$
A, C, F, L, N	Material constants of the transversely isotropic strata defined by Love (Pa)
$b_{if}(i=1,\cdots,10)$	Hydraulic constants defined in equations (15a) to (15j) (Pa^{-1})
$b_{it}(i=1,\cdots,10)$	Thermal constants defined in equations (16a) to (16j) $(^\circ C^{-1})$
$b_{it}^*(i=1,2,3)$	Thermal constants defined in equations (22a) to (22c) $(Pa^2/^\circ C)$
E_r, E_z	Young's modulus in horizontal/vertical direction (Pa)
$f_i(i=r,z)$	Body forces of the strata (N/m^3)
G	Shear modulus of the isotropic strata (Pa)
G_{rz}	Shear modulus for planes to the plane of isotropy of the normal isotropic strata (Pa)
k_r, k_z	Permeability of the strata in the horizontal/vertical direction (m/s)
$m_{if}(i=1,2)$	Hydraulic constants defined in equations (23a) and (23b) (Dimensionless)
$m_{it}(i=1,2)$	Thermal constants defined in equations (24a) and (24b) (Dimensionless)

p — Excess pore fluid pressure of the strata (*Pa*)

Q_f — Fluid volume of the injected hot fluid per unit time (*m³/s*)

Q_t — Thermal strength of the injected hot fluid (*J/s*)

(r,θ,z) — Cylindrical coordinates system (*m, radian, m*)

$R_{if}\,(i=1,\cdots,10)$ — Distance parameter defined in equation (19a) to (19 j) (*m*)

$R_{it}\,(i=1,\cdots,10)$ — Distance parameter defined in equation (20a) to (20 j) (*m*)

$S_{if}\,(i=1,\cdots,4)$ — Hydraulic constants defined in equations (25a) to (25d) (Dimensionless)

$S_{it}\,(i=1,\cdots,4)$ — Thermal constants defined in equations (26a) to (26d) (Dimensionless)

$u_i\,(i=r,z)$ — Displacement components of the strata (*m*)

α_s — Linear thermal expansion coefficient of the solid skeleton of the isotropic strata (*°C⁻¹*)

$\alpha_{sr},\,\alpha_{sz}$ — Linear thermal expansion coefficient of the cross-anisotropic strata in horizontal/vertical direction (*°C⁻¹*)

$\beta_r,\,\beta_z$ — Thermal expansion factors of the cross-anisotropic strata (*Pa/°C*)

γ_f — Unit weight of injected hot fluid (*N/m³*)

$\delta(x)$ — Dirac delta function (*m⁻¹*)

ϑ — Temperature changes of the strata (*°C*)

λ — Lame constant of the isotropic strata (*Pa*)

λ_t — Thermal conductivity of the isotropic thermoelastic strata (*J/sm°C*)

$\lambda_{tr},\,\lambda_{tz}$ — Thermal conductivity of the cross-anisotropic thermoelastic strata in the horizontal/vertical direction (*J/sm°C*)

$\mu_{3i}(i{=}f,t)$ — Characteristic root defined in equations (18a) and (18b) (Dimensionless)

$\mu_{ji}(i=1,2; j=f,t)$ — Characteristic roots of the characteristic equation (17) (Dimensionless)

ν — Poisson's ratio of the isotropic strata (Dimensionless)

ν_{rz} — Poisson's ratio for strain in the vertical direction due to a horizontal direct stress (Dimensionless)

$\nu_{r\theta}$ — Poisson's ratio for strain in the horizontal direction due to a horizontal direct stress (Dimensionless)

ν_{zr} — Poisson's ratio for strain in the horizontal direction due to a vertical direct stress (Dimensionless)

$\sigma_{ij}(i,j=r,\theta,z)$ — Total stress components of the strata (*Pa*).

Environmental science and environmental engineering

Advances in Energy, Environment and Materials Science – Wang & Zhou (Eds)
© 2017 Taylor & Francis Group, London, ISBN 978-1-138-03600-0

An analysis of the low-carbon consumption mode in Beijing

Ru Liu & Menghui Li
Beijing Municipal Institute of Science and Technology Information, Beijing, China

ABSTRACT: Low-carbon economy has gradually become the ideology and the mainstream value of the world. With its distinctive advantage and huge market, low-carbon economy has become the focus of the world's economic development. Developing low-carbon economy is an international revolution which is concerned with manufacturing mode, lifestyle, value, and national power and interests. To protect our earth, it is necessary to bring out the low-carbon manufacturing mode, consuming style, and lifestyle.

1 INTRODUCTION

Overall Urban Planning of Beijing (2004–2020), which is approved by the State Council in 2005, divides Beijing into the following 4 districts by laying different emphases on the development strategy: Core Districts of Capital Function, Extended Districts of Urban Function, New Districts of Urban Development, and Ecological Preservation Development Districts.

The economic output and the GDP per capita of Beijing are relatively low and it is a long and arduous task to develop low carbon economy; however, the rising Beijing economy has a broad development prospect. From 2006, Beijing has restructured more than 190 enterprises that caused serious pollution and shut down or suspended 23 cement production lines, 149 clay brick plants, and a batch of high energy consumption, high water consumption, and heavy-polluting industries such as Beijing Coking Plant, Beijing Organic Chemical Plant and Beijing Hua Er Co., Ltd, etc. It phased out nine industries, including small-scale casting, chemical, cement, chrome plating, printing and dyeing, paper-making and closed small coal mines in Men Tougou and Fang Shan successively. Beijing has bid farewell to the period of using small coal mines. Besides, it has shut down small thermal power generator units with a total capacity of 52,000 kilowatts, including Yan Hua Power Plant and Jing Gong Power plant, and relocated the Shougang Group.

After many years of efforts, Beijing effectively promoted its transformation to low carbonization economy, which lays a solid foundation for developing a low carbon economy and constructing a low carbon city. Beijing is the most capable of building the first low carbon city nationwide. When faced with China's target that the CO_2 emission levels per unit of GDP by 2020 will be reduced to 40%–45% when compared with that of 2005

levels, Beijing has taken this historic opportunity to play an exemplary role in tackling climate change and controlling carbon emissions.

2 THE ENERGY RESOURCE CONSUMPTION STRUCTURE OF BEIJING

2.1 *The general situation of energy resources in Beijing*

Firstly, the consumption level is high and increases quickly. The total amount of energy consumption of Beijing is just second to Shanghai. As the second largest city of high energy consumption, its total amount of energy consumption was about 58.5–65 million tons of standard coal in 2010. It is predicted that Beijing will consume 11,000 gigawatts of electricity, 11 billion cubic meters of gas, and 600 thousand of liquefied petroleum gas by 2020.

Secondly, the proportion of good-quality energy is enhanced, but the energy composition still needs to be optimized. Coal still plays the primary role and renewable resources take a small proportion.

Thirdly, energy-conservation work has gained certain effective results and still has great potential. The business and product structure of the industry can be further adjusted; motor vehicles consume 25% more petroleum gas per hundred kilometers; the average amount of energy consumption of central heating is twice or three times higher than the developed countries which have the similar climate.

2.2 *The situation of energy consumption*

The factors affecting energy consumption are the population, energy structure, and energy efficiency. To sum up, the weight of coal and its products in Beijing are reducing, while clean energy resources are increasing. Energy efficiency is enhancing, but it still has a long way to go to meet the requirement of energy-conservation and emission-reduction.

In Beijing, the private cars stock that relies heavily on fossil fuels is an indicator of car use. This has important impacts on energy consumption in Beijing. The developmental trend in China involves a continuous growth of the automobile industry. The government considers the automobile industry as one of the seven "pillar industries" of the economy. Private car use is increasing most rapidly in Beijing. Beijing has also undergone a noticeable climate change due to the increasing consumption of fossil fuels.

To implement a low-carbon economy, a developing strategy of our country, the "12th 5-year plan" explicitly sets energy-conservation and emission-reduction as a goal. Up to 2015, the emission of carbon dioxide has reduced to 17% per GDP and energy consumption has reduced to 16% per GDP; the proportion of non-fossil energy to primary energy resources has increased by 3.1% (from 8.3% to 11.4%); the total amount of the emission of major pollutants has reduced to 10%. In addition, the "12th 5-year plan" also specifies the major pollutants to be controlled. Besides, chemical oxygen demand and carbon dioxide which are specified in the "11th 5-year plan", two more categories of pollutants are added to the list of controlling objects, that is, ammonia, nitrogen, and oxynitride. The binding indicators proposed in the "12th 5-year plan" indicate our country's determination to conserve energy and reduce pollutant emission.

Beijing has performed a lot of efficient jobs as shown in the following aspects for conserving energy and reducing emission in Beijing: Beijing issued rules and regulations including "Beijing Energy-Conservation Supervision Measures", "Resolution of Developing Cycling Economy and Constructing a Resource-Saving City", and "Detailed Regulations of the Hygienic Manufacturing Auditing Interim Measures of Beijing"; Beijing also formulated a series of policies including "Opinions of the People's Government of Beijing in Implementing the Decisions of the State Council about Strengthening Energy-Conservation", the "Comprehensive Work Plan of Energy-Conservation and Emission-Reduction of Beijing", "Reform Opinions of Improving the Cities' Domestic Waste Operational Mechanism", and "Guideline for Industrial Energy Efficiency and Water Efficiency of Beijing"; the city established the "Action Plan of Developing Cycling Economy and Constructing a Resource-Conservation City" in a consecutive four years.

3 LOW-CARBON CONSUMPTION MODE IN BEIJING

3.1 *Developing low-carbon buildings with effort*

According to statistics, over 80% of the newly built houses in our country every year are high energy-consumption ones and over 95% of the existing buildings are high energy-consumption ones, whose carbon-emission amount is about 50% of that of the whole country. To change this situation, Beijing should learn from the zero-carbon hall of the Shanghai World Expo and strongly encourage low-carbon buildings; it should set up the energy-saving criteria for low-carbon emission buildings and carry out the rule of achieving energy-consumption; it should improve the energy efficiency of the newly built buildings and implement "energy-saving and carbon-reduction" in the constructing process of comprehensive residential areas, public buildings, and business buildings by taking energy-saving, water-saving, material-saving, land-saving, and environment protection into consideration.

The Real Estate Chamber of Commerce of All-China Federation of Industry and Commerce spread low-carbon techniques in the four systems (green, water-saving, energy-saving, and transportation) of the modeling projects in the low-carbon building areas. The techniques help in reducing 20.1 kilograms of carbon per square meter per year and saving the operating expenditure of about 20 yuan. Based on the above criteria, we could reduce 0.2 billion tons of carbon and save an operating expenditure of 100 billion yuan, if we assume that 10 billion square meters can be converted into green low-carbon residential areas by 2020. Business buildings occupy 25.4% of the total buildings in our country, but the energy they consume is 10–15 times as much as that of residential buildings do. If low-carbon business buildings amount to 5 billion square meters, carbon will reduce by at least one billion tons per year and save the operating expenditure of 500 billion yuan by 2020. It is also optimistic to expect the energy-consumption and carbon-emission amount of real estates in the process of manufacturing and construction. If we assume that one ton of carbon is emitted per square meter within the one billion square meters of the constructed buildings, and energy should be saved 20% per GDP, carbon should be reduced 0.2 billion tons per year, and an operating expenditure of 100 billion yuan will be saved per year.

With reducing carbon emissions in residential areas, public buildings, and business buildings, and manufacturing and constructing process as mentioned above, the amount of carbon emission will be reduced to 1.4 billion tons per year in the future and the expenses saved will amount to 700 billion yuan. 60% of emitted carbon in cities is from the building's maintenance function. To set up the green building technique system and develop low-carbon building is of vital importance and the key is the optimization of low-carbon control in the whole process of the building plan design, construction, use, implementation, maintenance, dismantling, and reuse. For example, at the phase of construction, the roof photovoltaic electricity generating technology can be utilized to combine natural light and lighting efficiently and air current and air speed can be adjusted and the air blower is driven to generate electricity by establishing the unpowered roof ventilating facility; solar energy, especially natural lighting, should be spread, and energy-saving heating and freezing systems should be selected; heat-preserving materials should be selected, appropriate decorations should be advocated, and new blank houses should be stopped. At the stage of use, a "green roof" can be made by planting grasses and flowers on the roof, which can not only lower down the temperature and save the electric power of the air-conditioners, but also absorb the pollutants in the air. At the stage of dismantling, secondary pollution can be prevented by recycling and reusing building wastes.

3.2 Construction of the green traffic system

The development of public traffic as a dominant means of transportation will be taken as a priority and the construction of track traffic should be strengthened for the purpose of forming a fast traffic system which is favorable for environment protection. The government should give financial support to the development of new energy autos to be firstly used as business cars, taxis, as well as public traffic. Financial subsidy should also be given to the purchasers of cars that use new energy. Automobiles, bicycles, and pedestrians will work together with harmony for the development of a multi-traffic system. Lanes for bikes should be well-maintained and using bikes is encouraged by perfecting both anti-theft facilities in parking places for bikes and bike-to-bus shift system by providing cheap bike-lending services near larger bus stops. The electric sensor net should be extended for the coverage of control of traffic lights, traffic emergency, flow rate of both one and two direction roads, and regular adjustment of time between red and green lights for relieving jam. The sensor net will help in making the fullest use of traffic information for reducing the rate of traffic idle load and unnecessary trips. The development of automobiles consuming blending fuel, electricity, diesel fuel, as well as hydrogen will help in reducing the stress caused by traffic upon the environment.

Target-setting is by no means a prerequisite for, or determinant of, successful EV deployment, but it is useful for understanding the level of ambition and support from national policymakers. Key areas were targeted to reduce GHG emission by the strengthening of institutional, legal, economic, and technological instruments and hence to achieve energy conservation, energy structure optimization, as well as ecological improvement. Supervision and monitoring will be enhanced on the spread of Electric Vehicles. Preferential policies for Electric Vehicles products will be formulated. Tax policies will be made favorable for the sale of Electric Vehicles. The government will also guide the public to purchase energy-saving and low carbon emission vehicles and disseminate the idea of conservation-orientation. Scientific research and technological development shall be promoted; constructions of talents and financial support are to be strengthened. Finally, public awareness will increase by using the promotion function of the government. Awareness will penetrate through all levels through a top-down procedure. Besides, publicity, public participation, education, and training will be reinforced by the government. International cooperation and communication shall be reinforced in order to promote public awareness on climate change issues as well as exchange experience of other countries on climate change publicity and spread of Electric Vehicles.

3.3 Boosting the low-carbon economy through low-carbon consumption

For the time-being, some hindrances exist against the life of low carbon consumption in China. First, the government officials need to take the leadership in energy conservation by showing how to solve their own problems of energy waste to the public. Second, people should set aright the wrong ideas of consumption that low consumption would cause low life quality and low consumption would result in low production. Some people even think that low consumption is just a problem of personal preference. Third, deficiency of low-carbon products has impeded low-carbon consumption, for the energy-saving products are the bulk of low-carbon consumption goods. The market for energy-saving products is undeveloped because of the lack of low-carbon consumption goods of fine quality.

4 CONCLUSIONS

To boost the development of low-carbon consumption, first, the government should set an example in implementing low-carbon consumption. On one hand, the government should form a new working style taking energy conservation as an important administrative job. On the other hand, the government should implement strict supervision on the use of public money and corruption of government fund must be eliminated. Only by doing so can the public believe that the government is credible in advocating low-carbon consumption and the policy will be effectively carried out. Second, the government should help the people form a habit of low-carbon consumption according to the requirements of public frugality. Third, the government should draft an "action plan for developing low-carbon society" as soon as possible, which will be feasible and specific in the aspects of the goal of low-carbon life style and the strategic industry planning for the output of low carbon consumption goods. The action plan will draw a new standard for low-carbon products according to SO14060 requirements, and at the same time implement strict supervision on markets of low-carbon products by examining the market admittance. Fourth, the government should intensify low-carbon education and propaganda to make people use low-carbon products.

Moreover, energy system is also the main driver to promote the development of low-carbon economy. At present, gasoline and diesel provide the most of the transport energy. EV can use electricity to provide power for transportation. Electricity has already provided power for some rail networks. In future, electricity will be used to charge batteries in fully electric vehicles. The demand for mitigating climate offered by future electricity-based EV systems would therefore be dependent on clean energy sources, such as renewable energy, nuclear energy, and coal energy with carbon capture and storage. Renewable energy is widely distributed in different parts of the China, which can form a mix grid. And EV might be recharged principally at night when the electricity demand is lower. Therefore, an EV system has less impact on the requirement for additional electrical capacity and EV can be used to absorb some surplus electricity from discontinuous forms of power generation. In the future, electricity based on the EV system could have important impacts on the China's electricity demand. The government shall also intervene in the energy system, so that electricity power generations are moving to lower carbon electricity generations.

Beijing has begun to prioritize the reality of climate change and attempted to modify their power framework towards a more environmentally friendly pattern. They will sooner or later compel the perfect low-carbon consumption patterns in their market and society.

REFERENCES

Chen, Z. Y. Cities of the world and low-carbon economy. www.lwxcw.com 07/07/2011.

Guo, J. (2010). Harmonizing cities' industrial structure adjustment and space optimization to achieve the goal of low-carbon: A study of Hangzhou. City Development Research, 17 (7).

Li Yang, The Development Path Options and Policy Proposals of Low-carbon Economy in China, Urban Studies, 2010, 17(12).

Lian, Y. M. Beijing should lead the construction of low-carbon city. Web issuing date: 07/26/2010. From China Business Times.

Liu, S. J. The key point of the current low-carbon economy and suggestions for policy. Web issuing date: 09/29/2010.

National Bureau of statistics of China, China Statistical Yearbook (2010), China Statistics Press.

Qiao, S., et al. (2010). Choices of low-carbon economic development routes and policy design studies in Jiangsu. Science Development, 9.

The route and policy of the development of low-carbon economy in China. www.southcn.com 12/20/2009.

Wang, H. T. (2010). The trend of total amount of energy consumption, structure and carbon emission in Beijing. City Development Research, 17 (9).

Wang, S. Yin, et al. Rational thoughts about low-carbon economy. www.chinacity.org.cn 03/14/2011.

Advances in Energy, Environment and Materials Science – Wang & Zhou (Eds)
© 2017 Taylor & Francis Group, London, ISBN 978-1-138-03600-0

Validation of scientificity of meteorological proverbs by statistical analysis

Fei Qian & Yuehong Zhang
South China Sea Standards and Metrology Center of S.O.A, Guangzhou, Guangdong, China

Su-an Xu
China Jiliang University, Hangzhou, Zhejiang, China

ABSTRACT: In this paper, topography, climatic conditions of Hangzhou, and the historical background of meteorological proverbs are described. By analyzing meteorological data during 1951–2013 in the Hangzhou area, the scientificity of meteorological proverbs are validated in this paper. The results show that the probability of occurrence of the "east wind in autumn, high temperatures" is 80% and therefore, this meteorological proverb has some credibility. And the probability of occurrence of "east wind in spring, rainy", "east wind in summer, dry", "east wind in winter, snowy" is 23.5%, 46.4%, and 21.9%, respectively. According to the above research findings, the reasons why weather forecast based on meteorological proverbs is different from the actual weather conditions are analyzed and the application and effect of meteorological proverbs in weather observation and forecast are summarized.

1 INTRODUCTION

With the development of science and technology, there are more and more means of weather prediction and forecast, such as satellite cloud picture, weather charts, and other advanced technology. Therefore, the speed and accuracy of weather forecast has been further improved. However, for those complex, terrain, and special geographical locations, even with the advanced technology, the weather forecast is less precise and less spatially accurate. If meteorologists are able to make the appropriate determinations according to the weather proverbs and use the advanced technology, the weather forecast would be perfect.

In response to this situation, scholars have carried out a lot of research and tried to solve accuracy issues of weather forecast. Some researchers use the monthly average temperature, the average temperature of ten days, monthly total precipitation, and other weather elements to analyze meteorological proverbs of the Haiyan area. The results show that local weather proverbs have great reference value in the weather forecast, but it requires a certain degree of flexibility when they are used (Ma & Chen 2007). Weather proverbs of the Hunan area used in meteorological observation and weather forecast are analyzed and summarized. The results show that seasonal factors should be considered. And weather proverbs should combine with real-time weather and local climate when they are used (Yang & Liu 2012). Other researchers study classified explanation of meteorological proverbs. By analyzing the application of weather proverbs in observation, the

results note that the quality of weather observations can be improved when vanguard technology is combined with proverbs (Deng 2004).

2 METEOROLOGICAL PROVERBS

The wisdom of farmers about the weather and farming is collected in many popular proverbs and passed on from generation to generation.

Weather proverbs are usually divided into the proverbs of clouds, the proverbs of wind, the proverbs of lightning, the proverbs of sleet, and so on. Such as "When cirrus appears on the sky, the weather will be windy and rain"—the reason for this phenomenon of the meteorological proverb is because the clouds of low pressure are moving from west to east. The wind speed is generally increased, and clouds are thickened. Therefore, the probability of rain is increased (Huang 2011). "When altocumulus castellanus appears on the sky, there'll be a thunder shower"—it is because there is instability caused by the warm air and cold air at high altitude when altocumulus castellanus appears. Although these weather proverbs are not necessarily correct, they can provide relatively credible forecast of the weather changes for people. Therefore, people can make appropriate preparations in advance (Liu & Hu 2011).

Also, in the meteorological proverbs, there exists the problem of applicability. The meteorological proverbs of Hangzhou do not apply to the Northeast. The latitude of human's main active area is 30° to 60° north (south). These zones don't sustain the same climate and weather variations are obvious. Although today is sunny and humid, tomorrow may be a cold day. Climate in the equatorial area does not have significant changes, such as the Sahara Desert which is hot and dry all year. And the weather of the West Pacific and the Indian subcontinent is rainy season. Therefore, weather proverbs in Chinese are mainly applied in the mid-latitude zone (Mu & Chen 2011, Yang 2011).

3 VALIDATE THE SCIENTIFICITY OF METEOROLOGICAL PROVERBS

As science and technology make rapid progress, the weather proverbs still have a certain reference value, but its science is yet to be verified. Based on the statistical analysis of meteorological data of Chinese ground stations, the scientificity of meteorological proverbs is validated and the results of the study provide a reference for weather forecast in this area.

3.1 Geography and climate of Hangzhou

Hangzhou is located in the southeast coast of China and the northern part of the Zhejiang Province. The terrain of Hangzhou is varied, including hills and mountains, and these occupy 66 percent of the total area. The mountainous region is mostly in the southwest and the elevations are generally below 500 meters. The Plain of Hangzhou occupies 26 percent of the total area and is located in the eastern region. Also, the plains are generally at an elevation of 3 to 10 meters. Eight percent of the total area in Hangzhou is made of rivers and lakes. Thus, the climatic resources, such as temperature, light, water, wind, and so on, are distributed unevenly. Therefore, the resources of the microclimate are abundant in the Hangzhou area.

Hangzhou has subtropical monsoon climate, with four distinct seasons: the full sunshine, with plenty of rainfall and the average temperature around 16°C. The annual average temperature of Hangzhou is from 15.9°C to 17.0°C with high temperatures in the south and lower temperatures in the north. Extreme maximum temperature is from 39.8°C to 42.9°C. Extreme minimum temperature is from −7.1°C to −15.0°C. The annual average relative humidity is from 76% to 81%. With controlled winter and summer winds alternately, the atmospheric circulation background, weather systems, and climatic conditions of Hangzhou will be changed seasonally and therefore, it exhibits the climatic conditions of rainy spring, hot summer, dry autumn, and cold winter.

3.2 "East wind in spring, rainy"

The scientificity of the meteorological proverb is verified by using the monthly meteorological data of

Hangzhou during 1951–2013. The three meteorological factors such as wind direction of maximum speed, wind direction of extreme speed, and anomaly percentage of precipitation are analyzed and the wind codes of 09–14 are classified as east wind. The result shows that the probability of "east wind in spring, rainy" occurring is 23.5 percent, as shown in Table 1.

In the case of east wind in spring, the probability of heavy rainfall is low and actual weather conditions are greatly different from this meteorological proverb. But this meteorological proverb still has a certain reference value.

3.3 "East wind in summer, dry"

The two meteorological elements of wind direction of maximum speed and anomaly percentage of precipitation are analyzed and the wind codes of 09–14 are classified as east wind. The result shows that the probability of "east wind in summer, dry" occurring is 46.4 percent, as shown in Table 2. The analytical process only considers two factors while ignoring other factors that may affect rainfall. Therefore, it follows that the probability of occurrence of this proverb is not very high.

3.4 "East wind in autumn, high temperatures"

The two weather elements of wind direction of maximum speed and temperature are analyzed and the wind codes of 09–14 are also classified as east wind. The result shows that the probability of "east wind in autumn, high temperature" occurring is 80 percent, as shown in Table 3. Therefore, this meteorological proverb has characteristics of

Table 1. Rainfall in spring.

	Frequency	Percent	Effective percent	Cumulative percent
Light rainfall	13	76.5	76.5	76.5
Heavy rainfall	4	23.5	23.5	100.0
Total	17	100.0	100.0	

Table 2. Rainfall in summer.

	Frequency	Percent	Effective percent	Cumulative percent
Light rainfall	15	53.6	53.6	53.6
Heavy rainfall	13	46.4	46.4	100.0
Total	28	100.0	100.0	

160

scientificity and reliability and it can help to improve the accuracy of weather observation and prediction.

3.5 *"East wind in winter, snowy"*

As shown in Table 4 and obtained by statistical analysis, the probability of "east wind in winter, snowy" occurring is 21.9 percent. Also the wind codes of 09–14 are classified as east wind. Rain and snow are just two different forms and therefore, these two factors are included in the analysis of snowfall.

The result shows that the probability is low and forecast obtained by using the proverb is different from actual weather conditions. It means that this meteorological proverb may be not scientific. Also, the factors considered in this analysis may be incomplete.

3.6 *Comparative analysis of the probability of occurrence of weather proverbs*

Through the above statistical analysis of weather data of the four seasons for Hangzhou, we finally obtained the probability of occurrence of meteorological proverbs, as shown in Table 5. From the table, we can find the probability of occurrence of

Table 3. Temperature in autumn.

	Frequency	Percent	Effective percent	Cumulative percent
Low temperature	1	20.0	20.0	20.0
High temperature	4	80.0	80.0	100.0
Total	5	100.0	100.0	

Table 4. Snowfall in winter.

	Frequency	Percent	Effective percent	Cumulative percent
Snowless	25	78.1	78.1	78.1
Snow	7	21.9	21.9	100.0
Total	32	100.0	100.0	

Table 5. The probability of occurrence of meteorological proverbs.

Meteorological proverbs	Probability of occurrence
"East wind in spring, rainy"	23.5%
"East wind in summer, dry"	53.6%
"East wind in autumn, high temperatures"	80%
"East wind in winter, snowy"	21.9%

"east wind in spring, rainy", "east wind in summer, dry", "east wind in autumn, high temperatures", and "east wind in winter, snowy" is 23.5%, 46.4%, 80%, and 21.9%, respectively.

The probability of occurrence of "east wind in summer, dry" and "east wind in autumn, high temperatures" is high. Weather forecast, which is according to the two weather proverbs, is relatively consistent with the actual situation and therefore, these two meteorological proverbs are comparatively dependable. Weather proverbs are the summary of people's life and production experience, but proverbs which are observed continuously and checked repeatedly can also provide useful information and clues and therefore, proverbs have a certain degree of scientificity. But when the weather proverbs are used to forecast long-term weather, weather changes will be the objective limitations.

When most weather proverbs were generated, science was not developed. Weather proverbs are just used to describe the phenomenon and they are not made to provide scientific explanation. Therefore, weather proverbs do not have theoretical support and they lack a scientific spirit. Therefore, the probability of occurrence of "east wind in spring, rainy" and "east wind in winter, snowy" is low. According to these two weather proverbs, weather forecast is different from the actual situation.

Since data were acquired due to the severe constraints, the reliability of above analysis results may be greatly reduced. Therefore, by improving the methods of observation of meteorological elements, improving the precision of observation instruments, strengthening skills of observation personnel, and so on, the reliable meteorological data can be obtained. Thus, meteorological proverbs can be analyzed more scientifically.

4 CONCLUSIONS

When weather data are used to verify meteorological proverbs, we need flexibility in the use of meteorological data. For example, statistics of rainfall should include not only precipitation anomaly percentage, but also other factors, such as soaking rain, and so on. Only in this way, the weather proverbs can be explained and validated more reasonably and they can provide a more reliable basis for weather forecast. Through the above validation, it suggests that these meteorological proverbs have a certain reference value and they can provide some clues for the Hangzhou weather forecast.

Meteorological proverbs have strong regional applicability. When regional weather proverbs are applied to other regions, they are not necessarily fit. Therefore, if meteorological proverbs are used to predict the weather, they cannot be analyzed in isolation. Meteorological proverbs should be combined with the local weather and climate background and then, accuracy of forecast can be guaranteed. Thereby, weather proverbs can play their due role in disaster prevention and reduction.

REFERENCES

Deng, X. 2004. The application of meteorological proverbs in observation. Journal of Sichuan Meteorology (87): 57–58.

Han, Y. & Pu, X. 2010. China's meteorological services and its effectiveness evaluation. Scientia Meteorologica Sinica 30(3): 420–426.

Huang, X.Y. 2011. Meteorological knowledges from common weather proverbs. Modern Agricultural Science and Technology (21): 37–38.

Huang, Y. 2012. Design and implementation of meteorological data quality control system. Chengdu: University of Electronic Science and Technology of China.

Liu, L.C. & Hu, R. 2011. A general description of the study of meteorological culture. Advances in Meteorological Science and Technology 1(2): 46–50.

Luo, D. 2009. Research on Hunan meteorological disaster warning management system of the city. Changsha: National University of Defense Technology.

Ma, X.H. & Chen, J. 2007. Weather proverbs verification for prediction in Haiyan area. Journal of Qinghai Meteorology (2): 26–27.

Mu, M. & Chen, B.Y. 2011. Methods and uncertainties of meteorological forecast. Meteorological Monthly 37(1): 1–13.

Shen, J. 2011. Research on data process based on automatic weather station. Changsha: Central South University.

Yang, F.Q. & Liu, Y.H. 2012. Generation of weather proverbs in Hunan and Its application in meteorological operation. Office Operations.

Yang, Y.B. 2011. A study on meteorological disaster warning reformation of Shanghai (2000–2010). Shanghai: East China Normal University.

Yuan, L. 2009. Study of meteorological disasters emergency management. Tianjin: Tianjin University.

Advances in Energy, Environment and Materials Science – Wang & Zhou (Eds)
© 2017 Taylor & Francis Group, London, ISBN 978-1-138-03600-0

An analysis of an asphalt layer's shear stress distribution under a large longitudinal slope

Min Luo
China Aiport Construction Group Corporation of CAAC, Beijing, China
Beijing Super-Creative Technology Co. Ltd., Beijing, China

Liang Cheng
Wuqi Branch of Yan'an Road Administration Bureau of Shaanxi Province, Yan'an, China

ABSTRACT: An asphalt layer's shear stress distributions under the standard design state (state 1) and the heavy-load and high-temperature state (state 2) with different levels of horizontal load coefficients were studied by using Bisar3.0; and then, the asphalt layer shear stress distribution's changing rule with different longitudinal slope gradients under heavy-load and high-temperature state conditions was analyzed. The results show that the level of horizontal load has a significant impact on the asphalt layer shear stress, which must be considered in the check process of shear strength. Its sphere of influence increases with an increase in the horizontal load coefficient, with the main effect sphere being in the range of 0~12 cm. On the contrary, the effect of the longitudinal gradient on the asphalt layer's shear stress is very limited and its sphere of influence increases with an increase in the slope, with the main effect sphere being in the range of 0~3cm. Thus the effect of the longitudinal gradient can be neglected during the check calculation of shear strength. It can be seen from the calculating results that, when considering horizontal loads, the asphalt layer's stress state is more serious under heavy loads and high temperature condition thans that in the standard design state and are more subjected to rutting formation due to shear damage in the pavement.

1 INTRODUCTION

Rutting is one of the main damage types. Under the modern traffic, the speed and universality of rutting occurrence are greater than expected. A large number of literature surveys indicate that the rutting disease occurred more often in large longitudinal slope road sections than that in normal sections. Therefore, the asphalt layer's stress is calculated under different longitudinal slopes by using bisar 3.0 in this article, and the shear stress distribution rule of a large longitudinal slope road asphalt layer is found, to provide a theoretical basis for the establishment of rutting prevention measures.

2 PAVEMENT STRUCTURE MODEL AND PARAMETER SELECTION

The road structure, design parameters, the working conditions, and loading mode used in this research work are shown Table 1.

A 65°C modulus was calculated by using the following equation in the book of strength variation rule and maintenance of asphalt pavement:

Table 1. Pavement structure and calculation parameters.

Structure	Thickness (cm)	Elasticity modulus (Mpa) 20°C	65°C	Poisson's ratio
Top asphalt layer	5	1200	279	0.35
Middle asphalt layer	6	1000	232	0.35
Bottom asphalt layer	7	1000	232	0.35
Base	18	1400	1400	0.25
Subbase	36	1200	1200	0.25
Subgrade	–	35	35	0.4

$$\frac{E_t}{E_c} = 1.0 - 1.5 \lg \frac{T}{C}$$

where Et—T is the temperature modulus; Ec is the modulus under the standard temperature C, with the value of C generally being 20°C.

The load scheme uses the experience relationship between the contacting area and the axle load in Belgian, (A = 0.008P + 152 where A is the contacting area, cm^2 and P is the tire load, N).

Table 2. Loading mode.

Ground pressure (Mpa)	Tire pressure (KN)	Radius (cm)	Grounding area (cm²)	Temperature (°C)	Note
0.7	25	10.65	356	20	The standard design state
1.0	76	15.55	760	65	Heavy load and high temperature state

This scheme wheel-pressure and the contacting area increase with an increase in the axle load in this scheme. Keeping the distance of circle center as 31.95 cm stable is the most in line with the actual situation for the fact that the heavier transport vehicles are more statistically likely to have a tendency to adopt high strength high pressure tires.

3 DIFFERENT HORIZONTAL LOAD COEFFICIENTS OF ASPHALT LAYER'S SHEAR STRESS DISTRIBUTION

When the horizontal load coefficient f is 0, 0.1, 0.2, 0.3, 0.4, and 0.5 respectively, the asphalt layer shear stress distribution under both the standard design state and the heavy load and high temperature state is calculated. The calculation results are shown in Figures 1–3.

With an increase in the horizontal load coefficient, the asphalt layer's shear stress increases sharply. When coefficient f increases from 0 to 0.5, the maximum shear stress under the standard design state increases from 0.2276 Mpa to 0.8947 Mpa, with the growth rate of 293%, while maximum shear stress under heavy load and high temperature increases from 0.3226 Mpa to 1.318 Mpa, with an increase rate of 309%.

Maximum shear stress appears at the edge of the wheel load and the outer edge is greater than the inner edge.

Overloading of the influence sphere of the horizontal force under heavy load and high temperature is 0~12 cm while under the influence range of horizontal force under standard design state is 0~6 cm. The influence of horizontal force increases greatly under heavy load and high temperature state.

When the horizontal load coefficient is less than 0.2, the maximum shear stress appears in the road under 1 cm while when the horizontal load coefficient is greater than or equal to 0.2, the maximum shear stress appears on the road surface.

When the horizontal load coefficient is less than 0.2, the maximum shear stress increases slowly while when the horizontal load coefficient is greater than or equal to 0.2, the maximum shear

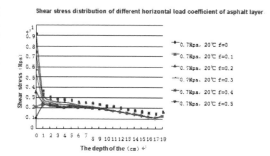

Figure 1. Graph showing asphalt layer's shear stress calculation under standard design conditions.

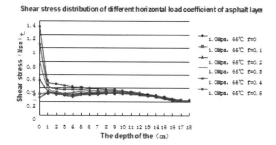

Figure 2. Graph showing asphalt layer's shear stress calculation under heavy load and high temperature state conditions.

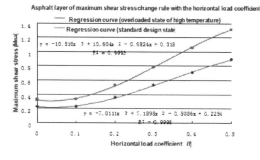

Figure 3. Graph showing the rule of asphalt layer's maximum shear stress with the horizontal load coefficient.

stress increases sharply, and the increase rate under heavy load and high temperature is greater than that in the standard state.

Table 3. The asphalt layer's shear stress calculation results under different longitudinal slope degrees.

Depth (cm)	Longitudinal slope degree (%)								
	0	1	2	3	4	5	6	7	8
0	1.318	1.344	1.37	1.396	1.422	1.448	1.473	1.499	1.524
1	0.5152	0.5218	0.5283	0.5349	0.5415	0.5481	0.5547	0.5612	0.5678
2	0.4605	0.4677	0.4749	0.482	0.4892	0.4963	0.5034	0.5104	0.5174
3	0.4389	0.4447	0.4505	0.4563	0.4624	0.4686	0.4747	0.4808	0.4869
4	0.4235	0.4281	0.4332	0.4382	0.4433	0.4483	0.4534	0.4584	0.4634
5	0.4131	0.4171	0.4211	0.4252	0.4292	0.4332	0.4372	0.4412	0.4452
6	0.3972	0.4006	0.404	0.4073	0.4107	0.414	0.4174	0.4207	0.424
7	0.3898	0.3926	0.3953	0.398	0.4008	0.4035	0.4062	0.4089	0.4116
8	0.382	0.3842	0.3865	0.3887	0.391	0.3932	0.3954	0.3976	0.3998
9	0.373	0.3748	0.3767	0.3785	0.3804	0.3822	0.384	0.3858	0.3876
10	0.3622	0.3638	0.3653	0.3669	0.3684	0.3699	0.3714	0.3729	0.3744
11	0.3494	0.3508	0.3521	0.3534	0.3547	0.356	0.3572	0.3585	0.3597
12	0.3344	0.3355	0.3367	0.3378	0.3389	0.34	0.3411	0.3422	0.3432
13	0.3168	0.3179	0.3189	0.3199	0.3208	0.3218	0.3228	0.3237	0.3246
14	0.2967	0.2976	0.2985	0.2994	0.3003	0.3011	0.302	0.3028	0.3036
15	0.2736	0.2745	0.2753	0.2761	0.277	0.2777	0.2785	0.2793	0.2800
16	0.2475	0.2483	0.2491	0.2499	0.2507	0.2514	0.2522	0.2529	0.2536
17	0.2262	0.2267	0.2271	0.2274	0.2278	0.2282	0.2285	0.2289	0.2292
18	0.2258	0.2263	0.2268	0.2272	0.2277	0.2282	0.2286	0.2291	0.2295

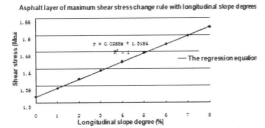

Figure 4. The distribution of the asphalt layer's maximum shear stress with longitudinal slope degrees.

4 THE ASPHALT LAYER'S SHEAR STRESS DISTRIBUTION UNDER DIFFERENT LONGITUDINAL SLOPE DEGREES

When the longitudinal slope degree is 0%, 1%, 2%, 3%, 4%, 5%, 6%, 7%, and 8%, respectively, the asphalt pavement structure layer's shear stress was calculated under heavy load and high temperature conditions with a horizontal load coefficient of 0.5 and the results are shown in Table 3.

From the calculation results, we can see that

1. The asphalt layer's shear stress increases with an increase of longitudinal slope degrees.
2. When the longitudinal slope degree increases from 0% to 8%, the asphalt layer's maximum shear stress increases from 1.318 Mpa to 1.524 Mpa, with an increase rate of 15.6%, and the increase in the amplitude is small.

3. It can be seen from Figure 4 that the asphalt layer's maximum shear stress increases with an increase in the longitudinal slope degree linearly.
4. The influence of the longitudinal slope's degree on shear stress is under 0~3 cm of the surface.

5 CONCLUSIONS

1. The horizontal load coefficient has a significant influence on the shear stress of the asphalt layer. This must be considered when designing a large longitudinal slope road pavement structure.
2. Maximum shear stress appears at the edge of the wheel load and the outer edge is greater than that of the inner edge.
3. The horizontal load coefficient has an influence range of 0~12 cm under heavy load and high temperature state conditions; therefore, it should be checked whether the asphalt mixture satisfies the shear requirements when designing a pavement structure
4. The maximum shear stress of the asphalt layer increases linearly with an increase in the longitudinal slope degree and the increase in the amplitude is small.
5. The longitudinal slope degree has an influence range of 0~3 cm under heavy load and high temperature state conditions; therefore, this effect can be ignored in the pavement design.
6. In a large slope road, the reason leading to rutting seriously is not slope, but slower driving speed and load for a long time.

REFERENCES

Li-jun sun. Asphalt pavement structural behavior theory [M]. Beijing: people's traffic press, 2005.

Sha Qing-lin. Highway asphalt pavement early damage phenomenon and prevention [M]. Beijing: China Communications Press, 2001.

Shen Ai-qin, Zhuang Chuan-yi. Mountain highway asphalt pavement rutting causes and prevention measures [J]. Road Machinery & Construction Mechanization, 2007, Sixth: 1–4.

Shen Jin-an. Asphalt and road performance of asphalt mixture [M]. Beijing: China Communications Press, 2001.

Sun Li-jun. Asphalt pavement structural behavior theory [M]. Beijing: China Communications Press, 2005.

Zhang Deng-liang. Asphalt and asphalt mixture [M]. Beijing: China Communications Press, 1993.

Zhou liang; Jian-ming ling; Xiao-ping Lin; Greeny; Temperature correction coefficient of asphalt layer modulus back calculation [J]; Journal of tongji university (natural science edition); 2011, 11.

Research status and prospect of rock wool in China

Pengqi Wang, Liang He, Danjun Tan & Ying Wang
Beijing New Building Materials Public Limited Company, Beijing, China

ABSTRACT: With the improvement of buildings' energy conservation and heat preservation, various types of insulation materials have been developed rapidly in our country, which are mainly divided into four categories: inorganic, organic, metal, and composite insulation. In this paper, the research of the rock wool in composite materials, integration decorative board, the recycling, and life cycle evaluation are overviewed and the forecast to the rock wool research direction in the future is carried out. Function, efficient insulation, decorative integration, reutilization, and selection of insulation material by the whole life cycle assessment will be the research focus.

1 INTRODUCTION

As China's buildings' energy consumption level of urbanization continues to increase, buildings' energy consumption accounted for about 30% of the total energy consumption, of which approximately half the energy loss occurs in between the wall and the external atmosphere. The various types of insulation materials in China have been rapidly developed with an increase in the buildings' energy-saving insulation requirements, which including inorganic, organic, metal, and composite insulation. The inorganic insulation materials include fibrous, porous, glass, and other materials; organic insulation materials include polyurethane foam, polystyrene board, phenolic foam board, and other materials; metal materials include aluminum foil insulation waveform paper insulation panels, reflective insulation web, and other reflective materials; composite insulation materials are the combination of above four kinds of materials.

2 COMPOSITE OF ROCK WOOL

The polyurethane/rock wool composite insulation board proposed by Guozhong Lu (2015) is prepared by using a high temperature foam mold by spraying polyurethane resin on the rock wool board's surface. Its pull-out strength is greater than 7.5 KPa and other properties meet the requirements of JG149-2003 (Table 1). Qiang Ren (2007 & 2008) studied the influence of rock wool fibers for the wood fiber/mineral wool fiber composite material, the research results show that the flame retardant properties of the composites is improved, but will reduce its MOR

and within bonding strength, there may even be very low intensity defective area. Xiaodong Wang (2006) studied the insulation properties of composite materials, which were made of high silica glass, aluminum, cotton, basalt, mullite, and aerogel by using a wet process (Table 2). The cold temperature of rock wool/ aerogel is lower than that of other composite materials, while the hot face is between 400°C and 800°C, even with a low 30°C, when compared with mullite/aerogel at 800°C. Hong Wang (2009) prepared composite

Table 1. List of properties of rock wool/polyurethane.

Index name	Detection value
Density (kg/m^3)	≥100
Conductivity [W/(m.K)] 25°C	Polyurethane ≤0.024 Rook wool ≤0.040
Compressive strength (kPa)	≥60
Tensile strength (kPa)	≥7.5
Water absorption (kg/m^2) (24 h)	≤1.0
Dimensional stability (%) (70°C ± 2°C, 48 h)	≤1.5
Combustion performance	B1 (Polyurethane)

Table 2. Thermal conductivity of rock wool/silicon dioxide composite material.

Temperature of hot surface	Thermal conductivity
25°C	0.0184 W/m • K
200°C	0.017 W/m • K
300°C	0.019 W/m • K
400°C	0.020 W/m • K
500°C	0.019 W/m • K
650°C	0.022 W/m • K

materials by using glass fiber, rock wool, aluminum silicate fiber, and aerogel which had good insulation properties that the thermal conductivity coefficient were lower than 0.025 W/(m • K).

At present, inorganic insulation materials have a relatively high thermal conductivity while the combustion performance of organic insulation materials cannot reach the A-level and therefore, the situation will lead to the rapid development of an efficient composite insulation material, such as rock wool/aerogel composite material.

3 DECORATIVE INTEGRATION BOARD

Guangqing Liu (2012) prepared a metal/rock wool board in which the upper and lower surface materials were color steel and the core material was rock wool. The bond strength of the composite insulation board is greater than 0.06 MPa and fire resistance is more than 1 hour under the thickness of 80 mm. Pingping Wang (2014) studied the insulation and sound performances of the metal/rock wool sandwich board, which the 67 mm metal/rock wool sandwich panels can achieve the insulation properties of 250 mm aerated concrete block; and sound insulation was greater than 30 dB when the thickness was 80 mm.

Zhengang Geng (2015) carried out a study on materials selection, engineering design, and construction elements of the insulation decorative integration board, which is made up mainly by rock wool and calcium silicate boards, and also proposed the performance of the G-type rock wool insulation decorative plate and rock wool (Tables 3 and 4), calcium silicate board, and decorative coatings. The engineering example analysis showed that the decorative integration system had improved water resistance, corrosion resistance, and thermal stability. Jintai He (2015) had studied the rock wool insulation decorative board, which consisted of rock wool and a calcium silicate board. The main production processes included producing decorative panels and molding of rock wool and a calcium silicate board, and the wall construction process including primary treatment, decorative plates, and fixed installation.

Development of mineral wool decorative board integration will simplify the external insulation process of construction, save energy, and improve its application. In addition, with the advancement of industrialization, building, and construction parts, prefabricated walls are rapidly advancing, the development of antibacterial, antifungal, anti-electromagnetic radiation, temperature, and humidity control, and other functions of composite materials as interior fillers will be the focus of future research directions.

Table 3. The performance of a G-type decorative board.

Item		Index
Appearance		Uniform color, no damage
Surface density		≤20 kg/m^2
Tensile bond strength between skin panel and insulation board	Primary strength	≥7.5 kPa, failure surface in the insulation board
		Water resistance
		Freezing thaw
Combustion		No less than level A
Thermal resistance		Meet the design requirement

Table 4. The performance of an external insulation system with G-type decorative board.

Item	Index	
Weather resistance	Appearance shock strength	No bubbling, falling off, crack 10.0 J
Thermal resistance		Meet the design requirement
water absorption in 24 h		<500 g/m^2
Resistance to wind load		No less than design
Impermeability in 2 h		Anti-permeate in the medial
Anti-extension intensity of single anchor bolt		Meet the design requirement and no less than 0.3 kN

4 REUSE OF WASTE ROCK WOOL

With the demolition of the building, a large amount of rock wool, used for exterior insulation, are generally discarded after a single use, which will cause greater impact on the environment if abandoned rock wool is thrown away. Therefore, the waste rock wool processing has attracted people's attention. At present, scholars use waste rock wool after modification for processing water pollution.

Chenxu Liu (2014) used ultrasonic and Polymerized Ferrous Sulfate (PFS)-modified waste rock wool and studied the treatment of the modified rock wool to waste water. The results showed that the modified rock wool in sewage COD and turbidity removal with good results, which is one of the best conditions for the modified wool accounted for 3.5% of PFS, ultrasonic oscillation frequency is 60 KHz, the processing time of 20 min (Figs. 1 and 2). This method can be achieved by building secondary use of waste rock wool, while improving the adsorption efficiency and service life of

Figure 1. Picture of waste rock wool with no modified PFS.

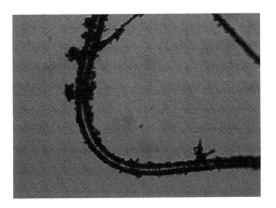

Figure 2. Picture of waste rock wool with the modified by 3.5% PFS.

Table 5. The removal rates of COD and turbidity.

Number	Shock time/min	PH	Rook wool/g	Removal rate of COD/%	Removal rate of turbidity/%
1	15	6	15	72.56	97.72
2	15	7	20	73.97	98.25
3	15	8	25	73.49	97.56
4	20	6	15	73.56	97.96
5	20	7	20	73.86	98.27
6	20	8	25	74.36	98.67
7	25	6	15	73.31	97.76
8	25	7	20	73.7	98.19
9	25	8	25	74.12	98.51

PFS. Ping Shang (2010) used waste rock wool that is modified by PFS and ultrasonic oscillation to remove COD and turbidity of landfill leachate. The removal rates of COD and turbidity will reach maximum values, respectively, 74.36% and 98.67% when the content of PFS was 7%, the ultrasonic frequency was 60%, time was 20 min, pH value was 8, and the dosage of the modified waste rock wool was 15 g (Table 5). Encheng Sun (2007 & 2008) had studied the removal of ammonia nitrogen and phosphorus by waste rock wool, which modified by acid and alkali. The results show that the best modified method was by using hydrochloric acid with a concentration of 15% and heating with ultrasonic oscillation for removing ammonia, while using the sodium hydroxide with a concentration of 15% and heating with ultrasonic oscillation for removing phosphorus.

It will generate a lot of solid waste in the waste phase of rock wool for the lack of reasonable recycling measures. But the analysis of the material itself proves that it has the rock wool waste recovery and recycling capacity, which will further enhance its environmental friendliness.

5 LIFE CYCLE ANALYSIS OF ROCK WOOL

The market of insulation materials is occupied by Expandable Polystyrene (EPS), extruded Polystyrene (XPS), Polyurethane (PU), phenol formaldehyde resin and rock wool, expanded perlite, and foamed ceramic/glass. These products formed respective competition with combustion or insulation performance, except the fire protection design requirements of a certain building that must use A-level external fire insulation materials. The whole life evaluation of insulation material was not included in the user's account. At present, scholars have launched a detailed and in-depth research on the field during the 12th "five year plan", and find out rock wool had significant advantages compared to other materials.

Table 6. Energy consumption of 50 years for each stage of thermal insulation materials for building life.

Thermal insulation	$E_{manu}/$ (kJ/m^2)	$E_{trans}/$ (kJ/m^2)	$E_{use}/$ (kJ/m^2)	$E_{Tot}/$ (kJ/m^2)
Rock wool	3.15×10^5	2901	6.23×10^6	6.55×10^6
EPS	3.95×10^5	765	6.23×10^6	6.63×10^6
XPS	3.95×10^5	870	6.23×10^6	6.58×10^6
Polyurethane	5.30×10^5	812	6.23×10^6	6.77×10^6

Table 7. Energy consumption ratio of 50 years for each stage of thermal insulation materials for building life.

Thermal insulation material	E_{manu}	E_{trans}	E_{use}
Rock wool	4.8	0.044	95.156
EPS	5.96	0.012	94.028
XPS	6.00	0.013	93.987
Polyurethane	7.82	0.012	92.168

Jun Liu (2014) evaluated the life cycle of EPS, expanded perlite, rock wool board, insulation mortar by inventory analysis and life cycle of resources, energy consumption, and environmental impact. The results showed that energy consumption and industrial waste gas emission of rock wool were in the second and carbon emission was the lowest, while solid waste emission was the biggest. Rock wool as an insulation material is a better choice under comprehensive consideration. Lianbin Zhu (2014) used the principle of life-cycle analyzed the total and the stage energy consumption on rock wool, XPS, EPS, and polyurethane in a building within 50a life (Table 6 and Table 7) and calculated the recovery period of capital, environmental benefits, and energy consumption. The results showed that the total energy consumption of rock wool was the lowest, and financial, environmental benefits, and energy recovery period were also the lowest, respectively, for 3.6 years, 2.2 years, and 2.09 years.

ACKNOWLEDGMENT

This research is sponsored by the Beijing Nova Program (Z141103001814029).

REFERENCES

Chenxu Liu, Nuting Yan & Jin Li (2014). Applied research of modified waste rock wool in sewage treatment. J. Environmental Pollution and Control, 36, 15–18.

Encheng Sun & Ping Shang (2007). Study on Phosphorus Removing from Eutrophication Water by Modified Basaltic Mineral Wool. J. Non-Metallic Mines, 30, 47–49.

Encheng Sun, Ping Shang & Ruihua Zhao (2008). Experimental study on ammonia nitrogen removal from water by modified basaltic mineral wool. J. Tianjin Chemical Industry, 22, 52–54.

Guangqing Liu (2012). Sealing Side Rock Wool Metal Surface Sandwich Board. J. Building energy efficiency, 6: 61–66.

Guozhong Lu, Xuesong Zheng & Lijuan Zhou (2015). Preparation and Properties of Polyurethane/Composite rock wool insulation board. J. Wall materials innovation & energy savingbuildings, 5, 48–50.

Hong Wang, Xingyuan Ni & Renwang Tang (2009). The thermal insulation material with ultra-low thermal conductivity made by using fiber felt to compound with SiO_2 aerogels and ambient pressure drying. J. Technical textiles, 11, 9–12.

Jintai He, Guozhong Lu & Xuesong Zheng (2015). Rock-wool insulation decorative panels used in the building's exterior wall decoration. J. Construction science and technology, 9, 70–71.

Jun Liu, Chen Chen & Wei Qi (2014). Life Cycle Analysis on Insulation Materials for Village Roof in Northern China. J. Journal of Shenyang Jianzhu University (Social Science), 16, 263–267.

Lianbin Zhu, Xiangrong Kong & Xian Wu (2014). Life cycle assessment and environmental & economic benefits research of important building external insulation materials in Beijing. J. Acta ecologica sinica, 34, 2155–2163.

Ping Shang, Taoli Liu & Xiangjun Kong (2010). Pretreatment of landfill leachate by modified basaltic men-made asbestos. J. Environmental Engineering, 28, 218–221.

Pingping Wang & Yunming Xu (2014). The application of metal rock wool sandwich board exterior wall in building energy saving. J. Shanxi architecture, 40(23): 236.

Qiang Ren, Jianzhang Li & Zhenyou Lu (2007). Studies on the wood fiber-rock wool composite. J. Journal of beijing forestry universtiy, 29, 161–164.

Qiang Ren, Jianzhang Li & Jinping Zhao (2008). Effeet of Coupling Agent on Properties of Wood-Rock Wool Composite MDF. J. Chemistry and adhesion, 30, 1–5.

Xiaodong Wang (2006). Base Research on the Application of nanoporous SiO_2 aerogel based thermal insulation composite. D. ChangSha, National University of Defense Technology.

Zhengang Geng, Wei Wang & Yang Yang (2015). On external wall outer insulation system of G-type rock wool insulation decoration integrated plate. J. Shanxi architecture, 41, 195–197.

Advances in Energy, Environment and Materials Science – Wang & Zhou (Eds)
© 2017 Taylor & Francis Group, London, ISBN 978-1-138-03600-0

Characteristics of the particle number concentration during the heating period in Harbin

Likun Huang
School of Food Engineering, Harbin University of Commerce, Harbin, China

ABSTRACT: The characteristics of the atmospheric particle number concentration were studied during the early heating period. The data of $PM_{2.5}$, relative humidity, temperature, wind scale, and atmospheric visibility were obtained in meteorological monitoring performed online. The results show that, the number of the days, of which the visibility was less than 10 km, was 3, 11, and 9, respectively. The days on which the $PM_{2.5}$ concentrations were higher than 100 $\mu g/m^3$ were 1, 9, and 9 in October, November, and December respectively. The average of the particle number concentration were 17775, 36345, and 34640, respectively; the average of visibility were 23.6, 8.5, and 9.7 km, it shows that an increase in the number concentration is mainly caused by coal-burning heating in November and December. The range of number concentration has a large fluctuation in December. One part of particles was adsorption in the snow, thereby falling to the ground, so that the content of the suspended particulate matter in air is reduced. The particle number concentration was significantly correlated with $PM_{2.5}$ concentration, visibility, and relative humidity. This shows that the number concentration of $PM_{2.5}$ in the atmosphere is higher and the visibility is lower.

1 INTRODUCTION

The main reasons for atmospheric particulate pollution were the increase of suspended fine particles and gaseous pollutants (Dai et al. 2013, Dong et al. 2012, Huang 2013). Due to the growth of urban population, industrial development, and the surge of motor vehicles, the concentration of the suspended fine particulate matter $PM_{2.5}$ and gaseous pollutants such as sulfur dioxide and nitrogen oxides increased sharply. The soluble particles (e.g. sulfate, nitrate, ammonium, and organic acid salt), with strong water absorption in $PM_{2.5}$, combined with water vapor to form haze weather. Furthermore, sulfur dioxide, nitrogen oxide emissions from coal combustion and motor vehicle exhaust were the precursors of sulfate and nitrate in atmospheric fine particles $PM_{2.5}$ (Huang 2014). Through a series of complex chemical reactions, sulfur dioxide, nitrogen oxides, and other pollutants in the air formed sulfate, nitrate, and other secondary particles. Transformation of gaseous pollutants into the homomorphic pollutants becomes the main composition of $PM_{2.5}$. The process of secondary pollution became the leading cause of rising $PM_{2.5}$ levels, which aggravated the formation of continuous haze weather (Mejia et al. 2006, Hiroaki et al. 2005, Tareq et al. 2006). Accordingly, the influence of the particle number concentration should also be observed (Yan et al. 2004).

In this study, the atmospheric particle number concentration was monitored from October to December 2015. We analyzed these data thereby concluding the variation characteristics and the main influence factors to provide the scientific basis for the prevention of the related air pollution in Harbin.

2 SAMPLES AND METHODS

By using the TSI number concentration meter, the particle number concentration is monitored online from October to December 2015 three times a day: 9:00, 12:00, and 15:00, and each time point was monitored for 5 minutes. The relative humidity and temperature were measured with a hygrometer and thermometer synchronously. Other meteorological factors were searched on the weather network and recorded.

3 RESULTS AND DISCUSSION

3.1 *Particle number concentration variation characteristics*

Figs. 1 and 2 show the daily variations of the particle number concentration curves from October to December 2015. The monitoring period was days 3–16 per month (three times a day: 9:00, 12:00, and

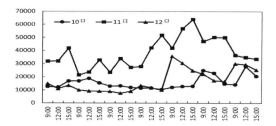

Figure 1. The daily variation of particle number concentrations at three times on every month 3rd to 9th.

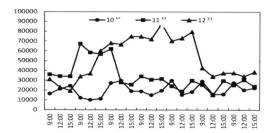

Figure 2. The daily variation of particle number concentrations at three times on every month 10th to 16th.

15:00). Fig. 1 shows that the number concentration in 3–9 November are higher than that in October and December; the change range is from 20000 to 40000 and its maximum is 63829, which is 2.26 times of that in October, and 1.79 times of the concentration in December. And the particle number concentration in October and December basically changes between 10000 and 30000. Fig. 2 shows that, during 12–14 December, the number concentration is high for the maximum value which is 87689 and the minimum value is 68082. Since mid-December is the medium of the heating period, the number concentration increased due to coal consumption. And the number concentration of 15, 16 down to about 40000, which may be related to the snowfall and temperature reduction. As can be seen from Figs. 1 and 2, the change of the number concentration in October is more stable and the fluctuation is not big; this is because, about on October 20th heating began. At this time, number concentration took less effect. The number of November and December change apparently, one of the reasons is that the fly ash and fine particulate matter entered the atmosphere when heating. Another reason is the influence of temperature, humidity, wind, snow, light, and other factors.

3.2 Particle number concentration daily variation characteristics

The daily changes are shown in Table 1 of the particle number concentration, the concentration of $PM_{2.5}$, and meteorological factors in October. The

particle number concentration on 3–4 October increases mainly because the wind scale reduced from 2 to 1 and wind speed decreased to below 1.6 m/s, which was not conducive to the pollutants' diffusion, so that the number concentration increased, thereby enhancing the atmospheric scattering effect and the visibility reduced from 34 km to 20 km. The number concentration of particulate matter in 5–7 is relatively stable between 11000 and 12800, the visibility is in the 9–13 km range, and the relative humidity is slightly increased. On days 8–13, the particle number concentrations remain over 20000. The visibility is above 20000 m, while the concentration of $PM_{2.5}$ swings around 20 ug/m³. All the above parameters were negatively correlated and the AQIs of these days were 52, 40, 45, 43, 28, and 29. The air quality was superior. The number concentration of 14 to 16 increases slightly, while the $PM_{2.5}$ concentration significantly increases and the maximum concentration is 138 ug/m³. The visibility decreased and the minimum value is 7.1 km. The relative humidity of days 3–16 is above 50% and the effect of the particle number concentration was not obvious.

The changes of the particle number concentration, the $PM_{2.5}$ concentration, and meteorological factors in November were analyzed by using Table 2. The higher the wind speed, the greater is the carrying distance of the pollutants, more air is mixed and the concentration of pollutants is lower, which is conducive to the diffusion of pollutants. The particle number concentration in November is all above 20000, the wind scale reaches grade 2 to 3 in 8 days, the visibility is less than 5 km for 6 days, and the $PM_{2.5}$ concentration for 9 days is higher than 100 ug/m³, with the maximum concentration being equal to 578 ug/m³. But these changes are not consistent with the rules before, which is possibly due to the effect of the inversion layer. When appearing on the ground, the inversion layer restrains the atmospheric surface layer turbulence and the atmosphere moves downward vertically. If it appears at a certain level, it will hinder the development of the vertical air movement, which diminished the mixture of pollutants and air exchange, eventually causing the diffusion of pollutants and diluted the ability to slow down; pollutants remain for a long time. As can be seen in the table, the change of the temperature was not stable and the particle number concentration can be affected. On days 9–12, the relative humidity values were 85%, 88%, 93%, and 88% and the values of visibility were 6.9, 8.1, 4.3, and 5.6 km, the higher relative humidity could decrease the visibility.

The changes of the particle number concentration, the $PM_{2.5}$ concentration, and meteorological factors in December were analyzed in Table 3. In December, the number concentration curve

172

Table 1. The daily change of number concentration, $PM_{2.5}$ concentrations, and meteorological factors in October 2015.

Date	Number concentration	$PM_{2.5}$ (ug/m³)	Visibility (km)	Temperature (°C)	Relative humidity (%)
10.3	13884	18	34.3	6	57
10.4	17030	30	20.0	8	59
10.5	12726	53	12.8	14	63
10.6	11192	85	9.2	17	69
10.7	12633	67	11.2	17	67
10.8	20868	26	22.6	13	61
10.9	20993	17	26.7	7	55
10.10	20580	22	30.8	6	84
10.11	11137	15	52.2	6	89
10.12	25287	13	48.0	5	59
10.13	17759	16	33.3	3	55
10.14	21115	59	12.4	11	50
10.15	20277	84	9.7	11	67
10.16	23364	138	7.1	17	57

Table 2. The daily change of number concentration, $PM_{2.5}$ concentrations, and meteorological factors in November 2015.

Date	Number concentration	$PM_{2.5}$ (ug/m³)	Visibility (km)	Temperature (°C)	Relative humidity (%)
11.3	35247	578	1.9	6	66
11.4	25965	157	5.0	10	43
11.5	28217	96	9.7	0	37
11.6	40572	392	3.4	−3	54
11.7	54171	244	4.0	−2	48
11.8	49073	251	3.9	−1	44
11.9	35045	148	6.9	−2	85
11.10	34689	140	8.1	−2	88
11.11	60943	232	4.3	−5	93
11.12	38519	164	5.6	−5	91
11.13	32105	97	9.8	2	89
11.14	24258	51	18.2	7	75
11.15	23452	48	17.6	2	69
11.16	26576	36	21.1	−6	55

Table 3. The daily change of number concentration, $PM_{2.5}$ concentrations, and meteorological factors in December 2015.

Date	Number concentration	$PM_{2.5}$ (ug/m³)	Visibility (km)	Temperature(°C)	Relative humidity (%)
12.3	13125	57	12.4	−12	87
12.4	9380	38	23.1	−9	68
12.5	8426	35	25.5	−10	73
12.6	11375	61	15.4	−17	62
12.7	30319	199	4.6	−19	76
12.8	18634	103	8.1	−11	75
12.9	28287	196	4.8	−4	89
12.10	24425	118	6.3	−4	92
12.11	43727	112	8.2	−15	76
12.12	69908	268	4.1	−15	83
12.13	78144	371	2.5	−17	85
12.14	74350	299	3.0	−15	84
12.15	38045	154	5.9	−10	85
12.16	36811	66	12.6	−17	79

changes greatly, the value of the number concentration in days 3–6 is about 10000 and the visibility is more than 10 km, which are 12.4, 23.1, 25.5, and 15.4 km. The wind scale is 2 and the high wind speed conducing to the speed of pollutants. The number concentrations in days 12–14 are 69908, 78144, and 74350, which are the largest among three months. At this point, the visibility in 3–6 December is minimal and its values are as follows: 4.1, 2.5, and 3.0 km and the value of $PM_{2.5}$ is the highest among these days. The three days' wind scales are 1, 0, and 0; the wind speed is low, less able to dilute the pollutants and therefore, pollution is more serious (AQI: 318, 414, and 349).

At the same time, it can be seen that the higher the concentration of $PM_{2.5}$ in the atmosphere, the lower is the visibility. Therefore, atmospheric particles' pollution and meteorological conditions are closely related to each other. When the emissions and the distribution of pollution sources were relatively stable, atmospheric particles' concentration mainly depends on particles' transport and diffusion in all kinds of weather.

4 CONCLUSION

In October and December, the average particle number concentrations were higher than that in the rest of the months of this year and the values of visibility were lower, mainly due to coal-burnt heating in winter.

In October, the number concentration fluctuated around in 10000–20000; there were 7 days of 2 wind scale, the number concentration was not high and stable. In November, the number concentration was above 20000 and the visibility was less than 5 km for 6 days. And in December, the number concentration curve fluctuated greatly, because of the snow. One part of particles was absorbed in the snow, and then down to the ground, so that the concentration of suspended particulate matter in air decreased.

The particle number concentration is significantly related to the $PM_{2.5}$ concentration, the visibility, and the relative humidity; the higher the concentration of $PM_{2.5}$ in the atmosphere, the lower is the visibility.

ACKNOWLEDGMENT

This study was supported by the National Natural Science Foundation of China (No. 51408168). The authors would like to express their sincere appreciation for the financial support to accomplish this study.

REFERENCES

Dai W., Gao J. Q., Cao G. (2013). Chemical composition and source identification of PM2.5 in the suburb of Shenzhen, China. *Atmospheric Research, 122*:391–400.

Dong X., Liu D., Gao S. (2012). Characterization of $PM_{2.5}$ and PM_{10} bound polycyclic aromatic hydrocarbons in urban and rural areas in Beijing during the winter. *Advanced Materials Research, 518–523*:1479–1491.

Huang L. K. (2013). Study on Concentration Characteristics of TSP, PM10 and PM2.5 in Harbin, *Advanced Materials Research, 777*: 412–415.

Huang L. K., Wang G. Z. (2014). Chemical characteristics and source apportionment of atmospheric particles during heating period in Harbin, China. *Journal of Environmental Sciences, 26*(12): 2475~2483.

Hiroaki M., Hideto T. (2005). Observation of number concentrations of atmospheric aerosols and analysis of nanoparticle behavior at an urban background area in Japan. *Atmospheric Environment, 39*: 5806–5816.

Mejia J. F., Wraith D. (2006). Mengersen K. Trends in size classified particle number concentration in subtropical Brisbane, Australia, based on a 5 year study. *Atmospheric Environment, 40*: 1064–1079.

Tareq H., Ari K., Jaakko K., Jari H. (2006). Meteorological dependence of size-fractionated number concentrations of urban aerosol particles. *Atmospheric Environment, 40*: 1427–1440.

Yan F. Q., Hu L., Yu T. (2004). Analysis of particularte mass concentration, aerosol number concentration and visibility in Beijing. China *Particuology, 2*(1): 25–30.

Advances in Energy, Environment and Materials Science – Wang & Zhou (Eds)
© 2017 Taylor & Francis Group, London, ISBN 978-1-138-03600-0

A study on the distribution characteristics of atmospheric particulates during a dust period in Harbin

Likun Huang
School of Food Engineering, Harbin University of Commerce, Harbin, China

ABSTRACT: This study investigates the distribution characteristics of atmospheric particulates during a dust storm episode, including mass concentration of particulate matter and the size distribution of the particles. According to the results of the particle size distribution, $PM_{2.5}/TSP$, $PM_{2.5}/PM_{10}$, and PM_{10}/TSP begin to rise after the end of the dust period; $PM_{2.5}/TSP$ and PM_{10}/TSP rise from the lowest points of 0.17 and 0.44 to the highest points of 0.43 and 0.87 respectively, which show that PM_{10-100} dominated in the dust period. After the dust weather period, wind speed decreased, the majority of the PM_{10-100} subsided, and particulate matter returned to the normal levels.

1   INTRODUCTION

Dust weather includes four categories, which are as follows: floating dust, blowing sand, sandstorms, and strong sandstorms. Floating dust, the horizontal visibility of which is less than 10 km, includes dust and sand that evenly floats in air; blowing sand, the horizontal visibility of which varies between 1 km and 10 km, is caused by dust blowing from the ground; while the horizontal visibility of the sandstorm is less than 1 km; and the horizontal visibility of the strong sandstorm is less than 500 m.

In the dust weather mentioned above, the sandstorm is considered to be a kind of severe weather phenomena accompanied with the interaction of wind and sand, which is more dangerous than other dust weather conditions and occurs more frequently, so that it has received more and more attention of many countries Jayaratne et al., 2011; Lee et al., 2010; Vanderstraeten et al., 2008). Its formation has an inseparable relationship with global warming, El Nino phenomenon, loss of forests, destruction of vegetation, species extinction, climate anomalies, and other factors. The formation of sandstorms or strong sandstorms is inevitably connected with three conditions; the weather situation is conducive to produce strong winds, distribution of sand and dust, and an unstable air condition.

The statistics show that, in China, there are 8 strong sandstorms in the 1960s, 13 times in the 1970s, 14 times in the 1980s, and 20 times in the 1990s. As we can see, the incidence of sandstorms is on the rise, thereby causing more and more heavy losses (Fu et al., 2008). In recent years, there is at least one sandstorm occurring in Harbin annually, and mostly occurrs in spring (April and May). When sandstorm occurs, the number of traffic accidents and attendance rates of respiratory diseases increase correspondingly and therefore, it is necessary to ascertain the chemical composition of the storm particles, types, and contribution of pollution sources, as well as its long-distance transmission pathways. However, there are different air quality characteristics between other cities and Harbin due to its high-latitude, cold climate, and frequent inversion phenomena in the winter season and no detailed investigation has been conducted in Harbin so far. (Chi et al., 2008; Zhao et al., 2010).

According to the characteristics of sandstorms and the reasons for their formation, this research carries out a dust storm tracking and monitoring during the dust period and no-dust period, and the sampling points are set in Harbin University of Commerce, which is in the upwind. By intensive sampling, three types of particulate matter (TSP, PM_{10}, and $PM_{2.5}$) are collected once in every 24 hours in continuous 10 days and their mass concentration and elemental composition analysis should be analyzed.

2   MATERIALS AND METHODS

Serious dust storm episodes frequently occur in spring every year in Harbin. In order to investigate the impacts of dust storm episodes on ambient air quality, daily atmospheric particulate samples (from 8:00 am to 8:00 pm) were consecutively collected from May 11 to May 20, including dust days

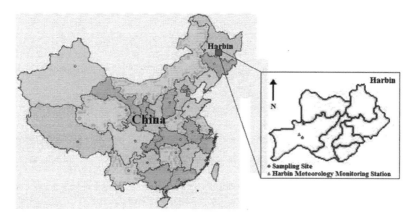

Figure 1. Daily atmospheric particulate samples from the South Campus of Harbin University of Commerce.

and no-dust days. The sampling site was located at the South Campus of Harbin University of Commerce, as illustrated in Fig. 1. The filters used in this study were 80 mm quartz filters, which were initially heated at 900°C before sampling in order to eliminate the interference of residual organic compounds. Three same medium-flow air samplers (Wuhan Tianhong Intelligent Instrument, Model TH-150) were used to collect TSP, PM_{10}, and $PM_{2.5}$ with a flow rate of 100 L/min and automatically recorded an overall sampling volume and duration. The sampling height was 5 m above the ground. Moreover, ambient temperature, dew point, wind speed, relative humidity, and atmospheric visibility were obtained from the Harbin Meteorology Monitoring Station situated at Jianguo Street, which is approximately 3 km northwest from the sampling site (Huang, Yuan, & Wang, 2011).

3 CHARACTERISTIC ANALYSIS OF ATMOSPHERIC PARTICULATE DURING THE DUST PERIOD

3.1 Mass concentration of particulate and weather

Sandstorms, which are related to special environmental characteristics, affect the local weather conditions. Variations of ambient temperature, dew point, relative humidity, atmospheric pressure, visibility, and wind speed during the sampling period are listed in Table 1. It can be seen from the table that wind speed is relatively higher on May 14, 15, and 16, when there is a significant reduction in visibility; on the other hand, the temperature of the entire sampling period is higher than the dew point and its humidity is less than 80%; during the days of low visibility, yellow dusts cover everything outside. All these phenomena illustrate

the same conclusion that the dust is the chief culprit to reduce the visibility, rather than haze or fog. To further confirm whether it is dust or not, TSP, PM_{10}, and $PM_{2.5}$ will be collected together and their changes of mass concentration will be analyzed as well and their results are shown in Table 1.

From the table, we can see that the concentration of the particulate matter reached a peak value on May 14~16, when the visibility is relatively low, which means that the increased wind speed resulted in foreign sand continuously pouring into Harbin and causing increased concentration of particles these days; the highest concentration, maximum wind speed, and the lowest visibility all occurred on May 15, indicating that there must have a significant correlation among these three phenomena. On May 16, the particle concentration began to decrease and the dust began to subside after showers; until May 17, dust weather disappeared and particle concentrations in the atmosphere returned to normal levels. Although the dust weather did not last for a long time, the condensation, duration, and disappearance of dust weather were captured, which can help to investigate characteristics and hazards of particulate matters of dust weather.

3.2 Distribution characteristics of particle size

As most of the particulate matter is introduced from outside sources during the dust weather period, the distribution of particles are essentially different from local sources. In Table 2, relative proportions of particles of various sizes are listed: $PM_{2.5}$/TSP, $PM_{2.5}$/PM_{10}, and PM_{10}/TSP decreased first and increased later; and in the dust weather period, the ratio dropped to the lowest value, indicating that the PM_{10-100} induced the increased concentrations of particles; although these particles could not float in the air for a long time. It still can maintain a certain concentration under the large

Table 1. Characteristics of the atmospheric environment during the sampling period.

Factor	5/11	5/12	5/13	5/14	5/15	5/16	5/17	5/18	5/19	5/20
Ambient temperature (°C)	13	12	10	8	16	14	13	15	13	22
Dew point (°C)	1	−5	−6	−7	2	11	0	−1	0	6
Relative humidity (%)	44	28	26	28	41	73	41	32	41	32
Atmospheric pressure (hPa)	1001	1003	1006	997	999	1002	1008	1005	1001	1006
Atmospheric visibility (km)	10	10	10	6	5	7	9	9	10	10
wind speed (km/h)	5	13	10	11	21	18	14	6	17	14
Weather	Cloudy	Showers	Clear	Clear	Cloudy	Showers	Showers	Cloudy	Cloudy	Showers

Table 2. Mass concentrations and relative proportions of particulate matter ($\mu g/m^3$).

Date	TSP	PM_{10}	$PM_{2.5}$	$PM_{2.5}/TSP$	$PM_{2.5}/PM_{10}$	PM_{10}/TSP
5/11	235.26	129.61	62.47	0.27	0.48	0.55
5/12	321.56	148.61	71.88	0.22	0.48	0.46
5/13	268.06	121.81	58.04	0.22	0.48	0.45
5/14	386.67	192.28	79.18	0.20	0.41	0.50
5/15	497.61	215.83	83.21	0.17	0.39	0.43
5/16	436.44	189.72	72.43	0.17	0.38	0.43
5/17	326.81	152.08	60.28	0.18	0.40	0.47
5/18	220.61	159.17	70.53	0.32	0.44	0.72
5/19	180.67	158.06	75.83	0.42	0.48	0.87
5/20	151.12	130.13	66.21	0.44	0.51	0.86

wind speed. Additionally, these three ratios began to rise after the end of the dust on May 17; $PM_{2.5}/TSP$ and PM_{10}/TSP raised from the lowest points of 0.17 and 0.44 to the highest points of 0.43 and 0.87, respectively; these changes showed once again that PM_{10-100} dominated in the dust weather period. After the dust weather period, the wind speed decreased, the majority of the PM_{10-100} subsided, and particulate matter returned to normal levels. The rise of $PM_{2.5}/PM_{10}$ levels is due to the consecutive showers on May 16 and 17 and their degradation capacity to coarse and therefore, the ratio after the sampling period is greater than that before sampling.

4 CONCLUSIONS

All these phenomena illustrate the same conclusion that the dust is the chief culprit in reducing the visibility, rather than haze or fog. According to the results of the particle size distribution, $PM_{2.5}/TSP$, $PM_{2.5}/PM_{10}$, and PM_{10}/TSP begin to rise after the end of the dust on May 17; $PM_{2.5}/TSP$ and PM_{10}/TSP rise from the lowest points of 0.17 and 0.44 to the highest points of 0.43 and 0.87 respectively, which show that PM_{10-100} dominated in the dust period. After the dust weather period, the

wind speed decreased, the majority of the PM_{10-100} subsided, and particulate matter returned to the normal levels. The rise of $PM_{2.5}/PM_{10}$ levels is due to the consecutive showers on May 16 and 17 and their degradation capacity to coarse and therefore, the ratio after the sampling period is greater than that before sampling.

ACKNOWLEDGMENT

This study was supported by the National Natural Science Foundation of China (No. 51408168). The authors would like to express their sincere appreciation for the financial support to accomplish this study.

REFERENCES

Chi, K.H., Hsu, S.C., Wang, S.H., & Chang, M.B. (2008). Increases in ambient PCDD/F and PCB concentrations in Northern Taiwan during an Asian dust storm episode. Science of the Total Environment, 401(1–3), 100–108.
Fu, P., Huang, J., Li, C., & Zhong, S. (2008). The properties of dust aerosol and reducing tendency of the dust storms in northwest China. Atmospheric Environment, 42(23), 5896–5904.

Huang, L.K., Yuan, C.S., & Wang, G.Z. (2011). Chemical characteristics and source apportionment of PM_{10} during a brown haze episode in Harbin, China. *Particuology, 9*(1), 32–38.

Jayaratne, E.R., Johnson, G.R., McGarry, P., Cheung, H.C., & Morawska, L. (2011). Characteristics of airborne ultrafine and coarse particles during the Australian dust storm of 23 September 2009, *Atmospheric Environment, 45*(24), 3996–4001.

Kim, N.K., Park, H.J., & Kim, Y.P. (2009). Chemical composition change in TSP due to dust storm at gosan, Korea: Do the concentrations of anthropogenic species increase due to dust storm? *Water, Air, and Soil Pollution, 204* (1–4), 165–175.

Lee, Y.C., Yang, X., & Wenig, M. (2010). Transport of dusts from East Asian and non-East Asian sources to Hong Kong during dust storm related events 1996–2007. *Atmospheric Environment, 44*(30), 3728–3738.

Shen, Z., Cao, J., Arimoto, R., Han, Z., Zhang, R., Han, Y., & Liu, S. (2009). Ionic composition of TSP and $PM_{2.5}$ during dust storms and air pollution episodes at Xi'an, China. *Atmospheric Environment, 43*(18), 2911–2918.

Tsai, H.H., Yuan, C.S., Lin, H.Y., & Huang, M.H. (2006). Physicochemical characteristics of suspended particles and hot-spot identification at a highly polluted region. In *99th Air & Waste Managment. Association Annual Meeting, New Orleans, Louisiana.*

Vanderstraeten, P., Lénelle, Y., Meurrens, A., Carati, D., Brenig, L., Delcloo, A., Offer, Z.Y., & Zaady, E. (2008). Dust storm originate from Sahara covering Western Europe: A case study. *Atmospheric Environment, 42*(21), 5489–5493.

Yuan, C.S., & Sau, C.C. (2004). Mass concentration and size-resolved chemical composition of atmospheric aerosols sampled at the Pescadores Islands during Asian dust storm periods in the years of 2001 and 2002. *Terrestrial, Atmospheric & Oceanic Sciences, 15*(5), 857–879.

Zhao, Q., He, K., Rahn, K.A., Ma, Y., Jia, Y., Yang, F., & Duan, F. (2010). Dust storms come to Central and Southwestern China, too: Implications from a major dust event in Chongqing, *Atmospheric Chemistry and Physics, 10*(6), 2615–2630.

Advances in Energy, Environment and Materials Science – Wang & Zhou (Eds)
© 2017 Taylor & Francis Group, London, ISBN 978-1-138-03600-0

3D ore grade estimation using artificial intelligence approach

Liansheng Han, Dehui Zhang & Guanghai Fan
AnSteel Groupe Mining Corporate, Gongchangling Open-pit Iron Mine, China

Jianfei Fu & Zhiqiang Ren
Northeastern University, Shenyang City, Liaoning Province, China

ABSTRACT: The demand for a new ore grade predictor originates from the limitation of conventional methods. Typically, artificial intelligent methods such as Genetic Programming (GP) and Support Vector Regression (SVR) have ability to capture non-linear relationships between the input and output data under extreme conditions. Their application in 2D ore grade estimation is quite mature, but on the other hand their application in 3D ore grade estimation is very poor. This paper introduces the Multi-Gene Genetic Programming (MGGP) model for 3D ore grade estimation and also compares it with the well-known SVR method. The results obtained from case study show that artificial intelligence approaches have high estimation accuracy and strong generalization ability in 3D ore grade estimation.

1 INTRODUCTION

Accuracy of ore grade estimation is one of the most important aspects of assessing mineral deposit. The reliability of a spatial grade model depends greatly on the type of modeling technique used as well as their geological complexity and adequacy of the data. Generally, applicable methods used to estimate spatial grade can be classified in three categories: traditional methods including geometrical, inverse distance weighted and others, geostatistical method and artificial intelligent techniques. Over the last four decades, geostatistics became the most well-established grade estimation technique because it makes best use of the spatial variation through variograms obtained from structural analysis (Oliver, 2010). Recently artificial intelligent techniques, such as expert system, fuzzy logic, genetic algorithm, neural network, SVR and MGGP, have been employed to solve ore grade estimation problems (Li, 2009; Li, 2013; Garg, 2014). The attraction of artificial intelligent methods is their flexibility in nature and their ability to capture non-linear relationships between the input and output data under extreme conditions.

But no matter which approach comes out on top, they mainly has been applied in 2D ore grade estimation (Oliver, 2010; Li, 2009; Li, 2013; Garg, 2014). Geological phenomenon essentially happens in 3-dimensional space and ore grade are changed with geological spatial coordinates.

Aimed at this problem, we present the use of Multi-Gene Genetic Programming (MGGP) method for 3D ore grade estimation. In this paper,

we explore the use of proposed model for 3D ore grade estimation and compare the obtained results with SVR method.

2 METHOLOGY

Genetic Programming (GP) developed by Koza (1992) is considered to be the most famous for solving symbolic regression problems and is widely used in modeling processes of varying nature. GP based Darwin's theory of "survival of the fittest" finds the best/optimal solution by mimicking the process of evolution in nature. GP generates both of model types and its coefficients automatically based on the given data (Garg, 2014; Koza, 1992; Searson, 2010).

Multi-Gene Genetic Programming (MGGP) is a robust variant of GP, which effectively combines the model structure selection ability of the standard GP with the parameter estimation power of classical regression by using a new characteristic called multi-gene. In traditional GP method, the model is a single tree/gene expression whereas in MGGP, the model formed is a linear combination of several trees/genes which each of them is a traditional GP tree. MGGP generates mathematical models of set of data by linear combinations of low order non-linear transformations of the input-output variables. Recently, the MGGP have been used successfully for engineering modeling problems (Garg, 2014). It has been shown that MGGP regression can be more accurate and efficient than the standard GP for modeling non-linear problems.

The key difference between GP and MGGP is that, in the latter, the model participating in the evolution is a combination of several sets of genes/trees.

For a system with u input of dimension $R^{n \times m}$ to produce a model output y with dimension $R^{n \times 1}$, where n is the number of observations taken and m is the number of input variables, we could produce a tree structure which introduces the mathematical relationship:

$$\hat{G} = f(u_1, ..., u_i) \quad (1)$$

In multi-gene symbolic regression GP, each prediction of the output variable $\hat{G}$ is formed by a weighted output of each of the trees/genes in the multi-gene individual plus a bias term. Each tree is a function of zero or more of the i input variables $u_1, ..., u_i$. Mathematically, a multi-gene regression GP model can be written as:

$$\hat{G} = d_0 + d_1 \times tree_1 + \cdots + d_M \times tree_M \quad (2)$$

where d_0 represents the bias or offset term while $d_1, ..., d_M$ are the gene weights and M is the number of gene (i.e. trees) which constitute the available individual. The weights (i.e. regression coefficients) are automatically determined by a least squares procedure for each multi-gene individual. In multi-gene symbolic regression each symbolic model is represented by number of GP trees weighted by linear combination. Each tree is considered as a gene by itself. The typical example of MGGP model and its mathematical expression are shown in Figure 1.

Comparing with kriging, MGGP for spatial grade prediction do not require the assumptions about the spatial distribution of the grade data. MGGP only require training inputs and outputs to estimate the underlying their relationship. When MGGP is used to estimate spatial grade, input data can normally come in the form of 3-dimensional space coordinates at sampled locations and the output data should be the spatial grade at these locations. MGGP treat the spatial grade prediction as problem of function approximation in the data coordinate space.

That is, the mathematical relationship represented by MGGP model for spatial grade prediction is as follows:

$$\hat{G} = f(x, y, z) \quad (3)$$

where (x, y, z) represents the 3D space coordinate and $\hat{G}$ is the grade value at these locations. At that time, the fitness function used for performance evaluation of population is root mean square error

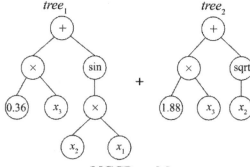

MGGP model:

$Z = d_0 + d_1(0.36x_3 + \sin(x_2 \times x_1)) + d_2(1.88x_3 + \text{sqrt}(x_2))$

Figure 1. Example of MGGP model.

between actual measured grade values and predicted values, given by:

$$fitness = \sqrt{\frac{\sum_{i=1}^{N} |G_i - \hat{G}_i|^2}{N}} \quad (4)$$

where G_i is the value predicted by the MGGP model at ith data coordinate space, $\hat{G}_i$ is the actual measured value at ith data coordinate space and N is the number of data samples.

The steps generally followed in MGGP are:

1. Creation of an Initial population of individuals (i.e. programs or equations).
2. Evaluation of individuals based on fitness function.
3. Selection of the fittest individuals as parents.
4. Creation of new individuals (also called the offspring) through the genetic operations of crossover, mutation, and reproduction.
5. Replacing the weaker parents in the population by the stronger ones.
6. Repetition of steps 2 through 5 until the user defined termination criterion is satisfied.

The termination criterion can be completion of a specified number of generations or fitness criterion such as minimum error reached.

3 CASE STUDIES

This section aimed to evaluate the performance of the artificial intelligence approaches including MGGP and SVR in 3D spatial grade estimation. For this illustration, we have selected data collected from a BIF-hosted iron ore deposit located in the Anshan-Benxi area, northeastern part of China. This iron ore deposit with 23.60%~39.80% (overall average, 32.70%) of iron ore grades (Figure 2), has been sampled by means of 116 randomly positioned boreholes perpendicular to the dip of the ore body. Among them, we randomly selected 91 boreholes as training data for model estimation and used the rest 35 boreholes as validating data for cross-validation (shown in Figure 3).

In estimating ore grades, the implementation of MGGP method required adjustment of its parameters. The parameter selection is important since it affects the generalization ability of the MGGP model. The two input variables used for the MGGP model in this study are x, y and z spatial coordinates of study area and the output variable is the coal seam width at these locations. The parameters selected based on trial-and-error approach are shown in Table 1.

The parameters like population size and number of generations fairly depends on the complexity of the regression problem. In generally, the population size and number of generations should be fairly large for training data of smaller samples. Since a MGGP model is formulated from the set of genes, the model will have higher complexity i.e. greater number of nodes along with the evolution, and may result in over-fitting. The restriction on the maximum number of genes and depth of the gene exerts control over the complexity of the models and results in accurate and compact models. Therefore, in this study, the maximum number of genes and maximum depth of tree is kept at 9 and 6, respectively. The best MGGP models are chosen on the basis of providing the best fitness value (considering the spatial variation) on the training data as well as the simplicity of the models.

In this study, GPTIPS toolbox (Searson, 2010), in conjunction with subroutines coded in MATLAB, is used to implement MGGP. The best MGGP model selected based on minimum fitness on training data is used to cross validation, which is as follows:

$$
\begin{aligned}
&G(x,y,z) \\
&= 39.490 + 0.257 \times \exp\!\Big(\mathrm{plog}\big(\mathrm{power}\big(\cos(x \times \mathrm{plog}(z)), \mathrm{atan}\big(\sin(\mathrm{plog}(z))\big)\big)\big)\Big) - 3.329 \\
&\quad \times \sin\!\Big(\exp\big(\mathrm{plog}\big(\cos(\tanh(z)/\cos(x))\big)\big)\Big) + 0.834 \times \mathrm{plog}(\cos(z \times \mathrm{plog}(z))/ \\
&\quad \mathrm{power}\big(\mathrm{psqrt}(z), \mathrm{power}\big(\cos(x), \mathrm{power}\big(\cos(z) \times \mathrm{plog}(y), 11.855x/y\big)\big)\big)\big) \\
&\quad - 2.178 \times \mathrm{psqrt}\big(\cos\big(\mathrm{plog}(y) \times \mathrm{plog}(z)^2\big)\big) + 1.079 \times \cos(\mathrm{atan}(\tanh(\mathrm{plog}(y))/\cos(x)) \\
&\quad \times \mathrm{square}\big(\mathrm{plog}(y)\big)) + 1.303 \times \cos\big(\mathrm{psqrt}(12.774x/y)\big) + 533 \times \tanh \\
&\quad \Big(\exp\big(\mathrm{psqrt}(z)/\mathrm{plog}(x)\big)/\big(\mathrm{atan}\big(z \times \mathrm{square}(\sin(z)) \times \cos(z)\big) \times \mathrm{square}\big(\mathrm{square}(\mathrm{plog}(y))\big)\big)\Big) \\
&\quad + 0.257 \times \mathrm{power}\big(\mathrm{square}\big(-\mathrm{square}(\sin(z)) - \cos(y+7.394) + \mathrm{psqrt}(z)/11.855\big), \\
&\quad \times \mathrm{power}\big(\cos(x \times \mathrm{plog}(z)), \mathrm{atan}\big(\sin(\mathrm{plog}(z))\big)\big) - 18855 \times \mathrm{atan}\big(\sin\big(\exp\big(\mathrm{plog}(y \times z^2)\big)\big)\big) \\
&\quad /\mathrm{square}(\mathrm{psqrt}\big(z^2 \times \mathrm{plog}(z) \times \cos\big(\mathrm{psqrt}(z)/\mathrm{plog}(x)\big)\big) + 2.178 \times \cos(y) \times \cos(z) + 0.001 \times \mathrm{power} \\
&\quad \times \big((x \times z \times \mathrm{plog}(\mathrm{psqrt}(x/y)))/163.183, \mathrm{power}\big((y, \mathrm{power}\big(\mathrm{power}(1/\cos(x), \mathrm{power}(x \times \mathrm{plog}(az), 11.855x/y)), \\
&\quad -3.574y)\big)\big)
\end{aligned}
$$

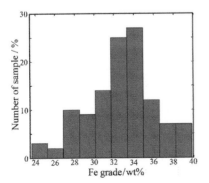

Figure 2.   Histogram of iron ore grade in study area.

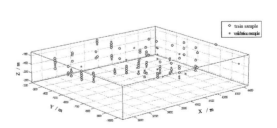

Figure 3.   Spatial distribution of iron ore sample points.

Table 1. Parameter Setting for MGGP Model.

Parameters	Selected values
Population size	500
Number of generations	200
Tournament size	20
Max depth of tree	6
Max genes	9
Functional set (F)	+, −, ÷, ×, sin, cos, tan, atan, tanh, exp, square, plog, psqrt, ppower
Terminal set (T)	$x, y, z; [-10\ 10]$
Crossover probability rate	0.85
Reproduction probability rate	0.10
Mutation probability rate	0.05

 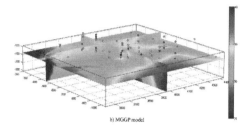

a) SVR model   b) MGGP model

Figure 4. Result of iron ore grade estimation in 3-demension space.

In this study, in order to construct SVR model for spatial grade prediction, we used the LS-SVM (Suykens, 2002). The selection of kernel function plays a key role in the learning and minimizing the loss function efficiently since it affects the generalization ability of the SVR model. In this work, Radial Basis Function (RBF) is chosen as kernel function for the performance of SVR models, since it is well known for its faster and efficient training process among researchers. The suitable regularization parameter C and σ of the chosen kernel function are determined using a combination of Coupled Simulated Annealing (CSA) and a grid search method. The CSA determines the good initial values of C and σ, and then, these are passed to the grid search method, which uses cross-validation to fine tune the parameters. Optimal parameters for iron ore deposit data are $C = 1.7392$ and $\sigma = 51.1855$.

Figure 4 show the actual observed values and the predicted values obtained by SVR and MGGP methods with validation data. Compared two prediction models with the map of actual observed values (Figure 4) can obviously resulted in that MGGP method is more robust than SVR.

Finally, correlation coefficient (R), Mean Absolute Prediction Error ($MAPE$) and Root Mean Square Prediction Error ($RMSPE$) are used to evaluate the cross-validation result of three models

for grade prediction. They are calculated using the following equations:

$$R = \frac{\sum_{i=1}^{N}(G(x_i)-\overline{G}(x_i))(\hat{G}(x_i)-\overline{\hat{G}}(x_i))}{\sqrt{\sum_{i=1}^{N}(G(x_i)-\overline{G}(x_i))^2 \sum_{i=1}^{N}(\hat{G}(x_i)-\overline{\hat{G}}(x_i))^2}} \quad (5)$$

$$MAPE = \frac{1}{N}\sum_{i=1}^{N}\left|G(x_i)-\hat{G}(x_i)\right| \quad (6)$$

$$RMSPE = \sqrt{\frac{1}{N}\sum_{i=1}^{N}[G(x_i)-\hat{G}(x_i)]^2} \quad (7)$$

where $G(x_i)$ is the observed value at sampling location x_i, $\hat{G}(x_i)$ is the estimated value at this location and N is the number of sample points for cross-examination.

The obtained comparison analysis results are shown in Table 2 and Figure 4.

From Table 2 and Figure 4, we can also see that $MAPE$ and $RMSPE$ of MGGP model are smaller than SVR and that R higher, indicating that proposed model is reliable in predicting the spatial grade. In addition, MGGP method overcomes the defect of SVR that does not provide explicit formulation for grade prediction model.

Table 2.   Comparison of Estimation Models.

Method	R	MAPE	RMSPE
SVR	0.86434	1.1703	1.7752
MGGP	0.83223	1.0633	1.5485

4   CONCLUSIONS

Determination of estimation method is essential for decreasing the prediction error and increasing the estimation accuracy of ore grade evaluation. We explore the application possibility of multi-gene genetic programming in 3D ore grade estimation. The obtained results show that the MGGP method as well as SVR improves remarkably the estimation accuracy of 3D spatial grade.

REFERENCES

Garg A., Garg A., Tai K. a Multi-Genetic Programming Model for Estimating Stress-dependent Soil Water Retention Curves. Computational Geosciences, 2014, 18(1): 45~56.

Koza J. R., Genetic Programming on the Programming of Computers by Means of Natural Selection. Cambridge: MIT, 1992: 1~609.

Li X. L., Xie Y. L., Li L. H. A nonlinear grade estimation method based on Wavelet Neural Network. IEEE conference Publications, 2009: 1~5.

Li X. L., Li L. H., Zhang B. L. Hybrid Self-adaptive Learning Based Particle Swarm Optimization and Support Vector Regression Model for Grade Estimation. Neurocomputing, 2013, 118: 179~190.

Oliver M. A. the Variogram and Kriging, Handbook of Applied Saptial Analysis: Software Tools, Methods and Applications. Berlin: Springer, 2010: 319~352.

Searson D. P., Leahy D. E., Willis M. J., GPTIPS: an Open Source Genetic Programming Toolbox for Multigene Symbolic Regression, Proceeding of the International Multiconference of Engineers and Computer Scientists. Hong Kong, 2010: 77~80.

Suykens J. A. K., De B. J., et al. Weighted Least Squares Support Vector Machines: Robustness and Sparse Approximation. Neurocomputing, 2002, 48(1–4): 85~105.

Advances in Energy, Environment and Materials Science – Wang & Zhou (Eds)
© *2017 Taylor & Francis Group, London, ISBN 978-1-138-03600-0*

Evaluation study on the environmental bearing capacity of China's scale pig breeding

Bi-Bin Leng
School of Economics and Management, Jiangxi Science and Technology Normal University, Nanchang, Jiangxi, China
Center for Central China Economic Development Research, Nanchang University, Nanchang, Jiangxi, China

Yue-Feng Xu
School of Economics and Management, Jiangxi Science and Technology Normal University, Nanchang, Jiangxi, China

Qiao Hu
Center for Central China Economic Development Research, Nanchang University, Nanchang, Jiangxi, China

ABSTRACT: By using the expert consultation theory, theoretical analysis theory, and frequency analysis theory, a scale pig breeding environment bearing capacity evaluation index system is built, which contains natural resources supply index and social economic conditions support and environmental pollution tolerance index. It effectively combines factor analysis and hierarchical analysis to evaluate the environmental bearing capacity of scale pig breeding in central regions of China. The evaluation results show some imbalance in the environmental bearing capacity among the regions, with Jiangxi Province the highest; followed by Hunan Province, Anhui Province, Shanxi Province, and Hubei Province; and Henan Province the lowest.

1 INTRODUCTION

Since 2004, the central committee of China has issued 13 "No.1 Documents" to direct "agriculture, farmer, and rural areas". They are the main projects for the government for a long time in the present and for the future. At the same time, scale pig breeding industry is an important part in the agriculture industry and for the prosperity of farmers. It plays an irreplaceable role in improving people's living conditions and realizing a well-off society. Pork production, according to data from the National Bureau of Statistics (NBS), is the principal part of animal husbandry in our country: the total output of the pigs, cattle, sheep, and poultry is 85.4 million tons in 2014, of which pork production constitutes 56.71 million tons (66.41%). Meanwhile, China is the world's largest pork producer and consumer. According to the USDA statistics, in 2014, the pork production of main country, which involved the global meat market, reached 110 million tons, while China's pork output is 56.71 million tons (51.3%). The total consumption of those countries is 110 million tons, while China's pork output is 57.17 million tons (52%). However, with the rapid development of pig industry and the rapid advance of intensification and scale in China, the problem of scale pig breeding pollution has been increasing seriously. This paper, from the perspective of bearing capacity of scale pig breeding, based on the construction of evaluation index system of scale pig breeding comprehensive carrying capacity, and the application of factor analysis and analytic hierarchy process (ahp), with the use of SPSS, carry out a statistical analysis on the scale pig breeding bearing capacity of composite index of six provinces in central China, and put forward policy suggestions.

2 CONSTRUCTION OF SCALE PIG BREEDING ENVIRONMENTAL CAPACITY EVALUATION INDEX SYSTEM

The design of this evaluation index system should fully consider the reality of China. Scale pig breeding is a complex nonlinear system. The measurement of complex systems using a single index is too difficult to reflect the main features. Information reflecting different aspects and characters, through the mathematical transformation or

Table 1. Scale pig breeding environmental capacity evaluation index system.

Target layer	Criterion layer	Indicator layer
Pig breeding environment bearing capacity index X	Natural resources supply index Y_1	Arable area per capita Z_1 Water resources per capita Z_2 Food per capita Z_3
	Social and economic conditions support index Y_2	Gross regional product per capita Z_4 The total proportion of industrial enterprises research and experimental development spending to the local GDP Z_5 The total output of pork Z_6
	Environmental pollution bearing index Y_3	Fertilizer use per unit area Z_7 (reverse target) Scale pig breeding density Z_8 (reverse target) Industrial waste water volume Z_9 (reverse target)

processing, has a comprehensive value with evaluation function. The numerical magnitude reflects the quantitative evaluation of the research object. That exactly is the essence of multi-index comprehensive analysis.

The current screening methods are expert consultation method, theoretical analysis, and frequency analysis. In this paper, we adopt the combination method of these three types, combining the current situation with principal contradiction of the scale pig breeding. It carries out the analysis, comparison, and combination, choosing those targeted strong indicators.

On this basis, the expert consultation method is further used to adjust the index. Considering the feasibility evaluation, further consulting experts, to replace a few incomplete data of indicators, the scale pig breeding environment bearing capacity evaluation index system was finally obtained. Scale pig ultimately meets the material needs of human, and only increasing the demand for pig production can promote its development. The industry development needs plenty of natural resources (land, water, etc.), to guarantee the material supply and pollutants given. The development of pig breeding technology and pollution treatment technology must have the economic basis. It seems that scale pig breeding is not an independent economic activity, and is related to natural resources, environment, social economic development, etc. Besides its condition reflecting index of pollutants producing ability, environment quality and natural resources can reflect the pollutants given ability indicator. Social and economic development reflects the support ability index and other factors. Therefore, we should fully consider the supply of natural resources, environmental protection, and social conditions support subsystems. Namely, the specific indicators should reflect the natural resources, environmental quality, social economy, scale pig breeding condition, and the environment bearing capacity evaluation index system shown in Table 1.

3 EMPIRICAL RESEARCH ON THE ENVIRONMENTAL BEARING CAPACITY INDEX OF SIX PROVINCES IN CENTRAL CHINA EMPIRICAL RESEARCH

3.1 Index measure of criterion layer

First, we obtain the corresponding standardized values of natural resources supply indices Z_{11}, Z_{22}, and Z_{33}, as shown in Table 2.

Then, the correlation matrix of the original variables is obtained.

Because of the correlation between these three original variables, the maximum correlation reached 0.862, further illustrating the original scalar suitable for factor analysis. According to the factor variance contribution rate from SPSS, the factor 1 variance contribution rate is 67.355%, and the characteristic value is 2.021. The factor 2 variance contribution rate is 28.293%, and the characteristic value is 0.849. These two measures show a variance of 95.648% (over 80%). Therefore, we take the first two components as the first and second principal components. According to the rotated component matrix from SPSS, component 1 maximum loads are 0.943 and 0.965, respectively, on the variable Z_{11} and Z_{33}, whereas that of component 2 on Z_{22} is 0.989. Ingredient 1 can be named F_{Y11}, and ingredient 2 F_{Y12}. The factors' score coefficients are shown in Table 4.

Expressions for various factors:

$$F_{Y11} = 0.514 * Z_{11} + 0.156 * Z_{22} + 0.558 * Z_{33}$$

$$F_{Y12} = 0.007 * Z_{11} + 1.025 * Z_{22} + 0.148 * Z_{33}$$

Normalizing the two factors' variance contribution rate (67.355%, 28.293%) results in the rates of 70.42% and 29.58%, respectively. Taking each factor variance contribution as the weight, the supply index of natural resources is:

$$Y_1 = 70.42\% * F_{Y11} + 29.58\% * F_{Y12}$$

Table 2. Standardized indices of six provinces.

Province	Z_{11}	Z_{22}	Z_{33}
Shanxi	0.26	0.11	0.60
Anhui	0.62	0.31	0.90
Jiangxi	0.39	1.00	0.77
Henan	1.00	0.07	1.00
Hubei	0.57	0.43	0.71
Hunan	0.60	0.75	0.72

Table 3. Correlation matrix of the original variables.

		Z_{11}	Z_{22}	Z_{33}
Correlation	Z_{11}	1.000	−0.336	0.862
	Z_{22}	−0.336	1.000	−0.232
	Z_{33}	0.862	−0.232	1.000

Table 4. Component score coefficient matrix.

	Component	
	1	2
Z_{11}	0.514	0.007
Z_{22}	0.156	1.025
Z_{33}	0.558	0.148

Similarly, through calculating and standardizing the social economy index, we can calculate the natural resources supply index standardization of corresponding values Z_{44}, Z_{55}, and Z_{66} in the six central provinces. According to the description of the social economic support index, GDPs of Hubei, Jiangxi, Shanxi, Anhui, Jiangxi, and Henan provinces are below average. Anhui Province has the maximum industrial enterprises research and experimental development spending to the local GDP ratio index. However, Jiangxi Province shows the minimum value. With an average value of 0.8, Shanxi, Henan, Anhui, and Jiangxi provinces are within the average level.

After testing the correlation and the applicability inspection over the social economic support indices, according to the analysis of the factor variance contribution rate, factor 1 variance contribution rate is 46.594% and its characteristic value is 1.398, while factor 2 variance contribution rate is 33.496% and its characteristic value is 1.005. These two indices showed a variance of 80.090% (over 80%), and hence we take the first two components as the first and second principal components. Component matrix shows that the first component results in the maximum load on variables Z_{44} and Z_{55}, and the corresponding correlation coefficients are 0.839 and

0.717. The second component results in the maximum load on Z66, and the correlation coefficient is 0.859. We can draw a conclusion that variables Z_{44} and Z_{55} are mainly composed of the first principal component, while Z_{66} is mainly composed of second principal component. After rotating, component 1 on variables Z_{44} and Z_{55} reaches a maximum load of 0.759 and 0.867, while component 2 on Z_{66} results in a maximum load of 0.958. The component 1 is named F_{Y21}, and component 2 F_{Y22}. In accordance with the component score coefficient matrix, the expressions for various factors are:

$$F_{Y21} = 0.543* Z_{44} + 0.681* Z_{55} - 0.081* Z_{66}$$
$$F_{Y22} = 0.256* Z_{44} - 0.253* Z_{55} + 0.904* Z_{66}$$

Through normalizing, we obtained the system index of social economy as

$$Y_2 = 58.18\%* F_{Y21} + 41.82\%* F_{Y22}$$

Similarly, we obtained the tolerance index for environmental pollution as:

$$Y_3 = 73.01\%* F_{Y31} + 26.99\%* F_{Y32}$$

From the expressions of natural resource supply index, social economic support index, and environmental bearing index, we can calculate their values, as shown in Figure 1.

3.2 Comprehensive estimate of the central region scale pig breeding environment bearing capacity level

Analytic hierarchy process (ahp) has the characteristics of systematic, flexibility, and practicability. Therefore, we decide to adopt analytic hierarchy process (ahp) to analyze the composite bearing capacity index. We invited more than 50 experts

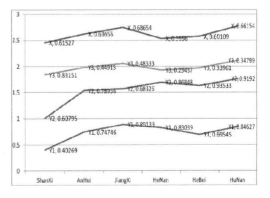

Figure 1. Environmental bearing capacity of six provinces in central comprehensive index figure.

Table 5. Judgment matrix table (A – B).

X	Y_1	Y_2	Y_3
Y_1	1	2	1
Y_2	1/2	1	1/2
Y_3	1	2	1

to score criterion layer weights, then consulted the experts repeatedly and revised it continuously. Finally, the judgment matrix is obtained, as shown in Table 5.

In the judgment matrix, the maximum eigenvalue $\lambda_{max} = 3$, and the characteristic vector $W = (0.4000, 0.2000, 0.4000)$, $CI = 0$, $CR = 0 < 0.1$. Therefore, we can determine that the judgment matrix is consistent, and then the environmental bearing capacity composite index is calculated as:

$$X = 0.4 * Y_1 + 0.2 * Y_2 + 0.4 * Y_3$$

The calculation of the environmental bearing capacity composite index is shown in Figure 1.

For the natural resources supply index, the supply index of Jiangxi Province is the highest, followed by Hunan, Henan, Anhui, and Hubei provinces. Meanwhile, Shanxi Province has the lowest corresponding value. This illustrates that Jiangxi and Shanxi province have the largest and least per capita natural resources, respectively, and hence the resource bearing capacity of the latter is low. In accordance with the environmental bearing capacity index of environmental pollution, Shanxi Province has the highest bearing index, followed by Jiangxi, Anhui, Hunan, Hubei, and Henan provinces. Figure 1 shows that the pollution bearing capacity of Shanxi Province is higher than other provinces. On the contrary, Henan Province has the minimum bearing capacity. In other words, because of the natural resources and social economic conditions to environmental pollution bearing of Jiangxi Province, the comprehensive bearing capacity of the scale pig breeding is the highest in central regions. The comprehensive carrying capacity of Henan Province is the lowest.

3 CONCLUSION

In recent years, the most serious smog pollution has warned us of environmental protection again. Therefore, as an important industry of animal husbandry and agriculture in China, scale pig breeding has an irreplaceable significance in the country's economic social development, improving people's living standard. On the basis of the scale pig breeding environmental capacity evaluation research of the central region of China, we found that imbalance exists in every place. The environmental bearing capacity index of Jiangxi Province is the highest, followed by Hunan, Anhui, Shanxi, Hubei, and Henan provinces.

In the whole evaluation process, because of the limitation of data, the evaluation index system is unable to adopt those excellent indicators accounting for a significant impact on the result of indicators. The evaluation result, as well, is dependent on the index system, which directs further research.

ACKNOWLEDGMENTS

The authors gratefully acknowledge the grant of project "Scale Pig Breeding Ecological Energy System Stability Feedback Simulation Study" (71501085) supported by the National Natural Science Foundation of China and the Project of Humanities and Social Sciences in Colleges and Universities by Jiangxi Province (GL1536).

REFERENCES

An Cui-juan, Hou Li-hua. Resources and environment carrying capacity by ecological civilization perspective evaluation research——In the case of guangxi beibu gulf economic zone [J]. Ecological economy, 2015, 11:143–147.

Li Zhi-gang, Leng Bi-bing, etc. The comprehensive evaluation research of scale pig breeding environmental capacity. [J]. An hui Agricultural Sciences, 2014, 27:9382–9385.

Lu Yuan-qing, Shi Jun. The construction of low carbon competitiveness evaluation index system [J] Statistics and Decision, 2013(1): 63–65.

Shi Jing, Xiao Hai-feng. The technical demand of farmers to animal husbandry and some influencing factors research—Based on the analysis of the questionnaire survey data to the wool-sheep farmers [J]. rural economy, 014(3): 56–60.

Song Fu-zhong. Livestock and poultry breeding environment bearing capacity of system and early warning research[D] Chongqing: Chongqing University, 2011.

Yang Xiu-ping. Based on the tourism environmental capacity analysis of DIAHP and the relevant countermeasures research. [J]. Journal of ecology and rural environment, 2008, 24 (1): 20–23.

The development of integrated management of water and fertilizer

L.M. Chuan, H.G. Zheng, J.J. Zhao, S.F. Sun & J.F. Zhang
Institute of Information on Science and Technology of Agriculture, Beijing Academy of Agriculture and Forestry Sciences, Beijing, China

ABSTRACT: Integrated management of water and fertilizer (fertigation) aims to control irrigation water and nutrient application comprehensively according to crop requirements, in order to promote the uptake of fertilizer nutrients by irrigation and enhance the effective utilization of water and nutrients in farmland. Beijing is one of the cities in the world that seriously lack water resource. This paper reviewed the current situation of fertigation in Beijing and analyzed the problems in the development of this technique. It is also suggested that fertigation is crucial to promote the efficient use of water and fertilizer resources. Beijing should pay more attention to develop the fertigation technology and research on water-soluble fertilizers and water–fertilizer integration equipment. Technical service, policies, and financial support should be focused on improving the utilization efficiency of water and fertilizer.

1 INTRODUCTION

Beijing is one of the cities in the world that seriously lack water resource (Yang et al 2016). Large population and less water is Beijing's basic situation. Agriculture is one of the major users of water in the city. In September 2014, the government issued the document of views on adjusting the structure, transforming the mode, and increasing the efficiency of water-saving agriculture. It is indicated that modern agriculture should pay more attention to develop with available water, and strictly limit the development of high water consumption industry. Irrigation consumes much of the limited agricultural water resources. Improving the efficiency of irrigation water is the key problem in modern agriculture. The integrated management of water and fertilizer (fertigation) is an inevitable choice to realize water-saving agriculture.

2 CURRENT SITUATION

2.1 *Various techniques of irrigation mode developed*

In recent years, Beijing has actively implemented integration technology of water and fertilizer to the planting of vegetables and fruits, and has gradually explored various techniques of irrigation mode including precise drip irrigation, surrounded drip irrigation, micro-spray irrigation, and plastic film mulching irrigation (Zhang 2015). Beijing also promoted the rainfed technology, random irrigation, and automatic moisture monitoring of the crops.

It is roughly estimated that 900 m^3 of water is saved per year in 1 ha, which saves 30% of water compared with the traditional irrigation. Because of the precise fertilization, we can reduce the amount of fertilizer input. Furthermore, the implementation of water–fertilizer integration technology in the greenhouse inhibits crop diseases largely, which significantly reduces the pesticide inputs. For example, Xinchengyuan, a planting park, located in Changping, around Beijing, reduced irrigation water from 300 to 120 m^3 on strawberry by drip irrigation rather than flood irrigation. The water-saving rate is up to 60% (Wang et al 2016).

2.2 *Much water has been saved*

In 2011, agricultural water saving reached about 41 million m^3 in Beijing, equivalent to 20 times of Kunming Lake water storage. Extensive farming is gradually replaced by the scientific water and fertilizer saving mode. It had reached the goal of water saving, fertilizer saving, and income rising, which is contributed by the integration of water and fertilizer.

2.3 *Water saving in agriculture achieved a great progress in the Twelfth Five Year*

During the "Twelfth Five-Year Plan" period, in order to realize the target of 35% of vegetable self-sufficiency, Beijing started to promote the integration technology in the suburban orchard, field and greenhouse. According to the statistic of Beijing Municipal Bureau of Agriculture in 2011, through the implementation of "comprehensive water-saving demonstration project in urban agriculture corridor", "the promotion project of water–fertilizer integration in fields and orchards", and "gravity drip irrigation–fertilization technology demonstration project", the city established 128 water–fertilizer integration bases, and up to 1800 ha. The monitoring results showed that after the application of irrigation–fertilizer integration technology, it saved 152 m^3 of water, 36 kg of fertilizer, and 927 yuan per year per 667m^2 for vegetables, which are 86 m^3, 26 kg, and 605 yuan for fruits, respectively, compared with the conventional technology (Gong et al 2013).

In 2012 and 2013, Beijing established three wheat demonstration areas of micro-spray water–fertilizer integration in Shunyi, Tongzhou, and Fangshan districts, totaling 62 031 m^2. Compared with the control group, it saved 160 m^3 of water and achieved a yield increase of 51 kg, which is 11.5% per 667 m^2. The water production efficiency is 1.61 kg/m^3, improved by 0.37 kg/m^3. The maximum production is 539 kg per 667 m^2. In 2015, Beijing actively promoted the water–fertilizer integration and plastic film mulching irrigation. By the end of May 2015, the high efficiency of water-saving technique application for vegetables covered 130.06 million m^2 and the coverage rate reached 45%. It saved 9.24 million m^3 of water, which was expected to reach 18.24 million m^3 at the end of the year.

2.4 Benefit of the integration management of irrigation and fertilization

The benefit of the integration management of irrigation and fertilization is not only saving water and fertilizer, but also the concept change of agricultural production. With the popularization and application of integration of water and fertilizer, the technology could create a "soil" environment for the plants and achieve planting modes of various three-dimensional styles, which expands the cultivation space and increases the yield. Therefore, it is also an effective method to increase the efficiency of agriculture (Gao et al 2015).

3 PROBLEMS EXISTED

Although good progress has been achieved in integration technology of irrigation and fertilization, there still exist some problems.

3.1 Imbalanced development

The imbalance is mainly revealed in different regions and different crops. The imbalanced development in different regions is mainly due to insufficient

understanding and enthusiasm for the technology in townships, which resulted in slow development. The imbalanced development in crops is due to the fact that most researches and applications are usually focused on economic crops, and hence more studies on the cereal crops should be conducted in future (Chen et al 2013).

3.2 High cost and not sustained

At present, there is lack of cost-effective equipment and automation management technology for irrigation–fertilization integration technology. Relative companies are less domestic and the R&D capability is not high. The equipment requires high investment first and has slow capital return. Furthermore, because of the poor quality, poor performance, pipeline blockage, and other reasons, the equipment is always gradually shelved and could not produce the maximum benefit. It dampens the enthusiasm of the farmers for the integration technology to a large extent.

3.3 Supporting products is not enough

The irrigation–fertilization integration technology needs a combination of irrigation equipment, water-soluble fertilizer, and corresponding system. However, so far, the combination of technology and products has not been close enough. Some people only focus on irrigation and fertilization equipment, while ignore the system optimization and the application of water-soluble fertilizer. The technology involves subjects including farmland water conservancy, irrigation engineering, fertilization, cultivation, soil, and so on. The discipline boundaries may limit its development. There is also lack of comprehensive specialized technical personnel (Zhao 2015).

3.4 Technical service is not adequate

The enterprises always pay more attention to production and sales of their products, while technical guidance and after-sale service for farmers are not adequate. Although there are a few of training programs, it is difficult for the farmers to grasp all the use method and performance parameters of various equipments accurately. Besides, service system is not very good. All these inhibit the farmers' motivation to adopt new equipment and technology.

3.5 Lack of water-soluble fertilizer

The production scale of water-soluble fertilizer, which the integration technology needs, is small and the price is relatively high. The research on product development and large-scale application of water-soluble fertilizer technology is still in the initial stage. The school–enterprise cooperative research is few and there are only few well-known fertilizer varieties suitable for irrigation fertilization system.

3.6 Supporting policy is not comprehensive

The financial subsidy mechanism is not fully established. The technology promotion funds are insufficient, and the subsidy standard is relatively low. Moreover, the standard system is not perfect. Some standards do not meet the requirements of the integration technology of irrigation and fertilization. The technology of water saving agriculture still needs further development. The industrial support system and the management and service support are not harmonious enough, which has become one of the main obstacles to the development of water saving.

4 DEVELOPMENT TREND OF THE IRRIGATION AND FERTILIZATION

4.1 Focus on the development of water-soluble fertilizer

The quality of water-soluble fertilizers has strict requirements, especially in insoluble content and particle size. At present, the technical services and supporting equipment are not perfect. Major factors restrict the development of water-soluble fertilizer, including the backward production technology, weak R&D foundation, high cost of production and logistics, serious product homogenization, lack of product standard, and inadequate household service input. Conceptual speculation and the absence of relevant supporting policy are also common phenomena currently. There are some problems in the links of R&D, production, distribution, and use of water-soluble fertilizer industry. It is urgent to establish an association of production enterprises, scientific research units, distribution businesses, and farmers to combine all aspects to solve the problems.

Production enterprises of water-soluble fertilizer and irrigation equipment should work together to research suitable fertilizer for crop field and actively cultivate the market. Combining production, teaching, and research is one of the ways to accelerate the technology innovation of water-soluble fertilizer production, reduce costs, and upgrade the product quality. It is also important to optimize the relevant product standards and strengthen the construction of the industrial integrity system.

4.2 Emphasizing the research on high use efficiency of water and fertilizer technology

The integration of irrigation and fertilization involves many disciplines including agriculture, information science, engineering, and management science. In the past, people only paid attention to the individual engineering technology, such as the waterproof canal, low-pressure pipeline, and sprinkler irrigation, and ignored supporting measures to form comprehensive technology. It resulted in water saving, but was less beneficial. We recommend using water-saving technology in integration of engineering, biological, agronomic, and management measures, such as water saving, drought-resistant varieties, fertilization, surface covering, and water-composing management. Attention should be paid not only to transforming the original surface irrigation systems, but also to large-scale development of spray and drip irrigation and other advanced technologies. Development of water-saving irrigation technique and the combination of agricultural practices and water management measures need the establishment of irrigation system according to the water requirement rules, monitoring soil moisture, root distribution, soil properties, facilities, and technical measures. About the nutrient management, fertilization technology and the amount of application should be scientifically determined in accordance with the objective yield, fertilizer requirement, soil nutrient content, and irrigation characteristics. More studies on water–fertilizer coupling mechanism are needed. We should take full advantage of the integration technology of irrigation and fertilization and improve the utilization efficiency of water and fertilizer.

4.3 Strengthening the R&D of water–fertilizer integration equipment

The performance of integration equipment made in China is less and the automation level is lower than that made in other countries. In general, senior technicians engaged in water-saving irrigation equipment manufacturing are less. At the same time, domestic enterprises producing integration equipment always have small scales, low benefits, and relatively low economic strengths, which limit product R&D funds and decreases technology strength. Globalization, engineering, and precision are basic trends of integration technology of irrigation and fertilization. Therefore, we should pay more attention to strengthening the development and application of water–fertilizer integration equipment in the future. It is urgent to line with international standards in many aspects, such as the diversification of specifications, the aging resistance of plastic pipe, and the uniformity degree of emitter effluent. Furthermore, improving the standardization system and quality of the supervision system is also of great importance.

4.4 Optimizing the subsidy policies for integration equipment

The high investment of water–fertilizer integration mode leads to poor enthusiasm of the farmers. Therefore, it is important to establish special foundation for water-saving agricultural development. Farmers who implement water–fertilizer integration technology can get material rewards from the foundation. When some water–fertilizer integration program needs major repair or renovation, the funds could be included in the annual financial budget. We can adopt more preferential water price policy for the water-saving agriculture projects. In regions where water resources are scarce, we should also improve the water subsidy proportion. The water-saving agriculture to ensure national food security should be taken as key financial support objective. Besides, it is also suggested to improve subsidies for the integral pipeline equipment and increasing support for water-fertilizer facilities, machinery, and the construction of storage tank to accelerate the promotion of integrated technology. All these steps are aimed to ensure that the farmers are able to afford advanced water-saving irrigation technologies, promote continuous transformation of agricultural production mode, and achieve high and stable yields (Yang et al 2015).

4.5 Establishing the standard system for water-saving agriculture

Agricultural water efficiency may or may not depend on its form, such as sprinkler, micro-irrigation, waterproof canal, pipelines, or small beds instead of large border. The benefits of some projects are little in spite of less irrigation water because of taking only one certain technical measure. In order to guide the healthy development and ensure the quality of water-saving agriculture, it is important to establish the standard index system of water-saving agriculture in China (Gao et al 2012).

Water-saving agriculture includes the source of water, conveying, irrigation, management, and crop consumption. The natural conditions are different from one region to another, and so do the form and scale of the water-saving agriculture. Therefore, water-saving agriculture can never be measured by a single indicator. It should be a system containing a set of interrelated indices. The level of water-saving agriculture is related the national economy and strength, which means that it is restricted by the ability of the country and the farmer's investment to develop the water-saving agriculture. Therefore, it is suggested to establish a consistent standard system in accordance with the three levels: water indicators, engineering indicators, and efficiency indicators.

5 CONCLUSIONS

In this paper, it is concluded that fertigation is an important way to promote the efficient use of water and fertilizer resources. Beijing should pay more attention to develop the fertigation technology or to make full use of the water and fertilizers. In the future, studies on water-soluble fertilizer, water–fertilizer integration equipment, and actives on technical service, policies, and financial support should be focused on to increase the utilization efficiency of water and fertilizer.

ACKNOWLEDGMENTS

This study is funded by the Innovation Program of Beijing Academy of Agricultural and Forestry Sciences "Structure evolution and driving forces analysis of agricultural water use in Beijing (KJCX20160503)", "Information research to promote the agricultural science and technology innovation ability (KJCX20140207)", and "Key Laboratory of Urban Agriculture (North), Ministry of Agriculture, P.R.China".

REFERENCES

Chen, G.F., Du, S. Jiang, R.F., et al. (2013). The present situation of research and application of water and fertilizer integration technology in our country. *China Agricultural Technology Extension* 05: 39–41.

Gao, P. Jian, H.Z., Wei, Y., et al. (2012). The application status and development prospect of integrative water and fertilizer. *Modern Agricultural Science and Technology* 8: 250, 257.

Gao, X.Z., Du, S., Zhong, Y.H., et al. (2015). Present situation and Prospect of the integrated development of water and fertilizer. *China Agricultural Information* 02: 14–19, 63.

Gong, J.J., Liu, Y.X., Liu, H., et al. (2013). The development and application of watr-fertilizer integration technology in China and abroad and its enlightenment to Beijing. *Reference of agricultural science and technology*, Oct. 2013.

Wang, Z.P., Wang, J.M., Cheng, K.W., et al. (2016). Water-fertilizer integration technology for strawberry. *Vegetables* 05: 27–29.

Yang, L.L., Wang, C.Z., Han, M.Q., et al. (2016). Reflections on the construction of high efficiency water saving agriculture in Beijing. *Beijing Agriculture* 01: 195–196.

Yang, L.L., Zhang, H.W.. Han, M.Q., et al. (2015). Analysis of techniques and application prospect of water and fertilizer integration technology. *Journal of Anhui Agri. Sci.* 43: 13–25, 28.

Zhang, M.F. (2015). High efficient water-saving development of agriculture in the new period of Beijing. *China Water Resources* 11: 49–50, 53.

Zhao, J. (2015). The problems and solutions in the adaption of fertigation technology. *Northwest A & F University*.

Advances in Energy, Environment and Materials Science – Wang & Zhou (Eds)
© 2017 Taylor & Francis Group, London, ISBN 978-1-138-03600-0

An integrated approach to analyze satisfaction on post-disaster recovery: A comparative study between Nanba Town, China, and Greensburg, US

Mimi Shi, Saini Yang, Shuangshuang Li, Jiayi Fang, Weiping Wang & Wenhui Zhang
State Key Laboratory of Earth Surface Processes and Resource Ecology, Beijing Normal University, Beijing, China
Academy of Disaster Reduction and Emergency Management, Ministry of Civil Affairs and Ministry of Education, Beijing Normal University, Beijing, China

ABSTRACT: Post-Disaster Recovery (PDR) is a complex social issue and is an untraditional topic for the Institutional Analysis and Development (IAD) framework. An integrated approach, which combines the IAD framework and a modified Capital Assets Framework (CAF), is proposed to provide a new perspective for analyzing complex social issues. A quantitative and qualitative comparative analysis on residents' satisfaction of PDR was conducted in Greensburg and Nanba Town, from the perspective of capital assets and institutions. Results indicate that the diverse exogenous driving factors of action situation and outcomes can be effectively represented by capital assets. Fewer losses of natural capital and higher physical, financial, human, and social capital improves residents' satisfaction towards PDR in Greensburg, as does the 'bottom-up' interaction mechanism and decentralized institution settings, which open channels for public participation in Greensburg. This integrated approach provides a new systematic conceptual framework for analysis on similar complex social issues.

1 INTRODUCTION

In recent decades, the world has experienced a climax of disastrous events, with an increase of remarkable social and economic consequences (the United Nations Development Programme 2004). Post-Disaster Recovery (PDR) is a crucial work as it plays a vital role in providing long-term developmental guidance for people, particularly for those in disaster-prone areas (Yi and Yang 2014). However, implementing PDR efficiently and obtaining high satisfaction from the affected households are still challenging to achieve. In practice, social satisfactions with PDR differ in different regions. Factors that lead to these differences are not well understood since PDR involves social, economic, political, and cultural factors and common resources—both natural and social resources. Among administrative regulators, academics, and some organizations, there is a growing consciousness that both short-term and long-term PDR are crucial to the future of the affected area (Ingram et al. 2006). A large amount of researches have concerned themselves with relevant social issues such as policy-making, program design, and experience summary of PDR (Hanes 2000; Yang et al. 2015). However, there is no systematic analytical framework aimed towards addressing the satisfac-

tion issue of PDR to the best of our knowledge. An integrated approach is needed to analyze this complex social issue. In this study, we propose an integrated approach combining the traditional Institutional Analysis and Development (IAD) framework and the modified Capital Assets Framework (CAF) to provide a new perspective for an unconventional social issue such as PDR.

As a typical approach to the systematic analysis of collective action and the commons (Ostrom 1986; Kiser and Ostrom 2000; Ostrom 2005), the IAD framework has been described as 'one of the most developed and sophisticated attempts to use institutional and stakeholder assessment in order to link theory and practice, analysis and policy' (Aligica 2006, pp. 89). As such, the IAD framework has proved useful in understanding a wide variety of institutional arrangements in both developed and developing countries (Imperial and Yandle 2005). It is a 'multi-tier conceptual map', (Ostrom 2011, pp. 9) which concerns how social-cultural, institutional, and biophysical contexts within which all such decisions are made affect the structure of action arenas and incentives of the resulting outcomes (Ostrom 2011). In the past three decades, scholars have applied the IAD framework to areas such as common pool resources and resource management (Feeny 1994; Lee 1994; Clement 2010) as well as irrigation systems (Araral 2005; Lam 2001). So far, however, there is no existing case study utilizing the IAD framework on PDR since it is an unconventional topic for the classic IAD framework.

The CAF was adapted from the Sustainable Livelihood framework, which was outlined by the Institute for Development Studies and widely used by the British Department for International Development (DFID) and the International Fund for Agriculture Development. The CAF considers five types of capital assets (human, social, natural, financial, and physical) to understand livelihood outcomes and risk. In recent years, the importance of institutions for disaster management has also been discussed (Barone and Mocetti 2014). Considering the importance of institution in PDR, we added institutional capital to the CAF to provide a simple and straightforward method for complex issues, without ignoring the exogenous factors such as contextual economic, political, and social elements.

In this research, a comparative case study on PDR was conducted to validate the validity of the integrated approach. Greensburg, Kansas (US) and Nanba Town, Pingwu County (China) were selected as study areas for a regional comparison to explore the possible causes of different degrees of satisfaction from residents' cognitive feedbacks.

2 CONCEPTUAL FRAMEWORK

The integrated approach, which combines the IAD framework and a modified CAF for PDR analysis, is shown in Figure 1. The modified CAF was inset to systematically describe the characteristics of a complex social problem.

For a detailed illustration of exogenous variables in this integrated approach, natural and physical capital are used to describe the biophysical condition, while social, human, and financial capital are applied to measure other attributes of the community.

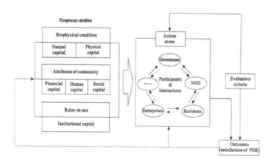

Figure 1.   The integrated framework for PDR analysis.

Institutional capital can be deemed as an indicator of rules-in-use. In the action arena of PDR, multiple stakeholders including government, NGOs, local enterprises, and residents interact with a common goal of post-disaster recovery. As a consequence, different satisfaction degrees of PDR are generated. Meanwhile, a series of evaluation criteria are set to assess interaction and outcomes. Section 3 provides a case study to illustrate the application of the proposed approach in the analysis of PDR from the perspective of social satisfaction.

3 CASE STUDY

3.1 *Study area*

Based on the proximity of geographical area, severity of hazards, disaster losses, and recovery time, we selected Greensburg, Kansas (US) and Nanba Town, Pingwu County (China) as study areas. Nanba Town is located in the southeast part of Pingwu County, Mianyang City, Sichuan Province with a 326 km² area. Nanba Town's construction area is 2.2 km², ranking as the second-largest town in Pingwu County. It consisted of 26 villages and one neighborhood committee with a population of 19,976 at the end of 2008[1]. Nanba Town represents the worst-hit areas affected by the Wenchuan Earthquake in May 2008. The infrastructures in Nanba were seriously damaged: 654,600 m² of housing collapsed, and another 190,000 m² of housing was severely knocked down[2]. In total, the earthquake killed 1,565 and injured 4,500 residents, which accounts for about 46% of the death toll in Pingwu County. Restoring the earthquake-affected areas requires large amounts of resources, including manpower, funds, and effective organizations and systems (Ge et al. 2010).

Greensburg is located in the southwest part of Kansas, US. Its size spans 3.83 km², and Greensburg was established to support the country's thriving railroad industry. Over the last four decades, the small rural farmtown's population has been declining from 1,988 in 1960 to 1,389 in 2006, with a struggling economic base (Berkebile Nelson Immenschuh McDowell Architects 2008). On May 4th 2007, an EF-5 tornado, which was estimated to be 2.74 kilometers wide with 329.92 km per hour winds, made its way through Greensburg at approximately 9:45 pm. The damage to Greensburg was significant, with more than 90% of the structures in the community severely damaged or destroyed and 11 people killed as a result of the tornado (Carbin and Schaefer 2008). Following the disaster, the community set forth to rebuild a prosperous future and a greener Greensburg (Paul and Che 2011).

3.2 Data and survey

The data used in this research consists of statistical data and survey data. The former were obtained from statistical yearbooks, reports, and press releases from government agencies, while the latter were acquired from interviews and questionnaires in study areas. A pre-test with a sample size of 15 was implemented in Nanba Town in December 2012. According to the pre-test results, we made modifications on the questionnaires.

Two surveys were conducted in Greensburg and Nanba Town. The first was carried out by a group of ten well-trained faculty and graduate students from January 9th to 15th, 2013 in the center area of Nanba Town. All households were surveyed with face-to-face instructions except those that were out of town or refused our visit. As a result, a total of 689 responses were collected. We obtained 513 valid responses after removing those that were invalid. Meanwhile, we conducted sampling surveys in villages located in the mountains. In total, 717 valid samples were obtained, 513 of which were obtained in the central district while 204 were gained in the mountainous villages.

A similar survey in Greensburg was conducted in April 2013 by the core group members of faculty and graduate students who conducted surveys in Nanba. There were almost 170 households in Greensburg, Kansas. Except for the households that were out of town or turned down our survey, we interviewed 122 households and received 117 valid samples.

When dealing with the samples, satisfaction of recovery, current life, education, medicare along with expectation of life are deemed as major contrast indicators. Moreover, educational attainment/ skill level, religious/collective activities, occupation and trust degree are all taken into account.

3.3 Analysis

From the statistics of our surveys, satisfaction towards PDR in two study areas shows significantly different. The degrees of satisfaction with recovery, current life, education, and medicare in Greensburg are higher than that in Nanba Town, as shown in Figure 2.

What are the potential driving factors for these differences? Taking residents' satisfaction degree as criteria of the interactions and outcomes in PDR process, we applied the proposed integrated approach to explore the answer to this question.

Exogenous variables
Disaster events significantly impact vulnerable communities through damages to natural, social, and economic conditions (Cutter et al. 2003).

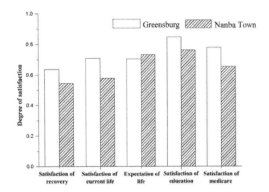

Figure 2. Average degree of satisfaction in Nanba Town and Greensburg.
Note: 4 and 5 choices were set respectively in original questionnaire of Greensburg and Nanba ("1" = very satisfied, "4" = very dissatisfied; "1" = very satisfied, "5" = very dissatisfied). For comparison and better understanding, data were normalized (modified degree = (max-original degree)/(max-min)).

Various capital assets are needed for PDR, while 'there is no consensus as to the precise factors that most contribute to recovery, and the relative weight of different factors is likely to vary according to the specific context and disaster region' (Shimada 2015, pp. 1). Some authors have claimed that the most critical factors in the process of PDR are the quantity of money flowing into a disaster-struck region (Ingram et al. 2006) and the socioeconomic and demographic conditions of the area (Katz 2006). Meanwhile, others have argued that a fundamental factor in any recovery is human capital rather than physical capital (Horwich 2000). Social capital plays a more significant role in the process of recovery than physical infrastructure (Aldrich 2011). Sensitive groups that have lower social capital stock (Wang et al. 2012) suffer from a longer recovery process. Therefore, we attempted to integrate six types of capital assets, namely natural, physical, financial, human, social, and institutional capital, to systematically analyze the potential causes of different satisfaction of PDR in Nanba Town and Greensburg.

Biophysical condition
Biophysical condition refers to the physical resources and capabilities to provide and produce goods and services (Polski and Ostrom 1999). Natural capital and physical capital are considered as biophysical condition in this case study.

Natural capital
Nanba Town was previously extremely rich in mineral resources, such as manganese ore and quartz. Its forest coverage reaches 61.8% with 302,575

acres, much higher than the national average forest coverage (20.36%) in 2008[3]. Fruit trees and walnut trees are the major types of cash tree. The catastrophe washed out ten thousand tons of minerals, about 34 hectares of sloped land, 28.25 hectares of farmland, and 17.06 hectares of woodland in Nanba Town alone[4]. In addition, frequent secondary disasters such as landslides, mud-rock flow, and barrier lakes changed the original landscape and destroyed regional ecological systems and consequently induced biodiversity loss. Large amounts of mineral and land resource losses have had long-term impacts on the production capacity and ecosystem function in Nanba Town. It is a tedious and arduous task to recover the mineral and land resources as well as the ecosystems. In this sense, the conditions are unfavorable for Nanba to recover and develop to rely on its own natural capital.

The tornado hit Greensburg was the strongest to hit the US since June (Edwards et al. 2013). Compared to the Wenchuan Earthquake disaster, this tornado brought similar damages to infrastructures and widespread crop failure, but the damage to land resources was much less. In addition, Greensburg is located in one of the windiest parts of the US, with consistently high wind speeds throughout the year. In addition, sufficient sunlight in Greensburg presents opportunities for energy-efficient buildings. For instance, solar panels on the roof convert sunlight directly to electricity, which can meet nearly ten percent of the building's electricity requirements (Judkoff 2008). Thus, the loss of natural capital in Greenburg is less than that in Nanba, which influences the recovery target determination in the action arena.

Physical capital

Hard infrastructure contributed positively to facilitate the recovery of community economic functions (Khew et al. 2015). Housings and the infrastructures in Nanba Town were mostly destroyed by the Wenchuan Earthquake. Similarly, over 90 percent of the structures in Greensburg were gravely destroyed.

Standards of housing, public facilities and infrastructures (e.g., schools, power stations, roads, and pipelines) reconstruction was addressed in the recovery plan of Nanba Town. Within three years, this town finished the reconstruction of houses and infrastructure in the central area. However, there are some underlying problems according to our surveys and interviews. For example, the infrastructures in the rural area are underdeveloped, and the water supply and sewer system are undeveloped, even in the central area. Further, the available services and equipment cannot meet the high-standard education and medical infrastructure in this town.

A detailed recovery plan was set in Greensburg, targeting a sustainable and greener future. As a result, a complete and environmentally friendly set of infrastructures (e.g., community meeting centers, schools, medical and emergency service facilities, transportation systems, and parks) has been established to foster economic development and provide convenient services to both the rural and urban residents. Particularly, all new city buildings larger than 4,000 square feet were built according to the U.S. Green Building Council LEED Platinum certification level, and they have all reduced energy consumption by 42% as compared to standard buildings (Paul and Che 2011). Thus, Greensburg owns higher physical capital compared to Nanba Town, making residents' lives easier and influencing their satisfaction degree.

Attributes of community
The attributes of a community include the demographic and social features of the community (Polski and Ostrom 1999). Financial capital, human capital and social capital are indicators of attributes of community in this case study.

Financial capital

Some scholars argue that it takes a longer amount of time for the people on lower economic levels to recover from natural disasters (Yasui 2007). GDP and income are commonly used to measure financial capital. Industrial structures also need to be taken into consideration since disaster impacts on various industry types may significantly influence the recovery cycle.

Led by mining industry and agriculture, Nanba Town has the major industrial clusters of Pingwu County. Since mineral and land resources are accumulated through gradual, long-term geological evolution, they cannot be recovered in a short amount of time, not even in decades, once they have been destroyed. Thus, the mining industry and agriculture have proved difficult to revive after the earthquake in Nanba Town. In addition, owing to the mountainous landscape and poor transportation system, non-agricultural industries cannot develop rapidly. According to the results of a questionnaire survey conducted in Nanba Town in 2013, the per capita income was 9,413.73 RMB ($1,520), which is about 51.4% ($2,986.6[5]) of the national average. The average annual household income was 33,656.2 RMB ($4,036.7). Since there is no similar survey conducted prior to 2008, the growth rate of economy in Nanba is not available directly. The growth rate of GDP in Pingwu between 2007 and 2010 is 0.32 times (17%[6]) of that of national level (52.6%). As a typical town in Pingwu County, Nanba's GDP growth rate is inferred to be significantly lower than national average.

Greensburg is an important agricultural production base in the US. Thus, agriculture is a significant contributor to the town's economic well-being. Meanwhile, large amounts of labor force are applied in non-agricultural industry as educational services, health care and social assistance, accounting for 28.75%[7]. The tornado on May 4th 2007 caused widespread crop failure, but did not ruin the land resource, which is the backbone of the agricultural industry. The focus on building a "green" town brought Greensburg a great opportunity to attract new environmentally friendly economies (Pless et al. 2010), and the non-agriculture industry benefited from the reconstruction and high-quality services of infrastructure. The per capita income in Greensburg in 2013 was $26,891[8], about 90% of the national average ($28,645)[9]. The median household income was $42,024, lower than that of Kansas State ($50,972)[10]. However, the growth rate of the median household income between 2000 and 2013 is 1.87 times (47.8%) that of Kansas State (25.5%)[11], which indicates an effective economic recovery. Above all, Greensburg's industrial structure suffered less impact by the tornado compared to the earthquake in Nanba Town, and the stock of financial capital is slightly better than that in Nanba, providing a better base for the action arena.

Human capital

A successful recovery depends on both individual and community-based capacities to respond to the impacts of a disaster. It is reasonable to expect that countries with low human capital are likely to poorly respond to disasters, leading to insufficient recovery. Educational attainment along with the skill levels of residents can be used to measure human capital.

A high percentage of Greensburg residents (82.6%) over the age of 25 have a high school education (Berkebile Nelson Immenschuh McDowell Architects 2008), and the total high school graduation rate is 86% in this area. In terms of education attainment and skill levels, Greensburg proves

much more successful than Nanba Town. In Greensburg, the average highest educational level of households is up to 6.67 ('6' = secondary completed, and '7' = vocational college completed), while in Nanba Town, it is only 2.50 ('2' = junior middle school completed, and '3' = senior high school/technical secondary school completed). With a view of householders' unique leading force in households, a comparison of their educational attainment in Nanba Town and Greensburg was conducted. Results showed that householders in Greensburg are better educated and trained, and this finding may make a difference in the interaction function and the degree of satisfaction towards PDR.

Social capital

Social capital has a significant influence on PDR at the individual, household, and community level. In general, residents in communities with great social capital show relatively high enthusiasm of recovery participation (Nakagawa and Shaw 2004). Households with richer social network resources are more likely to raise more funding (Narayan and Pritchett 1999). As for a community, frequent activities can enhance the recovery rate in the process of PDR (Tatsuki 2008). In addition, social network significantly affects the psychological recovery of survivors (Beaudoin 2007). In our study, based on the comparability of survey data, some indicators were used to measure social capital (see Table 2).

The above five indicators are deemed as equal weight variables, and the maximum score is five (the maximum of each indicator is regarded as one). As a result, the score of social capital in Greensburg is higher (4.40) than that in Nanba Town (3.41), mainly due to high participation of religious activities and collective activities, which influence the interaction degree of participants.

Rules-in-use

The rules-in-use are the "strategies adopted by participants within ongoing situations" (Ostrom,

Table 1.   Human capital contrast between Nanba Town and Greensburg.

	Nanba Town (n = 717)		Greensburg (n = 117)	
Indicators	Mean	Description of indicators	Mean	Description of indicators
The highest educational attainment of household	2.50	0–5: "0" = illiterate, "5" = graduate students	6.67	1–8:"1" = illiterate, "8" = university
The highest skill level of household	2.54	number of skill training items	7.10	1–8: "1" = farmer, "8" = skilled public sector
Educational attainment of householder	1.58	0–5: "0" = illiterate, "5" = graduate students	6.55	1–8: "1" = illiterate, "8" = university
Skill level of householder	0.33	number of skill training items	6.42	1–8: "1" = farmer, "8" = skilled public sector

Table 2. Social capital contrast between Nanba Town and Greensburg.

Indicators	Nanba Town (n = 717) Mean	Greens-burg (n = 117) Mean	Description of indicators
Heterogeneity of work	4.03	1.60	Work kinds among household members
Percentage of religious activity	29.15	69.64	Household participation rate of religious activity (%)
Frequency of collective activities	0.29	1.49	Average participation of collective activities (times/household/week)
Degree of consolidation	0.65	0.77	Residents in my town are more consolidated. Higher degree means a more consolidated town.
Degree of trust	0.61	0.64	Residents in my town are more trust worthy than that in other town. Higher degree means a more trustworthy town.

2005). Institutional capital is considered as the Rules-in-use in this case study since it decide the root mechanism of resource allocation and legislation in PDR.

To cope with reconstruction following catastrophe, a set of rules (policies, legislations, recovery plans) and even some formalized institutions were dedicated to these two study areas. More multi-level legislations were put forward to motivate the PDR process in Nanba Town, while Greensburg placed an emphasis on 'long-term sustainability' at the community level. When it comes to formalized institutions, results displayed a marked difference in that a systematic and integrated institutional setting from the national to county levels was established in China, while NGOs along with design enterprises played a significant role in the PDR process of Greensburg.

A conclusion can be drawn from these findings that the rules and institution settings of PDR in China are much more centralized than those in Greensburg. Previous researches indicate decentralized governments prepare for and respond to disaster more effectively in relation to more centralized systems (Drabck and McEntire 2003). This may help to explain the different interactions among participants, as well as the different outcomes of the action situation.

Participants and interactions

Disasters, especially catastrophes, demand the management and coordination of governments. NGOs and academic institutions also contribute through various channels (Agrawal and Gibson 1999). Central and local governments, NGOs, communities, academic institutions, and enterprises act as participants with different discourse power in the PDR process.

For emergency response and recovery response, 'partnership and trust between government agencies at all levels and between public and non-profit sector agencies' were significant (Kapucu 2006, pp. 217). Interactions among stakeholders, especially the incorporation of local decision-making, are helpful for a better understanding of the different points of view, criteria, preferences, and trade-offs involved in decision-making (Antunes et al. 2006).

After the Wenchuan Earthquake, the State Council set up a Post-Earthquake Reconstruction office. Under the guidance of an experts committee, a draft of an overall plan was proposed, followed by a revised plan after soliciting opinions to the society. Approved by a State Council executive meeting, the overall plan was localized by relevant ministries and subordinate administrative units. Overall, this approach shows a typical 'top-down' pattern in decision-making.

Greensburg decided to rebuild an eco-friendly town after the devastating tornado. From May to June 2007, with the invitation of the Kansas state government, Berkebile Nelson Immenschuh McDowell Construction Company together with local residents, local government, and the Federal Emergency Management Agency (FEMA) constructed the initial PDR plan. Based on the initial scheme, a long-term plan was put forward one year later after repeated community discussion. Approved by the city council, the long-term plan received funding and technological support from the federal government. This decision-making process is a typical 'bottom-up' pattern.

The strong central government control and weak participation by NGOs in China's PDR may be rooted in China's centralized political system where in central government controls a majority of national revenue and leaves limited political space for NGOs' operation (Huang et al. 2011). In addition, public participation in the PDR planning is still low based on our interviews: a majority of interviewees showed no passion or concern for the PDR plan (69.60%), or their feedback on the PDR plan was not appropriately addressed by local authorities (86.72%). Since all recovery strategies must pass a public vote in Greensburg, residents hold strong discourse power, while federal and state government plays the supporting role (Becker 2009).

Table 3. Institution capital dedicated to Nanba Town and Greensburg.

	Nanba Town, China	Greensburg, United States
Policies	Paired assistance program*	Green and Sustainable Greensburg
Legislations	**National level**	US Green Building Council LEED® Platinum standard
	"Regulation on restoration and reconstruction after Wenchuan Earthquake" (2008);	
	"Guiding Opinions of the State Council on Post-Wenchuan Earthquake Restoration and Reconstruction" (2008);	
	"Notice on paired assistance program of restoration and reconstruction after Wenchuan earthquake issued by the State Council Office"	
	Provincial level	
	"Reconstruction planning outline in urban and rural areas after the 5.12 earthquake in Sichuan Province"(2008)	
	Municipal level	
	"Management measures of village reconstruction planning in Mianyang City (provisional)" (2008);	
	"Guidelines for village reconstruction planning in Mianyang City (provisional)" (2008)	
Plans	**National level**	"Greensburg sustainable comprehensive plan"(2007);
	"The state Overall Planning for the Post-Wenchuan Earthquake Restoration and Reconstruction.";	"Long-term community recovery plan"(2007);
	"The overall plan for post-earthquake recovery and reconstruction scheme" (2008)	
	Municipal level	
	"Urban system planning for 5.12 post-earthquake recovery and reconstruction in Mianyang City" (2008)	
	Town level	
	"Post-disaster reconstruction planning of Nanba Town, Pingwu County (2008–2015)"	
Formalized institutions	**National level**	A Planning Team; The Public
	China National Commission of Disaster Reduction;	Square Steering Committee-Recovery Action Team;
	Post-earthquake recovery planning group of Earthquake Relief Headquarters of the State Council;	Sustainable Development Resource Office;
	Provincial level	Housing
	Emergency management office of the Sichuan Provincial Government	Resource Office;
	Municipal level	Greensburg GreenTown (NGO)
	Emergency management office of the Mianyang Municipal Government	
	County level	
	The earthquake relief headquarter of Pingwu County	

* Paired assistance program: 19 provinces or municipalities directly under the central government were assigned to make a one-to-one paired relationship with one of the 19 most severely earthquake-affected counties and donate annually no less than 1% of their financial income in a hardest hit county's reconstruction from 2008 to 2010. Nanba Town was assisted by Tangshan City (known for the Tangshan Earthquake on 28 July, 1976), Hebei Province.

Effective public involvement in the planning process provides a solid base for the implementation of a recovery plan. Further, NGOs (e.g., the American Red Cross, Greensburg GreenTown) are significant participants of the sustainable reconstruction in Greensburg.

4 RESULTS AND DISCUSSION

The quantitative and qualitative comparison results show significant differences in the exogenous variables and interaction mechanisms among participants. Gaps in natural, physical, financial, human,

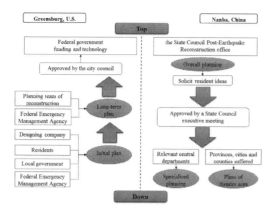

Figure 3. Programming flowchart in Greensburg and Wenchuan. (Adapted from Du et al, 2014).

social, and institutional capitals in Nanba Town and Greensburg lead to different action situations and interactions among participants in PDR. As outcomes, degrees of satisfaction in the two study areas differed significantly.

In particular, a less significant loss of natural capital in Greensburg presents opportunities for a 'Greener Greensburg', while ruins of land and mineral resources bring severe difficulty to Nanba. Based on better financial capital and less of an impact on the industrial structure, Greensburg has successfully pursued sustainable recovery. After five years, with the support from well-educated and skilled residents as well as a united community, Greensburg owns a set of well-developed infrastructures that can offer high-quality services for residents. What should not be ignored is that decentralized institution settings in Greensburg open channels of interaction among participants, which significantly improves the degree of satisfaction. The PDR in China was characteristic of strong central government control and relatively weak participation by NGOs and residents. This background is favorable for prompt approval and the implementation of policies and a recovery plan, but may not be advantageous in receiving residents' opinions on recovery. Therefore, the satisfaction of PDR is relatively low in Nanba. Due to the reality of severe losses in natural capital and low physical human, financial and social capital, the 'top down' might be a pragmatic approach.

From the perspective of social satisfaction and based on our analysis, the key recommendations to improve the PDR practices in Nanba or similar areas in China are as follows: to mitigate the impact of loss of natural capital, to integrate the existent resources, and to adjust the industrial structures. While it is difficult to increase financial

Table 4. Contrast of capital assets, participants and interactions in process of PDR.

		Nanba Town	Greensburg
Capital assets	Natural capital	More loss	Less loss
	Physical capital	Lower	Higher
	Financial capital	Lower	Higher
	Human capital	Lower	Higher
	Social capital	Lower	Higher
	Institutional capital	Centralized	Decentralized
Participants		Strong central government control and weak participation of the public and NGOs	Widespread public participation
Interactions		Centralized	Decentralized

capital and social capital in a short time, education and skill training should be addressed in PDR to increase human capital, as this is significant for financial capital accumulation and public participation. Partnership among governments, NGOs, residents, and enterprises should also be explicated and emphasized to promote public participation and effectiveness from the perspective of institutional capital.

5 CONCLUSION

In this paper, an integrated approach that combined the IAD framework and a modified CAF was proposed to provide a new systematic analytical framework for complex social issues such as PDR. Six capital assets, namely natural, physical, financial, human, social, and institutional capitals, are imbedded into the IAD framework as quantitative and qualitative measures for the analysis of biophysical conditions, community attributes, and rules-in-use.

Nanba Town and Greensburg were selected as study areas in this research. Based on statistical data, two field surveys, and a quantitative-qualitative comparison of capital assets, we conclude that gaps in these six capital assets in Nanba Town and Greensburg contribute to different action situations and interactions among participants in PDR. Ultimately, degrees of satisfaction in the two study areas differed significantly. What we can learn from the study is the establishment of a simple and practical strategy (i.e. a set of bottom-up rules) which can impact positively the lives of ordinary people, and therefore make a difference.

The comparative case study indicates that the integrated approach is valid to systematically analyze the satisfaction issue of PDR, and to identify causes of different outcomes. In practice, findings of this paper can help communities to better target the improvement of exogenous context and communication among participants during the PDR process. Furthermore, this approach contributes to the field of risk management in extending as a systematic conceptual framework for the analysis of similar social issues. In further research, how to build agent-based models and to simulate decision-making process quantitatively will be the keystone of our research.

REFERENCES

Agrawal A. and Gibson C. C. (1999) Enchantment and disenchantment: the role of community in natural resource conservation. *World Development* 27(4): 629–649.

Aldrich D. P. (2011) The power of people: social capital's role in recovery from the 1995 Kobe earthquake. *Natural Hazards* 56(3): 595–611.

Aligica P. D. (2006) Institutional and stakeholder mapping: frameworks for policy analysis and institutional change. *public organization review* 6(1): 79–90.

Antunes P., Santos R. and Videira N. (2006) Participatory decision making for sustainable development—the use of mediated modelling techniques. *Land use policy* 23(1): 44–52.

Araral, E. (2005) Bureaucratic Incentives, Path Dependence, and Foreign Aid: An Empirical Institutional Analysis of Irrigation in the Philippines. *Policy Sciences* 38(2–3):131–157.

Barone G. and Mocetti S. (2014) Natural disasters, growth and institutions: a tale of two earthquakes. *Journal of urban economics* 84: 52–66.

Beaudoin C. E. (2007) News, social capital and health in the context of Katrina. *Journal of health care for the poor and underserved* 18(2): 418–430.

Becker C. (2009) Disaster recovery: a local government responsibility. *Public Management* 91(2): 6–12.

BNIM Architects (2008) Greensburg Sustainable Community Comprehensive Plan. *BNIM Architects,* 2008. Accessed 8 November 2015.

Carbin G. W. and Schaefer J. T. (2008) Tornadoes of 2007: A Deadly Year." *Weatherwise* 61(2): 38–45.

Clement, F. (2010) Analysing Decentralised Natural Resource Governance: Proposition for a 'Politicised' Institutional Analysis and Development Framework. *Policy Sciences* 43(2):129–156.

Cutter S. L., Boruff B. J. and Shirley W. L. (2003) Social vulnerability to environmental hazards. *Social science quarterly* 84(2): 242–261.

Drabek T. E. and McEntire D. A. (2003) Emergent phenomena and the sociology of disaster: lessons, trends and opportunities from the research literature. *Disaster Prevention and Management: An International Journal* 12(2): 97–112.

Edwards R., LaDue J. G., Ferree J. T., Scharfenberg K., Maier C., Coulbourne W. L. (2013) Tornado intensity estimation: Past, present, and future. *Bulletin of the American Meteorological Society* 94(5): 641–653.

Feeny, D. (1994) Frameworks for Understanding Resource Management on the Commons. In *Community Management and Common Property of Coastal Fisheries in Asia and the Pacific: Concepts, Methods and Experiences.* R.S. Pomeroy, ed. Manila, Philippines: International Center for Living Aquatic Resources Management (ICLARM).

Ge Y., Gu Y. and Deng W. (2010) Evaluating China's national post-disaster plans: The 2008 Wenchuan earthquake's recovery and reconstruction planning. *International Journal of Disaster Risk Science* 1(2): 17–27.

Hanes J. E. (2000) Urban planning as an urban problem: the reconstruction of Tokyo after the Great Kanto earthquake. *Seisakukagaku* 7(3): 123–137.

Horwich G. (2000) Economic lessons of the Kobe earthquake. *Economic development and cultural change* 48(3): 521–542.

Huang Y., Zhou L. and Wei K. (2011) 5.12 Wenchuan Earthquake Recovery Government Policies and Non-Governmental Organizations' Participation. *Asia pacific journal of social work and development* 21(2): 77–91.

Imperial M. T. and Yandle T. (2005) Taking institutions seriously: using the IAD framework to analyze fisheries policy. *Society and Natural Resources* 18(6): 493–509.

Ingram J. C., Franco G., Rumbaitis-del Rio C., Khazai B. (2006) Post-disaster recovery dilemmas: challenges in balancing short-term and long-term needs for vulnerability reduction. *Environmental Science & Policy* 9(7): 607–613.

Judkoff R. (2008) Increasing building energy efficiency through advances in materials. *MRS Bulletin* 33(04): 449–454.

Kapucu N. (2006) Public-nonprofit partnerships for collective action in dynamic contexts of emergencies. *Public administration* 84(1): 205–220.

Katz B. (2006) Concentrated poverty in New Orleans and other American cities. *Chronicle of Higher Education* 52(48): B15.

Khew Y. T. J., Jarzebski M. P., Dyah F., San Carlos R., Gu J., Esteban M., Aránguiz R., Akiyama T. (2015) Assessment of social perception on the contribution of hard-infrastructure for tsunami mitigation to coastal community resilience after the 2010 tsunami: Greater Concepcion area, Chile. *International Journal of Disaster Risk Reduction* 13: 324–333.

Kiser L. L. and Ostrom E. (2000) The three worlds of action: A metatheoretical synthesis of institutional approaches. *Polycentric Games and Institutions* 1: 56–88.

Lam, W.F. (2001) Coping with Change: A Study of Local Irrigation Institutions in Taiwan. *World Development* 29(9):1569–1592.

Lee, M. (1994) Institutional Analysis, Public Policy, and the Possibility of Collective Action in Common Pool Resources: A Dynamic Game Theoretic Approach. (Ph.D. Dissertation, Indiana University). http://hdl.handle.net/10535/3591

McGinnis M. D. (2011) An introduction to IAD and the language of the Ostrom workshop: a simple guide to a complex framework. *Policy studies journal* 39(1): 169–183.

Nakagawa Y. and Shaw R. (2004) Social capital: A missing link to disaster recovery. *International Journal of Mass Emergencies and Disasters* 22(1): 5–34.

Narayan D. and Pritchett L. (1999) Cents and sociability: Household income and social capital in rural Tanzania. *Economic development and cultural change* 47(4): 871–897.

Ostrom E. (1986) An agenda for the study of institutions. *Public choice* 48(1): 3–25.

Ostrom E. (2005) Building a better micro-foundation for institutional analysis. *Behavioral and Brain Sciences* 28(06): 831–832.

Ostrom E. (2011) Background on the institutional analysis and development framework. *Policy studies journal* 39(1): 7–27.

Paul B. K. and Che D. (2011) Opportunities and challenges in rebuilding tornado-impacted Greensburg, Kansas as "stronger, better, and greener". *GeoJournal* 76(1): 93–108.

Paul, B. (2009) Reclaiming institutions as a form of capital. Paperpresented at the Pennsylvania Economic Association Conference, June 6.

Pless S., Billman L., Wallach D., GreenTown G.(2010) From Tragedy to Triumph: Rebuilding Greensburg, Kansas, To Be a 100% Renewable Energy City. Paper presented at the annual meeting for ACEEE Summer Study Conference, Pacific Grove, CA.

Polski M M, Ostrom E. (1999) An institutional framework for policy analysis and design. Paper was presented at the Workshop in Political Theory and Policy Analysis. Indiana University, Bloomington.

Shimada G. (2015) The role of social capital after disasters: An empirical study of Japan based on Time-Series-Cross-Section (TSCS) data from 1981 to 2012. *International Journal of Disaster Risk Reduction* 14(4): 388–394.

Tatsuki S. (2008) The role of civil society for long-term life recovery from a Megadisaster. Paper presented at the annual meeting of the American political science association.

United Nations Development Programme. (2004) Reducing Disaster Risk: A Challenge for Development–A Global Report. New York, US: UNDP.

Wang Y., Chen H. and Li J. (2012) Factors affecting earthquake recovery: the Yao'an earthquake of China. *Natural Hazards* 64(1): 37–53.

www.greensburgks.org/recoveryplanning/Greensburg%20Comprehensive%20Master%20Plan%2001-16-08%20DRAFT.pdf

Yang S., Du J., He S., Shi M., Sun X. (2015) The emerging vulnerable population of the urbanisation resulting from post-disaster recovery of the Wenchuan earthquake. *Natural Hazards* 75(3): 2103–2118.

Yasui, E. (2007) Community Vulnerability and Capacity in Post-disaster Recovery: The cases of Mano and Mikuraneighbourhoods in the wake of the 1995 Kobe earthquake. PhD thesis, University of British Columbia.

Yi H. and Yang J. (2014) Research trends of post disaster reconstruction: The past and the future. *Habitat international* 42: 21–29.

NOTE LIST

[1] Data source: Social and Economic Statistical Yearbook of Pingwu County 2009.

[2] Data source: The instruction of post-disaster reconstruction planning of Nanba Town (2008–2015).

[3] Data source: Main result of the seventh national forest resources inventory (2004–2008), http://www.forestry.gov.cn/portal/main/s/65/content-326341.html

[4] Data source: Seismological Bureau of Pingwu County, 2013. Wenchuan Earthquake relief records of Pingwu County (earthquake damage collection)

[5] Data source: Chinese Statistical Yearbook, 2014.

[6] Data source: Economic and social development program in Pingwu's 12th five-year plan.

[7] Data source: World Media Group, LLC website. http://www.usa.com/greensburg-ks-income-and-careers--historical-careers-data.htm

[8] Data source: Advameg, Inc website. http://www.city-data.com/city/Greensburg-Pennsylvania.html

[9] Data source: Department of Numbers website. http://www.deptofnumbers.com/income/us/

[10] Data source: Advameg, Inc website. http://www.city-data.com/income/income-Greensburg-Kansas.html

[11] Data source: Advameg, Inc website. http://www.city-data.com/income/income-Greensburg-Kansas.html

Advances in Energy, Environment and Materials Science – Wang & Zhou (Eds)
© 2017 Taylor & Francis Group, London, ISBN 978-1-138-03600-0

A compilation method of greenhouse gas emission inventory in ports of China

Xu Guo, Guoyi Li & Leilei Liu
Tianjin Research Institute for Water Transport Engineering, Key Laboratory of Environmental Protection Technology on Water Transport, Ministry of Transport, Tianjin, China

ABSTRACT: Greenhouse Gas (GHG) emission has become a global issue. In this regard, the port enterprises must compile an inventory of GHG emissions. By drawing on "2006 IPCC Guidelines for National Greenhouse Gas Inventories" and other literature, this paper determines the overall framework and main content of GHG emission inventory guidebook in a Chinese port.

1 INTRODUCTION

The port cargo throughput will continue to increase in China, thereby increasing energy consumption and Greenhouse Gas (GHG) emission. In order to control the GHG emissions effectively, relevant research on compilation of GHG emission inventory must be carried out.

At present, the GHG emissions have been studied in various fields.

E. H. Pechan & Associates Inc. developed 2006, 2007, and 2008 calendar year GHG and criteria pollutant emission inventories for PANYNJ facilities and operations, including the emissions of its tenants (e.g. airlines and shippers) and patrons (e.g. airport passengers and Port Authority Trans-Hudson (PATH) riders) (Colodner, 2011).[1]

G Villalba et al. put forward a method of GHG emission list based on the ports. They proposed the indicators of GHG emissions per ton of cargo for policy and emission prevention measures (Villalba, 2011).

HR Choi et al. proposed a system to estimate GHG emission from port facilities, and achieved a more efficient management of carbon through a dimensional analysis of its emission sources (Choi Hyung Rim et al., 2015).

H Sakai et al. proposed a system model to estimate the carbon dioxide emission at a container terminal. It consists of a decision process to select the evaluation level of accuracy according to the availability of information and data needed to evaluate it, and sub-models for calculating the carbon dioxide emissions. The model's effectiveness is verified by applying it to a Japanese port (Sakai Hiroshi et al., 2006).

YT Chang et al. calculated the GHG emission from the Port of Inchon, South Korea, based on the ship type and its activity to evaluate the emissions. Among all types of ship, the international car ferry is the most serious source of pollution (Chang Young Tae et al., 2013).

S Kim et al. designed the Network Design Problem (NDP), which determines investment alternatives to minimize the total system cost, including costs related to GHG emissions (Kim Suhyeon et al., 2013).

IL Marr et al. set up a network of 10 stations, with passive sampling for VOCs (including benzene), NO, and SO, over 2-week periods; grab sampling for CO; and 48 h pumped sampling for PM to conduct an air quality survey for 12 months around Aberdeen Harbour (Marr I L et al., 2007).

However, there is very few research on the compilation of GHG inventories in the port industries of China. By drawing on "2006 IPCC Guidelines for National Greenhouse Gas Inventories", "Guidebook for the compilation of provincial GHG emission inventory", and on the basis of investigating various types of large port in China, this study determines the overall framework and main content of GHG Emission Inventory Guidebook.

2 LIST TYPE AND SCOPE

2.1 *GHG emission types*

The main GHGs include CO_2, CH_4, N_2O, HFC_s, PFC_s, and SF_6. In order to simplify the statistical difficulty, only the emission data of CO_2 are selected.

2.2 *Scope of the GHG emission in the port*

The Greenhouse Gas Protocol (A Corporate Accounting and Reporting Standard) defines various emissions accounting scope:

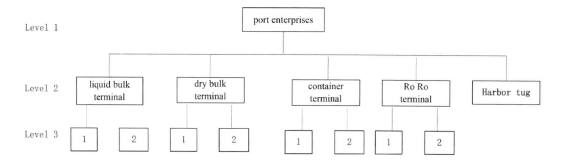

Figure 1. Three-level structure gHG accounting method.

1. Scope1: direct emissions of business activities;
2. Scope2: indirect emissions generated by power consumption of business activities;
3. Scope3: other indirect emissions generated by business activities.

Because the situation of each port area is not the same, it is difficult to define the power source and other indirect emission in the process of production. Therefore, the direct emission (scope1) is calculated within the list.

2.3 *Emission source analysis*

The main emission source of the port is the combustion of various types of fossil fuel. The main emission can be divided into three levels, whose structures are as follows:

Level 1: port enterprises (coastal, inland river);

Level 2: liquid bulk terminal, dry bulk terminal, container terminal, Ro Ro Terminal Harbor tug;

Level 3: loading and unloading production, auxiliary production.

3 GHG ACCOUNTING METHOD

On the basis of the current situation of administrative system and statistical data in China, the amount of GHG emission is calculated by the methods of bottom–up and layer-by-layer accumulation.

GHG emission is obtained by the product of the activity level data and the emission coefficient. Activity level data refer to the consumption of fossil fuels in the production of a port. Emission coefficient represents the amount of GHG from fossil fuel consumption. The calculation formula is as follows:

$$E_{i,j} = \sum EF_{i,j} \times Act_{i,j} \qquad (1)$$

where

$E_{i,j}$ = the GHG emissions (t);
$EF_{i,j}$ = the GHG emission coefficient (t-CO2/t);
$Act_{i,j}$ = the fuel consumption (t);
I = the fuel type;
j = the sector type.

4 THE EMISSION COEFFICIENT

The emission coefficient can be calculated as follows:

$$EF_{i,j} = Qdw_i \times CC_i \times OF_i \times 44\,/\,12 \qquad (2)$$

where

$E_{i,j}$ = the GHG emissions (t);
Qdw_i = the low calorific value of fuel "i" (GJ/t (GJ/ m^3);
CC_i = carbon content of unit calorific value of fuel "i" (t-C/TJ);
OF_i = oxidation rate of carbon of fuel "i";
$44/12$ = molecular weight ratio of carbon dioxide to carbon.

In actual calculation, the measured data (including "Qdw", "CC" *and* "OF") of fossil fuel are preferred. If there are no actual test data, the default data provided in various authoritative literature should be chosen.

5 ACQUISITION OF FUEL CONSUMPTION

In the collection of various types of energy consumption data, the good approach is to obtain the

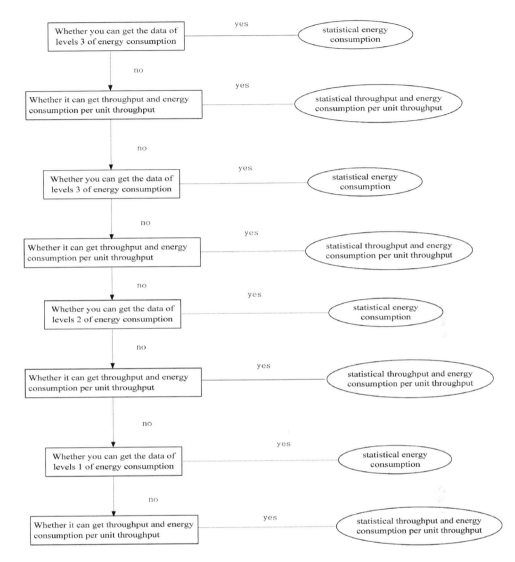

Figure 2.   Fuel consumption acquisition mode.

fossil fuel consumption data of level 3 (loading and unloading production, auxiliary production). The data can be obtained through the port of energy statistics and original accounting.

If there are still no corresponding data, the users should use the data of throughput and energy consumption per unit of energy consumption (level 3) to calculate the energy consumption.

If there are still no corresponding data, the users should use the data of level 2 (liquid bulk terminal, dry bulk terminal, container terminal, Ro Ro Terminal) to determine the energy consumption. The method is the same as above (level 3). If there are still not enough data, the users should use the data of level 1 (port enterprises).

6   CONCLUSIONS

According to the production characteristics of the port industry, this paper puts forward a compilation method of GHG inventory, which has reference significance for the future work of the GHG emission inventory in ports of China.

REFERENCES

Colodner Stephen, et al (E. H. Pechan and Associates, Inc., US). Port authority of New York and New Jersey criterion pollutant and greenhouse gas emission inventory[J]. Transportation Research Record, 2011(2233): 53–62.

Villalba Gara, et al (Universitat AutÒnoma de Barcelona, Spain). Estimating GHG emissions of marine ports-the case of Barcelona[J]. Energy Policy, 2011, 39(3): 1363–1368.

Choi Hyung Rim, et al (Dong-A University, KR). Development of ports carbon emission estimation system[J]. International Journal of Future Generation Communication and Networking, 2015, 8(8): 235–242.

Sakai Hiroshi, et al (Port and Airport Research Institute). The estimation model of carbon dioxide emission at container terminals[J]. Journal of Japan Industrial Management Association, 2006, 57(1): 68–79.

Chang Young Tae, et al (Inha University, KR). Assessing greenhouse gas emissions from port vessel operations at the Port of Incheon[J]. Transportation Research Part D: Transport and Environment, 2013(25): 1–4.

Kim Suhyeon, et al (Korea Institute of Construction Technology). Multimodal freight transportation network design problem for reduction of greenhouse gas emissions [J]. Transportation Research Record, 2013 (2340): 74–83.

Marr I L, et al (Aberdeen University, UK). An air quality survey and emissions inventory at Aberdeen Harbour[J]. Atmospheric Environment, 2007, 41(30): 6379–6395.

IPCC. 1996 IPCC Guidelines for National Greenhouse Gas Inventory[M]. Intergovemmental Panel on Climate Change, 1996.

IPCC. 2006 IPCC Guidelines for National Greenhouse Gas Inventory[M]. Imergovemmental Panel on Climate Change, 2006.

Writing Group of the Provincial Greenhouse Gas Inventories. The Provincial Greenhouse Gas Inventories[M], 2010.

Advances in Energy, Environment and Materials Science – Wang & Zhou (Eds)
© 2017 Taylor & Francis Group, London, ISBN 978-1-138-03600-0

Grey relational theory analysis of factors affecting haze pollution in Beijing City

Aixiang Tao

School of Management and Economy, HuaiYin Institute of Technology, Huai'an, China

ABSTRACT: In today's world, people's demand for environmental quality is increasing quickly. Haze governance has become an important issue. In this paper, the factors affecting the level of haze in Beijing are analyzed by using the gray correlation analysis method. The results show that, according to the degree of influence from high to low order, the factors affecting the level of haze in Beijing city is divided into three levels. The first level is the forest coverage, the resident population and energy consumption; the second level is the total amount of GDP and motor vehicle ownership; the third level is the investment in science and technology, education investment and energy saving and environmental protection. Therefore, the focus of Beijing haze governance is to enhance the forest coverage, the number of resident population to control and reduce energy consumption. At the same time, scientific measures should be taken to further enhance the quality of economic growth and control quantity of motor vehicle ownership. In addition, we should further increase the input of energy saving and environmental protection.

1 INTRODUCTION

With the government's efforts to strengthen environmental pollution control, haze governance has become an important issue. Many scholars have made scientific analysis on the factors that influence the level of fog and haze. They have drawn a series of scientific conclusions. Han Lijian and Zhang Shuping analyzed how meteorological factors affected PM2.5 in winter (Zhang, 2016). Feng Shaorong and Feng Kangwei analyzed the influence factors of haze by using statistical analysis method. They put forward the corresponding control measures (Feng, 2015). Jin Feifei and Ni Zhiwei evaluated factors affecting haze by using the theory of fuzzy preference relation (Jin, 2015). Chengting analyzed haze climate characteristics and influencing factors in nearly 50 years of Nanjing city. Results showed that fog and haze usually accompany each other and convert to each other as well (Chen, 2014). Using econometric empirical analysis method, Li Xiaoyan analyzed affecting factors of haze in Beijing, Tianjin and Hebei province. The results showed that, dust haze exerts the greatest influence on the smog problem in Beijing and Tianjin, while the car exhaust is the major source of the smog in Hebei Province (Li, 2016). Zhang Li (2015) analyzed the change characteristics of haze in Shandong province and the characteristics of the circulation and meteorological elements in the process of a one-time persistent fog and haze. The results showed that the distribution

of fog is related to geographical factors, while distribution of haze is related to the city industry and pollution emissions. Wang Huiqin and He Yiping (2014) studied the influence factors of the public participation in the haze governance, and proposed the path of public participation. Wangliping (2016) analyzed socio-economic factors influencing haze pollution. The result showed that, there exists spatial spillover effect of haze pollution in Chinese provinces and "high-high" polarization phenomenon in Jing-Jin-Ji area, Yangtze River Delta and middle-east region. Some factors have anti-disturbance impacts on the haze pollution, such as industrial structure, urban construction, energy consumption structure, vehicles ownership and population size (Wang, 2016).

Based on the above research, this paper analyzed the main factors that affected the haze pollution in Beijing, and drew scientific conclusion.

2 MODELING STEPS WITH GREY RELATIONAL ANALYSIS

A *The establishment of the original series and dependent variables refer to the number of columns and compare the number of independent variables listed*

Refer to the number of columns known as the dependent variable sequences recorded as the mother; $x_0^{(k)} : x_0^{(k)} = \left[x_0^{(1)}, x_0^{(2)}, x_0^{(3)}, \ldots, x_0^{(k)} \right]$;

Comparing the number of independent variables is also called the sub-sequence of the column, $x_i^{(k)} : x_i^{(k)} = \left[x_i^{(1)}, x_i^{(2)}, x_i^{(3)}, \ldots, x_i^{(k)} \right] (i = 1,2,3,\ldots,n)$

B The original sequence is to be treated of non-dimensional

The purpose is to eliminate the impact of different sizes and to facilitate calculation and comparison. initialize method and the average method Can be used. calculate formulas are $x_i^{(k')} = {x_i^{(k)}}/{x_i^{(1)}}$; or $x_i^{(k')} = x_i^{(k)} / \overline{x_i}$

C Calculate the absolute value between parent sequence and each sub-sequence at each time point to identify the biggest difference and minimum difference

difference sequence: $\Delta i(k) = | x_0^{(k')} - x_i^{(k')} | (i = 1,2,3, \ldots,n)$
The biggest difference: $\Delta_{max} = \max_i \max_i | x_0^{(k')} - x_i^{(k')} |$ the minimum difference: $\Delta_{min} = \min_i \min_i | x_0^{(k')} - x_i^{(k')} |$

D Calculate the Gray correlation coefficient

$L_{0i}^{(k)} = \frac{\Delta_{min} + \lambda \Delta_{max}}{\Delta i(k) + \lambda \Delta_{max}}$ Among these, $L_{0i}^{(k)}$ is Gray correlation coefficient between the number of sub-sequences and the parent sequence, λ is distinguish factors, usually between 0 and 1.

E Calculation of gray correlation degree

The overall correlation need to take the different observation points in the overall level of the importance of observation into account, therefore need to determine the weight of each point. Under normal circumstances, the arithmetic mean method is used to calculate the grey correlation degree.
$r_{0i} = \frac{1}{n} \sum_{k=1}^{n} r_{0i}(k)$ r_{0i} represent the correlation coefficient between x_0 and x_i.

F Sort the correlation degree

Correlation is sorted based on size of order. The bigger a correlation is, the bigger the relation degree between the mother sequence and sub-sequence. According to experience, when the correlation is greater than 0.6, it will be considered a significant association (Liu, 1991; Yi, 1992; Deng, 2002).

3 INDEX CHOOSE AND CALCULATION

A Index selection

The daily mean of the particulate matter and the daily average of SO2 are the two most important indicators reflecting the degree of haze. In his paper, the index are selected to reflect the level of haze pollution indicators are as follows: the daily average of particulate matter, denoted as A1, unit (mg/M); sulfur dioxide daily average, denoted as A2, unit (mg/m). Factors affecting haze pollution degree in Beijing city are as follows: the total GDP, remember B1 (unit: hundred million yuan); forest coverage rate, denoted B2 (unit:%); resident population, B3 (unit: ten thousand people); energy consumption and B4 (unit: ten thousand tons of standard coal); spending on energy conservation and environmental protection, remember to B5 unit (hundred million yuan); motor vehicle retains quantity, remember B6 (unit: ten thousand); education investment amount, remember to B7 (unit: hundred million yuan); the amount of investment in science and technology, remember as B8 (unit: hundred million yuan). Specific data are as follows:

	2009	2010	2011	2012	2013	2014
A1	0.121	0.121	0.114	0.109	0.108	0.116
A2	0.034	0.032	0.028	0.028	0.027	0.022
B1	12153.0	14113.6	16251.9	17879.4	19800.8	21330.8
B2	36.7	37.0	37.6	38.6	40.1	41.0
B3	1860.0	1961.9	2018.6	2069.3	2114.8	2151.6
B4	6570.3	6954.1	6995.4	7177.7	6723.9	6831.2
B5	54.0	60.9	94.5	113.5	138.2	213.4
B6	401.9	480.9	498.3	520.0	543.7	559.1
B7	365.7	450.2	520.0	628.7	681.1	742.0
B8	126.3	178.9	183.0	199.9	234.7	282.7

B Calculation

According the above steps, the results are as follows:
According the importance, the grey relation coefficient are as follows
B4>B2>B3>B1>B6>B7>B8>B5

Table 2. Grey relation coefficient between the daily average of particulate matter and its affecting factors.

	B1	B2	B3	B4	B5	B6	B7	B8
A1	0.799	0.937	0.916	0.941	0.720	0.792	0.732	0.727

Table 3. Grey relation coefficient between the sulfur dioxide daily average and its affecting factors.

	B1	B2	B3	B4	B5	B6	B7	B8
A2	0.771	0.893	0.874	0.892	0.712	0.814	0.731	0.718

According the importance, the grey relation coefficient are as follows

B2>B4>B3>B6>B1>B7>B8>B5

4 CONCLUSION AND ADVICE

According to the above calculation, the factors affecting haze pollution degree in Beijing city can be divided into three levels. The first level includes the following aspects: forest coverage rate, the resident population quantity, the energy consumption. The second level includes the following aspects: the total amount of GDP and motor vehicle ownership. The third level includes following: the amount of energy saving and environmental protection expenditure, education investment and science and technology investment.

Firstly, the most important indicators affecting the degree of haze pollution in Beijing city are as follows: the forest coverage, the resident population and the amount of energy consumption. Forests have the function of conserving water and absorbing carbon dioxide and other harmful gases. According to the results of the study, the forest coverage rate and the degree of haze pollution has a very large correlation. Therefore, Beijing city should increase the intensity of afforestation to increase the forest coverage rate. In this way, we can significantly reduce the degree of haze pollution.

The greater the population there is in a city, the more rubbish is produced. During this process, the haze pollution degree is becoming more serious. We should take practical measures to reduce the population over expansion in Beijing city. In this way, we can effectively reduce the level of haze pollution in Beijing city.

Energy consumption also has a direct relationship with the degree of haze pollution in Beijing. During the process of energy consumption, a large number of toxic and harmful gases are produced, which directly increase the amount of toxic and harmful gases in the air.in Beijing city, we should increase the structural adjustment to move the energy consumption of large units out of the city. At the same time, we should increase the proportion of clean energy consumption. Feasible measures should be taken to increase energy conservation technology research and development and promotion. This can effectively reduce the total amount

of energy consumption. It also can help to reduce haze pollution in Beijing.

Secondly, the total amount of GDP and motor vehicles also have a significant impact on the degree of haze pollution in Beijing. It is clear that the process of economic growth will inevitably produce a variety of industrial waste and domestic waste, which will release a large number of toxic and harmful gases. Therefore, Beijing should accelerate the pace of economic restructuring and upgrading. The proportion of the first and second industries should be gradually reduced, while the third industry should be vigorously developed. in this way, the economy can grow steadily while the production of haze can be reduced. The use of motor vehicles will release toxic and harmful gases and pollute the air. In Beijing city, we should control the number of motor vehicles. At the same time, we should enhance motor vehicle energy use standards to reduce pollution.

Thirdly, energy saving and environmental protection investment, investment in science and technology and education also have an important impact on the degree of haze pollution in Beijing. When energy saving and environmental input increases, we can use a large number of new technologies and new equipment. We can increase the environmental protection staff. These measures can effectively reduce the environmental pollution. By increasing investment in science and technology, we can upgrade environmental protection technology upgrade, which can indirectly enhance the level of haze pollution prevention.

REFERENCES

Chen Ting, Wei Xiaoyi, Zhai Lingli, et al. An analysis of climatic characteristics and influence factors fog and haze in Nanjing in recent 50 years [J], Environmental Science & Technology, 2014, 34(6):54–61.

Dengjulong.grey theoty base [M]. wuhan: Huazhong university of science and technology press, 2002, (2)

Feng Shaorong, Feng Kang-wei. Haze Pollution and Its Control Measures Based on Statistical Mathods. Journal of Xiamen University (Natural Science), 2015(1):114–116.

Jin Fei-Fei, Ni Zhi-Wei. Factors Evaluation of Fog-Haze Weather Based on Hesitant Fuzzy Preference Relations. PR & AI. 2015(9):840–843.

Li Xiaoyan. Empirical Analysis of the Smog Factors in Beijing-Tianjin-Hebei Region. Ecological Economy, 2016, 32(3):144–148.

Liusifeng, Guotianbang. The grey system theory and its application [M]. Zhengzhou: Henan university press, 1991(2).

Wang Huiqin, He Yiping. Influence Factors and Path Optimization of Fog Haze Public Participation Governance. Chongqing Social Science, 2014(12):41–45.

Wang Liping, Chen Jun. Socio-economic Influential Factors of China Haze Pollution-Empirical Study on EBA Model using Spatial Panel Data. Acta Science Circumstantiae, 2016(4):12–16.

Yidesheng, Guoping. grey theory and method [M]. Beijing: oil industry press, 1992(1).

Zhang Li, Gong Wu, Wei Zong Gu, Chen Yanchun, Chen Cui. Haze changes in Shandong Province and a persistent haze process analysis. Journal of Ocean University of China 2015, 45 (11): 10–14.

Zhang Shuping, Han Lijian, Zhou Weiqi, Zhang Xiaoxin. Relationships between fine particulate matter (PM2.5) and meteorological factors in winter at typical Chinese cities. Acta Ecologica Sinica, 2016(12):2–5.

Advances in Energy, Environment and Materials Science – Wang & Zhou (Eds)
© 2017 Taylor & Francis Group, London, ISBN 978-1-138-03600-0

Effects of humus soil on pollutant removal and activated sludge sedimentation of the SBR process

Yuting Zhang
Henan Zhibo Architectural Design Group Co. Ltd., Luoyang, P.R. China

Ke Zhao, Zhenling Lu, Nan Cao & Kuiquan Duan
School of Municipal and Environment Engineering, Jilin Jianzhu University, Changchun, P.R. China

ABSTRACT: This paper investigates the pollutant removal of SBR which contains humus soil and compared experiments of internal and external Humus Soil SBR (HS-SBR) with conventional SBR (cSBR), from the Loosely Bound Extracellular Polymer Substances (LB-EPS), to analyze the effect of increasing the humus soil on activated sludge sedimentation. The results indicate that the removal rate of NH_4^+-N in the external HS-SBR was 62.9%, and removal rates in cSBR and internal HS-SBR were increased by 5.9% and 8.7%, respectively. The removal rate of TP was as high as 89.2%, and removal rates in cSBR and internal HS-SBR were increased by 8.5% and 13.3%, respectively. The concentration of LB-EPS in external and internal HS-SBR was significantly less than that in cSBR. SVI value was lower than cSBR by 17 ml/g, so sedimentation was superior to cSBR. In conclusion, the external HSR-SBR increased the effect of nitrogen and phosphorus removal. Moreover, adding humus soil significantly improved activated sludge sedimentation.

1 INTRODUCTION

The effective separation of microbial flocs and mixed liquor is the key to the successful operation of the activated sludge treatment system, which also ensures effluent standards and enough microorganisms in SBR. Therefore, this paper combines humus soil with SBR process, to explore the pollutant removal in internal and external humus soil reactor and the effect on activated sludge sedimentation.

2 MATERIALS AND METHODS

2.1 Experiment devices

In order to examine the effect of humus soil on SBR process, the experiment sets up three-group SBR reaction devices. One set is cSBR. Another set is the aerobic/anaerobic external HS-SBR process, which increases the HSR on the basis of cSBR. The other set is internal HS-SBR. Devices are shown in Fig. 1.

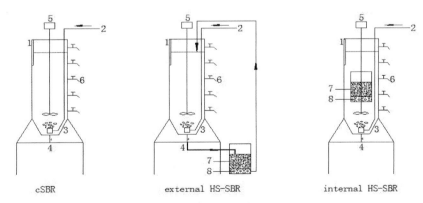

cSBR   external HS-SBR   internal HS-SBR

1-thermo; 2-compressed air; 3-aeration head; 4-mud mouth;
5-blender; 6-drain valve; 7-humus soil; 8-pumice.

Figure 1. cSBR, and external and internal HS-SBR.

2.2 Analysis indexes

COD uses 5B-1 fast tester for determination. NH_4^+-N is measured by Nessler's reagent spectrophotometry. Moreover, TP is detected by antimony molybdenum spectrophotometry. Sedimentation calls the Sludge Volume Index (SVI). LB-EPS is extracted by high-speed centrifugation and ultrasonic process. Polysaccharides, proteins, and DNA are all characterized against LB-EPS, which can be tested by the spectrophotometric method.

3 POLLUTANT REMOVAL

3.1 NH_4^+-N removal

The changes in concentration of NH_4^+-N in the typical operation cycle in the cSBR, and external and internal HS-SBR are shown, respectively, in Fig. 2. In the initial reaction stage, owing to enough air and predominant nitrifying bacteria, the NH_4^+-N content decreased, which was mostly transformed into NO_3^--N. Furthermore, owing to the degradation of polymer peptone, the NH_4^+-N content of external and internal HS-SBR increased slightly in 120 min and 240 min. As a whole, nitrification was stronger than ammoniation, which made NH_4^+-N content downtrend in the overall.

The results indicate that NH_4^+-N removal rates in the three reactors are 57%, 62.9%, and 54.2%, respectively. The removal effectiveness of NH_4^+-N in the external HS-SBR was better, which increased by 5.9% and 8.7% compared with those in the cSBR and internal HS-SBR. Thus, built-out humus soil improved the removal rate of NH_4^+-N.

3.2 TP removal

Fig. 3. shows that TP concentration were rising within the first 30 min during the aeration in the cSBR, and external and internal HS-SBR from

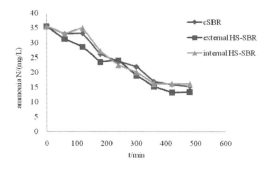

Figure 2. Variation in the concentration of NH_4^+-N in the cSBR, and external and internal HS-SBR.

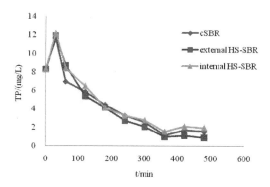

Figure 3. Variation in the concentration of TP in the cSBR, and external and internal HS-SBR.

8.3 mg/L to 11.6 mg/L, 11.9 mg/L, and 12.2 mg/L, respectively. The reason was that the active sludge carried out on the static settling for 14 h and its bottom was the anaerobic state, hence Phosphorus-Accumulating bacteria (PAOs) released phosphorus. At the same time, in the initial stage of aeration, organic matter oxidation needed plenty of DO, which maintained the reactors in hypoxia state, and therefore, the TP concentration increased. With the degradation of organic matter, the demand of DO was decreasing. As the reactors were in aerobic condition, PAOs took excessive phosphorus and TP content gradually decreased. The TP content dropped to 1.2 mg/L, 1 mg/L, and 1.6 mg/L, respectively, at the end of the aeration in reactors. When the aeration ended and mixing began, the systems entered the anaerobic phase. The TP concentration increased slightly, but not by much. After the mixing stopped, the TP contents of three-group reactors were 1.6 mg/L, 0.9 mg/L, and 2.0 mg/L, respectively, and the removal rates were 80.7%, 89.2%, and 75.9%. At the same time, the external HS-SBR system promoted the enrichment of PAOs, thus improving the phosphorus removal efficiency. It was concluded that the TP removal rate was relatively high in the external HS-SBR.

3.3 Comparison of sedimentation

In the study, SVI of three-group reactors are shown in Fig. 4.

Fig. 4 shows that addition of humus soil greatly reduced SVI of activated sludge. While SVI in the cSBR was 53.5 mL/g, SVI in the external and internal HS-SBR were 36.3 mL/g and 36.8 mL/g, respectively. This implies that SVI in the cSBR was higher than that in the external and internal HS-SBR process about 17 mL/g. The external and internal HS-SBR system had better sedimentation than the cSBR system. However, there was no obvious distinction between the sedimentation of

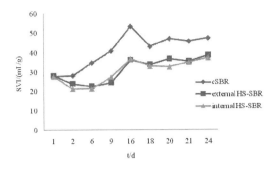

Figure 4. SVI comparison in the cSBR, and external and internal HS-SBR.

Table 1. LB-EPS concentration during startup (mg/g VSS).

Run Cycle (d)	3	6	14	17	18	21
cSBR	64	68	100	115	97	165
External HS-SBR	62	47	87	92	88	132
Internal HS-SBR	27	54	94	90	96	158

the external and that of the internal humus soil. The concentration of LB-EPS during startup in the three-group reactors are listed in Table 1.

LB-EPS concentration in the cSBR was higher than that in the external HS-SBR by 32.52 mg/g VSS, and higher than that of internal HS-SBR by 37.23 mg/g VSS. By adding humus soil, LB-EPS content became relatively small, and SVI was accordingly small. The reason was the physical and chemical properties of soil humus. Studies have demonstrated that EPS has a great influence on sedimentation, especially LB-EPS. Because LB-EPS with loose structure and liquid is located in the outer layer of EPS, and therefore, it directly affects the performance of activated sludge. In addition, LB-EPS and SVI have positive correlation.

In addition, the humus soil that makes contact with water can precipitate Ca, Mg, etc. Moreover, multivalent cations (especially Ca^{2+}) in salt bridge function are closely related to their roles in EPS bridge. The Ca^{2+} combines with anionic functional groups in EPS, which make sizes of sludge floc bigger and sedimentation better.

4 CONCLUSIONS

The removal rate of NH_4^+-N of external HS-SBR was 62.9%, which increased by 5.9% and 8.7%, respectively, compared with those of cSBR and internal HS-SBR. TP removal rate of external

HS-SBR system was as high as 89.2%, which increased by 8.5% and 8.5%, respectively, compared with those of cSBR and internal HS-SBR process. It was concluded that the effect of nitrogen and phosphorus removal in the external was better.

LB-EPS mass concentration of the external and internal HS-SBR was much smaller than that of the cSBR, and SVI of those were lower than that in the cSBR by 17 mg/L. Sedimentation of the external and internal HS-SBR was superior to that of cSBR. Moreover, there was no obvious distinction between the sedimentation of the external and that of the internal humus soil.

ACKNOWLEDGMENTS

This research was financially supported by the Natural Science Foundation of China (NSFC) (51478206).

REFERENCES

Haisong Li, Yue Wen. (2012). Research progress of the influence of EPS on the flocculation properties of activated sludge. Environmental Pollution & Control. 34(7):64–89.

Hongwu Wang, Xiaoyan Li, Qingxiang Zhao. (2004). Surface properties of activated sludge and their effects on settleability and dewaterability. J Tsinghua Univ. 44(6):766–769.

Huirong Ma, Min Wu, Yulei Jia. (2014). Effects of phosphorus removal and microbial community characteristics of humus activated sludge process. China Water & Wastewater. 30(17): 26–30.

Ke Zhao, Jun Yin, Lijun Wang, Baojun Jiang. (2009). Performance improvement of SBR process by addition of humus soil.Journal of Harbin Institute of Technology. 41(4):81–84.

Lanhe Zhang, Jun Li, Jingbo Guo. (2012). Effect of EPS on flocculation sedimentation and surface properties of activated sludge. CIESC Journal. 63(6):1865–1871.

Min Wu, Rui Zhu, Xiaohui Pan. (2009). Physicochemical characteristics of humus soil. Industrial water & wastewater. 40(1):61–63.

More T.T., Yadav J.S.S., Yan S., Tyagi R.D., Surampalli R.Y. (2014). Extracellular polymeric substances of bacteria and their potential environmental applications. Journal of Environmental Management. 144–169.

Advances in Energy, Environment and Materials Science – Wang & Zhou (Eds)
© 2017 Taylor & Francis Group, London, ISBN 978-1-138-03600-0

Influence of temperature on the decontamination efficiency and microbial community of the humus soil SBR process

Ke Zhao, Zhenling Lu, Yuting Zhang, Lizhu Chen & Chongyang Sun
School of Municipal and Environment Engineering, Jilin Jianzhu University, Changchun, China

ABSTRACT: The influence of temperature (10°C, 15°C, 25°C) on the decontamination efficiency of a humus soil Sequencing Batch Reactor (SBR) was investigated, and using high-throughput sequencing analysis, the effect of temperature on the microbial community was studied. The results indicated that temperature had little effect on the removal of COD; low temperature was not conducive to the removal of NH_4^+-N, and the removal rate of NH_4^+-N of 25°C reactor was higher than that at 10°C and 15°C by 13.4% and 11.7%, respectively. By contrast, low temperature was beneficial to the removal of TP; the removal rates of 10°C and 15°C reactors were higher than that of 25°C reactor by 11.4% approximately. High-throughput sequencing results show that, at the family level, the differences of 10°C, 15°C, and 25°C reactors' microbial community relative abundance mainly consists of *Rhodocyclaceae* (20.62%, 15.22%, and 11.34%), *Comamonadaceae* (7.65%, 4.16%, and 10.42%), *Planctomycetaceae* (1.29%, 2.79%, and 4.47%), *Aeromonadaceae* (2.12%, 0.81%, and 0.1%), and *Flavobacteriaceae* (2.9%, 1.08%, and 1.51%).

1 INTRODUCTION

Among various environmental factors that affect the metabolic activity of activated sludge, temperature is an important factor. In general, the microbial activity of activated sludge decreased with the decrease in temperature, and the sewage treatment efficiency will also decline in different degrees. Many operation results of sewage treatment plant in cold area show that with the decreasing temperature, activated sludge sedimentation going down and organic matter removal, nitrification/denitrification by a greater impact directly affects the water quality (Oleszkiewicz 2004, Knoop 2003). This experiment will combine humus soil fillers and SBR process, to improve the metabolic activity of microorganisms in the activated sludge, to enhance the efficiency of wastewater treatment, and to investigate the effects of different temperatures on the system performance and microbial community by experiments and provide technical basis for the practical engineering application.

2 MATERIALS AND METHODS

2.1 Configuration

This study was carried out with three parallel experiments, and the reactors device were the same (10°C named 1#, 15°C named 2#, 25°C named 3#) (Fig. 1). All reactors are made of plexiglass, and the relevant parameters are listed in table 1.

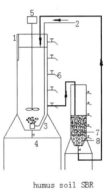

humus soil SBR

1–thermo; 2–compressed air; 3–aeration head; 4–mud mouth; 5–blender; 6–drain valve; 7–humus soil; 8–pumice.

Figure 1. Schematic diagrams of experiment.

A sludge discharge tube was installed at the bottom of the reactors, making porous stone as microporous aeration head, adopting blower aeration.

2.2 Wastewater and test methods

Test influent was synthetic to imitate the domestic sewage. It was composed of glucose, ammonium sulfate, potassium dihydrogen phosphate, sodium bicarbonate, and other experimental medicines. The characteristics of the influent and the test methods are given in table 1.

Table 1. Quality of influent and test methods.

Indicators	Quality of influent	Test methods
COD (mg/L)	300	Portable chemical oxygen demand (5B-1)
TP (mg/L)	5	Nessler's reagent spectrophotometry method
NH₄⁺-N (mg/L)	30	ammonium molybdate spectrophotometric method
Microbial community		High-throughput sequencing

2.3 Operation

The inoculation was collected from the secondary sedimentation tank of a wastewater treatment plant in the city of Changchun. MLSS of three SBR were maintained at about 4000 mg/L, temperature was set at approximately 10°C, 15°C, and 25°C, respectively. The SBR was operated at one cycle per day for 10 h. Aerobic and anaerobic operations took 6 h and 2 h, respectively, followed by a settling period of 2 h and a decanting period of 5 min, and the remaining time was used as an idle phase. MLSS of humus soil reactors were maintained at 8000 mg/L, operated with aeration for 12 h, and then settled for 12 h. At the beginning of the next cycle, moderate mixed liquor was changed among SBR and humus soil reactor of each other. Then, the next cycle was started.

3 RESULTS AND DISCUSSION

3.1 Removal efficiency

The removal effect during stable operation is given in table 2, from which it can be observed that the change in temperature had little effect on the removal of COD. The NH₄⁺-N removal rate of 3# reactor was higher than that of 1# and 2# reactors by 13.4% and 11.7%, respectively, but 1# and 2# reactors had little difference. The removal efficiencies of TP in 1# and 2# reactors were higher than that of 3# reactor by 11.4%. In terms of one cycle, the influent concentration of TP was 6.3 mg.L⁻¹, after 2 hours, measured TP were 3.22 mg/L, 5.12 mg/L¹, and 7.61 mg/L, respectively; after 4 hours, measured TP were 0.31 mg/L, 0.72 mg/L, and 2.33 mg/L. Although the removal efficiency of TP in 1# was very close to that of 2#, the phosphorus uptake rate in the former was faster than that of the latter in the aerobic stage. It can be inferred that low temperature is more conducive to the removal of phosphorus, but not to the removal of ammonia nitrogen.

3.2 Microbial community in reactors

In order to explore the principles of microbiology that cause different removal effects of nitrogen and phosphorus among these reactors, sludge samples on day 30 after the stable operation of the reactor were analyzed by high-throughput sequencing; microbial colonies' relative abundance at the family level is shown in Figure 2. Their dominant bacteria comprised the following similarly: unclassified, *Rhodocyclaceae*, *Comamonadaceae*, *Chitinophagaceae*, *Planctomycetaceae*, *Rhodobacteraceae*, *Xanthomonadaceae*, *Carnobacteriaceae*, *Saprospiraceae*, and *Flavobacteriaceae*. However, the relative abundance of some dominant bacteria was different as follows: *Rhodocyclaceae* (20.62%, 15.22%, 11.34%), *Comamonadaceae* (7.65%, 4.16%, 10.42%), *Planctomycetaceae* (1.29%, 2.79%, 4.47%), *Aeromonadaceae* (2.12%, 0.81%, 0.1%), and *Flavobacteriaceae* (2.9%, 1.08%, 1.51%).

Rhodocyclaceae belongs to the group of denitrifying bacteria, with a good phosphorus accumulating ability. Li et al (2013) detected that *Pseudomonas sp.* and *Rhodocyclus sp.* were dominant bacteria in enrichment of denitrifying phosphorus bacteria in MBR. *Rhodocyclaceae* concentration in 1# reactor was higher than that in 2# and 3# reactors by 5.4% and 9.28%, respectively, and *Rhodocyclaceae* concentration in 2# reactor was higher than that of 3# reactor by 3.88%. Studies have demonstrated that *Aeromonas* are polyphosphate bacteria, have anti-nitrification function (Xiuguang et al 2007), and are widely distributed in nature, aerobic or facultative anaerobic. *Aeromonadaceae* concentration in 1# reactor was higher than that in 2# and 3# reactors by 1.31% and 2.02%. *Flavobacteriaceae* has the ability to degrade organic phosphorus and polycyclic aromatic hydrocarbons (Yinshan et al 1985, Shuanxi 2010), and the concentration in 1# reactor was also higher than that of 2# and 3# reactors. Thus, the relative abundance of phosphorus-removing bacteria *Rhodocyclaceae*, *Aeromonadaceae, and Flavobacterium* all dropped with the increase in temperature. This is because the phosphorus-accumulating bacteria is a type of cold bacteria; the biochemical reaction rate under low-temperature conditions was fast, and therefore, under the condition of low temperature we can still get good biological phosphorus removal (Jun et al 2011).

Comamonadaceae is important for the composition of the bacterial micelles to adsorb the organic compounds in wastewater, but its denitrification pathway is more complex. Wengong et al. (2011) think that the cluster is a kind of heterotrophic nitrification bacteria. The concentration of these

Table 2. The effluent concentration and removal efficiency of two reactors.

Reactors	Effluent COD (mg/L)	COD removal efficiency(%)	Effluent NH$_4^+$-N (mg/L)	NH$_4^+$-N removal efficiency(%)	Effluent TP (mg/L)	TP removal efficiency(%)
1#	31.0	90.4	8.74	72.7	0.25	93.0
2#	20.29	92.9	7.99	74.4	0.38	93.9
3#	11.11	95.9	4.12	86.1	1.16	81.6

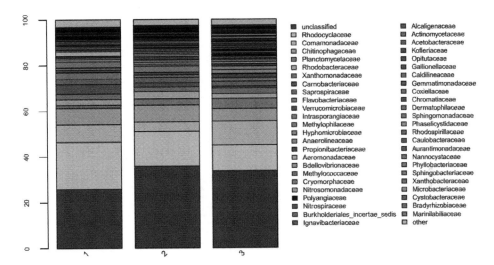

Figure 2. Distribution of the relative abundance of microbial communities in each reactor.

bacteria in 3# reactor were higher than that of 1# and 2# reactors by 2.77% and 6.26%, respectively. *Planctomycetes* are small door aquatic bacteria, a class of bacteria which is related to Planctomyces distantly. *Planctomycetes sp.* can survive under the hypoxia environment, using nitrite oxidizing ammonium ion formation of nitrogen to obtain energy; hence, they are known as anaerobic ammonia oxidation bacteria, which play an important role in denitrification (Haiyan et al 2005). The concentration of *Planctomycetaceae* in 3# reactor was higher than that of 1# and 2# reactors by 3.18% and 1.68%, respectively. This shows that low temperature is not conducive to the enrichment of nitrogen-removing bacteria, which is consistent with the low efficiency of nitrogen removal efficiency under low temperature conditions.

4 CONCLUSIONS

1. Under different temperature conditions, it had little effect on the removal of COD, but the removal rate of NH$_4^+$-N of 25°C reactor was higher than of 10°C and 15°C reactors by 13.4% and 11.7%, respectively, and the removal rates of TP in 10°C and 15°C reactors were higher than that of 25°C reactor by approximately 11.4%.

2. The relative abundances of some useful phosphorus-removing bacteria, such as *Rhodocyclaceae, Aeromonadaceae*, and *Flavobacterium*, all dropped with the increase in temperature, and some contribute to nitrogen-removing bacteria such as *Comamonadaceae and Planctomycetaceae* in 10°C reactor, whose abundance was lower than that of 15°C and 25°C reactors.

ACKNOWLEDGMENTS

This research was financially supported by the Natural Science Foundation of China (NSFC) (51478206).

REFERENCES

Haiyan Wang, Yuexi Zhou, Jinyuan Jiang (2005). Microbial population and characterization techniques for the enhanced biological phosphorus removal system [J]. Journal of Microbiology, 32 (1): 118–122.

Jun Yin, Liang Liu, Ke Zhao et al (2011). Effect of different temperature on the operation efficiency of SBR humic activated sludge system [J]. Journal of environmental engineering, 01:7–10.

Knoop S, Kunst S (2003). Influence o f temperature and sludge loading on activated sludge settling, especially on microthrix parvicella. W at. Sc. i Tech., 37(45): 2735.

Li Liu, Bing Tang, Shaosong Huang (2013), et al. Rapid enrichment culture and fluorescence in situ hybridization for the denitrification phosphorus accumulating organisms (J). Environmental science, 34 (7): 2869–2875.

Oleszkiewicz JA, Berquist SA (2004). Low temperature nitrogen removal in sequencing batch reactors. W at. R es., 22(9): 11631171.

Shuanxi Fan (2010). The study on performance of three strains of dominant bacteria for degradation of polycyclic aromatic hydrocarbons [J]. pollution control technology, 23 (1): 1–6.

Xiuguang Jiang (2007). The function of nitrogen removal microbial in aerobic granular sludge and community dynamics [D]. Harbin Institute of Technology.

Yinshan Wang, Xuejun Pang (1985). The isolation of Flavobacterium p3–2 and some properties in degradation of organophosphorus pesticide [J]. Journal of environmental science, 5 (3):315–320.

Wengong Si, Zhigang Lv, Chao Xu (2011). Screening of high concentration ammonia nitrogen heterotrophic nitrification bacteria and optimization of its nitrogen removal conditions [J]. environmental science, (11): 3448–3454.

Advances in Energy, Environment and Materials Science – Wang & Zhou (Eds)
© 2017 Taylor & Francis Group, London, ISBN 978-1-138-03600-0

Use of multi-oxidants produced by deep seawater brine for postharvest storage of custard apples (*Annona squamosa* L.)

J.X. Liu
Department of Life Science, National Taitung University, Taiwan

C.H. Yen
Department of Life Science, National Taitung University, Taiwan
Department of Horticulture, National Chiayi University, Taiwan

C.H. Chen & H.C. Huang
Department of Life Science, National Taitung University, Taiwan

Y.Y. Li & Y.H. Hsieh
Department of Environmental Engineering, National Chung-Hsing University, Taiwan

C.Y. Chang
Center for General Education, National Taitung College, Taiwan

ABSTRACT: Deep seawater, which exists in the depth below 200 meters, is low in temperature, clean, contains plentiful nutrient salts, and is stable. Deep seawater brine that is produced using a desalination device contains a high concentration of salt, and therefore, it is not widely used for commercial purposes. Custard apples (*Annona squamosa L.*) that are stored at a low temperature experience the dumb fruit phenomenon (the fruits turn hard and the surface is black) and these fruits are not easily exported. In this study, custard apples are treated with multi-oxidants, including chlorine dioxide (ClO_2), chlorine (Cl_2), ozone (O_3), and hypochlorous acid ($HOCl$), which are produced using a new electrolysis method that is modified from the diaphragm electrolytic processes of Model Hooker S–3. The experimental results demonstrate that the storage life of custard apples that are treated with multi-oxidants by adding newspapers and ethylene absorbents doubles, the rate of water loss decreases, and softening is retarded, which extends their shelf life. The green color of the fruit is also maintained. Therefore, chlorine dioxide that is manufactured by the electrolysis of deep seawater extends the shelf life of fruit and maintains the quality of fruits, postharvest, which allows their exportation.

1 INTRODUCTION

Deep seawater, which exists in the depth below 200 meters, is low in temperature, clean, contains plentiful nutrient salts, and is stable. Deep seawater can be used in different industries and products. Deep seawater brine that is produced by electrodialytic desalination to concentrate salt has a higher concentration of salt, and hence, it is not widely used industrially, except for food and biological additives.

Sugar apple (*Annona squamosa L.*), belongs to the *Annona* genus of the Annonaceae family, known as the custard apple (Nakasone & Paull 1998). This fruit is native to the tropical Americas and it has a high economic value in Taiwan. The export volume of sugar apples in 2015 was 12,392 tons and the

main species was custard apples. Custard apple is a type of climacteric fruit, and therefore, it needs 3 to 5 days and 5 to 7 days for postharvest ripening at room temperature in summer and winter, respectively (Yang 1996a). In addition, this type of fruit is not suitable for export because it cannot be stored at low temperatures and transported for a long time (Yang 1996b, Yang 1984, Lu et al. 2010, Lee et al. 2005, Cheng 2015). As storage at a low temperature causes the dumb fruit phenomenon and browning, the exportation of custard apples is difficult. In this study, multi-oxidants that are produced from the brine of deep seawater are shown to extend the shelf life of custard apples, raise the fruit quality, and increase the economic value. The results of this study show that the brine of deep seawater can be widely used in agricultural and commercial applications.

2 MATERIALS AND METHODS

2.1 *Material*

Custard apples (*Annona squamosa L.*) were harvested from Taitung in the morning on February 24, 2016. The winter fruits, ripened by electric supplemental lighting, were 70% mature and the fruit weight was 500–600 g. Each fruit was put in the double-layered fruit bag and packaged in the corrugated cartons for exportation. Each carton contained 11 fruits and the net weight was approximately 6 kg.

2.2 *Methods*

2.2.1 *Preparation of multi-oxidants containing chloride dioxide*

A new electrolysis protocol that is modified from the diaphragm electrolysis of Model Hooker S-3 was used in this study. Brine of deep seawater was electrolyzed to produce numerous strong oxidants, including Cl_2, ClO_2, O_3, and H_2O_2 (Pillai et al. 2009, Hsu et al. 2012). The positive pole, which was a titanium rod and titanium mesh coated with Pd, and an insoluble anode, which was made of Rh with ISA as a catalyst, were separated by using a diaphragm of Nafion N-2030 membrane in the electrolyte chamber (Liu 2015, H).

$$\text{Brine} \xrightarrow{\text{electrocatalyze}} Cl_2 + ClO_2 + O_3 + H_2O_2 \quad (1)$$

2.2.2 *Treatment with chloride dioxide and analysis*

Custard apples were given two concentrations of multi-oxidants (0 ppm and 50 ppm) using ultrasonic atomization (Sumo, Ultrasonic Nebulizer V-15) for 15 minutes and using various components (adding newspapers, N; ethylene absorbent, EA) and procedures (custard apples were put in a PE bag, which was inflated using multi-oxidants and then packaged in corrugated cartons; inflating multi-oxidants directly into the corrugated cartons), and then, the custard apples were stored at 11°C for 10 days. After 10 days, the custard apples were stored at room temperature. During the 17-day period of storage, total soluble solids, hardness, rate of water loss, storage life, and the appearance of the fruits were measured and then analyzed. In this experiment, 50 ppm of multi-oxidants and a storage temperature of 11°C were the optimal experimental conditions, as determined from previous studies. The experimental flow chart is shown in figure 1.

The concentrations of ClO_2, Cl_2, and O_3 were measured using a photospectrometer (PC MULTI-DIRECT O 5M 86 MODEL, AQUALYTIC Co. Ltd.). Reagents and protocol used (Method

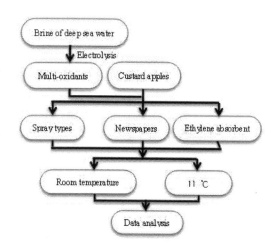

Figure 1. Flow chart for the experiments.

1.00608.0001, 1.00602.0001 and 1.00607.0001) were developed by Merck Co. Ltd. (HACH 2005).

3 RESULTS AND DISCUSSION

3.1 *Percentage of ripening of custard apples (Annona squamosal L.)*

Table 1 shows that the percentage of ripe custard apples with 50 ppm multi-oxidants is higher than that of apples with no added multi-oxidants. When comparing the different package components in the study, the percentages of ripe fruit in the groups that used added newspapers and added ethylene absorbents were higher than the other groups. The storage life doubles in some cases. This study enhances the production value of the fruits and creates the commercial opportunity of export.

3.2 *Variation in the hardness of custard apples (Annona squamosal L.)*

In studying the effect of multi-oxidants on the hardness of custard apples, figure 2 shows that custard apples are still ripen as well when multi-oxidants are combined with different package components. Comparing the effects of different package components on the hardness of custard apples, the hardness of the group with no added ethylene absorbents is greater than the other groups. It is speculated that the ethylene absorbents (1-methylcyclopropene, C_4H_6) absorb the ethylene that is produced by the custard apples, extend the shelf life of the custard apples and maintain the green color, but decrease the rate of softening of the custard apples and maintain hardness.

Table 1. The percentage of ripe custard apples on day 14 and day 17, when treated with multi-oxidants and packaged using different components at the experimental temperature.

Groups		Day 14	Day 17
A	The custard apples were bare and placed in a corrugated carton and ethylene absorbent was put in the carton.	63.6%	36.3%
B	The custard apples were bare and placed in a corrugated carton without adding any materials.	36.3%	63.6%
C	The custard apples were sprayed with 50 ppm of multi-oxidants and then placed in a carton and ethylene absorbent was put in the carton.	54.5%	45.4%
D	The custard apples were placed in the carton and were then sprayed with 50 ppm of multi-oxidants and ethylene absorbent was put in the carton.	27.2%	72.7%
E	The custard apples were placed in the carton and were then sprayed with 50 ppm of multi-oxidants. Newspapers were put on the bottom of the carton and ethylene absorbent was also put in the carton.	9.0%	90.9%
F	The custard apples were sprayed with 50 ppm of multi-oxidants and then placed in a carton, without adding ethylene absorbent or newspapers.	27.2%	72.7%
G	The custard apples were placed in the carton and were then sprayed with 50 ppm of multi-oxidants, without adding ethylene absorbent or newspapers.	36.3%	63.6%
H	The custard apples were sprayed with 50 ppm of multi-oxidants and then placed in a carton. Newspapers were put on the bottom of the carton but no ethylene absorbent was used.	27.2%	72.7%

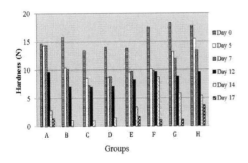

Figure 2. Hardness of custard apples that are treated with multi-oxidants and packaged using different components at the experimental temperature during the period of storage.

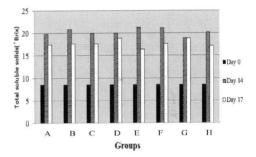

Figure 3. Soluble solids content of custard apples that are treated with multi-oxidants and packaged using different components at the experimental temperature on day 14 and day 17.

3.3   Variation in the soluble solids (Brix) content of custard apples (Annona squamosal L.)

Figure 3 shows that there is no significant difference in soluble solids content for differently packaged components, procedures, and concentrations of multi-oxidants. Most results for soluble solids content are around 18° to 21° Brix and the soluble solids content is not affected in the latter storage days. Therefore, the soluble solids content of custard apples that are stored at the experimental temperature is not affected by different packaging components or the addition of multi-oxidants.

3.4   Variation in the rate of weight loss for custard apples (Annona squamosal L.)

Storage at the experimental temperature retards breathing and evaporation, and hence, the rate of water loss for custard apples is less than that of those stored at room temperature. The results show that there is no significant difference in the rate of weight loss for all custard apple groups (shown in figure 4), but the groups that use added newspapers or ethylene absorbents have lower values than other groups.

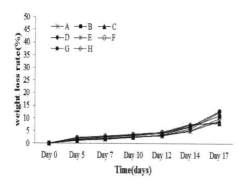

Figure 4. Rate of weight loss for custard apples that are treated with multi-oxidants and packaged using different components at the experimental temperature during the period of storage.

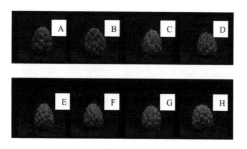

Figure 5. Appearance of custard apples that are treated with multi-oxidants and packaged using different components at the experimental temperature on day 17.

3.5 Variation in the appearance of custard apples (Annona squamosal L.)

Figure 5 shows that the custard apples that are treated with 50 ppm multi-oxidants are greener than the others, which implies that treatment with multi-oxidants maintains the green color of fruits. Different packaging components and procedures produce no significant difference in the appearance of fruits.

4 CONCLUSIONS

Custard apples that are stored at low temperature will be dumb fruits and will become brown easily, and therefore, their shelf life at room temperature is approximately 5 to 7 days. This study uses multi-oxidants and a lower experimental temperature to extend the shelf life of custard apples. The custard apples are stored at 11°C and then moved to an atmosphere of room temperature to ripen on day 10 of the experiment. The experimental results indicate that all of the custard apples ripen on day 17 with no dumb fruit phenomenon. The results of this research demonstrate that custard apples that are treated with multi-oxidants remain green.

The direct filling or spraying of multi-oxidants into corrugated cartons is a convenient technology for packaging large quantities of fruits for export. The addition of newspapers and ethylene absorbents increases their shelf life, decreases the rate of water loss, and retards softening of custards apples. This method allows custard apples to be exported and increases their economic value.

ACKNOWLEDGMENTS

This study was supported by a grant from the Ministry of Economic Affairs of Taiwan under contract no. 102-EC-17-A-32-S1–230.

REFERENCES

Chang, C.Y. (2007). Taiwan Patent M322947.

Cheng, Y.L. (2015). The world's first pineapple custard God sent. *Common Wealth Magazine* 578, 28–32. (in Chinese)

HACH (2005). Operation Menu DOC 022.53.00720.

Hsu, J.J., C.Y. Chang, & M.C. Wu (2012). Application of preservation and bacterial inhibition of electrolytic water on water convolvulus (Ipomoea aquatica Forsk). *Journal of Biobased Materials and Bioenergy* 6, 1–4.

Lee, J.X., C.S. Yang, & L.X. Ko (2005). 1-MCP and storage temperature effect on the physiology of atemoya after ripening. *The conference special issue for the research and application on the treatment technology of garden product postharvest*, Taiwan agricultural research institute, pp. 79–90. (in Chinese)

Liu, C.H., C.Y. Chang, Y.J. Shih, N.T. Chen, Y.L. Guan, Y.Y. Li, C.H. Yen, & H.C. Huang (2015). The study on the preparation of chlorine dioxide by using brine of deep sea water. *Green Science & Technology Journal* 5 103–116. (in Chinese)

Lu, B.S., S.W. Jiang, & C.S. Yang (2010). Introduction of a new Annona squamosa L. variety of Taitung No. 2. *Agricultural policy and agricultural situation* 211, 97–99. (in Chinese)

Nakasone, H.Y., & R.E. Paull (1998). Tropical fruits. *Annona*, pp. 71–75.

Pillai, K.C., T.O. Kwon, B.B. Park, & I.S. Moon (2009). Studies on process parameters for chlorine dioxide production using IrO_2 anode in an un-divided electrochemical cell. *Journal of Hazardous Materials* 164: 2–3, 812–819.

Yang, C.S. (1984). The management and outlook of Annona squamosa L. *Special issue of the tropic orchard management conference in the Taiwan area*, Taichung District Agricultural Research and Extension Station, pp. 119–133. (in Chinese)

Yang, C.S. (1996a). Common characteristics and production adjust technology of Annona squamosa L. *The production and distribution of Annona squamosa L.*, Taitung district agricultural research and extension station, pp. 1–11. (in Chinese)

Yang, C.S. (1996b). The flowering habit and storage after fruit ripening of Annona squamosa L. *The production and distribution of Annona squamosa L.*, Taitung district agricultural research and extension station, pp. 12–24. (in Chinese)

A non-intrusive load identification method based on transient event detection

Feifei Gan, Bo Yin & Yanping Cong
Ocean University of China, Qingdao, Shandong Province, China

ABSTRACT: In order to identify appliance load, this paper proposes a non-intrusive load identification method by analyzing sampled data of various domestic electric appliances, which is based on turn-on event detection. According to a periodic current state variation rule, transient event detection algorithm is performed. Then, turn-on waveform acquisition, in combination with load characteristics analysis, is applied to extract relevant transient features. Finally, Artificial Neural Network (ANN) is used for load classification and identification by comparing the performance of Back Propagation (BP) with its improved algorithm, that is, the Levenberg-Marquardt method (LM). As the experimental results suggest, the proposed method is feasible and practical for load identification.

1 INTRODUCTION

Non-Intrusive Appliance Load Monitoring (NIALM) technology plays a vital role in smart grid, which will help us to monitor electricity consumption and save power resource. Traditionally, various sensors and smart meters are installed in power distribution system to acquire real-time power grid data. In this way, data can be feed back to the control of electric system, and power consumption can be monitored in real time by users. When sensors are installed at the entrance of electric power supply, we call it Non-Intrusive Appliance Load Monitoring (NIALM) that is installed easily.

In the early 1980s, Professor George W. Hart from the Massachusetts Institute of Technology (MIT) has proposed NIALM. Hart (1992) put forward the most primitive MIT method with the support of Electric Power Research Institute (EPRI), using the cluster analysis algorithm to monitor the active power and reactive power, but it is difficult to identify appliance load with similar power. Since then, the MIT team (Norfold & Leeb 1996) increased the transient features to improve the MIT method, and Albicki (1998) extended the approach. Laughman (2003) used harmonic waves to recognize load, making the calculation simplified. Leeb (Lee et al. 2005) established the relationship between harmonic waves and the active power, the reactive power, and classified the appliance load based on the principle of minimum distance. Lin (Lin & Tsai 2015) employed genetic programming and pattern recognition technology to NIALM. Apparently, more and more techniques including FFT, wavelet transform, were applied to load identification, but features were relatively limited.

Considering that proper appliance characteristics have significant effect on the accuracy, this paper extracts features from transient event of raw voltage current, and then chooses highly recognizable ones to make up for the limitation of load features.

2 GENERAL SCHEME OF NIALM SYSTEM

Considering non-intrusive technology methods, there are several existing solutions. Although they have many differences in detail, the overall design framework for NIALM system is roughly the same. An overall design scheme is given in this paper.

Taking all the existing works into consideration (see Fig. 1), non-intrusive load monitoring methods can be mainly divided into four parts, including data acquisition and pro-processing, transient event detection, feature extraction and optimization, load classification and identification. It can be observed from Fig.1 that event detection and feature extraction are the key techniques for NIALM system, which is directly related to the accuracy of load identification.

3 PROPOSED KEY METHODS

3.1 *Transient event detection process*

Among all of load monitoring steps, event detection is of the utmost importance because of its fundamental role. Only turn-on event is detected correctly, can we reduce the mistaken probability and grasp the transient response accurately. Through analyzing the current state of appliances

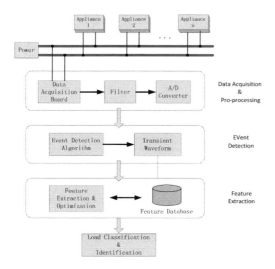

Figure 1. Overview scheme for NIALM.

in turn-on moment, load instantaneous current has evident variation tendency, where current curve rises instantly, and appliance is judged whether open or not according to this change.

3.1.1 Description of event detection algorithm

We assume that the current signal to be detected is I and the corresponding voltage is V, with a total of M sampling cycles and N sampling points in each cycle. I_i (see Equation 1) is the current state parameter of cycle i. I_{i_mean} (see Equation 2) is the current mean value of cycle i. ΔI_i (see Equation 3) is the current state variation of cycle i. ΔI_{i_p} (see Equation 4) is the difference in mean peak value between before and after cycle i. Similarly, ΔI_{i_v} (see Equation 5) is the difference of valley value. I_{on} (see Equation 6) is the final current matrix of load turn-on event, and V_{on} (see Equation 7) is the final voltage matrix accordingly.

$$I_i = \frac{\sum_{j=1}^{N} |I_{ij} - I_{i\_mean}|}{N} \quad (i = 1, 2, ..., n) \quad (1)$$

$$I_{i\_mean} = \frac{\sum_{j=1}^{N} I_{ij}}{N} \quad (i = 1, 2, ..., n) \quad (2)$$

$$\Delta I_i = I_{i+1} - I_i \quad (i = 1, 2, ..., n-1) \quad (3)$$

Where I_{ij} is the current value of the point j for cycle i.

$$\Delta I_{i\_p} = \frac{\sum_{m=i+2}^{i+6} I_{m\_p}}{5} - \frac{\sum_{m=i-1}^{i-5} I_{m\_p}}{5} \quad (4)$$

$$\Delta I_{i\_v} = \frac{\sum_{m=i-1}^{i-5} I_{m\_v}}{5} - \frac{\sum_{m=i+2}^{i+6} I_{m\_v}}{5} \quad (5)$$

where I_{m_p} is the peak value of current state parameter in cycle m and I_{m_v} is the valley value similarly.

$$I_{on} = \left[I_{kj} - I_{(k-1)j}, I_{(k+1)j} - I_{(k-1)j}, ..., I_{lj} - I_{(k-1)j} \right] \quad (6)$$

$$V_{on} = \left[V_{kj} - V_{(k-1)j}, V_{(k+1)j} - V_{(k-1)j}, ..., V_{lj} - V_{(k-1)j} \right] \quad (7)$$

where $j = 1, 2, ..., N$; k is the start cycle of transient turn-on event; and l is the end cycle of transient event.

Through theoretical research and analysis on transient current variation trend, the process of turn-on event detection algorithm based on current variation is put forward as follows:

1. Input current matrix (I) and voltage matrix (V), and initialize cycle $i = 0$. Update $i = i + 1$, calculate ΔI_i using equation (3). If $\Delta I_i > \alpha$, then go to step 4, where α is the given threshold. It is considered that current change of cycle i is relatively large. Otherwise, go back to step 2. Calculate ΔI_{i_p} and ΔI_{i_v} using equations (4) and (5), respectively. If $\Delta I_{i\_p} > \beta$ and $\Delta I_{i\_v} > \beta$ simultaneously, then $k = i$, where β is the given threshold. It is considered that load turns on in cycle k, and record k. Otherwise, go back to step 2.
2. Update $i = i + 1$, and calculate ΔI_i, ΔI_{i+1}, ΔI_{i+2}, ΔI_{i+3}, ΔI_{i+4}.
3. If ΔI_i, ΔI_{i+1}, ΔI_{i+2}, ΔI_{i+3}, $\Delta I_{i+4} < \gamma$ simultaneously, then $l = i$. It is considered that the load transient event ends in cycle l, and record l. Otherwise, go back to step 6.
4. Store I_{on} and the corresponding V_{on} matrix, that is, transient data of load turn-on event from cycle k to cycle l.
5. Let $i = l$ and go back to step 2 to detect the next transient turn-on event, until $i = $ end cycle of sample data.

3.1.2 Turn-on waveform acquisition

After performing the transient event detection algorithm, we can acquire the turn-on waveform of appliance load. Taking a piece of data, for example, the incandescent lamp is frequently switched on five times. Raw current waveform and periodic current state variation are shown below (see Fig. 2). The points that the arrows refer to represent the load turn-on points or cycles, and it turns out to be accurate when contrasting to the actual situation.

From the principle of the turn-on event detection, we can conclude that the transient period of different appliances in different events can be different and the transient period of the same appliance in different events may not be exactly the same. Therefore, the sampling number of each event can also be different. Transient current waveform of part residential appliances is given based on the event detection algorithm (see Fig. 3).

3.2 Feature extraction

Residential appliance load in the electricity power system varies, which can be roughly divided into

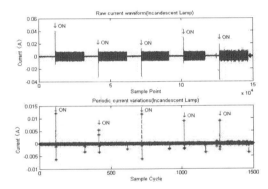

Figure 2. Turn-on event detection of incandescent lamp.

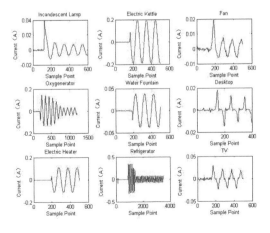

Figure 3. Transient current waveform of part appliances.

three categories, including resistive load such as water fountain and electric kettle, capacitive load such as TV and computer, and inductive load such as fan and refrigerator. Research of power system load characteristics is mainly from two aspects, namely the steady-state analysis and transient analysis. This paper focuses on the load transient turn-on event, and therefore, transient features are mainly discussed.

Considering the load signal transient characteristics, the feature parameters contain time domain and frequency domain characteristics. All features are extracted from I_{on} and V_{on} (see section 3.1). Potential features in time domain are given in Table 1.

Some of the interpretation or calculation method of the feature is shown below. I_{peak} is the maximum value between crest value and absolute value of I_{on} valley. *Slope* is the tangent value of first-wave crest point and valley point. R_{pp} is the relative error between first-wave peak value and stable mean value. I_{area} is defined as the product of the figure area enclosed with absolute value of I_{on} and cycle numbers. I_{rms} is given (see Equation 8). *AP* is the product of current and voltage virtual values.

$$I_{rms} = \sqrt{\frac{\sum\limits_{i=1}^{M}\sum\limits_{j=1}^{N}(I_{ij})^2}{M*N}} \tag{8}$$

It often takes harmonic problems into consideration, when analyzing the signal in the frequency domain. Potential part features related to harmonic waves in the frequency domain are given in Table 2. Frequency-domain issues will not be discussed in detail in this paper.

In order to identify appliance load in a more accurate way, some features are selected from potential ones to ensure the validity of the characteristic and the efficiency of recognition. Ten features in time domain are adopted to make up the

Table 1. Potential time-domain features.

Tag	Feature	Description
1	I_{peak}	Current peak value
2	*Slope*	Waveform slope
3	R_{pp}	Peak to peak ratio
4	I_{area}	Current area
5	I_{rms}	Root-mean-square value
6	I_{pp}	Peak to peak value
7	T_t	Transient time
8	TP	Transient power
9	TE	Transient energy
10	AP	Apparent power

Table 2. Potential frequency-domain features.

Tag	Feature	Description
1	$E_{fundamental}$	Energy of fundamental wave
2	$E_{harmonic}$	Energy of harmonic wave
3	Ratio	Ratio of fundamental and harmonic

eigenvalue matrix of sample, FEATURE = [I_{peak}, Slope, R_{pp}, I_{area}, I_{rms}, I_{pp}, T_t, TP, TE, AP, Class]. Class is the corresponding category label of the eigenvalue. Save the eigenvalue matrix in the sample feature database, respectively, and then set up characteristics model for each appliance to get ready for load identification.

3.3 Load identification

Classification and identification is the important part of the appliance load monitoring system, and the key is the design of classifier. Artificial Neural Network (ANN) has its unique advantages on self-adaption, capable of searching optimal solution, etc. Therefore, neural network is often used for solving classification problems. Generally speaking, learning rules, the characteristics of the neurons, and the topology of the network connection are taken into consideration when selecting and designing the neural network model. The neural network topology model includes the input layer, hidden layer, and output layer. Back Propagation Artificial Neural Network (BP-ANN) can achieve arbitrary linear or nonlinear function mapping based on the gradient descent method, and it is implemented to minimize the mean square error between expected value and neural network output by adjusting the weights. As the improved algorithm of BP, the Levenberg-Marquardt (LM) method can greatly improve the classification efficiency and accuracy. Neural network used for classification mainly includes several aspects as follows:

1. Input and expectant output. To improve the efficiency of the network, features as network input, the load category label as the expectant output, both need to be normalized.
2. Neural network structure. The input layer nodes are decided by input feature numbers, and the output layer nodes are up to network output. The hidden layer has no strict rules but only by experience.
3. Neural network training. If transfer function and learning algorithm are different, network training results will not be the same.
4. Neural network simulation. Through the simulation of trained net, the actual output is generated. Then, we can judge the load category label and compute the recognition accuracy.

4 LOAD IDENTIFICATION EXPERIMENT AND RESULT

4.1 Data selection

First, common residential appliances are selected to be used for the experiment (see Table 3). Second, voltage and current senor and A/D converter are adopted to collect real-time voltage and current data with a sampling rate of 5 KHz. Third, event detection program is executed to capture transient waveform (see section 3.2). Fourth, features are extracted from the acquired transient data. Finally, the sample feature values matrix is used for load identification as the train and test sample.

4.2 Comparison of experimental result

As shown in Table 3, the category tags of ten kinds of different appliances are set from 1 to 9, respectively, in this experiment. Sample matrix (see section 3) is made up of ten kinds of transient features and one corresponding appliance category label, having a total of 11 columns, and it is given as follows:

$$Feature = [I_{peak}, Slope, R_{pp}, I_{area}, I_{rms}, I_{pp}, T_t, TP, TE, AP, Class]$$

Training sample and test sample in each appliance have 25 sets of feature data respectively. As 9 kinds of appliance load data are selected, there are 225 sets of training sample and 225 sets of testing sample totally. The following experiments are implemented on the MATLAB platform by programming the load identification algorithm.

When training the Artificial Neural Network (ANN), transfer function selects tansig in hidden layer nodes and purelin in output layer nodes. Both standard BP and LM are used for identification, and the two methods adopt same sample and do multiple experiments. Maximum iterations are set to 1000 and the goal of Mean Square Error (MSE) is 0.001. One identification result of BP and LM (see Figs. 4, 5), performance, and accuracy contrasted (see Table 4) are given, respectively.

In comparison to standard BP and its improved method (Levenberg-Marquardt), the load identification results indicate that the Levenberg-

Table 3. Selected appliance category.

Tag	Appliance category		
1~3	Incandescent Lamp	Electric Kettle	Fan
4~6	Oxygenerator	Water fountain	Desktop
7~9	Electric heater	Refrigerator	TV

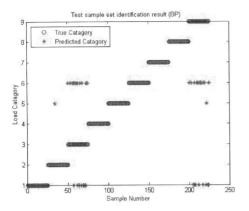

Figure 4.   Identification result of BP-ANN.

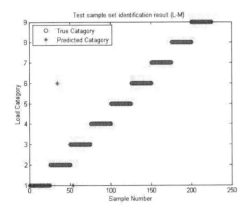

Figure 5.   Identification result of LM.

Table 4.   Performance and accuracy contract.

Algorithm	Iterations	Time (sec.)	MSE	Accuracy (%)
BP	1000	7	0.030110	81.778
	1000	12	0.028772	79.111
	1000	6	0.028315	80.889
	1000	8	0.027610	85.333
	1000	8	0.028267	83.111
LM	56	2	0.000462	99.111
	71	2	0.000498	99.111
	51	2	0.000481	99.111
	53	2	0.000497	99.556
	80	3	0.000485	99.111

Marquardt method recognizes appliance loads more effectively and the result owns higher mean accuracy of about 99.333% compared with 82.044% of the BP method.

5   CONCLUSIONS

This paper gives a method of non-intrusive load identification, which is based on turn-on transient event detection. Misclassification probability is greatly reduced because of the improved event detection algorithm based on current signal state parameter variation. By contrasting BP-ANN and LM method, Mean Square Error (MSE) is reduced from 0.0286148 to 0.0004846 on average, and the accuracy of appliance load identification result is increased by 21.07%. As the load result shows us, the given method has certain practicability and popularization for load identification.

ACKNOWLEDGMENTS

This work was financially supported by Qingdao innovation and entrepreneurship leading talent project (13-cx-2), Qingdao strategic industry development project (13-4-1-15-HY), and international cooperation project (2015DFR10490).

REFERENCES

Albicki, A.I. & Cole, A. 1998. Algorithm for non-intrusive identification of residential appliances. Proceedings of the IEEE International Symposium (3):338–341.
Albicki, A.I. & Cole, A. 1998. Data extraction for effective non-intrusive identification of residential power loads. IEEE Instrumentation and Measurement Technology Conference.
Hart, G.W. 1992. Nonintrusive appliance load monitoring. Proceedings of the IEEE.
Laughman, C. et al. 2003. Power signature analysis. IEEE Power & Energy Magazine.
Lee, K.D. et al. 2005. Estimation of variable-speed-drive power consumption from harmonic content. IEEE Transactions on Energy Conversion (20): 566–574.
Lin, Y.H & Tsai, M.S. 2015. The integration of a genetic programming-based feature optimizer with fisher criterion and pattern recognition techniques to nonintrusive load nonitoring for load identification. International Journal of Green Energy.
Norfold, L.K. & Leeb, S.B. 1996. Non-intrusive electrical load monitoring in commercial buildings based on steady-state and transient load-detection algorithms. Energy and Buildings (24): 51–64.
Shaw, S.R. & Leeb, S.B. et al. 2008. Nonintrusive load monitoring and diagnostics in power systems," IEEE Transactions on Instrumentation and Measurement (57), 1445–1454.
Wichakool, W. & Avestruz, A. 2009. Modeling and estimating current Harmonics of variable electronic loads. IEEE Transactions on Power Electronics (24):2803–2811.

Advances in Energy, Environment and Materials Science – Wang & Zhou (Eds)
© 2017 Taylor & Francis Group, London, ISBN 978-1-138-03600-0

Combining recency of ordering time and different sources from upstream manufacturers to predict the materials demand of downstream manufacturers

H.H. Huang
Department of Business Administration, Aletheia University, Taiwan

ABSTRACT: In this paper, the author combines the information of demand from different sources of upstream manufacturers and the information of ordering time recency to forecast the demand of downstream manufacturing. This paper proposes a new model which extends the materials demand model of Huang (2014b) and Huang (2016c) and conducts the empirical data to evaluate, respectively, the RMSD of a new model and that of Huang's. The results show that the new model has better fitness than the other.

1 INTRODUCTION

It is an important topic to predict the material demand in the manufacturing process (Liu & He 2014; Nielsen et al. 2016). The results of forecasting can provide the manufacturers or managers to evaluate their purchasing (Kilger & Wagner 2014), inventory and output. It can also avoid manufacturer's overproduction, which causes more cost of inventory (Sanni & Chukwu 2013; Varyani et al. 2014).

In material demand quantity prediction, the source of material demand quantity may purchase from different upstream manufacturers (Peng & Zhou 2013; Kilger & Wagner 2014). Thus, in this paper, we take this information into account of the material demand model. In Huang's (2016c) model, the "ordering quantity of past" and the "recency of ordering time" are two variables of predicting the material demand quantity. In the proposed new model of this paper, we consider the variable of "ordering quantities from different sources" to provide more information of the variable of "ordering quantities of past". Thus, this new model can demonstrate not only the information of downstream manufacturer itself but also reflect the various ordering inputs from upstream. It can detect the increase or decrease in quantities from different sources in advance. This can make the model more flexible to fit the dynamitic demand process (Peng & Zhou 2013; Sanni & Chukwu 2013).

For the variable of ordering quantities from different inputs (source), we depend on Huang (2016c), who uses characteristic functions to develop the different sources of total demand in

stochastic theory. Therefore, this paper combines recency of ordering time (Huang 2014c) and different sources from upstream (2014b) to develop a new model to predict the materials demand of downstream manufacturers.

In the following section, first we will briefly introduce Huang's model (2016c). The ordering quantity will be considered as different random variables which reflect different ordering sources. The recency of ordering time will be demonstrated by the characteristic function based on Huang (2013a; 2014b). Second, the cumulative distribution function (cdf) and the probability density function (pdf) of new materials demand quantity model will be derived. Third, the data analysis is introduced. Moreover, the empirical data analysis, parameters estimation, and model validation are shown in this section. In this part, we also compare the fitness between empirical data and simulation data in both Huang's model and the new model. Finally, the conclusions are drawn.

2 THE MATERIALS DEMAND MODEL

Based on Huang's (2016c) model, we also consider that the materials demand model is composed of the ordering quantity of past (denoted as Q) and the recency of ordering time (denoted as R).

$$D = R \cdot Q \tag{1}$$

2.1 The ordering quantity

We consider that the ordering quantities (denoted as Q) are from two different sources, which are denoted by upstream suppliers $U(Q_U)$ and $W(Q_W)$. The probability of ordering from supplier U is denoted by P and ordering from supplier W is denoted by $1 - P$ (Huang 2014b). Then, the order quantities Q_U and Q_W are the random variables u and w, which follow normal distribution with the parameters (μ_u, σ_u^2) and (μ_w, σ_w^2). According to Huang (2014b), the characteristic function can be used to divide the pdf of total ordering quantity as follows:

$$f_Q(q) = \frac{1}{2\pi} \times \int_{-\infty}^{\infty} \left(P\, e^{it(\mu_U - Q) - \frac{\sigma_U^2 t^2}{2}} \right)$$
$$\left(1 - e^{it\mu_W - \frac{\sigma_W^2 t^2}{2}} - P e^{it\mu_W - \frac{\sigma_W^2 t^2}{2}} \right)^{-1} dt \quad (2)$$

In which the total ordering quantities are considered as random variable Q. If the downstream company only orders materials form source U, the probability is P. Then $Q = Q_U$. If the downstream company orders materials form both sources U and W, the probability is $P(1 - P)$. Then, $Q = Q_U + Q_W$.

Thus, the probability is $P(1 - P)^{n-1}$, when $Q = Q_U + \sum_{j=1}^{n-1} Q_{W_j}$. Its characteristic function is

$$E\left(e^{itQ}\right) = P\,E\left(e^{itQ_U}\right) + P(1-P)E(e^{it(Q_U + Q_W)})$$
$$+ P(1-P)^2\, E\left(e^{it(Q_U + Q_W + Q_U)}\right)$$
$$+ \ldots + P(1-P)^{n-1} E\left(e^{it(Q_U + \sum_{j=1}^{n-1} Q_{W_j})} \right) \quad (3)$$

Let $E(e^{itQ}) = h(t)$, $E(e^{itQ_u}) = g_U(q_1)$, $E(e^{itQ_W}) = g_W(q_2)$.
And $|g_U(q)| < 1, |g_W(q)| < 1$.
Then,

$$h(t) = \frac{P g_U(q)}{1 - (1 - P)g_W(q)}.$$

2.2 The recency of ordering time

Based on the study by Huang (2013a; 2016c), the recency of ordering time (denoted as R) is demonstrated as a renew process and the purchase interval time is a random variable that follows exponential distribution. The pdf and cdf in the recency of ordering time (denoted as r) are computing as

$$f_R(r|\theta) = \frac{1}{\theta} \exp\left(-\frac{r}{\theta}\right) \quad (4)$$

$$F_R(r|\theta) = \frac{1 - \exp\left(-\frac{r}{\theta}\right)}{\theta} \quad (5)$$

2.3 The cumulative distribution function of materials demand model

According to equation (1), the materials demand model is composed of "the total ordering quantity (Q)" multiplying "the recency of ordering time (R)". In order to calculate the cdf of the full model, we denote the random variable Y as demand quantity. Then, its cdf is

$$F_D(y) = P(R \cdot Q < y) = \frac{P}{2\theta^2 \pi}$$
$$\times \int_0^y \int_0^r \int_{-\infty}^{\infty} \left(e^{it(\mu_U - Q) - \frac{\sigma_U^2 t^2}{2} - \frac{r}{\theta}} \right) \quad (6)$$
$$\left(\frac{1}{P} - \frac{1}{P} e^{it\mu_W - \frac{\sigma_W^2 t^2}{2}} - e^{it\mu_W - \frac{\sigma_W^2 t^2}{2}} \right)^{-1} dt\,dq\,dr$$

2.4 The probability density function of materials demand model

Based on equation (6), we can compute the pdf of the proposed model, as shown in equation (7).

$$f_D(y) = \frac{d}{dy} F_D(y) = \frac{P}{2\theta^2 \pi} \times \frac{d}{dy} \int_0^y \int_0^r \int_{-\infty}^{\infty} S\, dt\,dq\,dr \quad (7)$$

where

$$S = \left(e^{it(\mu_U - Q) - \frac{\sigma_U^2 t^2}{2} - \frac{r}{\theta}} \right)$$
$$\times \left(\frac{1}{P} - \frac{1}{P} e^{it\mu_W - \frac{\sigma_W^2 t^2}{2}} - e^{it\mu_W - \frac{\sigma_W^2 t^2}{2}} \right)^{-1}$$

3 THE ANALYSIS PROCESS

There are 20155 samples, which include the information about "ordering quantity of past", "recency of ordering time", and "ordering quantities from different upstream manufacturers". First, we use the empirical data to estimate the parameters of the proposed model. The estimation will be provided in the following section. Second, we use the results of parameter estimation to poll our simulation data according to the proposed model. Third, the comparison between empirical data and simulation data is shown with the Root-Mean-Square Deviation (RMSD). The analysis process is shown in Figure 1.

3.1 The parameters estimation

We use MLE (Maximum Likelihood Estimate) to estimate the parameters. Let y_j denote the materials demand quantities by upstream j, that is:

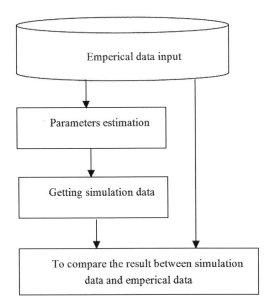

Figure 1. The analysis process.

$$L(\mu_U, \sigma_U^2, \mu_W, \sigma_W^2, P, \theta) = \left(\frac{P}{2\theta^2 \pi}\right)^n \frac{d}{dy} \int_0^\infty \int_0^r \int_{-\infty}^y S \, dt dq dr$$

(8)

We differentiate $L(\mu_U, \sigma_U^2, \mu_W, \sigma_W^2, P, \theta)$ respectively, considering $\mu_u^2, \sigma_u^2, \mu_w^2, \sigma_w^2, P, \theta$ and set them equal to zero, that is,

$$\begin{cases} \dfrac{\partial}{\partial \mu_U} L(\mu_U, \sigma_U^2, \mu_W, \sigma_W^2, P, \theta) = 0 \\[6pt] \dfrac{\partial}{\partial \sigma_U^2} L(\mu_U, \sigma_U^2, \mu_W, \sigma_W^2, P, \theta) = 0 \\[6pt] \dfrac{\partial}{\partial \mu_W} L(\mu_U, \sigma_U^2, \mu_W, \sigma_W^2, P, \theta) = 0 \\[6pt] \dfrac{\partial}{\partial \sigma_W^2} L(\mu_U, \sigma_U^2, \mu_W, \sigma_W^2, P, \theta) = 0 \\[6pt] \dfrac{\partial}{\partial P} L(\mu_U, \sigma_U^2, \mu_W, \sigma_W^2, P, \theta) = 0 \\[6pt] \dfrac{\partial}{\partial \theta} L(\mu_U, \sigma_U^2, \mu_W, \sigma_W^2, P, \theta) = 0 \end{cases}$$

(9)

The estimation results are shown in table 1.

3.2 *The model validation*

Based on the empirical data, we compute, respectively, Root-Mean-Square Deviation (RMSD) from the model of Huang (2016c) and the proposed new model.

Table 1. The results of parameters estimation.

Parameters	New model	Parameters	Huang's model (2016c)
μ_u^2	433	μ	975
σ_u^2	6.21	σ^2	6.08
μ_w^2	515	θ	2.31
σ_w^2	8.98		
P	0.67		
Θ	2.31		

Table 2. The results of parameters estimation.

	RMSD*
Huang's model (2016c)	0.9013
New model	0.8775

$$*\text{RMSD} = \frac{\sqrt{\sum_{j=1}^n (y_j - \hat{y}_j)^2}}{n}$$

The results show that the new model (RMSD = 0.8775) has better fitness than that of Huang (2016c) (RMSD = 0.9013).

4 CONCLUSION

This paper extends Huang's model to combine recency of ordering time and different sources from upstream to predict the materials demand of downstream manufacturers. This new model provides more information of ordering from different upstream manufacturers and considers dynamic ordering quantities. The results show good fitness of the new model. Thus, in the future, manufacturers or managers can use this new model to check the different sources ordering to ensure the materials demand.

ACKNOWLEDGMENTS

The author would like to thank Aletheia University (AU-AR-104-014) for supporting this research.

REFERENCES

Huang, H. H. 2013a. A detection model of customer alive in information management application. *Advanced Materials Research* 684: 505–508.

Huang, H. H. 2014b. A Predicting model for demand quantities of downstream manufacturer. *Applied Mechanics and Materials* 624: 694–697.

Huang, H. H. 2016c. A materials demand model with ordering quantity of past and recency of ordering time. *MATEC Web of Conferences* forthcoming.

Kilger, C. & Wagner, M., 2014. *Supply Chain Management and Advanced Planning*, Springer: New York.

Liu, X., Li, Z. & He, L. 2014. A Multi-objective stochastic programming model for order quantity allocation under supply uncertainty. *International Journal of Supply Chain Management* 3(3): 24–32.

Nielsen, I., Dung Do, N. A., Nielsen, P. & Khosiawan, Y. 2016. Material supply scheduling for a mobile robot with supply quantity consideration—A GA-based approach, *Advances in Intelligent Systems and Computing* 432: 41–52.

Peng, H. & Zhou, M. 2013. Quantity discount supply chain models with fashion products and uncertain yields. *Mathematical Problems in Engineering* 2013: 1–13.

Sanni, S. S. & Chukwu, W. 2013. An economic order quantity model for items with three-parameter weibull distribution deterioration, ramp-type demand and shortages. *Applied Mathematical Modelling* 37(23): 9698–9706.

Varyani, A., Jalilvand-Nejad, A. & Fattahi, P. 2014. Determining the optimum production quantity in three-echelon production system with stochastic demand. *The International Journal of Advanced Manufacturing Technology* 72 (1): 119–133.

Material science and engineering

Advances in Energy, Environment and Materials Science – Wang & Zhou (Eds)
© 2017 Taylor & Francis Group, London, ISBN 978-1-138-03600-0

The influence of rectangular grooved labyrinth seals on the flow characteristics of valve spools

Wenhua Jia
School of Mechanical Engineering, Nanjing Institute of Technology, Nanjing, China

Chenbo Yin
School of Mechanical and Power Engineering, Nanjing University of Technology, Nanjing, China

Guo Li & Menghui Sun
School of Mechanical Engineering, Nanjing Institute of Technology, Nanjing, China

ABSTRACT: According to the characteristics of the valve core and valve cavity relative motion, considering many factors such as wall slippage, sticky characteristics and so on, a mathematical model of the micro flow valve and spool valve's cavity clearance was established. On the basis of research on the corresponding relationship of the uniform pressure groove length, width, quantity, and flow characteristics, four kinds of configuration schemes of uniform pressure grooves were put forward and the multi-objective osculating value method was applied to perform a decision analysis to obtain the optimal effect of one of the four groups.

1 INTRODUCTION

On the macro scale, the effect of the wall slip[1–3] on internal leakage is insignificant. Sealing gaps exist in the micro/nano level and the wall slip is bound to lead to obvious changes in the leakage[4,5] amount and the flow field[6] characteristics, while existing research is mainly concentrated on the bearing capacity of friction and oil film force of the impact.

For construction machineries such as excavators, the excellent sealing performance of the valve core and valve cavity is a guarantee for proper functioning (Figure 1). The gap space between the valve core and valve cavity is in the micro-nano level. The design of gap space[7] should ensure the movement of the valve core and prevent the valve core from being stuck. The gap space between the valve core and valve cavity is designed to be a labyrinth seal structure, which combines the structure of the circumferential gap seal and axial face seal. Minimizing the radial clearance could reduce the leakage amount as well as contribute to sealing and maintenance of pressure.

A large number of experimental observations show that a wall slip exists on most of the liquid-solid interfaces with obvious scale effect. As shown in Figure 2, the micro gap flow driven by pressure difference is a laminar flow and the parameters can be denoted as follows: gap height – h, length – L,

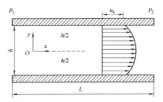

Figure 1. Structure of the valve core and valve cavity.

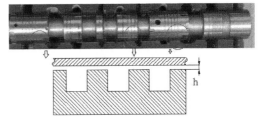

Figure 2. The model of the gap flow under pressure difference.

valve core radius – r_0, pressure difference between the ends of the gap – $P_2 - P_1$, slip coefficient – b, and fluid dynamic viscosity – μ.

Due to the fact that the size of gap height h is much smaller than the valve core radius r_0, the circular micro gap flow can be simplified to an equivalent micro gap flow of parallel plates. The valve core circumference, $2\pi r_0$, can be regarded as the width of a parallel plate.

$$Q = \int_{-h/2}^{h/2} 2\pi r_0 u\, dy = -\frac{\pi r_0 h^3}{6\mu}\frac{dp}{dx}\left(1+6\frac{b}{h}\right) \qquad (1)$$

Then, $\bar{u} = \dfrac{Q}{2\pi r_0 h} = -\dfrac{\pi r_0 h^3}{6\mu}\dfrac{dp}{dx}\left(1+6\dfrac{b}{h}\right)$

The gap flow characteristics driven by pressure difference in the wall-slip condition are expressed as equation (1). From the equation, it can be found that the pressure difference is proportional to the flow field velocity at a certain length of the model, aggravating the internal leakage in the meantime.

2 MODELING OF THE LABYRINTH SEAL OF THE MULTIWAY VALVE GAP

Models were established and compared pairwise purposefully to obtain the influence of relevant parameters and structures on the performance of hydraulic systems. In the models, the direction of axis X is selected as the axial direction and Plane YZ as the axial section, as shown in Fig. 3. The parameters of the model are as follows: the valve core radius R is 14 mm, gap width h is 0.02 mm, and model length L is 4 mm.

2.1 Grid division and setting of boundary conditions

As shown in Figure 4, there are two inflection points: point A and point B in the intersection between the gap flow channel and the pressure groove.

The seal groove of the model (Figure 4) can lead to a higher pressure drop at point A, causing more cavities and higher impact on point B. The

mesh around the inflection point needs to be densified and the grid division of the model is illustrated in Figure 3.

Face a, which belongs to the seal groove of the interface, is denoted as interface-a, and Face b, which belongs to the gap of the interface, is denoted as interface-b. Merge the two faces and set the type of the other boundaries as "wall". The boundary types of the model are shown in Figure 5.

The micro channel equivalent diameter d, also known as hydraulic diameter, is the ratio of the quadruple area of the wetted cross section to the wetted circle perimeter (equation 2).

$$d = \frac{4A}{l} = \frac{4\frac{1}{4}\pi(D^2 - d^2)}{\pi(D+d)} = D - d = 2\delta \qquad (2)$$

$$\mathrm{Re} = \frac{\rho V d}{\mu} = \frac{\rho V d}{\rho v} = \frac{V d}{v} = \frac{2\delta V}{v} = \frac{2\cdot10^{-5}V}{40\cdot10^{-6}} = 0.5V$$

According to previous experiment's results, for the micro gap flow in the actual situation, $V_{max} \leq 10$ m/s and $Re_{max} \leq 5$. Compared with the data in Table 1, Reynolds numbers[14,15] of the concentric micro gap flows is far below the critical value 1100, and for this reason, the laminar flow model is selected.

The boundary definition of the inlet and outlet is pressure type. As shown in Figure 6 (SY215C8M, Sany Heavy Industry, China), the inlet pressure is set as 22 MPa and the outlet pressure is 0.5 MPa. The wall boundary setting is based on the UDF added previously and the loading path is given as follows: Fluent → Define → Boundary conditions

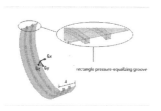

Figure 3. Schematic of the model chart.

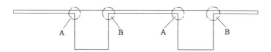

Figure 4. Schematic of a section of the gap flow model.

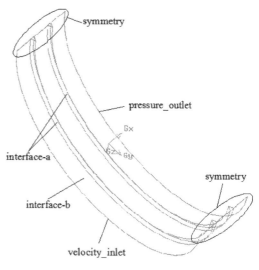

Figure 5. Schematic of boundary types.

Table 1. Critical Reynold numbers.	
Runner shape	Numbers
Concentric ring seam	1100
Eccentric ring seam	1000
Concentric with average pressure groove	700
Eccentric with average pressure groove	400

Figure 6. Picture showing pressure acquisition.

→ wall → Moving Wall → Components → X—Velocity. The filename is udf b_U.

3 INFLUENCE OF RECTANGULAR LABYRINTH SEAL PARAMETERS ON INTERNAL LEAKAGE AND RADIAL UNBALANCED FORCE

3.1 Depth of the rectangular groove

The model investigated is also under circumstances of parallel eccentricity and an inverted cone. The only changing parameter is the depth of the groove and all the other conditions remain unchanged. The depth is set as follows: 0.02 mm, 0.18 mm, 0.26 mm, and 0.42 mm. The length is 0.4 mm, the number of grooves is 2, and the position remains unchanged. The results are shown in Figures 7 and 8.

From Figure 7, it can be clearly observed that as the depth of the rectangular groove increases, the pressure change tendency of face Y+ and face Y− is almost the same, while the quantity of the pressure change is quite different. The area enclosed by the two curves is actually the area enclosed by the radial unbalanced force on the valve core. As a result, it can be seen that the area enclosed by the radial unbalanced force on the valve core constantly reduces with an increase in the rectangular groove depth. When the depth of a pressure groove is equal to the gap width, its pressure equalizing capability will get considerably poor and the groove will not show the same pressure in the X direction which can be seen from Figure 7(a), where the slope of the pressure line at x = 1–1.5 mm is negative. When

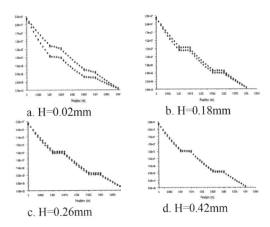

a. H=0.02mm   b. H=0.18mm

c. H=0.26mm   d. H=0.42mm

Figure 7. Graphs showing the spool pressure on section XY (Y+ & Y−).

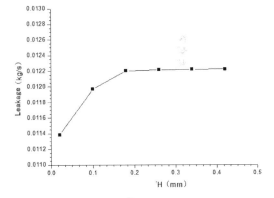

Figure 8. Graph showing the inner leakage under different depths.

rectangular groove depth reaches 0.42 mm, the pressures on face Y+ and face Y− are almost the same and the two pressure curves in Figure 7(d) almost coincide with each other. Under this circumstance, the area enclosed by the radial unbalanced force on the valve core can be taken as zero approximately, achieving the expected results.

As seen in Figure 8, the internal leakage also hikes with an increase in the rectangular groove depth. At this point, the line slope is relatively high. As the depth H increases, the internal leakage amount grows faster and the slope also gradually becomes smaller. And then, with an increase in the depth H to a larger scale, the internal leakage increases slowly and the tendency is likely to be smooth and steady.

3.2 Length of the rectangular groove

In order to investigate the influence of the groove length on the gap flow characteristics, the length is

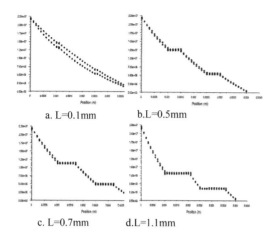

a. L=0.1mm   b.L=0.5mm

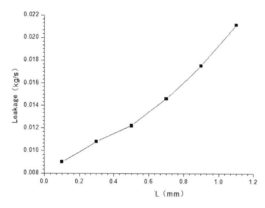

c. L=0.7mm   d.L=1.1mm

Figure 9. Graphs showing the spool pressures on section XY (Y+ & Y−).

Figure 10. Graph showing the inner leakage under different lengths.

set as 0.1 mm, 0.3 mm, 0.5 mm, 0.7 mm, 0.9 mm, and 1.1 mm, respectively, for comparison. The groove depth is 0.3 mm and two grooves are designed in the same position for all the cases. The results are shown in Figures 9 and 10.

It is observed from Figure 9 that, as the length L of the rectangular groove increases, the area enclosed by the radial unbalanced force on the valve core al decreases. When the groove length is higher than 0.5, the pressures on face Y+ and Y− almost become equal to each other and the two pressure curves coincide with each other approximately as well. In the meantime, the area enclosed by the radial unbalanced force on the valve core is equal to zero accordingly. The first part and the second part of the groove for all the models in this figure are different, and the general tendency is to descend gradually; the corresponding reason could be the length of the pressure groove, where the

pressure is close to the pressure of the same position without the groove.

From Figure 10, it can be observed that the increasing length can contribute to a higher internal leakage amount as well as a higher increasing rate of leakage.

3.3 *Number of rectangular grooves*

Effects of the groove number on gap flow characteristics are also studied. The groove number is set as 0, 1, 2, and 3 respectively. The length and depth are 0.5 mm and 0.2 mm, respectively.

With an increase in the groove number, the area enclosed by the curves from the non-horizontal lines gradually decreases. The curves from horizontal lines correspond to the position of the rectangular groove. By increasing the depth of the pressure groove, the pressures of Y+ and Y− on the XY cross section become close to each other and the area enclosed by horizontal lines is approximately close to zero. As result, the area enclosed by horizontal lines needs to be ignored in comparison with the area enclosed by the radial unbalanced force on the valve core.

4 CONCLUSIONS

1. Under the circumstance of parallel eccentricity and inverted cone, the addition of the pressure groove on the shaft shoulder of the valve core could reduce the radial unbalanced force, but also could increase the internal leakage amount. The influence of different types of pressure grooves on flow characteristics has been analyzed and the rectangular groove has been evaluated to have the optimal effect.
2. The influence of length, depth, and number of rectangular pressure grooves on flow characteristics has been investigated. Based on the correlations revealed above, we proposed four groove allocation options in terms of radial unbalanced force and internal leakage amount according to the simulation results. The multi-objective osculating value method has been employed to obtain the optimal case of groove allocation, ensuring the smooth and efficient operation of the valve core under the circumstance of parallel eccentricity and inverted cone.

ACKNOWLEDGMENTS

This work was supported in part by The National Natural Science Fund of China, Jiangsu Natural Science Foundation, University of Jiangsu Natural Science Foundation, and SANY Co., Ltd.

in Jiangsu and Open Subject of Provincial Key Construction Subject of School of Mechanical Engineering. Lecturer Support Fund Nos. are as follows: 51505211, 51505212, 51405222, 11302097 and BK20130741, BY2015005-15, and JXKJ201513.

REFERENCES

Bair S, McCabe C. A study of mechanical shear bands in liquids at high pressure [J]. Tribology International, 2004, 37: 783–789.

Henry C L, Craig V S J. Measurement of no-slip and slip boun dary conditions in confined Newtonian fluids using atomic force microscopy [J]. Physical Chemistry Chemical Physics, 2009, 11: 9514–9521.

Inaguma Y. A practical approach for analysis of leakage flow characteristics in hydraulic pumps [J]. Proceedings of the Institution of Mechanical Engineers, Part C: Journal of Mechanical Engineering Science, 2013, 227(5): 980–991.

Inaguma Y, Nakamura K. Influence of leakage flow variation on delivery pressure ripple in a vane pump [J]. Proceedings of the Institution of Mechanical Engineers, Part C: Journal of Mechanical Engineering Science, 2014, 228(2):342~357.

Khaled M B, Moh'd A.A N. 2D navier-stokes simulations of microscal viscous pump withSlip flow. Journal of fluids engineering, 2009, 131:051105-1~051105-7.

Neto C, Evans D R, Bonaccurso E, Butt H J, Craig V S J. Boundary slip in Newtonian liquids: a review of experimental studies [J]. Reports on Progress in Physics, 2005, 68: 2859–2897.

Experimental and simulation study on the interlaminate properties of fiber reinforced thermoplastic resin composites

S.C. Zhao & Z.Q. Wang
College of Aerospace and Civil Engineering, Harbin Engineering University, Harbin, Heilongjiang, China

B. Yang
School of Mechanical and Power Engineering, East China University of Science and Technology,
Shanghai, China

ABSTRACT: To study the effect of manufacture processing on the novel fiber reinforced Cyclic Butylene Terephthalate (CBT) thermoplastic resin composite materials, this paper firstly used two approaches that named Vacuum Assisted hot-Press Molding (VAMP) and Vacuum Assisted Prepreg (VAP) to manufacture unidirectional Glass Fiber reinforced Polymerized Cyclic Butylene Terephthalate (PCBT) composites (GF/PCBT). Subsequently, the effect of manufacture processing on the interlaminate properties of the obtained composites was analyzed by Double Cantilever Beam (DCB) test. FEM technology was further used to analyze the stress distribution in the material in DCB experiment. The experiment results show that compared with VAP, composites obtained by VAMP that use the limit displacement method has better interlaminate properties. Compared with VAP processing, VAMP processing could avoid the appearance of poor adhesive in the composites effectively. The obtained composite laminates in VAMP have better interlaminate strength, and Mode-I strength increased by 35% while the failure displacement decreased by 40%. Simulation results show that the initial crack in the material has a limited value, and the specimen will completely damaged by delamination once the crack tips formed with increasing of loading time.

1 INTRODUCTION

With the rapid development of aviation and astro-navigation, automobile, shipbuilding industries, composite materials have been applied more and more extensively in actual engineer fields (Du, 2008; Du, 2007). This is mainly due to the merits of Fiber-Reinforced Plastics (FRP), for example, FRPs possess excellent design ability, higher specific strength and specific modulus comparing with traditional materials (Zhang, 2000). However, composite materials are facing problems in the practical application due to its low interlaminar strength and the excessive dependence of mechanics performance on preparation processes. Therefore, it is necessary to study the interlaminar strength of composite materials prepared by different preparation processes.

Compared with thermosetting resin, thermoplastic resin, as recyclable materials, has attracted a lot of attention due to its favorable environmentally friendly features. Unfortunately, thermoplastic resin is difficult to prepare the products with complex shapes and high backfill body content; this also limits the application of thermoplastic

resin. Cyclic oligomer is a kind of low molecular polymer with cyclic structure and can be melt under low temperature during processing. Moreover, the melt not only possess low viscosity, but also could react with appropriate amount of ring opening agents and generate thermoplastics polymer with high molecular weight. At present, the most representative one of this kind of materials is Cyclic Butylenes Terephthalate (CBT) produced by Cyclics Corporation Schenectady, NY USA. CBT can be obtained by the depolymerization of Polybutylene Terephthalate (PBT) (Tripathy, 2007; Ishak, 2006). This resin have the processing characteristics of liquid thermosetting resin, such as the low viscosity during processing, easy to infiltrate the reinforced materials and easy to be modeled by various methods in the future; on the other hand, this resin is more environmentally friendly that can be processed repeatedly after its polymerization reaction (Philip, 2005; Ishak, 2006). Meanwhile, the main characteristics of CBT are low viscosity, high-rate polymerization, thermoformable, no reaction heat, no VOC and release, which render it more environmental friendliness. In addition, the PBT that was generated after polymerization of

CBT have extraordinary properties. Hence, CBT have gained great interests of both researchers and manufactures, especially in the area of composite materials, because it has the characteristics of both thermosetting and thermoplastic resins. The delamination of composite materials is a significant failure mode of fiber reinforced resin composites because of the lower strength of laminate along the vertical direction (Meo, 2010). Defects made by impact and manufacture would result in the poor interlayer performance of laminates, and then further decrease the overall mechanical properties of composites, especially the compression performance. The failure modes of high performance composites are extremely complex, it involves not only interlaminar delaminations, but also other delamination modes, such as matrix cracks, fiber breakage. Double-cantilever beams (DCB) tensile test is one of the effective means to evaluate the interlayer performance of laminated composite. Considerable efforts have been made to study the DCB testing technology numerically and experimentally (Schön, 2000; Choi, 1999; La, 2002). Yu et al. (1996) studied the crack size effect of composites in DCB test, provided detail results of the effects of precrack length, shapes of crack tip and measuring error of crack length on DCB test. Liu et al. (2014) used the continuity boundary conditions to solve the integral constant and obtained the load-displacement curve with of crack length and the cohesive as variables, and then obtained the DCB specimen crack propagation process. They also made some comparisons with the existing theoretical results to verify presented theoretical analysis.

In this paper, the unidirectional fiber reinforced PCBT resin composite laminates were fabricated by Vacuum Assisted Molding Process (VAMP) and Vacuum bag Assisted Preimpregnating molding process (VAP), and the interlayer performance of laminates fabricated by different molding processes were evaluated by DCB testing technology.

2 MATERIALS AND METHODS

2.1 Materials

The resin used in this paper is thermoplastic CBT resin. The resin has characteristics of low melt viscosity and can generate PBT after its ring opening polymerization acting under catalytic agent. The catalyst used in experiment is the chlorides of butyltin oxide (PC-4101). Besides, polyimide vacuum bags, steel molds with apertures on its sides, high temperature resisting sealants, and isopropyl alcohol solution are required for the experiment. All the specimens were fabricated with the assist of thermocompressor. It should be mentioned that all

the materials are baked in vacuum drying oven at 100°C for 10h before use to eliminate the effect of damp on ring opening polymerization of CBT.

2.2 Preparation processing

The unidirectional glass fiber reinforced PCBT resin composite laminates were fabricated by VAMP. In order to ensure the sufficient reaction of catalyst (PC-401) and CBT, the catalyst were dispersed thoroughly with isopropyl alcohol solution under stirring at 70°C for 20 min. After that, the unidirectional glass fibers were immersed in mixed solution, after baking at 100°C the glass fibers attached with catalyst were obtained. The flow chart of VAMP is illustrated in Figure 1. The as prepared glass fiber were placed in steel modes, the steel modes were heated to 220°C by thermocompressor and the PCBT resin was heated by oil bath at 180°C. PCBT resin was poured into fiber fabrics immediately with the assist of vacuum after completely melt. The whole system was heat treated at 200°C for 1 h to ensure the sufficient reaction of resin and catalyst. And then, the steel molds were heat treated at 190°C for 1h, and demolding after natural cooling. The unidirectional glass fiber reinforced PCBT resin laminates ($280 \times 180 \times 2.4$ mm^3) with 12 layups were successfully prepared by the above preparation process.

There are two kinds of pretreatments that need to be down on fiber fabrics before preparing composite laminates by VAP. One is the preparation of unidirectional glass fiber fabrics containing CBT resin with preimpregnating method. In detail, CBT resin in impregnating bath was heat treated to 190°C for melting, and then the prepared fiber fabrics with size of 40×40 mm^2 were immersed in CBT to infiltrate completely, the prepreg after impregnation was used as a standby. Note that fibers are easily infiltrated due to the low viscosity in the experiment. The other one is the preparation of fiber fabrics containing catalyst. In order to ensure the sufficient reaction of catalyst and CBT resin, solvent evaporation method were adopted to adhere catalyst to fiber fabric

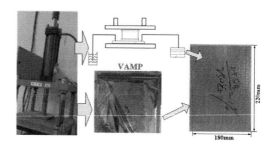

Figure 1. Flow chart of VAMP and VAP processing.

surfaces. The preparation process is as follows: we added 0.6% catalyst into isopropanol and mixed at 70°C. Fiber fabrics were immersed in the mixed solution for 2 h, and then baked at 100°C for drying. Two kinds of unidirectional glass fiber reinforced PCBT resin laminates with 12 layups were prepared as well. Note that the fiber fabrics with catalyst and prepreg should be tiled alternatively for sufficient reaction of catalyst and CBT resin. The preparation process of fiber reinforced PCBT composite with VAP are depicted in Figure 1. The reaction was at 220°C, temperature field was provided by thermocompressor, the high-temperature vacuum bag was prepared by polyimide film (300°C upper temperature limit) and high temperature resisting sealants. Vacuum state should be maintained to expel gas during the whole preparation process. Meanwhile, appropriate press was provided by thermocompressor to expel inside gas and excessive resin. The composite materials have step by step curing temperature: curving at 220°C and 190°C for 1h, respectively. The vacuum bag can be peeled off after the system cooled at room temperature. The as prepared laminate with thickness 2.4 mm and effective size of 30 × 30 mm² is shown in Figure 1. It's worth mentioning that the fiber volume content of the obtained composite by VAP can be as large as 68%.

2.3 Mechanical testing

The mechanical test in this paper were performed by using ZwickZ010 servo-electric testing machine. The specimens used for test were obtained by cutting the prepared laminate with the assist of diamond cutting machine according to the national standards, as shown in Figure 2. It is worthy to point out that the precrack was obtained by

setting polyimide films between sixth and seventh fiber fabrics. The thin film with thickness less than 0.01 mm meets requirements of this study. The length and width of precrack are 25 mm. Each experiment run comprises of 5 specimens, and mean value will be used to evaluate the prepared composite. The cross-head speed is 2 mm/min, and load-deformation curves can be obtained by the computer connected with testing machine.

2.4 Finite element emulation analysis

In order to analyze the stress variation trend with time, finite element software ABAQUS was adopted in this paper to simulate the cracking process.

Figure 3 shows the FEM model adopted in DCB test. The model size was determined by experimental setting, and element type is C3D8I. Interlayer performance of the laminates was based on cohesive zone model. Elastic and strength parameters of GF/PCBT composites are given in Tables 1 and 2. All the simulation parameters in Tables 1 and 2 was obtained by experimental measures. Where E_{ii}, G_{ij}, and V_{ij} are elastic properties of materials, and X_{1t} is axial tensile strength, X_{2t} is transverse tensile strength, S is interlaminar shear strength. In addition, axial and transverse tensile tests were adopted to obtain axial elastic modulus (E_1), Poisson's ratio (v_{12}), axial tensile strength (X_{1t}), transverse elastic modulus (E_2), and transverse tensile strength (X_{2t}), while three point bending test for specimens with precrack was used to obtain interlaminar shear Strength (S). Furthermore, E_1 and E_2 can be calculated by the initial linear stages of load-deformation curves, while X_{1t} and X_{2t} can be calculated by finding the ratio between failure load and the cross section area of specimen.

Figure 2. Dimension of the test specimens used in DCB test.

Figure 3. FEM model adopted in DCB test.

Table 1. Elastic parameters of GF/PCBT composites (GPa).

Material	E_1	E_2	E_3	G_{12}	G_{13}	G_{23}	μ_{12}	μ_{13}	μ_{23}	v_f	ρ (Kg/m³)
E-glass/PCBT	28	15.3	15.3	5.1	5.1	5.8	0.3	0.3	0.332	0.6	2002

Table 2. Strength parameters of GF/PCBT composites (MPa).

Material	X_t	Y_t	Z_t	S_{12}	S_{13}	S_{23}	X_c	Y_c	Z_c
E-glass/PCBT	997	26.8	26.8	35.2	35.2	33	520	150	150

3 RESULTS AND DISCUSSION

3.1 *DCB test results*

Figure 2 shows the photos of GF/PCBT composite laminate used in DCB test. The sheet metals were clamped by testing machine to immobilize specimens. It should be noted that the suspended part of specimens need to remain level when lamping. Figure 4 illustrates the typical load-displacement curves of the specimens obtained by VAMP and VAP processing in DCB tests. One can see that overall trends of the two curves remain almost the same. Load increases first and then decreases until crack are unstable with the increase of displacement. The difference between the two curves is that the maximum load (74N) of load-displacement curve obtained by VAMP is at the displacement of 12 mm, while the maximum load (48N) of load-displacement curve obtained by VAP is at the displacement of 25 mm. The contrast shows that unidirectional glass fiber reinforced PCBT resin composite laminates obtained by VAMP

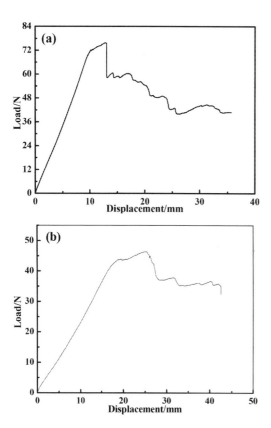

Figure 4. Typical load-displacement curves of the specimens obtained by VAMP and VAP processing in DCB tests.

have the better interlaminar strength, however, the crack will propagate very fast and damage once the interlaminar load exceeds the limit strength of composite laminates. When it comes to VAP, despite of the lower interlaminar failure strength, the laminates obtained by VAP have better interlaminar toughness, even if load exceeds the limit strength of composite laminates, the system will come to failure slowly instead of delamination failure immediately.

3.2 *DCB test results analysis*

Mode I fracture toughness can be expressed by

$$G_{IC} = \frac{3P_c\delta_c}{2ba} \tag{1}$$

where G_{Ic} is mode I fracture toughness, a the total length of crack, P_c and δ_c are the corresponding load and displacement for crack length, respectively, and b is the length of specimens.

Table 3 shows fracture toughness values of laminates fabricated by VAMP and VAP according to Eq. (1). One can see that there are no significant differences in fracture toughness of the two kinds of laminates, but differences of limit strength is obvious. Compared with VAP, VAMP can effectively prevent the lack of resin. The laminates prepared by VAMP have better interlaminar strength, the mode I interlaminar strength increased by 35%, and failure displacement decreased by 40%. This is mainly due to that the constant volume of steel mold in VAMP, resulting in relatively. Therefore, the laminates possess better strength due to the area with abundant resin. As for VAP, the inside resin of laminates was squeezed out by thermocompressor, resulting the laminates in the stage of lacking resin. In this case, fiber volume content is increased, but the laminate shows low strength in DCB test. For failure displacement, because of the bad toughness of PCBT resin and the abundant resin located in the interlayer of laminates caused by VAMP, crack propagate very fast in resin and failure displacement is small. While the crack of laminates prepared by VAP have the chance to propagate in fibers due to the relatively less resin

Table 3. Fracture toughness values calculated from the experimental test.

Process	Fracture toughness (KJ/m²)	Failure load (N)	Initial crack deformation (mm)
VAMP	2.1 ± 0.31	74.62 ± 8.7	15.7 ± 1.37
VAP	1.96 ± 0.1	48.62 ± 6.79	25.3 ± 0.29

inside. This would limit the crack propagation and finally increase the critical failure displacement.

3.3 *FE simulation results analysis*

Through comparison, the composite laminates prepared by VAMP have better interlaminar strength than that prepared by VAP. Figure 5 shows the comparison of experiment and simulation results obtained in DCB tests. It can be found that the deformation of specimens in DCB test is very big, but the length of corresponding delamination cracks is small. This further confirm the extraordinary toughness of the composites. On the other hand, Simulation results and experimental results show excellent agreement, and the displacement values and the corresponding crack length of simulation are nearly equal to that of experimental results. Hence, the finite element method can be further used for predicting the failure process of laminates in DCB test.

Figure 6 shows stress distribution in the specimens as a function of loading time. One can see that the initial debonding starts at 0.5769s, and significant debonding appears at 0.5781s. Note that the toughness of composites is obvious in the whole delamination process, the maximum load always located in crack tip. Figure 7 shows the variety of max-stress value as a function of loading time. It can be found in Figure 7 that the maximum load increases first from 500 MPa to 4000 MPa, and then decreases to 4500 MPa over time. This trend signifies that the initial crack forming is relatively difficult. However, once crack tip forms in the interlayer of composite, it will propagate immediately along the interface of two adjacent layers, and not stop until delamination failure.

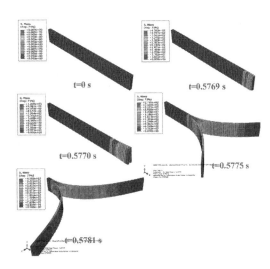

Figure 6. Stress distribution in the specimens as a function of loading time.

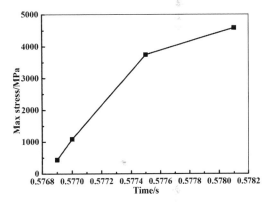

Figure 7. The variety of max-stress value as a function of loading time.

4 CONCLUSIONS

In this paper, unidirectional glass fiber reinforced CBT resin composites were fabricated by VAMP and VAP. The DCB test and finite element simulation were used to study the effect of preparation process on interlaminar performance. The results show that VAMP can effectively prevent the lack of resin. The laminates prepared by VAMP have better interlaminar strength, the mode I interlaminar strength increased by 35%, and failure displacement decreased by 40%. The simulation results show that the initial debonding is relatively difficult. Once crack tip forms in the interlayer of composite, it will propagate immediately along the interface of two adjacent layers, and not stop until delamination failure.

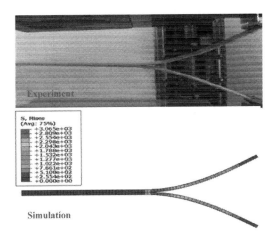

Figure 5. Comparison of experiment and simulation results obtained in DCB tests.

REFERENCES

Choi, N.S. Kinloch, A.J. & Williams, J.G. (1999). Delamination fracture of multidirectional carbon-fiber/epoxy composites under mode I, mode II and mixed-mode I/II loading. J. Compos. Mater. 33, 73–100.

Du, S.Y. & Guan Z.D. (2008). Strategic considera tions for development of advanced composite technology for large commercial aircraft in China. Acta Materiae Compositae Sinica. 2, 1000–3851.

Du, S.Y. (2007). Advanced composite materials and aerospace engineering. 1, 1–12.

Ishak, Z.A.M. Shang, P.P. & Karger-Kocsis, J. (2006). A modulated DSC study on the in situ polymerization of cyclic butylene terephthalate oligomers. J. Therm. Anal. Calorim. 84, 637–641.

Ishak, Z.A.M. Gatos, K.G. & Karger-Kocsis, J. (2006). On the in-situ polymerization of cyclic butylene terephthalate oligomers: DSC and rheological studies. Polym. Eng. Sci. 46, 743–750.

La Saponara, V. Muliana, H. Haj-Ali, R. & Kardomateas, G.A. (2002). Experimental and numerical analysis of delamination growth in double cantilever laminated beams. Eng. Fract. Mech. 69, 687–99.

Meo, M. & Thieulot, E. Delamination modelling in a double cantilever beam. Compos. Struct. 71, 429–434.

Philip, H. Howard, M.C. (2005). Recent work on entropically-driven ring-opening polymerizations: some potential applications. Polym. Advan. Technol. 16, 84–94.

Schön, J. Nyman, T. Blom, A. & Ansell, H. (2000). A numerical and experimental investigation of delamination behaviour in the DCB specimen. Compos. Sci. Technol. 60, 173–184.

Tripathy, A.R., Farris, R.J., & MacKnight, W.J. (2007). Novel fire resistant matrixes for composites from cyclic poly (butylene terephthalate) oligomers. Polym. Eng. Sci. 47, 1536–1543.

Yu, Z.C. & Jiao, G.Q. (1996). The size effects of crack in DCB test of composite materials. J. Aeron. Mater. 4, 46–53.

Liu. W.X. Zhou, G.M. Wang, X.F. & Gao, J. (2014). Theoretical analysis of crack propagation in composite DCB specimen. Acta Materiae Compositae Sinica. 1, 207–212.

Zhang, B.M. Wang, D.F., Du, S.Y. & Li, C.S. (2000). Investigation of multi-functional fiber optic smart composite. Acta Materiae Compositae Sinica. 1, 37–41.

Advances in Energy, Environment and Materials Science – Wang & Zhou (Eds)
© 2017 Taylor & Francis Group, London, ISBN 978-1-138-03600-0

The lubrication state analysis of automobile drive axles

Yongcong Wang, Youkun Zhang & Yanhui Lu
Jilin University, Changchun, China

ABSTRACT: The vehicle drive axle is one of the main sources of power loss in the drivetrain system and its improvements can have a significant impact on vehicle fuel economy. Gears churning loss, bearing friction loss, and engaging friction loss all make a great contribution to the heat generation. In order to avoid direct friction and to protect gears, we should add the lubricants to the tooth surface to form an oil film. One part of the heat convects with the surface of the box and another part is absorbed by lubricants, resulting in its high temperature. If the temperature is too high, the lubricants' viscosity would decrease, which reduces the film thickness and results in gear damage. If the temperature is too low, the lubricants viscosity would increase, thereby leading to an increase in the churning loss. Therefore, it is important to understand the heat generation and dissipation in automotive drive axles. However, the depth of understanding of drive axle temperature is limited and published information is deficient.

1 INTRODUCTION

In recent years, automotive original equipment manufacturers and suppliers strive as much as possible to improve the vehicle's driveline efficiency. This is the response to consumers' desire for greater fuel economy and the environment's objective need to be protected. A drive axle is part of the automobile drivetrain system, the main function of which is changing the rotational speed and torque as designed. The drive axle transports power by meshing of gears and heat will be generated while meshing due to the power loss. In order to avoid direct friction and protect gears, we should add the lubricants to the tooth surface to form an oil film. One part of the heat convects with the surface of the box and another part is absorbed by lubricants, resulting in its high temperature. If the temperature is too high, the lubricants' viscosity would decrease, which reduces the film thickness and results in gear damage. If the temperature is too low, the lubricants' viscosity would increase, thereby leading to an increase in the churning loss.

By analyzing the source of the drive axle power loss, a power loss model is established in this paper. Meanwhile, tests to investigate the relationship between the temperature of drive axle and influent factors like rotational speed, torque, and oil depth are also conducted. According to the final temperature, lubricants viscosity is calculated to analyze lubrication; this can provide the basis for the selection of the amount of lubricants.

2 HEAT ANALYSIS

The drive axle generates heat because of the power loss during the transmission process, which mainly includes gears churning loss, bearing friction loss, and engaging friction loss between gears. The power loss mechanism of the drive axle is complex and currently its calculation can only be determined by using the semi-empirical formula with experimental parameters.

Gear churning loss is mainly related to the size and rotational speed of gears and the quantity and quality of lubricants. Its calculation using ISO/TR 14179 is given as equation 1.

$$P_c = \frac{1.474 f_g v n^3 D^{5.7}}{A_g 10^{26}} + \frac{7.37 f_g v n^3 D^{4.7} F \dfrac{R_f}{\sqrt{\tan \beta}}}{A_g 10^{26}} \quad (1)$$

where v is the oil viscosity; D is the pitch circle diameter of the gear; and n is the rotational speed.

Bearing friction loss is mainly caused by friction torque of bearing, which is calculated by using ISO/TR 14179-1, as shown in equation 2.

$$P_f = \frac{\pi n M_f}{30} \quad (2)$$

where n is the bearing's rotational speed; f_1 is the friction coefficient; P is the friction dynamic load; and d_m is the bearing's diameter.

Engaging friction loss happens when gears are engaging because of friction. Its calculation is carried out by using ISO/14179, as shown in equation 3.

$$P_g = \frac{fTn(\cos\alpha)^2}{9549M} \quad (3)$$

where f is the friction factor of engaging; T is the torque of the drive gear; n is the tational speed of drive gear; α is the pressure angle of tooth surface; and M is the mechanical efficiency of engagement.

The total power loss during the transmission process transports into the heat power of the drive axle. The heat power integrated over time is the heat of the drive axle. Drive axle heat analysis can obtain the equilibrium temperature of different speeds, torques and oil immersion factors and this will provide a basis for the lubrication analysis.

$$P_e = P_c + P_f + P_g \quad (4)$$

The drive axle heat dissipation is divided into heat conduction, heat convection, and heat radiation. The transfer of heat from one molecule to another is known as conduction. Convection is defined as the heat transfer between fluids. When heat is transferred by electromagnetic waves, it is called radiation.

Heat conduction follows Fourier's law, which is calculated by using equation 5. Convection follows Newton's law of cooling, which is as shown in equation 6. Radiation transports the energy by electromagnetic waves, as shown in equation 7.

$$\varnothing = -\lambda A \frac{dt}{dx} \quad (5)$$

$$Q_a = \alpha A\left(T_w + T_f\right) \quad (6)$$

$$\rho c \frac{\partial t}{\partial \tau} = \frac{\partial}{\partial x}\left(\lambda\frac{\partial t}{\partial x}\right) + \frac{\partial}{\partial y}\left(\lambda\frac{\partial t}{\partial y}\right) + \frac{\partial}{\partial z}\left(\lambda\frac{\partial t}{\partial z}\right) + \dot{\phi} \quad (7)$$

When the fluid flows through the wall at a constant speed, due to the presence of viscous forces, the speed gradually decreases close to the wall. The fluid is in a non-slip statement. The energy transfer should pass through the wall of this layer in a thermally conductive manner and the amount of convective heat is equal to the heat of conduction.

In this paper, the method of numerical heat transfer theory is used, the loss power of drive axle is the source of heat, and the drive axle is seen as a whole. Conduction will be considered only inside the drive axle and it loses heat outside by convection. Without considering gears' running,

we would increase the actual thermal conductivity because the drive axle is filled with a mixture of oil and gas.

Finally, we get the conclusion that heat power and oil immersion factor is in a linear relationship when speed and torque are determined. And then, when the oil immersion factor is constant, heat power generally increases with increasing speed. At low speed, the churning loss is far less than engaging loss, but when speed is high and torque is low, the churning loss will be much larger than the engaging loss.

3 EXPERIMENT

To explore the drive axle heat, we carried out the drive axle oil temperature experiments. In this study, input speed, torque, and oil temperature were measured by sensors and the oil temperature was observed at different speeds, torques, and oil immersion depths, which varied with time.

Drive axle experiments are performed by using an used an electricity-enclosed bed. AC motors provided power to offset the loss of internal friction and other losses. The test is benched by using a motor, drive axle, inverter, and a variety of sensors. The temperature sensor used is a thermocouple temperature sensor PT100, whose measuring range is −50°C ~ 450°C.

The bed bench used a hall speed sensor to test the input speed. The bed bench is shown in Figure 1. Experimental conditions are as shown in Table 1.

Through testing, we obtained the drive axle temperature curve at different speeds, torques and oil immersion factors, as shown in Figure 2.

From the results of the experiment, we arrive at the conclusion that when the speed is high and torque is low, there is much difference between different oil immersion factors. And when speed is

Figure 1.   Pictures of the test.

Table 1. List of input speeds, input torques, and oil volumes.

Number	Oil (ml)	Speed (r/min)	Torque (Nm)
1	1500	100	900
2	1500	150	800
3	1500	200	600
4	2000	150	600
5	2000	100	800
6	2000	200	900
7	3000	150	900
8	3000	200	800
9	3000	100	600

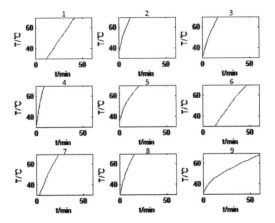

Figure 2. Graphs showing the temperature rise curve.

low and torque is high, the difference between different oil immersion factors is little.

4 LUBRICATION ANALYSIS

By studying the dissipation of the drive axle, the final equilibrium temperature at different speeds, torques, and oil immersion factors can be obtained. Firstly, the oil viscosity is calculated according to the formula of viscosity and temperature. Secondly, the minimum film thickness of oil can be obtained and this will help in determining the gears' state in the drive axle. The gears' lubrication state is divided into boundary lubrication and mixed lubrication. This means that, if the speed and torque are constant, we can see the impact on the amount of lubricants to the lubrication state of gears.

$$\log\left(\log\left(v+0.7\right)\right) = a + b \times \log T \qquad (8)$$

This is the formula of viscosity and temperature. In this formula, v is the oil viscosity; T is the absolute temperature of oil; and a and b are constants. The engagement of gears is in line contact; the lubrication region is small and therefore, the pressure will be high and the deformation is elastic deformation. This belongs to Elastohydrodynamic Lubrication (EHL). EHL features are as follows: pressure at the outlet area has a second peak pressure, the film begins to shrink in the second pressure peak; two gears start meshing when the film thickness is the smallest. Therefore, lubrication should be used here to determine the entire gear's lubrication. The minimum film formula is shown as equation 9.

$$h_0 = 6.76 \times U^{0.75} \times \alpha^{0.53} \times \eta_t^{0.75} \times R^{0.41}$$
$$\times E^{-0.06} \times \frac{B_1}{P_{max}}^{0.61} \qquad (9)$$

where U is the rolling speed of gears; R is the equivalent radius of curvature; α is the lubricant pressure viscosity coefficient; η_t is the oil viscosity; E is the equivalent elastic modulus; P_{max} is the tooth face maximum load in the normal direction; and B_1 is the driving gear width.

From equation 10, it can be observed that material properties have little effect on the film thickness, but the speed and viscosity have much impact on that.

$$\lambda = \frac{h_0}{\sqrt{R_{a1}^2 + R_{a2}^2}} \qquad (10)$$

where λ represents the ratio of film thickness and surface roughness, h_0 is the minimum oil film thickness, and R_{a1}^2 and R_{a2}^2 are the two arithmetical mean deviations of surface roughness.

If $\lambda > 3$, gears are at the whole film EHL state; if $\lambda < 1$, gears are at the boundary lubrication state; if $\lambda < 1$, gears are at the mixed lubrication state.

The whole film EHL state is the ideal lubrication station, and when gears are in boundary lubrication, they get easily worn and glued. When speed is high and torque is low, the amount of lubricants should be lesser; when speed is low but torque is high, the amount of lubrication oil should be larger.

5 CONCLUSION

By analyzing the source of the drive axle power loss, power loss is established in this paper. Heat power will increase when the oil immersion factor increases or speed and torque increases. And when the oil immersion factor is constant, heat power generally increases with increasing speed. At low speed, the churning loss is far less than the

249

engaging loss, but when speed is high and torque is low the churning loss will be much larger than the engaging loss.

We also carried out an experiment to explore the temperature of the drive axle. When speed is high and torque is low, there is much difference between different oil immersion factors. And when speed is low but torque is high, the difference between different oil immersion factors is little.

The lubrication state can be judged by oil viscosity, which is obtained by calculating the equilibrium temperature. Lubrication state and oil immersion factor can provide the basis for the amount of lubrication oil used. When speed is high and torque is low, the amount of lubricants used should be lesser; when speed is low but torque is high, the amount of lubrication oil used should be larger.

REFERENCES

Arup Gangopadhyay, Sam Asaro, Michael Schroder, Ron Jensen & Jagadish Sorab. 2002. Fuel Economy Improvement Through Frictional Loss Reduction in Light Duty Truck Rear Axle. *SAE Technical Paper* 2002–01–2021.

Akucewich E. S., B. M. O'Connor, J. N. Vinci & C. Schenkenberger. 2009. Next Generation Axle Fluids, Part III: Laboratory CAFE Simulation Test as a KeyFluid Development Tool. *SAE Technical Paper* 2003–01–3235.

ChristophWincierz. 2000. Influence of VI Improvers on the Operating Temperature of Multi-Grade Gear Oils. *SAE Technical Paper* 2000–01–2029.

Hai Xu, Avinash Singh & Don MaddockGeneral Motors Company Ahmet Kahraman and Joshua Hurley. 2011. Thermal Mapping of an Automotive Rear Drive Axle. *SAE Technical Paper* 2011–01–0718.

Ju Tonghui. 2015. Research on the Lubrication Problems of Driveline Gearbox. Jilin University, 2015.

Masanori Iritani & Hiroshi Aoki. 1999. Prediction Technique for the Lubricating Oil Temperature in Manual Transaxle. *SAE Technical Paper* 1999–01–0747.

Richard J. The Effect of Viscosity Index on the Efficiency of Transmission Lubricants. 2009. *SAE Technical Paper* 2009–01–2632.

Electrocatalytic dechlorination of 2,3,6-trichlorophenol in aqueous solution on graphene modified palladium electrode

Jinwei Zhang & Zhirong Sun
College of Environmental and Energy Engineering, Beijing University of Technology, Beijing, PR China

ABSTRACT: Graphene (Gr) modified palladium electrode (Pd/Gr/Ti electrode) was prepared by electrochemical deposition in this paper. The electrode was used for electrochemically reductive dechlorination of 2,3,6-trichlorophenol (2,3,6-TCP) in aqueous solution. Effects of the dechlorination current and the initial pH value of the solution on the removal efficiency, current efficiency and dechlorination efficiency were studied in the 2,3,6-TCP dechlorination process on Pd/Gr/Ti electrode. The results indicate that, under the condition of dechlorination current of 5 mA and initial pH value of 2.3 within 80 min, the removal efficiency, current efficiency and dechlorination efficiency on Pd/Gr/Ti electrode could be up to 100%, 36.6% and 100%, respectively. The intermediate products included phenol, 2-CP, 2,6-DCP and 2,3-DCP and the final products were mainly phenol. The reaction followed the pseudo-first-order kinetics. The modification of graphene provides the high surface area for the deposition of Pd catalyst and improves the deposition morphology of catalyst, which is favorable to electrochemically reductive dechlorination, and the electrode exhibits promising potential for dechlorination with high catalytic activity.

1 INTRODUCTION

Chlorophenols (CPs), which have been widely used for the production of dyes, drugs, pesticides and preservatives, are one kind of the hazardous organic compounds (Huang & Chu 2012). CPs can cause serious environmental pollution and health problems due to their high toxicity, recalcitrance, bioaccumulation and suspected carcinogenicity even at low concentration (Liu et al. 2012, Briois et al. 2007), which have been listed as priority pollutants by The United States Environmental Protection Agency (USEPA) (Liu et al. 2001, Vallejo et al. 2013). They are recalcitrant to biodegradation and persistent in the environment. Therefore, it is of great importance to develop effective methods to remove them from the environment.

A variety of treatment methods including biological, thermal and chemical treatments have been developed for the degradation of chlorinated organic pollutants (Laine & Cheng 2007, Weber 2007, Sze & McKay 2012). Electrochemically reductive dechlorination has been suggested as a promising method due to its rapid reaction rate, low apparatus cost, mild reaction conditions, and low environmental contamination (Huang et al. 2012, He et al. 2011, Cui et al. 2008). The mechanism of electrochemically reductive dechlorination is known as Electrocatalytic Hydrogenolysis (ECH) (YangYu & Huang 2007, Chen & Wang 2004, Cheng & Fernando 1997, Mahdavi et al. 1994).

Briefly, it's a process in which chemisorbed hydrogen atoms (Hads) generate on the electrode surface by electrolysis of water and then exchange with chlorine atoms (Kim et al. 2015, Li et al. 2013, Knitt et al. 2008, Dabo et al. 2000).

Among the factors affecting the ECH efficiency, the nature of the electrode material and the feature of the electrode surface are of a prime importance (Meng et al. 2011). Titanium (Ti) is one of the most general metal materials in the electrolytic industry (Chu et al. 2008, Yi et al. 2008, Peng et al. 2004) and can be selected as the cathode substrate. Due to its strong ability to intercalate hydrogen in its lattice, Palladium (Pd) catalyst is usually selected to be applied in the hydrodechlorination of CPs in liquid phase (Andersin et al. 2012, Matsumoto et al. 2011, Xia et al. 2009, Yuan & Keane 2004). High dispersity and large surface area of Pd particles are desirable for the electrochemically reductive dechlorination process. Therefore, it's necessary to make research on the modification of electrode. Graphene (Gr), a two-dimensional sheet of sp2-hybridized carbon atoms, was discovered by Novoselov et al. in 2004 (Novoselov et al. 2004). This material has many unique properties (Liu et al. 2014, Balandin et al. 2008, Nair et al. 2008, Geim & Novoselov 2007), such as the large theoretical specific surface area, high electron mobility at room temperature, excellent thermal conductivity and good optical transparency.

In this work, aiming at the full use of excellent properties of Gr, especially its' large specific surface area and good electrical conductivity, Pd/Gr/Ti electrode modified with Gr was prepared. 2,3,6-trichlorophenol (2,3,6-TCP) was selected as model compound and electrochemically reductive dechlorination of 2,3,6-TCP on Pd/Gr/Ti electrode was investigated in aqueous solution.

2 EXPERIMENTAL

2.1 Experimental chemicals and materials

Experimental chemicals including $PdCl_2$ powder, H_2SO_4 (98%), HCl, $H_2C_2O_4$, Na_2CO_3, Na_2SO_4 and Isopropanol were analytic purity and were supplied by Beijing Chemical Works. Graphene was supplied by Chengdu Organic Chemicals Co., Ltd. 2,3,6-TCP was from AccuStandard Inc., USA. Meshed Ti was from Anping Wire Screen Mesh Plant, China. Cation-exchange membrane Nafion-324 (DuPont) was from Sigma-Aldrich Chemical Co. Solutions were prepared with Millipore-Q water.

2.2 Experimental methods

2.2.1 Electrode preparation
The meshed Ti plate was used as substrate, Gr and Nafion solution was drop-casted evenly on the surface of the meshed Ti substrate. Gr/Ti supporting electrode was used as cathode and the platinum foil was used as anode for Pd deposition to prepare Pd/Gr/Ti electrode.

2.2.2 Dechlorination of 2,3,6-TCP
Dechlorination experiments were carried out in a two-compartment cell separated by a cation-exchange membrane with stirring, which was used to prevent the chloride atom generated on cathode during dechlorination process from transporting to the anode surface to form Cl_2, and to avoid the dechlorination products being rechlorinated further (Sun et al. 2010). The Pd/Gr/Ti electrode was used as cathode and a platinum foil was used as anode for constant current electrocatalytic reduction of 2,3,6-TCP. The catholyte was 30 mL of 0.05 mol/L Na_2SO_4 solution containing 100 mg/L 2,3,6-TCP. Catholyte pH was adjusted by addition of H_2SO_4 when the pH effect was studied. The anolyte was 30 mL of 0.05 mol/L Na_2SO_4 solution.

2.3 Analysis methods

The concentrations of 2,3,6-TCP, intermediates and products were determined by High Performance Liquid Chromatography (HPLC, Waters, USA).

2.4 Calculation of removal efficiency and current efficiency

The removal efficiency (η) and current efficiency (φ) of 2,3,6-TCP were used to evaluate the process performance and efficiency. The removal efficiency (η) was expressed as:

$$\eta(\%) = \frac{C_0 - C_t}{C_0} \times 100 \tag{1}$$

where C_0 = the initial concentration of 2,3,6-TCP; and C_t = 2,3,6-TCP concentration at different electrolysis time t.

The current efficiency (φ) was calculated as that part of current (or charge) passed to convert 2,3,6-TCP to phenol, 2,3-dichlorophenol (2,3-DCP), 2,6-dichlorophenol (2,6-DCP), 2-chlorophenol(2-CP), which were detected by the HPLC analysis, during the whole dechlorination process:

$$\varphi(\%) = \frac{(m_1 \times n_1 - \sum(m_2 \times n_2)) \times F}{I \times t} \times 100 \tag{2}$$

where m_1 = the quantity of the conversed 2,3,6-TCP (mol); n_1 = the number of electrons in forming phenol from 2,3,6-TCP (n_1 = 6); m_2 = the quantity of the residual 2,3-DCP, 2,6-DCP and 2-CP (mol); n_2 = the number of electrons in forming phenol from 2,3-DCP, 2,6-DCP and 2-CP (n_2 = 4, 4, 2, respectively); F = the Faraday constant (96,500 C/mol); I = dechlorination current (A); and t = dechlorination time (s).

3 RESULTS AND DISCUSSION

3.1 Effect of electrolysis current on 2,3,6-TCP dechlorination

The effects of electrolysis current on the removal efficiency and current efficiency of 2,3,6-TCP dechlorination were investigated under initial pH value of 2.5 with different dechlorination currents (3 mA, 4 mA, 5 mA, 6 mA and 7 mA, separately).

Figure 1 shows the removal efficiencies of 2,3,6-TCP on Pd/Gr/Ti electrode with different dechlorination currents and times. Dechlorination current has a significant influence on the conversion efficiency of 2,3,6-TCP on Pd/Gr/Ti electrode in aqueous solution. Under the condition of the dechlorination current of 3 mA and the dechlorination time of 90 min, the conversion efficiency was very low (71%), which was probably because the low current could not offer enough Hads for the ECH. The conversion efficiency reached 100% at dechlorination current of 7 mA with dechlorination time of 80 min, and dechlorination current of

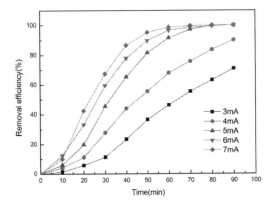

Figure 1. Effect of current on removal efficiency of 2,3,6-TCP.

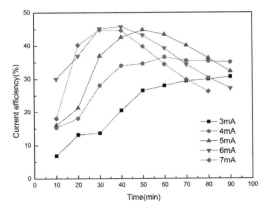

Figure 2. Effect of current on current efficiency of 2,3,6-TCP.

5 mA and 6 mA with dechlorination time of 90 min. The removal efficiency reached 99% at dechlorination current of 5 mA with dechlorination time of 80 min. That is to say, the generation of active hydrogen on the electrode surface was accelerated with the increase of dechlorination current, which promoted the reaction of dechlorination. In addition, the removal efficiency of 2,3,6-TCP improved with the increase of dechlorination time at the same dechlorination current.

Figure 2 shows the current efficiencies on the Pd/Gr/Ti electrode at different dechlorination currents in aqueous solution. At the time of 80 min, the maximal current efficiency of 36.2% was obtained at the dechlorination current of 5 mA, while at the time of 90 min, and the maximal current efficiency of 35.0% was obtained at the dechlorination current of 4 mA within the investigated dechlorination current. From overall, the current efficiencies increased with the dechlorination time at relatively low dechlorination current (3 mA and 4 mA).

And the current efficiencies at the dechlorination current of 5 mA, 6 mA and 7 mA also increased with the dechlorination time at initial stage (0–40 min), and then decreased with the dechlorination time going by. According to the mechanism of ECH, side reaction of Hydrogen Evolution Reaction (HER) will occur inevitably. Both the main reaction dechlorination and the side reaction HER will be speeded up. With current increasing, a greater proportion of Hads took part in the reaction for H_2 generation, which caused the decrease of the current efficiency.

With the integrated analysis of removal efficiency and current efficiency, the optimum dechlorination current was 5 mA with the initial pH value of 2.5 under the investigated range of experimental research. At this current value, the removal efficiency of 2,3,6-TCP and the current efficiency on the Pd/Gr/Ti electrode within 90 min reached 100% and 32.3%, respectively.

3.2 Effect of initial pH value

The effect of initial pH value on the electrocatalytic dechlorination activity was investigated under constant current of 5 mA and time of 80 min at different initial pH values (1.9, 2.2, 2.3, 2.4 and 2.5).

Figure 3 shows the removal efficiencies on the Pd/Gr/Ti electrode at different initial pH value in aqueous solution. It can be seen that the removal efficiencies increased with the dechlorination time in the investigated initial pH value (1.9–2.5). The removal efficiency of 96% could be obtained after dechlorination time of 80 min, and the complete removal was obtained with the initial pH value of 2.3. The reason is that, under the high pH value, insufficient H in the solution restricted the dechlorination reaction, which resulted in the poor removal efficiency of 2,3,6-TCP, while under the

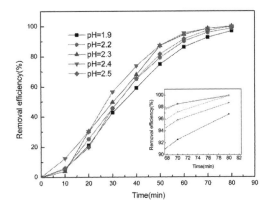

Figure 3. Effect of pH on removal efficiency of 2,3,6-TCP.

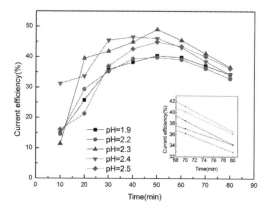

Figure 4.   Effect of pH on current efficiency of 2,3,6-TCP.

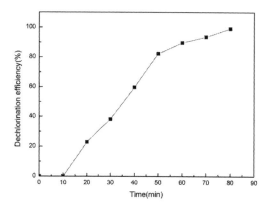

Figure 5.   Dechlorination efficiency of 2,3,6-TCP.

low pH value, abundant H^+ in the solution can promote the side reaction of hydrogen evolution reaction, which also resulted in the low removal efficiency.

Figure 4 shows the current efficiencies on the Pd/Gr/Ti electrode at different initial pH values in aqueous solution. From overall, the current efficiencies increased at initial stage, and then decreased with the dechlorination time going by. This is because, there were sufficient 2,3,6-TCP at initial stage, while the concentration of 2,3,6-TCP decreased and HER increased at late stage, and the current efficiencies decreased consequently. The current efficiency of 36.6% could be obtained after dechlorination time of 80 min with the initial pH value of 2.3, which is the highest one among the investigated dechlorination pH value.

We also investigated the dechlorination efficiency under constant current of 5 mA and time of 80 min at the initial pH value of 2.3, as shown in Figure 5. From overall, the dechlorination efficiency

increased with the dechlorination time going by, and the dechlorination efficiency of 100% was obtained after dechlorination time of 80 min when the 2,3,6-TCP was removed completely. However, the chloride ion was not detected in the first 10 min when the 2,3,6-TCP was partly removed, and it was probably because a small quantity of chloride ion generated at initial stage absorbed on the electrode surface without diffusing into the solution.

With integrated analysis of removal efficiency, current efficiency and the dechlorination efficiency, the initial pH value of 2.3 was selected for 2,3,6-TCP dechlorination. With the pH value of 2.3 and current of 5 mA, the removal efficiency, current efficiency and dechlorination efficiency on the Pd/Gr/Ti electrode at 80 min reached 100%, 36.6% and 100%, respectively.

3.3   The kinetics of 2,3,6-TCP dechlorination

The kinetics of electrocatalytic dechlorination of 2,4,6-TCP was investigated in this paper. The dechlorination experiments under 2,4,6-TCP initial concentration of 100 mg L-1, dechlorination current of 5 mA and initial pH of 2.3 were conducted at four temperatures of 288 K, 298 K, 308 K and 318 K, respectively.

Table 1 shows the variation of 2,3,6-TCP concentration with time at the four temperatures, where C0 is the initial concentration of 2,3,6-TCP, Ct is the residual concentration of 2,3,6-TCP at a certain time. The reaction rates correspond to the slopes of the fitting lines. Good linear fits were obtained between ln(C0/Ct) and time(t), meaning the reaction followed the pseudo-first-order kinetics and the apparent reaction rate constant k was approximately equal to 0.0507, 0.0676, 0.0723 and 0.0835 min-1, respectively. The removal of 2,3,6-TCP increasing with the rise of temperature indicated that the dechlorination of 2,3,6-TCP on the Pd/Gr/Ti electrode was endothermic reaction.

3.4   Product analysis

Under initial pH value of 2.3, products from the dechlorination of 100 mg/L 2,3,6-TCP with current of 5 mA and time of 80 min were analyzed.

Table 1.   Linear equations of 2,3,6-TCP dechlorination at different temperatures.

Temperature	Linear equations
288 K	$\ln (C_0/C_t) = 0.0507t - 0.6864$
298 K	$\ln (C_0/C_t) = 0.0676t - 1.0811$
308 K	$\ln (C_0/C_t) = 0.0723t - 1.1302$
318 K	$\ln (C_0/C_t) = 0.0835t - 1.2340$

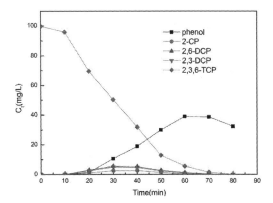

Figure 6. Concentration variation of 2,3,6-TCP and intermediate products.

Figure 6 indicates that 2,3,6-TCP was degraded gradually in the entire process and finally removed completely at 80 min. The intermediate products of 2,3,6-TCP dechlorination included phenol, 2-CP, 2,6-DCP and 2,3-DCP. The concentrations of the generated 2,6-DCP and 2,3-DCP increased during the dechlorination time of 0–30 min, and then decreased to 0 during 40–80 min. The concentration of the generated 2-CP increased during the dechlorination time of 0–40 min, and then decreased to 0 during 50–80 min. It was because the generation rates of the intermediate products were higher than the degradation rates at initial stage. With the concentration of 2,3,6-TCP decreasing, the generation rates decreased and were lower than the degradation rates of CPs to phenol. The maximum concentration of 2,6-DCP, 2,3-DCP and 2-CP were 5.5, 4.9 and 2.6 mg/L, respectively. 3-CP was not detected by HPLC in the entire dechlorination process. It was probably because the reactivity of meta-chlorion was higher than that of ortho-chlorion due to the steric hindrance caused by the neighbouring functional groups (Tsyganok et al. 1999).

However, the intermediate products were not detected in the first 10 min when the concentration of 2,3,6-TCP decreased slightly, and it was probably because a small amount of intermediate products generated at initial stage absorbed on the electrode surface without diffusing into the solution. Moreover, it also might be attributed to the adsorption of 2,3,6-TCP on the electrode surface from the solution without being degraded yet.

The amount of phenol increased with the dechlorination time (0–60 min) and the maximum amount of phenol reached 39.0 mg/L at the dechlorination time of 60 min, and then decreased to 32.4 mg/L at the dechlorination time of 80 min. Thus, it implied that the generated phenol could be further reduced to other reductive products partly. Cyclohexanone was the possible substance generated from phenol reduction(Sun et al. 2012, Sun et al. 2012).

3.5 Electrocatalytic degradation pathways of 2,3,6-TCP

Considering of intermediates and products, the possible electrocatalytic degradation pathways of 2,3,6-TCP are as following: As the intermediate products were not detected in the first 10 min, but detected at 20 min, it's possible that one or two chlorines are replaced by the active hydrogen to generate 2-CP, 2,3-DCP and 2,6-DCP. Further, except phenol, 2-CP, 2,3-DCP and 2,6-DCP were not detected at dechlorination time of 80 min when 2,3,6-TCP was removed completely. It can be inferred that three chlorines were replaced simultaneously to generate phenol directly.

According to the relevant literature (Roy et al. 2004, Shin & Keane 1999, Koet al. 2007), there were two possible pathways of 2,3-DCP degradation, meta-chlorion was replaced first by the active hydrogen to generate 2-CP and then 2-CP was dechlorinated to phenol, or ortho-chlorion and meta-chlorion were replaced simultaneously to generate phenol directly. Similarly, there were two possible pathways of 2,6-DCP degradation, one ortho-chlorion was replaced to generate 2-CP and then 2-CP was dechlorinated to phenol, or two ortho-chlorions were replaced simultaneously to generate phenol directly.

In summary, the electrocatalytic degradation pathways of 2,3,6-TCP are illustrated in Figure 7.

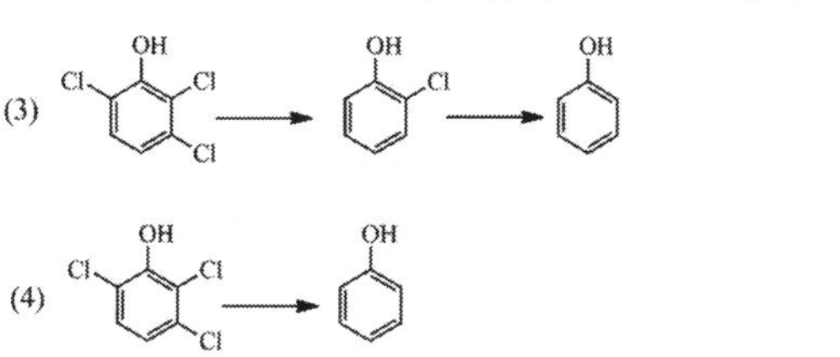

Figure 7. The electrocatalytic degradation pathways of 2,3,6-TCP.

4 CONCLUSIONS

Pd/Gr/Ti electrode with high catalytic activity was successfully prepared under modification of Gr via the facile drop-casting/ electrodeposition process and applied to the dechlorination of 2,3,6-TCP in aqueous solution. CV tests indicate that the maximal hydrogen adsorption peak current of -170 mA at about -650 mV was obtained. Gr played a crucial role in providing the high specific surface area and improving the deposition morphology of Pd catalyst, which enhanced the hydrogen adsorption capability and catalytic activity of the electrode. With the integrated analysis of removal efficiency and current efficiency, the dechlorination current of 5 mA and initial pH value of 2.3 are the optimum dechlorination conditions of 2,3,6-TCP. Complete dechlorination of 2,3,6-TCP with a concentration of 100 mg/L in aqueous solution at 298 K could be achieved under the optimum conditions within 80 min, and dechlorination efficiency of 100% and current efficiency of 36.6% could be obtained. The analysis of HPLC identifies that the intermediate products included phenol, 2-CP, 2,6-DCP and 2,3-DCP and the final products were mainly phenol. The reaction followed the pseudo-first-order kinetics. The Pd/Gr/Ti electrode exhibits a promising prospect of catalytic capability for dechlorination.

REFERENCES

Andersin, J., P. Parkkinen & K. Honkala (2012) Pd-catalyzed hydrodehalogenation of chlorinated olefins: Theoretical insights to the reaction mechanism. Journal of Catalysis, 290, 118–125.

Balandin, A. A., S. Ghosh, W. Bao, I. Calizo, D. Teweldebrhan, F. Miao & C. N. Lau (2008) Superior Thermal Conductivity of Single-Layer Graphene. Nano Letters, 8, 902–907.

Briois, C., S. Ryan, D. Tabor, A. Touati & B. K. Gullett (2007) Formation of Polychlorinated Dibenzo-p-dioxins and Dibenzofurans from a Mixture of Chlorophenols over Fly Ash: Influence of Water Vapor. Environmental Science & Technology, 41, 850–856.

Chen, G., Z. Wang & D. Xia (2004) Electrochemically codeposited palladium/molybdenum oxide electrode for electrocatalytic reductive dechlorination of 4-chlorophenol. Electrochemistry Communications, 6, 268–272.

Cheng, I. F., Q. Fernando & N. Korte (1997) Electrochemical dechlorination of 4-chlorophenol to phenol. Environmental Science & Technology, 31, 1074–1078.

Chu, D., M. Xu, J. Lu, P. Zheng, G. Qin & X. Yuan (2008) Electrocatalytic reduction of diethyl oximinomalonate at a Ti/nanoporous TiO2 electrode. Electrochemistry Communications, 10, 350–353.

Cui, C., X. Quan, H. Yu & Y. Han (2008) Electrocatalytic hydrodehalogenation of pentachlorophenol at palladized multiwalled carbon nanotubes electrode. Applied Catalysis B: Environmental, 80, 122–128.

Dabo, P., A. Cyr, F. Laplante, F. Jean, H. Ménard & J. Lessard (2000) Electrocatalytic Dehydro—chlorination of Pentachlorophenol to Phenol or Cyclohexanol. Environmental Science & Technology, 34, 1265–1268.

Geim, A. K. & K. S. Novoselov (2007) The rise of graphene. Nature Materials, 6, 183–191.

He, Z., Q. Wang, J. Sun, J. Chen & S. Song (2011) A Silver-Modified Copper Foam Electrode for Electrochemical Dehalogenation Using Mono-Bromoacetic Acid as an Indicator. Int. J. Electrochem. Sci., 2932–2942.

Huang, B., A. A. Isse, C. Durante, C. Wei & A. Gennaro (2012) Electrocatalytic properties of transition metals toward reductive dechlorination of polychloroethanes. Electrochimica Acta, 70, 50–61.

Huang, C. P. & C. Chu (2012) Indirect Electrochemical Oxidation of Chlorophenols in Dilute Aqueous Solutions. Journal of Environmental Engineering-Asce, 138, 375–385.

Kim, S., T. Park & W. Lee (2015) Enhanced reductive dechlorination of tetrachloroethene by nano-sized mackinawite with cyanocobalamin in a highly alkaline condition. Journal of Environmental Management, 151, 378–385.

Knitt, L. E., J. R. Shapley & T. J. Strathmann (2008) Rapid Metal-Catalyzed Hydrodehalogenation of Iodinated X-Ray Contrast Media. Environmental Science & Technology, 42, 577–583.

Ko, S., D. Lee & Y. Kim (2007) Kinetic studies of reductive dechlorination of chlorophenols with Ni/Fe bimetallic particles. Environmental Technology, 28, 583–593.

Laine, D. F. & I. F. Cheng (2007) The destruction of organic pollutants under mild reaction conditions: A review. Microchemical Journal, 85, 183–193.

Li, J., H. Liu, X. Cheng, Q. Chen, Y. Xin, Z. Ma, W. Xu, J. Ma & N. Ren (2013) Preparation and characterization of palladium/polypyrrole/foam nickel electrode for electrocatalytic hydrodechlorination. Chemical Engineering Journal, 225, 489–498.

Liu, L., F. Chen, F. Yang, Y. Chen & J. Crittenden (2012) Photocatalytic degradation of 2,4-dichlorophenol using nanoscale Fe/TiO2. Chemical Engineering Journal, 181–182, 189–195.

Liu, L., Z. Niu, L. Zhang & X. Chen (2014) Structural Diversity of Bulky Graphene Materials. Small, 10, 2200–2214.

Liu, Y. H., F. L. Yang, P. L. Yue & G. H. Chen (2001) Catalytic dechlorination of chlorophenols in water by palladium/iron. Water Research, 35, 1887–1890.

Mahdavi, B., P. Los, M. J. Lessard & J. Lessard (1994) A Comparison of Nickel Boride and Raney-Nickel Electrode Activity in the Electrocatalytic Hydrogenation of Phenanthrene. Canadian Journal of Chemistry-Revue Canadienne De Chimie, 72, 2268–2277.

Matsumoto, I., K. Sakaki, Y. Nakamura & E. Akiba (2011) In situ atomic force microscopy observation of hydrogen absorption/desorption by Palladium thin film. Applied Surface Science, 258, 1456–1459.

Meng, Y., L. Aldous, B. S. Pilgrim, T. J. Donohoe & R. G. Compton (2011) Palladium nanoparticle-modified carbon nanotubes for electrochemical hydrogenolysis in ionic liquids. New Journal of Chemistry, 35, 1369–1375.

Nair, R. R., P. Blake, A. N. Grigorenko, K. S. Novoselov, T. J. Booth, T. Stauber, N. M. R. Peres & A. K. Geim (2008) Fine Structure Constant Defines Visual Transparency of Graphene. Science, 320, 1308–1308.

Novoselov, K. S., A. K. Geim, S. V. Morozov, D. Jiang, Y. Zhang, S. V. Dubonos, I. V. Grigorieva & A. A. Firsov (2004) Electric field effect in atomically thin carbon films. Science, 306, 666–669.

Peng, X., K. Koczkur, S. Nigro & A. Chen (2004) Fabrication and electrochemical properties of novel nanoporous platinum network electrodes. Chemical Communications, 2872.

Roy, H. M., C. M. Wai, T. Yuan, J. Kim & W. D. Marshall (2004) Catalytic hyrodechlorination of chlorophenols in aqueous solution under mild conditions. Applied Catalysis A: General, 271, 137–143.

Shin, E. & M. A. Keane (1999) Detoxification of dichlorophenols by catalytic hydrodechlorination using a nickel/silica catalyst. Chemical Engineering Science, 54, 1109–1120.

Sun, Z., G. Hui, H. Xiang & Y. Peng (2010) Preparation of foam-nickel composite electrode and its application to 2,4-dichlorophenol dechlorination in aqueous solution. Separation & Purification Technology, 72, 133–139.

Sun, Z., K. Wang, X. Wei, S. Tong & X. Hu (2012) Electrocatalytic hydrodehalogenation of 2,4-dichlorophenol in aqueous solution on palladium–nickel bimetallic electrode synthesized with surfactant assistance. International Journal of Hydrogen Energy, 37, 17862–17869.

Sun, Z., X. Wei, X. Hu, K. Wang & H. Shen (2012) Electrocatalytic dechlorination of 2,4-dichlorophenol in aqueous solution on palladium loaded meshed titanium electrode modified with polymeric pyrrole and surfactant. Colloids and Surfaces A: Physicochemical and Engineering Aspects, 414, 314–319.

Sze, M. F. F. & G. McKay (2012) Enhanced mitigation of para-chlorophenol using stratified activated carbon adsorption columns. Water Research, 46, 700–710.

Tsyganok, A. I., I. Yamanaka & K. Otsuka (1999) Dechlorination of chloroaromatics by electrocatalytic reduction over palladium-loaded carbon felt at room temperature. Chemosphere, 39, 1819–1831.

Vallejo, M., M. Fresnedo San Roman & I. Ortiz (2013) Quantitative Assessment of the Formation of Polychlorinated Derivatives, PCDD/Fs, in the Electrochemical Oxidation of 2-Chlorophenol As Function of the Electrolyte Type. Environmental Science & Technology, 47, 12400–12408.

Weber, R. (2007) Relevance of PCDD/PCDF formation for the evaluation of POPs destruction technologies— Review on current status and assessment gaps. Chemosphere, 67, S109-S117.

Xia, C., Y. Liu, S. Zhou, C. Yang, S. Liu, J. Xu, J. Yu, J. Chen & X. Liang (2009) The Pd-catalyzed hydrodechlorination of chlorophenols in aqueous solutions under mild conditions: A promising approach to practical use in wastewater. Journal of Hazardous Materials, 169, 1029–1033.

Yang, B., G. Yu & J. Huang (2007) Electrocatalytic Hydrodechlorination of 2,4,5-Trichlorobiphenyl on a Palladium-Modified Nickel Foam Cathode. Environmental Science & Technology, 41, 7503–7508.

Yi, Q., W. Huang, X. Liu, G. Xu, Z. Zhou & A. Chen (2008) Electroactivity of titanium-supported nanoporous Pd–Pt catalysts towards formic acid oxidation. Journal of Electroanalytical Chemistry, 619–620, 197–205.

Yuan, G. & M. A. Keane (2004) Liquid phase hydrodechlorination of chlorophenols over Pd/C and Pd/Al2O3: a consideration of HCl/catalyst interactions and solution pH effects. Applied Catalysis B: Environmental, 52, 301–314.

Advances in Energy, Environment and Materials Science – Wang & Zhou (Eds)
© 2017 Taylor & Francis Group, London, ISBN 978-1-138-03600-0

Research on the technology of preparation of modified starch with a high degree of substitution

Yanjun Tan, Hua Wang, Jiali Ma, Shurui Liu, Qian Huo & Wenyan Li
Functional Fabric Key Laboratory of Xi'an Polytechnic University, China

ABSTRACT: Ethanol and isopropyl alcohol are used as solvents in the preparation of carboxymethyl corn starch. The traditional method is to use the single factor and orthogonal experiments of the dosage of ethanol to determine the optimization range of synthetic process parameters and use the range to determine results of statistical analysis of the DS and viscosity. However, the size of the trial of this synthetic process is very large and there is a certain defect in the process and cannot be analyzed or predicted. This experiment adopts the alkalization method step by step. Therefore, by designing the above-mentioned experiment through the Box-Behnken method that can create a mathematical model to analyze and predict the experimental results, the optimum synthesis process is determined and the high DS of modified starch is prepared; and then, the change in the molecular structure of modified starch is analyzed by using IR, X-Ray Diffraction (XRD), and Scanning Electron Microscopy (SEM), to provide the theoretical basis of carboxymethylated modified starch.

1 INTRODUCTION

Sodium alginate can be used as the main printing thickener; inside its macromolecular structure, many carboxyl anions, which have strong sodium alginate hydration properties that can produce high Newton viscosity and the electrostatic repulsion, are present; these result in the net structure not being easily formed in the macrostructure chain, and the structure viscosity of sodium alginate being relatively low. Therefore, sodium alginate cannot satisfy the high quality printing requirements (He, 2010), while the COO⁻ group of Carboxymethyl Starch (CMS) has great hydratability that leads to a higher Degree of Substitution (DS) and accumulates more of the negative charge in CMS molecules. The CMS molecules can repel with each other, and lead to the water molecules and carboxyl groups coming in contact. As a result, the thickener can hold more water and perform well in the printing process (Zhao, 2009).

Nowadays, researchers focus on the development of carboxymethyl starch with high a degree of substitution. For example, Ragheb uses ethyl alcohol and methylbenzene as solvents to produce CMS with DS of 1.23, and later discusses the production procedure (Ragheb, 1997). Scholars including Zelijko Stojanovic study the development mechanism of CMS and indicate that factors such as the type of starch, the dosage of the solvent, the temperature of the reaction, the time spent, and other factors can influence the process (Stojanovic, 2000).

In CMS solvent method preparation, the kind of organic solvent used has great influence on the DS and Reaction Effect (RE). A Professor in the University of Maribor, Čeh, M., uses acetone as the organic solvent, explaining its mechansim and produces the CMS such that the DS is 0.81 (M.Ceh, 1972). A Professor in Wuhan Univeristy, Fan lihong, published an article ttled "Synthesis, characterization and properties of carboxymethyl kappa carrageenan" that studies the preparation process of CMS with high substitution (Fan, 2011) and he uses ethanol and carrageenan starch. The experimental results show that chloroacetic acid has a remarkable effect on the DS; Another researcher in the National Research Centre of Egypt, A.A. Ragheb (1997), used an ethanol and toluene mixture to assemble the CMS with a DS of 1.23; A Researcher in Chiang Mai University, Ornanong S. Kittipongpatana (2006), had carried out a thorough research on the preparation of carboxymethyl mung bean starch, and Ornanong prepared the CMS with 15 different DSs which were between 0.06 and 0.66. A graduate student in Dalian University of Technology, Fan Qingsong (2005), used a dry method to improve the process and prepared the CMS with a DS of 0.73 and RE of 73%. The above-mentioned studies are merely focused on the CMS preparation mechanism; nevertheless, authors of these studies have not demonstrated details about designing and conducting experiments based on mathematical software to optimize the experimental mechanism or use the results to analyze and predict the experiment and optimize the mechanism. This topic uses isopropyl alcohol (ISO alcohol) as the solvent, adopts the alkalization method step by step, and then designs the above-mentioned experiment through the Box-Behnken method. This work implements a mathematical model to predict the result of this experiment and eventually optimizes the CMS preparation mechanism.

2 INSTRUMENTS AND MATERIALS

2.1 *Instruments and equipment*

The following are the instruments and equipment used in this work: an electric air blowing drying box, Shanghai Experimental Instrument Factory; a HJ-5 digital thermostat magnetic stirrer, Changzhou Guohua Electric Appliance Co., Ltd.; a HH-4 digital thermostat water bath, Jintan Danyangmen Quartz Glass Factory; a SHB-III-type multi-use recycled water pump, Zhengzhou Great Wall Science Industry and Trade Co., Ltd.; a Fourier transform infrared spectroscope (FT-IR), Nicolette 5700, American Nicolet Instrument Co., Ltd.; a KYKY-2008 type scanning electron microscope, KYKY Technology Development Co., Ltd.; a synchronous thermal analyzer, sta 449F3 Jupiter, Germany Netzsch Instruments Manufacturing Co., Ltd.; and a Dmax-Rapid type II X-ray diffractometer, Japan Science.

2.2 *Materials and agents*

Corn starch, food grade, was purchased from Jining, Shandong Province; chloroacetic acid, sodium hydroxide, ethanol, isopropanol, hydrochloric acid, and phenolphthalein were of analytical grade and were obtained from Xian Sanpu Chemical Reagent Co., Ltd.

3 EXPERIMENTAL METHODS

3.1 *Alkalization reaction*

A certain amount of starch and solvent are added into three-neck flasks under stirring. After few minutes, add NaOH and a certain amount of solvent and the alkalization reaction occurs at a certain temperature under high-speed stirring.

3.2 *Etherification reaction*

After heating to a certain etherification temperature, Monochloroacetic Acid (MCA) and NaOH are added. After allowing the reaction to proceed for some time, industrial alcohol is used with a small amount of glacial acetic acid to wash to

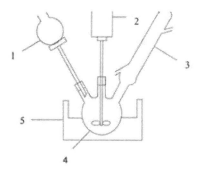

Figure 1. Schematic of experimental installation. 1 Filtrate funnel, 2 Motor stirrer, 3 Condenser tube, 4 Flask with three necks, 5 Thermostat water bath.

neutral, filtration is performed, and the filter residue is dried in a vacuum drying oven for some time, and the product is obtained. The experimental installation is shown in Fig. 1.

3.3 *Optimization of the preparation process for carboxymethyl corn starch with high DS*

The main difficulty in preparing a high substitution degree of modified starch is the presence of starch in an alkaline medium, and small molecules reagent in an alkali solvent can fully penetrate into the starch granules, and then react with the hydroxyl group of the structural unit, produce starch sodium salt, and thus performs the carboxymethylated reaction with chloroacetic acid under alkaline conditions; as a result, carboxymethyl starch sodium is obtained, but at the same time, the process is accompanied by the reaction of sodium chloroacetate hydrolyzing glycolic acid sodium, thereby inhibiting the reaction of carboxymethyl starch, resulting in a decrease in the degree of substitution. This experiment adopts the alkalization method step by step, in which the alkali agent leads to the swelling of starch molecules and loosed starch particles firstly and this will help in improving the carboxy methylation reaction rate. Therefore, starch crystallinity should be decreased as much as possible, so that a large number of alkali molecules will be present on the inside of the starch molecules, and react with the hydroxyl group of the starch-generated active center $[C_6H_9O_4(O^- Na^+)]_n$. If the crystalline region of the starch particle is difficult to be destroyed, sodium hydroxide and chloroacetic acid will be retained in the reaction medium and then, this will result in generating more by-products. A lot of active centers, $[C_6H_9O_4(O^- Na^+)]_n$, are combined with chloroacetic acid and then, the S_N2 double molecular nucleophilic substitution reaction takes place, thereby generating carboxymethyl starch sodium. The reaction process is as shown in reaction equations (1) and (2):

$$Starch\text{-}OH + NaOH \rightarrow Starch\text{-}ONa + H_2O \quad (1)$$

$$Starch\text{-}ONa + ClCH_2COOH + NaOH$$
$$\leftrightarrow Starch\text{-}O\text{-}CH_2COONa + NaCl + H_2O \quad (2)$$

While the above-mentioned equations denote primary reactions, there are side effects, such as the reaction depicted by equation (3).

$$ClCH_2COOH + 2NaOH \rightarrow HOCH_2COONa$$
$$+ NaCl + H_2O \quad (3)$$

Therefore, the more number of active centers $[C_6H_9O_4(O^-Na^+)]_n$ in the alkalization reaction, the greater the utilization rate of chloroacetic acid will be; the smaller the probability of side reactions, the DS of modified starch will be higher.

Meanwhile, this experiment uses the Box-Behnken plan to carry out the test. Six effective factors are selected—the concentration of isopropyl alcohol, the dosage of isopropyl alcohol, the temperature of the reaction, the time of the reaction, the ratio of n_{NaOH}/n_{AGU}, and the ratio of n_{MCA}/n_{AGU}. The response value Y is the DS of the prepared carboxymethyl corn starch (Yao, 2005).

Computation method of *DS* is given as follows:

$$DS_t = \frac{n_{MCA,0}}{n_{AGU,0}} \ (n_{NaOH,0} \geq n_{MCA,0}) \quad (4)$$

$$or \ DS_t = \frac{n_{NaOH,0}}{n_{AGU,0}} \ (n_{NaOH,0} < n_{MCA,0}) \quad (5)$$

(The variables $n_{NaOH,0}$, $n_{MCA,0}$ and $n_{AGU,0}$ in these equations represent the total number of moles of sodium hydroxide, acetic acid, and the hydroxyl unit of Anhydrous Glucose Unit (AGU) in the whole reaction, respectively.)

3.4 *Box-Behnken test design*

According to the design principles of Box-Behnken, six effective factors including the concentration of isopropyl alcohol, the dosage of isopropyl alcohol, the temperature of the reaction, the time of the reaction, the ratio of nNaOH/nAGU, and the ratio of nMCA/nAGU which have a great effect on the DS are chosen to design the three levels of analyzing the experiment, add seven zero points, and in order to reduce the error, the zero point experiment is repeated seven times. The main influence factors are determined preliminarily. According to the preparation process, design each factor and level, as shown in Table 1.

3.5 *The establishment of a mathematical model*

The final preparation process is based on the best preparation method predicted by the multivariate two times equation and validation tests of the DS are performed. The response value Y is the DS of the prepared carboxymethyl corn starch and optimization of the experiment is characterized by using a quadratic polynomial.

4 RESULTS AND ANALYSIS

4.1 *Analyzing the optimization process by factors and three levels*

According to the experimental design, SAS V (9.0) is used to perform regression fit of the results of the experiment, and the response value Y is the DS of the prepared carboxymethyl corn starch. The regression equation obtained is as follows:

$$Y = 715.71 + 3.22A + 29.74B + 36.84C$$
$$+ 10.05D + 30.98E - 9.55F - 22.35A^2$$
$$- 0.97AB - 42.37B^2 - 1.22AC - 4.295BC$$
$$- 19.24C^2 + 12.44DA - 2.29DB + 12.44DA$$
$$- 2.29DB + 12.44DC - 66.12D^2 - 13.38EA$$
$$- 3.5EB - 4.94EC - 3.91ED - 32.09E^2$$
$$+ 6.84FA - 11.399FB - 3.27FC - 6.201FD$$
$$- 13.996FE - 48.45F^2$$
$$(6)$$

The significance of the effect of each variable on the response value is judged by performing the

Table 1. Design of factors and levels.

Factors		Concentration of ISO alcohol (%)	Dosage of ISO alcohol (ml·g⁻¹)	Reaction temperature (°C)	Reaction time (min)	n_{NaOH}/n_{MCA} (mol·mol⁻¹)	n_{MCA}/n_{AGU} (mol·mol⁻¹)
Level	−1	90	2.5	50	200	2.0	0.9
	0	94	4	60	250	2.5	1.2
	1	98	5.5	70	300	3.0	1.5

Notes: A represents the concentration of ISO alcohol, B, C, D, E, and F represent the dosage of isopropyl alcohol, the temperature of the reaction, the time of the reaction, the moles ratio of NaOH/MCA (n_{NaOH}/n_{MCA}), and the dosage of etherifying agent (n_{MCA}/n_{AGU}), respectively.

261

F test in the regression equation. From equation 6, the preparation process is analyzed by the lack of fit and the significance of the influence factor of the test is determined. The loss of test $F_1 = 2.56 < F_{0.05}$ (17,9), indicates that the difference is not significant and shows that the impact factor is very comprehensive; there are no factors that cannot be ignored. The fitting test is able to examine whether the regression equation can reflect the actual situation. The fitting test $F_2 = 3.37 > F_{0.05}$ (27,26) shows that the difference is significant and indicates that the regression equation in the test is significant and can reflect the actual situation.

According to the analysis and prediction of SAS v9.0, when A = −0.125000, B = −0.100000, C = 0.000000, D = 0.300000, E = 0.200000, and F = −0.166667, the best theoretical value of DS is 1.08, carboxymethyl corn starch with a maximum DS can be prepared under this condition.

4.2 Determination and verification of optimal conditions for preparation of carboxymethyl corn starch

In order to test the correlation between the predicted value of the model and the actual test value that is to test the reliability of the mathematical model, the DS value of carboxymethyl corn starch that is prepared under the optimal conditions is verified by using the testing result. The optimal values of each factor in the experiment are that the concentration of ISO alcohol is 93.5%, the dosage of ISO alcohol is 3.85 ml/g, the reaction temperature is 60°C, the reaction time is 265 min, $n_{NaOH}/n_{AGU} = 2.6$ and $n_{MCA}/n_{AGU} = 1.25$, respectively, the DS values of the three parallel experiments are 0.974, 0.986 and 0.992, respectively, and the average value is 0.984, which is very close to the predicted value.

4.3 Infrared spectrum analysis of carboxymethyl corn starch prepared by using an optimal process

The surface structure of carboxymethyl corn starch by performing the infrared spectrum test is shown in Fig. 2.

Fig. 2 indicates that not only the characteristic absorption peak of two types of starch appear at around 3000 cm^{-1}, but also for carboxymethyl corn starch, a strong absorption peak of the stretching vibration of C-O-C (ether bond), symmetric stretching vibration weak absorption peak of -COO^{-}(carboxylates), and stretching vibration of strong absorption peaks of asymmetric carboxylate appear at around 1017 cm^{-1}, 1422 cm^{-1}, and 1640 cm^{-1}, respectively. Therefore, it can be proved that corn starch has undergone the carboxymethyl

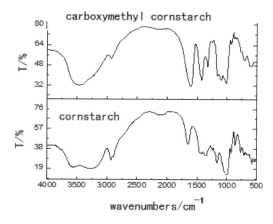

Figure 2. Infrared spectrum of original corn starch and carboxymethyl corn starch.

reaction and the product is carboxymethyl corn starch.

4.4 XRD analysis of carboxymethyl corn starch prepared by using an optimal process

The crystal structures of carboxymethyl corn starch obtained by using XRD analysis is depicted in Fig. 3.

Fig. 3 indicates that, in the XRD spectrum of corn starch, there are obvious diffraction peaks at around 15°, 17°, and 23° of the characteristic diffraction angles in the crystalline region, and for that of carboxymethyl corn starch, the intensity of these peaks around these characteristic diffraction angles decrease, and the peak of 15° disappears, the diffraction curve drifts downward, the crystal structure loosens, and the molecular structure changes. It shows that the crystallinity of the corn starch is decreased by carboxymethylation (Zhong, 2011).

4.5 Thermo gravimetric curve analysis of carboxymethyl corn starch prepared by using an optimal process

Fig. 4 shows the TG-DSC curve of carboxymethyl corn starch and the situation of weight loss in which the heat effect can be analyzed from the figure.

TG-DSC curves shown in Fig. 4 indicate that at 100°C, there is a small variation trend caused by weight loss (around 2%) of water and other volatile small molecules, there is a greater transition observed at around 250°C, then the phenomenon of weight loss increases; there is a higher peak of DSC which is shown at around 300°C, after which the trends become slower gradually.

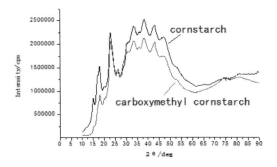

Figure 3. XRD pattern of carboxymethyl corn starch.

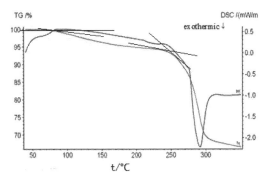

Figure 4. TG-DSC curves of carboxymethyl corn starch.

Carboxymethyl corn starch comes into splitting around 250°C, about 30% of weight loss is observed at around 300°C, and the weight loss thereafter tends to slow down, which shows that the speed of structure splitting for carboxymethyl corn starch reduces and the splitting tends to be completed. The final weight loss of carboxymethyl corn starch is 46.71%.

5 CONCLUSIONS

An optimal process for preparing carboxymethyl corn starch is advanced by Box-Behnken. It concludes that the six effective factors which have influence the DS of production are as follows: ISO alcohol with the concentration of 93.5%, the dosage of 3.85 ml/g, temperature of the reaction is 60°C, 265 min of reaction time, $n_{NaOH}/n_{AGU} = 2.6$, and $n_{MCA}/n_{AGU} = 1.25$. The result is that the DS of the prepared product is 0.984.

From the analysis of the infrared spectrum and XRD pattern for carboxymethyl corn starch, we find that the hydroxyl groups of the corn starch are carboxylated, the crystal structure changes and the diffraction peak at 15° disappears.

ACKNOWLEDGMENT

This work was financially supported by the Shaanxi Province Key Laboratory of functional fabrics.

REFERENCES

Ceh, M., Starke. Vol. 24 (1972), p. 124.
Fan L., L. Wang, S. Gao; Carbohydrate Polymers, Vol. 86 (2011), p. 1167.
Fan Q.S., B.Z.Ju, S.F.Zhang, J.Z. Yang; Fine Chemicals Vol. 22(S1) (2005), p. 112.
He B., J.S. Guo; Progress in Textile Science & Technology. Vol. 06(2010), p. 27.
Kittipongpatana O.S., J. Sirithunyalug, R. Laenger; Carbohydrate Polymers. Vol. 63 (2006), p. 105.
Ragheb A., HS. EI-Sayiad, A. Hebeish; Starch. Vol. 49 (1997), p. 238.
Stojanovic Z., K. Jeremic; Starch. Vol. 53 (2000), p. 413.
Yao J., H.You, ZH. Bao, J.She, J. G. Klaassien, J. H. Marsman, H. J. Heeres, W.R.Chen; Chinese Journal of Analytical Chemistry. Vol. 02 (2005), p. 201.
Zhao T., Dyeing Technology and principle. (China Textile & Apparel Press. 2009).
Zhong Z.H., Y.P. Zhong, A. Sun; Journal of South China University of Technology (Natural Science Edition), Vol. 03 (2011), p. 22.

Study of Shen-625 block underground string anti-corrosion technology under reduced oxygen air drive

Feng Yan

D&P Technology Research Institute, Panjin, Liaoning, China

ABSTRACT: The method of oil displacement in Shen-625 block is injecting reduced oxygen air. There has been serious corrosion of casing and tubing in the actual production process and packer has sealing leakage problems due to corrosion. To solve this problem, we have studied oxygen-containing corrosion mechanism, inhibiter mechanism and corrosion rate variation of field gas injection well under different conditions; we have carried out the string and packer corrosion resistant material screening, optimized corrosion resistant L80 material; at the same time, we filter out the inhibitor which can reduce the N80 material corrosion rate to 0.0687 mm/a; we also evaluate several materials of cathode protection nipple, screen and develop a cathode protection tubing nipple.

1 OXYGEN-CONTAINING CORROSION TYPE IN SHEN-625

The oxygen corrosion form in air injection mainly includes the following:

1. Crevice corrosion. The essence of the crevice corrosion is oxygen concentration cell. The width of the gap in the range of 0.025–0.1 mm is the sensitive seam width, which can form occluded cell. Almost all of the corrosive media are prone to crevice corrosion, and containing Cl—solution most likely lead to crevice corrosion.
2. Pitting corrosion. Corrosion concentrated in individual points or small area on the metal surface, and deep into the metal matrix, appears narrow and deep pits.
3. Groove corrosion. Groove corrosion is a form of localized corrosion, it is formed by the pitting corrosion of the groove shape and its cross section is V-shaped.
4. Interlayer corrosion. It is forging, rolling metal inner corrosion, sometimes leads to separation, peeling off generally occur along the lamellar structure such as rolling, extrusion or deformation direction.
5. Corrosion fatigue. Metallic materials have a brittle fracture under the combined action of cyclic stress or pulse stress and electrochemical corrosion.

2 DEVELOPMENT OF REDUCED OXYGEN AIR CORROSION EVALUATION SYSTEM

The determination of the corrosion rate of the reduced oxygen air needs a set of apparatus for measuring the corrosion rate under high temperature and high pressure conditions. In order to achieve this goal, we have developed a LHZCY-I type dynamic corrosion tester. The apparatus is an experimental device for testing the corrosion rate of materials as well as evaluating the corrosion resistance under high temperature and high pressure conditions. The corrosion rate and corrosion resistance evaluating of materials are tested by the methods of coupon corrosion test and high-temperature inductor-probe test, the test can be carried out at the set temperature and pressure conditions, and it is applicable to all types of corrosive media.

3 CORROSION FACTORS OF REDUCED OXYGEN AIR

The results of the experiment of oxygen concentration effect on the tubing corrosion are shown in Table 1.

From the experimental results in Table 1, coupon corrosion test and high-temperature inductor-probe test are basically consistent with the

Table 1. Effect of oxygen concentration on the corrosion rate.

Material	Corrosive media	Oxygen concentration (%)	Temperature (°C)	Pressure (MPa)	Average corrosion rate (mm/a)	
					Coupon corrosion test	Inductor probe test
N80	Formation water	10	110	20	1.3925	1.3798
		21			1.4517	1.4385

Table 2. Effect of temperature on the corrosion rate.

Material	Corrosive media	Oxygen concentration (%)	Temperature (°C)	Pressure (MPa)	Average corrosion rate (mm/a)	
					Coupon corrosion test	Inductor probe test
N80	Formation water	10	80	20	1.1906	1.1857
			100		1.3274	1.3011
			120		1.4954	1.4788

Table 3. Effect of pressure on the corrosion rate.

Material	Corrosive media	Oxygen concentration (%)	Temperature (°C)	Pressure (MPa)	Average corrosion rate (mm/a)	
					Coupon corrosion test	Inductor probe test
N80	Formation water	10	110	10	1.2941	1.2846
				20	1.4517	1.4469
				25	1.5354	1.5272

measured data. The corrosion degree of the reduced oxygen air containing 10% oxygen is slightly lower than that of ordinary air, but it is still more than 1.3 mm/a and the corrosion is very serious.

Through the above analysis results in Table 2, the corrosion rate of N80 material rises with the increase of temperature under the condition of constant oxygen concentration and pressure, in the temperature range of 80°C–120°C.

Through the above analysis results in Table 3, we can see the corrosion rate of N80 material rises with the increase of pressure under the condition of constant oxygen concentration and temperature, in the pressure range of 10 MPa–25 MPa.

According to the above experiments, it is found that with the increase of temperature, oxygen concentration and pressure, the corrosion rate of N80 material is on the rise. But the rise is not large. Temperature is the most influential factor of corrosion rate.

4 SCREENING OF THE MATERIAL OF THE CORROSION RESISTANT TUBING AND PACKER

Corrosion rate measurements are carried out at 110°C, 20 MPa, 10% Oxygen, using the methods of coupon corrosion test and high-temperature inductor-probe test. We use 12 kinds of materials, taking the average of the results of the two methods as a final result. The experimental results are in the Table 4.

Through the above analysis results in Table 4 we can see, N80 material corrosion rate reach 1.3935 mm/a, the situation of corrosion is very serious. L80 material corrosion rate is 0.0182 mm/a. After the experiment, the surface of the coupon is bright and corrosion point is not obvious. The corrosion resistance of L80 material is better than other experimental materials. It can meet the needs of the site.

Table 4. Corrosion rate of different materials.

Material	Corrosion rate (mm/a)
40CRMO	1.8032
J55	0.8470
35CRMO	2.4693
N80	1.3925
P110	2.9083
L80	0.0182
L80–3CR	1.2492
BG80H-13CR	0.0318
BG80H-3CR	1.0859
BG110H	2.2647
25CRMO	2.4510
TP110H	1.2383

Table 5. Corrosion evaluation test results.

Inhibitor no.	Dosing concentration	Corrosion rate (mm/a)
1#	200 ppm	1.7481
	500 ppm	1.2482
2#	200 ppm	1.3055
	500 ppm	1.1344
3#	200 ppm	1.2067
	500 ppm	0.9847
4#	200 ppm	1.5626
	500 ppm	0.0687
5#	200 ppm	1.6891
	500 ppm	1.4084

5 SCREENING OF CORROSION INHIBITOR

Under the condition of 110°C, 10 MPa, 10% Oxygen, 5 kinds of corrosion inhibition results are in the Table 5.

Through the above analysis results in Table 4 we can see, 1# inhibitor and 5# inhibitor play a counter effect after the addition, corrosion rate has increased in varying degrees. 2# inhibitor and 3# inhibitor play a slight effect after the addition. 4# inhibitor plays a counter effect when the dosing concentration is 200 ppm, but when the concentration increases to 500 ppm, the corrosion rate decreases to 0.0687 mm/a. This is lower than the corrosion rate indicator of oilfield water injection (0.076 mm/a), it completely meets the needs of the site.

6 CONCLUSIONS

1. We develop a LHZCY-1 type dynamic corrosion tester, maximum pressure 65 MPa, maximum temperature 500°C.

2. According to Shen-625 site conditions and corrosive environment, we have studied the general law of corrosion rates under different temperatures, different pressures and different oxygen concentrations. The results show that the corrosion rate of N80 material is on the rise with the increase of temperature, oxygen concentration and pressure, but the rise is not large. Among them temperature is the most influential factor of corrosion rate.
3. The material of string and packer are selected, the corrosion rate of L80 material is low to 0.0182 mm/a under Shen-625 site conditions, at the same time the mechanical performance of the material can meet the needs of the site.
4. We screen the 4# inhibitor which is suitable for the site. When the dosing concentration of inhibitor is 500 ppm, the corrosion rate fall to 0.0687 mm/a, which is less than oilfield water injection corrosion rate indicators (0.076 mm/a), it completely meets the needs of the site.

ACKNOWLEDGEMENTS

This work was financially supported by the Research and experiment on key technologies of heavy oil fire flooding in Liaohe oilfield.

REFERENCES

Bin Wu, Junxia Liu, Wenhua Man (2010). Column corrosion mechanism and prevention measures. Inner Mongolia Petrochemical Industry, 6: 9–11.
Dugstad, A. (1992). The importance of $FeCO_3$ super saturation of carbon steel in Corrosion [J]. California, 136–140.
Heuer, J.K., J.F.S. (1999). An XPS characterization of $FeCO_3$ films from CO_2 corrosion [J]. Corrosion Science, 41(7): 1226–1231.
Ikeda, A., M. Ueda, and S. Mukai (2003). CO_2 corrosion behaviour and mechanism of carbon steel and alloy steel [J]. NACE Corrosion, 21(5):394–402.
Jasinski, R. (1987). Corrosion of N80-type steel by CO_2/water mixtures [J]. Corrosion, 43(4):210–214.
Jun Ren (1998). Changqing Oilfield corrosion and protection. Gas Insutry, 18(5):63–67.
Lee, M.H., Roberts L.D. (1980). Effects of heat of reaction on temperature distribution and penetration in a facture. SPEJ, Dec: 501–507.
Shadldy J.R., C.A. P.a. (1993). CO_2 corrosion of N-80 steel at 71°C in a two-phase flow system. Corrosion Engineering Science and Technology, 49(8): 681–686.

Advances in Energy, Environment and Materials Science – Wang & Zhou (Eds)
© 2017 Taylor & Francis Group, London, ISBN 978-1-138-03600-0

A model of calculating radon and its daughters' concentrations in the chamber stope of underground uranium mines

Yong-jun Ye
Key Discipline Laboratory for National Defense for Biotechnology in Uranium Mining and Hydrometallurgy, University of South China, Hengyang, Hunan, China
School of Environmental Protection and Safety Engineering, University of South China, Hengyang, Hunan, China

Xin-tao Dai
School of Environmental Protection and Safety Engineering, University of South China, Hengyang, Hunan, China

De-xin Ding
Key Discipline Laboratory for National Defense for Biotechnology in Uranium Mining and Hydrometallurgy, University of South China, Hengyang, Hunan, China

ABSTRACT: In underground uranium mines of China, there are many chamber stopes that can generate and accumulate radon and its daughters, the concentrations of which are about 3–5 times higher than other countries. Therefore, it is very urgent to improve the level of radiation protection of the chamber stopes in China. Ventilation is one of the effective ways to control the concentrations of radon and its daughters. First, this study established the mathematical models that calculated the concentration of radon and its daughters in the chamber stope for designing and optimizing ventilation systems by using the theories of turbulent mass transfer and radioactive decay. Second, the models were used to analyze radon and its daughters' concentrations with the change of airflow rate, air volume, and radon generation rate in chamber stope. Third, the ventilation conditions of chamber stopes of four typical uranium mines in China were investigated and the models to calculate radon and its daughters' concentrations were used to verify its accuracy. Finally, the difference is analyzed between the measured values and the theoretical values of the total Potential Alpha Energy Concentration (PAEC) of radon daughters and a linear correction formula was established.

1 INTRODUCTION

Nuclear power has become an effective way to solve the increasingly serious problem of energy shortage and environmental pollution in China (Yan et al., 2011). At present, underground mining methods are used for exploitation of the uranium ore resources in China (Li, 2002). The mining methods have a common feature that can produce a lot of chamber stopes in underground uranium mines. Meanwhile, chamber stopes have become one of the main places where radon and its daughters generate and accumulate because of the existence of radium in uranium ores. Radon and its daughters are important carcinogens and their excessive inhalation can cause cancer (IAEA, 2015; ICRP, 2014; WHO, 2009). In order to reduce the personal radiation dose in underground mines, technical regulations for radon exhaustion and ventilation in underground uranium mines (regulatory standard of China) (EJ/T, 359-2006) stipulate that radon concentration and the PAEC of radon daughters should be controlled below 2.7×10^3 Bq/m^3 and $5.4\,\mu$J/m^3 in the chamber stope or other workplaces respectively and provide some strategies to reduce the radon generation rate in stopes. Because these strategies have not reasonably considered the factors such as airflow rate, radon release character, and the dimensional size of the ventilation space, the ventilation effect was unsatisfactory for reducing radon and its daughters' concentrations in the past. Currently, radon and its daughters' concentration levels in underground stopes are about 3–5 times higher than other countries and have greatly exceeded the reference level in China (Hu et al., 2012). Therefore, it is very urgent to reduce the concentration levels of radon and its daughters in underground stopes in China.

In this paper, two mathematical models were proposed for the PAEC of radon and its daughters in the chamber stope respectively, and based on these mathematical models, the strategies were discussed and summarized for exhausting and reducing the concentrations of radon and its daughters in the chamber stopes. Furthermore, the models were verified.

2 MATHEMATICAL MODELS

2.1 *The mathematical model for radon concentration*

The concentration of radon in the atmosphere of underground uranium mines mainly depend on radon emissions from the surface of uranium ore, heap of the blasted uranium ore, groundwater, and inlet air. The quantity of radon generated in the chamber stope comes from the sources except for the inlet air. The cross-sectional area at the ventilation space is much bigger than the area of the air inlet. The fresh air enters the chamber stope through an air inlet and can generate a strong turbulent mass exchange with harmful gases. Meanwhile, fresh air quickly mixes with harmful gases in the ventilation space with the help of turbulent diffusion and the mixed gas is exhausted through an air outlet. The ratio of the radon concentration in exhaust air to the average radon concentration in the ventilation space of the chamber stope is called the turbulent diffusion coefficient K_t. Figure 1 shows the schematic diagram of the calculation model for radon concentration in the chamber stope.

By using the continuity equation, the differential equation for the quantity of radon in the chamber stope can be written as follows:

$$V\frac{dC}{dt} = R - K_t(Q - vS)C - \lambda VC + QC_0 \qquad (1)$$

where C is the average radon concentration of the ventilation space, Bq/m³; C_0 is the radon concentration of the inlet air, Bq/m³; Q is the airflow rate, m³ /s; R is the radon generation rate of the ventilation space, Bq/s; λ is the radon decay constant, $\lambda = 2.1 \times 10^{-6}$ s⁻¹; V is the ventilation volume of the chamber stope, m³; K_t is the turbulent diffusion coefficient; v is the seepage velocity, m/s; and S is the surrounding rock area of the chamber stope, m².

Generally speaking, the quantity of air permeated from the surrounding rock is far less than the quantity of ventilation in the stope ($vS \ll Q$) in the chamber stope of the underground mine. When the ventilation is stable, $V(dC/dt) = 0$. Then, equation (1) becomes

$$C = \frac{R + QC_0}{K_t Q + \lambda V} \qquad (2)$$

As can be seen from equation (2), the average concentration of radon in the ventilation space mainly depends on the radon generation rate in the ventilation space, ventilation air inflow, radon concentration in inlet air, ventilation volume, and turbulent diffusion coefficient. Because the turbulent diffusion coefficient is influenced by geometry size and spatial distribution of the stope, which is determined by using a specific mining method, in view of the specific type of the chamber stope, we only consider the factors such as radon generation rate, airflow rate, radon concentration of the inlet air, and the ventilation volume. As $K_t = 1$, the radon concentration of the exhaust air is equal to the average radon concentration in the ventilation space in the chamber stope. We assume $K_t = 1$ to simplify the calculations in this paper, and equation (2) becomes

$$C = \frac{R + QC_0}{Q + \lambda V} \qquad (3)$$

2.2 *The mathematical model for the total PAEC of radon daughters*

As one of the radioactive gases, radon can produce daughters by way of decay. ²²²Rn can produce its short-lived daughter products such as ²¹⁸Po, ²¹⁴Pb, ²¹⁴Bi, and ²¹⁴Po with time. The regression equation of the total PAEC of radon daughters can be written as follows (Ye et al., 2015):

$$E_\alpha = \beta(1 - e^{-\lambda_e t})C_0 \qquad (4)$$

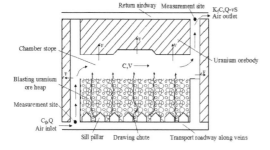

Figure 1. Schematic diagram of the calculation model for radon concentration in the chamber stope.

where β is the conversion constant, $\beta = 5.37\,\mathrm{uJ/kBq}$ and λ_e is the constant of equivalent decay, $\lambda_e = 3.3 \times 10^{-4}\mathrm{s}^{-1}$.

There are mainly three sources of radon daughters in the ventilation space of the chamber stope. These sources include the remnant radon daughters decayed from the radon in the inlet air, radon daughters decayed from the radon in the inlet air, and radon daughters decayed from the radon generated in the chamber stope. Without considering the importance attached to the wall and deposited of the radon daughters, the PAEC of the remnant radon decayed from the radon daughters in the inlet air $E_{\alpha 1}$, the PAEC of the radon daughters decayed from the radon in the inlet air $E_{\alpha 2}$, the PAEC of the radon daughters decayed from the radon generated in the chamber stope $E_{\alpha 3}$ and the total PAEC of the radon daughters in the outlet air E_α are respectively calculated as follows:

$$E_{\alpha 1} = E_{\alpha 0}e^{-\lambda_e V/Q} \tag{5}$$

$$E_{\alpha 2} = \beta C_0 (1 - e^{-\lambda_e V/Q}) \tag{6}$$

$$E_{\alpha 3} = \frac{R}{2Q}\beta(1 - e^{-\lambda_e V/Q}) \tag{7}$$

$$\begin{aligned} E_\alpha &= E_{\alpha 1} + E_{\alpha 2} + E_{\alpha 3} \\ &= E_{\alpha 0}e^{-\lambda_e V/Q} + \beta C_0 (1 - e^{-\lambda_e V/Q}) \\ &+ \frac{R}{2Q}\beta(1 - e^{-\lambda_e V/Q}) \end{aligned} \tag{8}$$

where $E_{\alpha 0}$ is the PAEC of the radon daughters in the inlet air, $\mu\mathrm{J/m}^3$.

3 RESULTS AND DISCUSSION

According to technical regulations for radon exhaustion and ventilation in underground uranium mines (EJ/T, 359-2006), the reference level of radon concentration is 1000 Bq/m³ in the inlet air of the workplace at an underground uranium mine and the total PAEC of radon daughters is 2 μJ/m³. For obtaining the general rules of exhausting radon, we assume that C_0 is 1000 Bq/m³ and $E_{\alpha 0}$ is 2 μJ/m³. Figure 2 shows the relationship between radon concentration and airflow rate in the chamber stope. Figures 3 and 4 show the relationships between the PAEC of radon daughters and airflow rate.

If the radon generation rate in the chamber stope is a constant, the influence of the change in the ventilation volume on the radon concentration can be ignored because equation (1) is almost equal to zero. In order to ensure that the radon concentration will not exceed the reference level, the airflow rate should be increased with an increase in the radon generation rate. If other conditions are

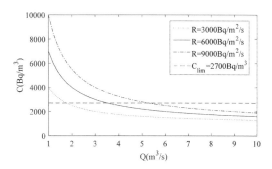

Figure 2. C versus Q at different values of R when V is 1000 m³.

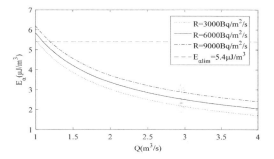

Figure 3. E_α versus Q at different values of R when V is 1000 m³.

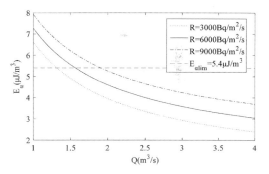

Figure 4. E_α versus Q at different R values when V is 2000 m³.

kept unchanged, the radon concentration decreases with an increase in the airflow rate, and increases with an increase in the radon generation rate.

Figures 3 and 4 show that the influence of the ventilation volume V is significant if the radon generation rate and airflow rate in the chamber stope are kept unchanged and the PAEC of radon daughters increases with an increase in the volume of air supply. The PAEC of radon daughters decreases with an increase in the airflow rate and increases with an increase in the radon generation rate, if other conditions are kept unchanged.

It is found that the airflow rate that maintain the concentrations of radon and its daughters in the chamber stope below the reference level should be designed separately by comparing Fig. 2 and Fig. 3. Under the same conditions given in this paper, the airflow rate for reducing radon is greater than that for radon daughters. Consequently, we should calculate the demand of ventilation air inflows for reducing the concentrations of radon and radon daughters below the reference level in the chamber stope respectively, and select the biggest of them as a reference to design a ventilation system.

From the results of the previous discussion, two optimization studies should be conducted. First, the radon generation rate in the ventilation space should be optimized, so that the measures for controlling radon exhaling from the surfaces of the chamber stope can be taken. Second, the airflow distribution in the chamber stope should be optimized, so that the radon and its daughters can't reside in the chamber stope for a long time. For an underground uranium mine, it is important to control the radiation hazard of radon daughters effectively. Therefore, the volume of the chamber stope should be minimized to reduce the total PAEC of radon daughters. In addition, reducing concentrations of radon and its daughters in the inlet air are important measures.

Because the mathematical model for the total PAEC of radon daughters does not consider the effect of radon daughters' settlement and adhesion in the chamber stope, we monitored the ventilation conditions in chamber stopes at four typical underground uranium mines in China in order to test the equation (8). The radon concentration and total PAEC of its daughters at the inlet air and outlet air of the chamber stope were measured with the KF602A type radon and its daughters' moni-

tor, which has been widely used in underground uranium mines of China and was calibrated at the Radon Laboratory of the University of South China. The average wind speed was measured at the air inlet of the chamber stope by using the EY3-2A type anemometer and the airflow rate of the chamber stope is the product of the average wind velocity and the cross-sectional area of the air inlet.

The measured and the theoretical values for the total PAEC of radon daughters are shown in Table 1. There are 8 samples whose theoretical values exceed the measured values and the maximum relative error is 37.82% and 3 samples whose theoretical values are smaller than the measured values and the maximum relative error is 11.81%.

From Table 1, the linear fitting formula for corre-lating the measured values with the theoretical val-ues of the total PAEC of radon daughters is ob-tained as follows:

$$E_{\alpha \cdot Measure} = 0.9775 E_{\alpha.Theoretical} - 0.1417 \quad (9)$$

The linear relationship between the measured and the theoretical values of the total radon daughters PAEC is obtained from the 11 chamber stope samples and is shown in Fig. 5. The linear correlation coefficient square (R-square) is 0.9712 between the measured values and the theoretical values, and a good overall correlation is obtained. It can be seen from equation (9) that the theoretical value is larger than the measured value. The main reason for this may be that radon daughters' particles will reduce because of the adsorption by the surrounding rock and their deposition and these effects are not considered in equation (8), when transporting along with the ventilation airflow in the chamber stope. Therefore, when using equation (8) to estimate the total PAEC of radon daughters in the chamber stope and the estimated results should be modified with equation (9).

Table 1. Comparison between the measured and the theoretical values of the total PAEC of radon daughters.

Stope number	Q (m³ s⁻¹)	V (m³)	R (kBq·s⁻¹)	C_0 (kBq·m⁻³)	$E_{\alpha 0}$ (uJ·m⁻³)	$E_{\alpha Measured}$ (uJ·m⁻³)	$E_{\alpha Theroetical}$ (uJ·m⁻³)	Relative error (%)
1	1.97	1600	5.106	4.377	4.346	7.616	10.4959	37.82
2	2.23	1600	8.499	3.515	4.250	10.56	9.4990	−10.05
3	2.59	1600	21.468	7.289	9.536	19.20	19.1161	−0.44
4	1.60	217	0.555	0.881	1.523	1.933	1.7047	−11.81
5	1.94	217	2.701	0.814	0.960	1.216	1.2195	0.29
6	6.65	637	27.942	0.303	0.198	0.550	0.5942	8.12
7	6.48	576	80.212	0.459	0.947	1.901	1.9549	2.85
8	4.28	576	53.972	0.613	0.486	1.504	1.9797	31.66
9	5.42	720	9.915	1.245	2.134	2.379	2.5406	6.79
10	4.75	633	7.166	1.893	3.716	3.997	4.1684	4.29
11	4.02	744	5.945	1.546	3.547	3.571	4.0649	13.83

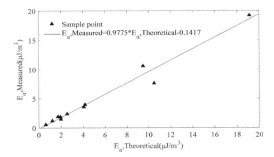

Figure 5. Graph showing the relationship between the measured and the theoretical values of the total potential alpha energy concentration of radon daughters.

4 CONCLUSIONS

The calculation models of the concentration of radon and its daughters in the chamber stope of an underground uranium mine can be used to analyze the effect of the airflow rate, the ventilation volume of the chamber stope, and the radon generation rate on the radon concentration and the total radon daughters' PAEC. The approaches for reducing the concentration of radon and its daughters can be proposed. The linearly modified formula for correlating the measured values and the theoretical values proposed in this paper is helpful to improve the calculation accuracy of the total PAEC of radon daughters and especially to decide the calculated amount of the radon or radon daughters, which is more suitable to design and optimize the ventilation scheme of the chamber stope.

ACKNOWLEDGMENTS

This work was financially supported by the National Natural Science Foundation of China (No.11105069), Hunan Provincial Natural Science Foundation of China (No. 09JJ6078), and Hunan Province Science and Technology Program (No. 2011SK3085).

REFERENCES

EJ/T., 359-2006, Technical regulations for radon exhaustion and ventilation in underground uranium mine [S] (Doctoral dissertation).

Harrison, J.D., & Marsh, J.W., 2012. Effective dose from in haled radon and its progeny. Annals of the ICRP, 41(3), 378–388.

Hu, P.H., & Li, X.J., 2012. Analysis of radon reduction and ventilation systems in uranium mines in China. Journal of radiological protection, 32(3), 289.

IAEA, 2015. Protection of the public against exposure indoors due to radon and other natural sources of radiation specific safety guide. International Atomic Energy Agency, Vienna.

ICRP, 2014. Radiological Protection against Radon Exposure. ICRP Publication 126. Ann. ICRP43 (3).

Li K.W., 2002. Mining technological properties and develop ment level of Chinese uranium ore mines. China Mining Magazine. 11(1), 23–27.

Miles, J., 1998. Development of maps of radon-prone areas us ing radon measurements in houses. Journal of Hazardous Materials, 61(1), 53–58.

Tomášek, L., Kunz, E., Darby, S.C., Swerdlow, A.J., & Placek, V., 1993. Radon exposure and cancers other than lung cancer among uranium miners in West Bohemia. The Lancet, 341(8850), 919–923.

Who, 2009. Handbook on indoor radon. World Health Organization, Geneva.

Yan, Q., Wang, A.J., Wang, G.S., Chen Q.S., Yu, W.J., & Li, R. P., 2011. Uranium resources survey and demand prediction for 2030. China Mining Magazine, 2, 1–5.

Ye, Y.J., Ding, D.X., Wang, L.H., Li, X.Y., Xie, D., Zhong Y.M., Zhao Y.L., 2014. Calculation model of radon and its daughters concentration in blind roadway with forced ventilation and their distribution rule, Journal of Central South University (Science and Technology), 46(5), 1800–1801.

Advances in Energy, Environment and Materials Science – Wang & Zhou (Eds)
© *2017 Taylor & Francis Group, London, ISBN 978-1-138-03600-0*

The structure and thermal properties of Nd and Ce-doped $Gd_2Zr_2O_7$ pyrochlore for nuclear waste forms

Shunzhang Chen, Di Huang, Long Fan, Xirui Lu & Dengsheng Ma
Subject Laboratory of National Defense for Radioactive Waste and Environmental Security, Southwest University of Science and Technology, Mianyang, Sichuan, P.R. China

Wei Ren
Laboratory of Electronic Information Materials and Devices, Chinese Academy of Sciences, Urumqi, Xinjiang, P.R. China

Liang Bian
Heibei Huicong Ecommerce Company Limited, Shijiazhuang, Hebei, P.R. China

ABSTRACT: Pyrochlore-structured ceramics, which are candidate hosts for nuclear wastes, are sintered by using the solid-state reaction method at 1500°C for 72 h in air. $Gd_{2-x}Nd_xZr_2O_7/Gd_2Zr_{2-x}Ce_xO_7$ $(0.0 \le x \le 2.0)$ compounds were investigated in respect of their overall performances. The phase structures of these compounds were identified by X-ray diffraction and their microstructures were observed by using SEM. Moreover, the thermal properties of these oxides were examined. In both kinds of solid solutions, all of the three observed objects increased with an increasing value of x from $x = 0$ to 2.0. The specific heat capacity of the two systems drew near, while the thermal expansion coefficient and the thermal conductivity of $Gd_{2-x}Nd_xZr_2O_7$ were relatively lower than those of $Gd_2Zr_{2-x}Ce_xO_7$.

1 INTRODUCTION

Formation and disposal of nuclear waste is one of the hardest challenges for the sustainable nuclear energy growth. In the study of artificial rock solidification of radioactive waste, pyrochlore oxides $A_2B_2O_7$ have attracted significant interest in recent years due to their desirable physic-chemical properties (Su, 2015), especially for their potential use as Thermal Barrier-Coating Materials (TBCs) (Cao, 2004; Bansal, 2007; Xiang, 2012), buffer layer for YBCO-coated conductors (Kato, 2004; Knoth, 2007; Qiu, 2014), and solidifying materials for irradiation wastes (Lu, 2014; Mandal, 2010). The general formula of the pyrochlore structure is $A_2B_2O_7$. The site A is mostly substituted by trivalent actinides and the B-site is generally substituted by Ti, Zr, Sn, Hf, and so on (Lang, 2010; Zhao, 2015), which leads pyrochlore compounds to large loading amounts for radioactive wastes.

Previous studies of gadolinium zirconium pyrochlore systems on fabrication, chemical durability, electrical properties, and radiation stability have yielded remarkable progress (Lu, 2015; Xia, 2011; Wang, 1999). And some researchers who are involved in searching for thermal properties of $Gd_2Zr_2O_7$ have also made an attempt (Wang, 2013). High-Level Radioactive Wastes (HLW) are widely known as heavy exothermal matters after immobilization of nuclear waste. Thus, the waste forms should not only perform well in terms of chemical durability and electrical properties, but also exhibit excellent thermal properties in consideration of the fixation of HLW. However, there are a few reports on whether the thermal properties of gadolinium zirconium pyrochlore materials meet the standards.

We majored in analyzing the phase transformation of gadolinium zirconium pyrochlore systems in our previous studies (Fan, 2015). In order to provide overall features of these interesting compounds, a series of substituted pyrochlore samples at either site A or site B have been synthesized and investigated in this study. We obtained the value of research parameters such as the specific heat capacity, the thermal expansion coefficient, and the thermal conductivity separately as a function of the value of x in the $Gd_{2-x}Nd_xZr_2O_7$ and $Gd_2Zr_{2-x}Ce_xO_7$ systems, which can be comprehensively used to evaluate their thermal properties.

2 EXPERIMENTAL PROCEDURE

Nd^{3+} and Ce^{4+} have so far gained widespread recognition and become representative substitutes for An^{3+} (An: actinide) and An^{4+} owing to their similar ionic radii to some actinide elements. Therefore, in this work, Nd is selected as an alternative to the trivalent actinides' substitutes for Gd, while Ce is

Table 1. The contents of starting materials for synthesizing the series of compounds.

Target compounds	Additive amount of raw materials / g		
	Gd_2O_3	ZrO_2	Nd_2O_3
$Gd_2Zr_2O_7$	0.7739	0.5261	—
$Gd_{1.8}Nd_{0.2}Zr_2O_7$	0.6995	0.5284	0.0721
$Gd_{1.6}Nd_{0.4}Zr_2O_7$	0.6244	0.5307	0.1449
$Gd_{1.4}Nd_{0.6}Zr_2O_7$	0.5488	0.5329	0.2183
$Gd_{1.2}Nd_{0.8}Zr_2O_7$	0.4724	0.5353	0.2923
$Gd_{1.0}Nd_{1.0}Zr_2O_7$	0.3954	0.5376	0.3670
$Gd_{0.8}Nd_{1.2}Zr_2O_7$	0.3177	0.5400	0.4423

Target compounds	Additive amount of raw materials / g			
	Gd_2O_3	ZrO_2	Nd_2O_3	CeO_2
$Gd_{0.6}Nd_{1.4}Zr_2O_7$	0.2393	0.5423	0.5183	—
$Gd_{0.4}Nd_{1.6}Zr_2O_7$	0.1603	0.5448	0.5950	—
$Gd_{0.2}Nd_{1.8}Zr_2O_7$	0.0805	0.5472	0.6723	—
$Nd_2Zr_2O_7$	—	0.5496	0.7504	—
$Gd_2Zr_{1.8}Ce_{0.2}O_7$	0.7617	0.4660	—	0.0723
$Gd_2Zr_{1.6}Ce_{0.4}O_7$	0.7498	0.4078	—	0.1424
$Gd_2Zr_{1.4}Ce_{0.6}O_7$	0.7383	0.3514	—	0.2103

Target compounds	Additive amount of raw materials / g		
	Gd_2O_3	ZrO_2	CeO_2
$Gd_2Zr_{1.2}Ce_{0.8}O_7$	0.7272	0.2966	0.2762
$Gd_2Zr_{1.0}Ce_{1.0}O_7$	0.7164	0.2435	0.3401
$Gd_2Zr_{0.8}Ce_{1.2}O_7$	0.7059	0.1919	0.4022
$Gd_2Zr_{0.6}Ce_{1.4}O_7$	0.6957	0.1418	0.4624
$Gd_2Zr_{0.4}Ce_{1.6}O_7$	0.6858	0.0923	0.5210
$Gd_2Zr_{0.2}Ce_{1.8}O_7$	0.6762	0.0460	0.5779
$Gd_2Ce_2O_7$	0.6668	—	0.6332

employed as the surrogate of tetravalent actinides' substitute for Zr in gadolinium zirconium pyrochlore solid solutions to simulate ceramics solidifying radioactive wastes.

In this study, AR grade Gd_2O_3, ZrO_2, CeO_2, and Nd_2O_3 powders were selected as reagents. $Gd_{2-x}Nd_xZr_2O_7$ and $Gd_2Zr_{2-x}Ce_xO_7$ ceramics were prepared by performing the solid-state reaction method. Before weighing, all the raw powders were heated at 800°C for 24 h to remove adsorptive water. The powders were compacted in a pellet form at a pressure of 10 MPa. The pellets were sintered at 1500°C for 72 h in the atmosphere to fabricate dense bulk ceramics. The sintered compounds should be taken out of the furnace when they have been naturally cooled to 200°C. The substituting rates were set as 0–100%. Tab. 1 presents the contents of starting materials for synthesizing the series of $Gd_{2-x}Nd_xZr_2O_7$ and $Gd_2Zr_{2-x}Ce_xO_7$ ($0 \leq x \leq 2.0$).

3 RESULTS AND DISCUSSION

We have proved in our previous study (Fan, 2015) that the gadolinium zirconium pyrochlore ceramics were fabricated by performing the solid state reaction at 1500°C for 72 h and $Gd_2Zr_2O_7$ exhibits a pyrochlore structure, while the gadolinium zirconium pyrochlore solid solutions display a defective fluorite structure. Some representative XRD patterns are shown in Fig. 1. It could be summarized that gadolinium zirconium pyrochlore ceramics were successfully fabricated in a single phase.

Fig. 2 shows the SEM results of $Nd_2Zr_2O_7$ and $Gd_2Ce_2O_7$ ceramics. The figures obviously display that the $Gd_2Ce_2O_7$ samples are relatively dense, while the $Nd_2Zr_2O_7$ samples have some pores, which may attribute to sintering. The grain boundary of the ceramics is clear and no other phase is found, which is in good accordance with the XRD pattern in Fig. 1.

Fig. 3. presents the specific heat capacity of $Gd_{2-x}Nd_xZr_2O_7$ and $Gd_2Zr_{2-x}Ce_xO_7$ ceramics as a function of the value of x. Both the specific heat capacity of $Gd_{2-x}Nd_xZr_2O_7$ and $Gd_2Zr_{2-x}Ce_xO_7$ gradually increase when the compositions change from $x = 0$ ($Gd_2Zr_2O_7/Gd_2Zr_2O_7$) to

$x = 2.0$ ($Nd_2Zr_2O_7/Gd_2Ce_2O_7$). And the specific heat capacity of the two discussed compounds does not exhibit much difference before Nd/Ce completely takes the place of Gd/Zr. Finally, when compared with $Gd_2Zr_2O_7$, the specific heat of $Nd_2Zr_2O_7$ is just 0.098 $MJ·m^{-3}·K^{-1}$ higher, while $Gd_2Ce_2O_7$ increases by 0.227 $MJ·m^{-3}·K^{-1}$. The specific heat capacity should increase with the decrease of density according to the following equations:

$$c = \frac{Q}{m \cdot \Delta t} \tag{1}$$

$$m = \rho \cdot V \tag{2}$$

$Gd_{2-x}Nd_xZr_2O_7$ compositions are in good agreement with this rule, whereas $Gd_2Zr_{2-x}Ce_xO_7$ compositions seem to be so not just by taking density into account.

Fig. 4 shows the thermal expansion coefficients of $Gd_{2-x}Nd_xZr_2O_7$ and $Gd_2Zr_{2-x}Ce_xO_7$ ceramics. Both thermal expansion coefficients of the two systems monotonically increase when the value of x varies from $x = 0$ to $x = 2.0$, which may be partly due to the increase of temperature. And the thermal expansion coefficient of $Gd_{2-x}Nd_xZr_2O_7$ is relatively lower than that of $Gd_2Zr_{2-x}Ce_xO_7$ all along. Ultimately, as the value of x rises to 2.0, the thermal expansion coefficient of $Gd_{2-x}Nd_xZr_2O_7$ and $Gd_2Zr_{2-x}Ce_xO_7$ systems respectively increase by 0.0740 $mm^2·s^{-1}$ and 0.0466 $mm^2·s^{-1}$.

These phenomena can be explained by the crystal energy. As Madelung constant was reduced, the crystal energy also decreased. Thus, when the structure becomes more disordered, the Madelung constant of $Gd_2Zr_2O_7$ is conformed to be reduced (Kutty, 1994). Therefore, in $Gd_2Zr_{2-x}Ce_xO_7$ and $Gd_{2-x}Nd_xZr_2O_7$ series, the degree of disorder of the structure increases with increasing CeO_2 and Nd_2O_3 contents, thereby leading to an enhanced thermal expansion coefficient. In addition, the substitution of Nd^{3+} for Gd^{3+} and Ce^{4+} for Zr^{4+} can cause point defects in $Gd_2Zr_2O_7$, which would create stress filed in the lattice and enhance lattice vibration, thereby leading to a larger thermal expansion coefficient (Zhou, 2008).

The thermal conductivity of $Gd_{2-x}Nd_xZr_2O_7$ and $Gd_2Zr_{2-x}Ce_xO_7$ solid solutions are plotted in Fig. 5.

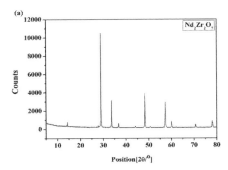

(a)

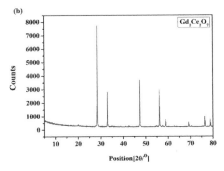

(b)

Figure 1. XRD patterns of typical samples: (a) $Nd_2Zr_2O_7$ and (b) $Gd_2Ce_2O_7$.

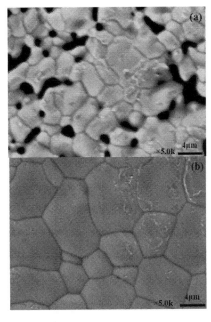

Figure 2. SEM micrographs of the ceramics: (a) $Nd_2Zr_2O_7$ and (b) $Gd_2Ce_2O_7$.

The measured thermal conductivities of these ceramics display an ascent curve as the value of x increases. And the thermal conductivity of $Gd_{2-x}Nd_xZr_2O_7$ is always lower than that of

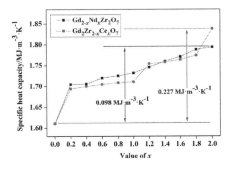

Figure 3. Graph showing the composition-dependent specific heat capacities of $Gd_{2-x}Nd_xZr_2O_7$ and $Gd_2Zr_{2-x}Ce_xO_7$.

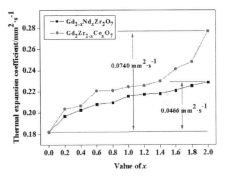

Figure 4. Graphs showing the composition-dependent thermal expansion coefficient of $Gd_{2-x}Nd_xZr_2O_7$ and $Gd_2Zr_{2-x}Ce_xO_7$.

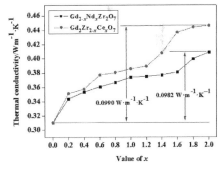

Figure 5. Graphs showing the composition-dependent thermal conductivity of $Gd_{2-x}Nd_xZr_2O_7$ and $Gd_2Zr_{2-x}Ce_xO_7$.

$Gd_2Zr_{2-x}Ce_xO_7$ within the scope of the discussion. The curves conform well to the following equation:

$$\kappa = \lambda \cdot \rho \cdot C_p \tag{3}$$

where κ stands for the thermal conductivity, λ is the thermal diffusivity, ρ denotes the density of the compositions, and Cp represents the specific heat capacity of the ceramics.

The excellent thermal and physical properties of $Gd_2Zr_2O_7$ are closely related to its unique crystal structure. In the pyrochlore structure ($Gd_2Zr_2O_7$), except a lot of oxygen vacancy defects, there are also two kinds of point defects; one is the Gd^{3+} ion and Zr^{4+} ions exchange positions with each other to form a reverse defect while another is the oxygen O (O_{48f}) ions that migrate to the O_{8b} position is not occupied by the formation of the Kerr defect. When the temperature change is relatively low, the three point defects work together to perform phonon scattering and reduce the average free path of the phonon, so that the thermal conductivity of $Gd_2Zr_2O_7$ became low. However, with the increase of temperature, the concentration of the defect and the defect of Kerr decrease and the sensitivity of oxygen vacancies to the temperature also decreases, so that their effect on the thermal conductivity is reduced.

4 CONCLUSIONS

In this study, it could be found that the Nd and Ce-doped $Gd_2Zr_2O_7$ ceramics were successfully fabricated in a single phase. The XRD patterns further show that Nd and Ce were successfully incorporated into the crystal lattice. The thermal properties were observed severally over the process and all of them displayed marked variation with an increased incorporation of Nd/Ce. The specific heat capacity increased with an increase in the value of x, which suggests that the gadolinium zirconium pyrochlore compounds will remain stable and safe while the temperature of the whole system changes distinctly. The thermal expansion coefficient of $Gd_2Zr_{2-x}Ce_xO_7$ is always higher than that of $Gd_{2-x}Nd_xZr_2O_7$ along the discussed range; it indicated that $Gd_2Zr_{2-x}Ce_xO_7$ ceramics will turn into larger volume than $Gd_{2-x}Nd_xZr_2O_7$ systems do under the same conditions of temperature variation. The thermal conductivity of $Gd_{2-x}Nd_xZr_2O_7$ and $Gd_2Zr_{2-x}Ce_xO_7$ also increased. It follows that, $Gd_2Zr_2O_7$ used as the radioactive wastes' solidification matrix may be appropriate for heat dissipation in case of unexpected explosion induced by thermal expansion within a short time.

ACKNOWLEDGMENTS

This work was supported by the National Natural Science Foundation of China (No. 21507105, No. 41302029), Thousand Youth Talents Plan (Y42H831301), Key Project of Sichuan Education Department(No.15ZB0116), Foundation of Laboratory of National Defense Key Discipline for Nuclear Waste and Environmental Safety, and Southwest University of Science and Technology (No. 15yyhk10).

REFERENCES

Bansal N P, Zhu D M. Effects of doping on thermal conductivity of pyrochlore oxides for advanced thermal barrier coatings. Mat. Sci. Eng., 2007, 459: 192–195.

Cao X Q, Vassen R, Stoever D. Ceramic materials for thermal barrier coatings. J. Eur. Ceram. Soc., 2004, 24: 1–10.

Fan L, Shu X Y, Ding Y, et al. Fabrication and phase transition of $Gd_2Zr_2O_7$ ceramics immobilized various simulated radionuclides. J. Nucl. Mater., 2015, 456: 467–470.

Kato T, Muroga T, Iijima Y, et al. Transmission electron microscopy studies of a $CeO_2/Gd_2Zr_2O_7$ buffer layer on an Ni-based alloy for YBCO coated conductor. Physica. C., 2004: 813–818.

Knoth K, Huhne R, Oswald S, et al. Detailed investigation on $La_2Zr_2O_7$ buffer layers for YBCO-coated conductors prepared by chemical solution deposition. Acta Mater., 2007, 55: 517–519.

Kutty K V G, Rajagopalan S, Mathews C K, et al. Thermal expansion behaviour of some rare earth oxide pyrochlores [J]. Mater. Res. Bull., 1994, 29(7): 759–766.

Lang M, Zhang F X, Zhang J M, et al. Review of $A_2B_2O_7$ pyrochlore response to irradiation and pressure. Nucl. Instr. Meth. B., 2010, 268: 2951–2959.

Lu X R, Ding Y, Dan H, et al. High capacity immobilization of TRPO waste by $Gd_2Zr_2O_7$ pyrochlore. Mater. Lett., 2014, 136: 1–3.

Lu X R, Fan L, Shu X Y, et al. Phase evolution and chemical durability of co-doped $Gd_2Zr_2O_7$ ceramics for nuclear waste forms. Ceram. Int., 2015, 41: 6344–6349.

Mandal B P, Pandey M, Tyagi A K. $Gd_2Zr_2O_7$ pyrochlore: Potential host matrix for some constituents of thoria based reactor's waste. J. Nucl. Mater., 2010, 406: 238–243.

Qiu W B, Fan F, Lu Y M, et al. Thickness effect of $Gd_2Zr_2O_7$ buffer layer on performance of $YBa_2Cu_3O_{7-\delta}$ coated conductors. Physica. C., 2014, 507: 81–84.

Su S J, Ding Y, Shu X Y, et al. Nd and Ce simultaneous substitution driven structure modifications in $Gd_{2-x}Nd_xZr_{2-y}Ce_yO_7$. J. Eur. Ceram. Soc., 2015, 35(6): 1847–1853.

Wang L, Eldrige J I, Guo S M. Thermal radiation properties of plasma-sprayed $Gd_2Zr_2O_7$ thermal barrier coatings. Scripta Mater., 2013, 69: 674–677.

Wang S X, Begg B D, Wang L M, et al. Radiation stability of gadolinium zirconate: A waste form for plutonium disposition. J. Mater. Res., 1999, 14(12): 4470–4473.

Xia X L, Liu Z G, Ouyang J H, et al. Effect of Ce substitution for Zr on electrical property of fluorite-type $Gd_2Zr_2O_7$. Solid State Sci., 2011, 13: 1328–1333.

Xiang J Y, Chen S H, Huang J H, et al. Phase structure and thermophysical properties of co-doped $La_2Zr_2O_7$ ceramics for thermal barrier coatings. Ceram. Int., 2012, 38: 3607–3612.

Zhao M, Ren X R, Pan W. Mechanical and thermal properties of simultaneously substituted pyrochlore compounds $(Ca_2Nb_2O_7)_x(Gd_2Zr_2O_7)_{1-x}$. J. Eur. Ceram. Soc., 2015, 35: 1055–1061.

Zhou H M, Yi D Q. Effect of rare earth doping on thermo-physical properties of lanthanum zirconate ceramic for thermal barrier coatings [J]. J. Rare. Earth., 2008, 26(6): 770–774.

Advances in Energy, Environment and Materials Science – Wang & Zhou (Eds)
© 2017 Taylor & Francis Group, London, ISBN 978-1-138-03600-0

Based on nanofiltration membrane separation of monovalent salt and divalent salt performance study

Daoyou Hu, Jinghuan Ma, Ying Liu & Xuejing Kang
Department of Environmental and Chemical Engineering, Tianjin Polytechnic University, Tianjin, China

Kai Zhang
Tianjin TangDa Environmental Protection Productivity Promotion Center, China

Fang Bian
Tianjin Tianrenhe Environmental Technology Research and Development Center, China

ABSTRACT: For the study of the separation performance of nanofiltration membrane of monovalent salt and divalent salt in seawater and decalcification seawater, the DL nanofiltration membrane of GE was selected to separate seawater and decalcification seawater respectively. Meanwhile, the different result of nanofiltration running under constant pressure and the effect of ions trapped under different pressure were examined. The membrane under optimal conditions exhibited SO_4^{2-} average intercept of 95.34%. Although the maximum retention rate of Ca^{2+} did not reach 85%, we got the effluent water which calcium ion content was less than 30 mg/L by using the decalcification seawater as influent water. These results indicated that compared with the seawater filtration process, the antifouling property of nanofiltration membrane improved apparently in decalcification seawater treatment.

1 INTRODUCTION

At present, the fresh water resources are in short supply, but the seawater is rich in resources. One of the effective ways to solve the shortage of fresh water and realize the sustainable development of economy and society is the development and utilization of seawater resources. However, the seawater contains a large number of salt and other components. It makes the seawater into a high salinity, high hardness and high corrosive mixed electrolyte solution system, which increased the difficulty of its utilization. Nanofiltration has been broadly used in wastewater reclamation and desalination because of its separation effect. In terms of seawater resource exploitation and utilization, nanofiltration is mainly used for pretreatment of seawater and optimization of quality of seawater, to provide better quality of water for reverse osmosis. Furthermore, low-temperature multi effect evaporation etc., reduce membrane fouling and scaling and enhance the recovery rate of the system.

Although the nanofiltration technology has good effect to optimize the quality of seawater, its application also has the problem of membrane fouling. The most common is the calcium salt scaling, including calcium carbonate and calcium sulfate. It is difficult to remove calcium sulfate with acid cleaning solution. Therefore, a further investigation of the nanofiltration membrane performance

is necessary for the application of nanofiltration technology. In this paper, the seawater and decalcification seawater were regarded as the test water and the separation performance of nanofiltration membrane of monovalent salt and divalent salt was studied. Additionally, the membrane surface fouling at different conditions were investigated through scanning electron microscopy.

2 THE EXPERIMENT PART

2.1 *The membrane used*

The nanofiltration membrane we using was produced by GE company, DL1812C-34D, with the intercept molecular weight of $1.5 \times 10^5 \sim 3 \times 10^5$, effective membrane area of 0.32 m², minimum of $MgSO_4$ intercept rate was 96% and the membrane material of TFM.

2.2 *Experiment materials*

The original seawater from Bohai bay in Tianjin had been pretreated before the nanofiltration. Decalcification seawater was the original seawater which the Ca^{2+} was removed by chemical precipitation method. The Ca^{2+} content of decalcification seawater was less than 60 mg/L. The main properties and components of the two water inlet were listed in Table 1.

Table 1. raw seawater properties and main components.

Parameters	Seawater	Decalcification seawater
pH	7.5	7.49
baume degree	3.3	3.3
TDS (mg/L)	34015	34068
conductivity (ms/cm)	34.6	34.9
Ca^{2+} (mg/L)	460.8	57.6
Mg^{2+} (mg/L)	1260	1236
SO_4^{2-} (mg/L)	2718	2598.4
Cl^- (mg/L)	17605.7	17962.3
K^+ (mg/L)	369.4	377.2
Na^+ (mg/L)	11429	11637.5
HCO^{3-} (mg/L)	137.5	143.8

2.3 Experimental device process

As shown in Figure 1, the original seawater was pretreated by the microfiltration and the larger particles, suspended solids, colloid and microorganisms were intercepted. The pretreatment reduced the water turbidity and enhanced the quality of water before nanofiltration process. The feed seawater was pumped into the nanofiltration membrane module by the plunger metering pump. The feed water was divided into two parts after nanofiltration, including concentrated water and production water. Additionally, the concentrated water would return to the original water while the concentrated water was collected to the raw water tank. Nanofiltration membrane module of the water inlet side and water production side were equipped with a pressure gauge, which could monitor operation pressure. In this case, the operating pressure and water production were effectively controlled by adjusting the reflux control valve. With nanofiltration experiments running constantly, seawater concentration in the raw water tank increased gradually.

2.4 Experimental detection and analysis method

Baume degree of Seawater was tested by Baume meter; pH was measured by using a pH meter; TDS instrument used in determination of TDS; conductivity was measured by using a conductivity meter. Chemical analysis methods included: EDTA complexometric titration method for the determination of Ca^{2+}, Mg^{2+} concentration; silver nitrate precipitation titration determination of Cl^- concentration; sodium tetraphenyl boron quaternary amine salt method the determination of K^+ concentration; EDTA complexometric titration method for the determination of SO_4^{2-} concentration; ion balance subtraction method to calculate the Na^+ concentration.

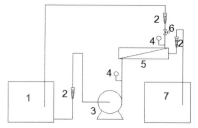

1. he raw water tank; 2. flowmeter; 3. plunger metering pump;
4. pressure gage; 5. nanofiltration membrane; 6. control valve;
7. Producing water tank

Figure 1.   Experimental process flow chart.

3   THE RESULTS AND DISCUSSION

3.1   Pure water flux and pure water permeability coefficient of membrane

Pure water permeability coefficient L_p is the intrinsic performance parameters of the nanofiltration membrane. It is an important symbol of membrane stability performance. Therefore, in order to ensure the accuracy and reliability of experimental results, the L_p value must firstly be determined of selected nanofiltration membrane. The L_p value can be calculated by the following equation:

$$L_P = J_V / \Delta P \qquad (1)$$

where J_V is the pure water flux; where ΔP is transmembrane pressure.

Under the pressure of 0.2~3.0 MPa, the pure water permeability coefficient of GE nanofiltration membrane remained stable, at the range of 42~47 L/(m²·h·MPa), and its average value was 44.89 L/(m²·h·MPa). The basic stability of pure water permeability coefficient showed that the GE nanofiltration membrane can be used in experimental study as it had good stability and reliability in the operating pressure for the range of 0.2~3.0 MPa.

3.2   Run at constant pressure of effluent flux changes

The operation pressure was kept constant by adjusting the concentration water return valve. Seawater and decalcification seawater were used as inflow water respectively and the change of effluent flux under a certain time was recorded.

As shown in Figure 2, under the condition of a certain pressure, the change of effluent flux showed a downward trend. The pressure was controlled at 1.2 MPa, filtering decalcification seawater and seawater, and the effluent flux respectively dropped from the initial 7.2 L/h and 7 L/h to 1.45 L/h and 1.5 L/h. Furthermore, the effluent flux respectively dropped from the initial 10.05 L/h and 9.9 L/h

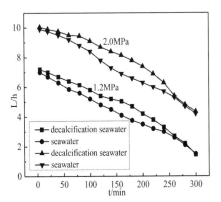

Figure 2. Water yield variation chart under a certain amount of pressure.

to 4.35 L/h and 4.2 L/h when the pressure was at 2.0 MPa. The reason for the phenomenon was that with the increase of influent concentration, the membrane fouling emerged in the blocked part of the membrane. At different pressures, the effluent flux of using decalcification seawater as influent water were slightly larger than directly using seawater as influent water in the whole process. It could be explained that the decalcification seawater more easily crossed nanofiltration membrane and reduced the form of calcium carbonate and calcium sulfate fouling layer in the surface of membrane.

3.3 The interception and separation effect of nanofiltration membrane for monovalent salt and divalent salt

The interception and separation effect of nanofiltration membrane for solute was expressed with retention rate. The retention rate is percentage of the membrane capsule intercept dissolving occupy the total quality of the solute. The retention rate R was calculated according to the following equation:

$$R = \left(1 - \frac{C_P}{C_0}\right) \times 100\% \qquad (2)$$

where R is the retention of a solute in the influent water; where C_P is the concentration of the solute in the effluent water; where C_0 is the concentration of the solute in the influent water.

Figure 3 and Figure 4 respectively reflected the retention rate of nanofiltration membrane for major ions in the seawater and decalcification seawater under different operating pressures. With the increase of pressure, the retention rate of nanofiltration membrane for major ions in the seawater and decalcification seawater first increased then decreased and tended to equilibrium. This phenomenon could be explained that at the beginning, with the increase of pressure, the water flux increased and the salt flux

changed slightly and the concentration of ions in the effluent water was low, then the main ion retention rate gradually increased. However, as the pressure continued to rise, ion concentration gap between the two sides of the membrane became lager, making increased ion flux and decreased rejection rate before the final balance.

As shown in Figure 3 and Figure 4, the interception rate of nanofiltration membrane for SO_4^{2-} and Mg^{2+} in seawater and decalcification seawater were high, all above 90%. For Ca^{2+}, the average retention rate of Ca^{2+} in seawater was 76.87% and the average retention rate of Ca^{2+} in the decalcification seawater was 57.68% while the Ca^{2+} concentration less than 30 mg/L in effluent water. The reason was that the Ca^{2+} concentration was low in decalcification seawater, less than 60 mg/L. For Cl⁻ and K⁺, the ion rejection rate was low, all below 18%. From the above data, it was observed that the interception and separation effect of nanofiltration membrane for monovalent salt and divalent salt in seawater and decalcification seawater were excellent, but the retention rate of Ca^{2+} was not very high, the maximum retention rate was 83.33%. However, the lower calcium ion content of effluent of decalcification seawater was achieved in the filtration.

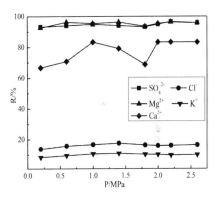

Figure 3. The interception rate of nanofiltration membrane for major ions in seawater.

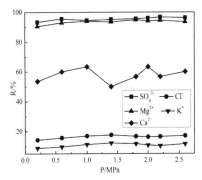

Figure 4. The interception rate of nanofiltration membrane for major ions in the decalcification seawater.

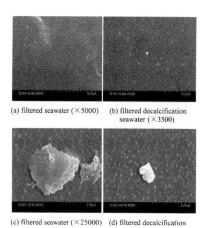

(a) filtered seawater (×5000)   (b) filtered decalcification seawater (×3500)

(c) filtered seawater (×25000)   (d) filtered decalcification seawater (×25000)

Figure 5. SEM diagram of GE nanofiltration membrane after being used.

3.4 Fouling analysis of nanofiltration membrane

Nanofiltration membrane will also appear fouling in use process, including membrane pore blocking and membrane surface fouling. In this experiment, seawater and decalcification seawater were used as influent water. After running for 7 days, the membrane sample was prepared first and observed under scanning electron microscopy then. The SEM diagram of GE nanofiltration membrane after filtered seawater and decalcification seawater respectively were shown in Figure 5.

As shown in Figure 5, it was observed that the membrane surface became rougher after being used and the fouling was accumulated in the membrane surface. At low magnification, the membrane surface (a) of the filtered seawater was coarser than that of the membrane surface (b) of the filtered decalcification seawater. At high magnification, it was observed that the membrane surface of (c) had larger sediment particles, with tens of microns in diameter, while the membrane surface of filter decalcification seawater (d) had smaller pollution particles, about 5 ~ 6 um in diameter. It was obvious that two kinds of membrane surface in high magnification were rougher than before.

4 CONCLUSIONS

Through the above experiments, the following conclusions were drawn:

1. Under the constant pressure of operating condition, the nanofiltration membrane possessed a better operation effect for decalcification seawater than seawater.
2. The interception and separation effect of nanofiltration membrane for monovalent salt and divalent salt in seawater and decalcification seawater was excellent, the size of the ion intercept rate order was: $SO_4^{2-} > Mg^{2+} > Ca^2 > Cl^- > K$. However, Ca^{2+} content was still high in the treated seawater, more than 100 mg/L, which would affect the further use about effluent water of nanofiltration, such as use the effluent water as influent water for reverse osmosis. These problems would be solved by using decalcification seawater as influent water for nanofiltration.
3. Compared with seawater, the membrane fouling caused by decalcification seawater was weaker, it was contributed to prolong the service life of the nanofiltration membrane and shorten the cleaning cycle of nanofiltration membrane. Although using decalcification seawater has many advantages compared with seawater for nanofiltration system, it was necessary to investigate its system economy in the next work.

ACKNOWLEDGEMENTS

This work was financially supported by the demonstration project of Tianjin Oceanic Administration (cxsf2014–30), Program of Tianjin Municipal Science and Technology Commission (14 JCQNJC02500) and Tianjin Tianrenhe environmental technology research and Development Center.

REFERENCES

Hassan, A. M. Farooque, A. M. Jamaluddin, A. T. M. (2000). Ademonstration plant based on the new NF–SWRO process. J. Desalination.131, 157–171.

Ryabchikov, B. E. Panteleev, A. A. Gladush, M. G. (2012). Performance testing of seawater desalination by nanofiltration. J. Petroleum chemistry. 52(7): 465–474.

Song Yuefei, Gao Xueli & Gao Congjie. (2013). Evaluation of scaling potential in a pilot-scale NF–SWRO integrated seawater desalination system. J. Journal of Membrane Science. 443, 201–209.

Subramanian, S. Seeram, R. (2013). New directions in nanofiltration applications—Are nanofibers the right materials as membranes in desalination? J. Desalination. 308, 198–208.

Wang Tao, Wang Shu, Wang Yi. (2015). Separation from sugar/salt mixtures by spiral wound nanofiltration membrane. J. Technology of water treatment. 41(1): 107–111.

Wang Wei, Li Guodong & Du Qiyun. (2007). Ionic selectivity of nanofiltration membrane. J. Journal of tianjin polytechnic university. 26(4): 5–7.

Zhou Dong, Zhu Lijing, Fu Yinyi. (2015). Development of lower cost seawater desalination processes using nanofiltration technologies—A review. J. Desalination. 376, 109–116.

Advances in Energy, Environment and Materials Science – Wang & Zhou (Eds)
© 2017 Taylor & Francis Group, London, ISBN 978-1-138-03600-0

Research progress in the application of nanomaterials for the adsorption of heavy metal ions in wastewater

Jianwen Gao

College of Information Technology and Engineering, Marshall University, Huntington, WV, US

ABSTRACT: Heavy metal contamination of water is a significant global issue. Removal of heavy metals from aqueous solution by nanoparticles, characterized by low cost, high efficiency, and easy separation and regeneration, is extensively studied in the field of wastewater treatment. The problems existing in the application of nanomaterials are described, and their prospects for environmental modification are discussed. Finally, suggestions for conducting further studies on the removal of heavy metals by nanoparticles are presented.

1 INTRODUCTION

Because of the development of industries, heavy metal contamination in water and soil have become more severe recently (Ozdes, 2009). Industrial wastewater and even seawater are often contaminated by poisonous impurities causing ecological imbalance and severe public health problems.

The term heavy metal refers to any metallic element with atomic weight in the range of 63.5–200.6 and specific gravity higher than 5.0. Heavy metals are natural components of the earth's crust and can enter water by means of industrial and consumer waste, or even from acid rain breaking down soils and releasing heavy metals into streams, lakes, rivers, and groundwater (Yakout, 2016; Raval, 2016).

Because of their toxicity, bioaccumulation, and biomagnification in the food chain, heavy metal contamination is a serious problem. The most common hazardous metals found in industrial wastewater are cadmium and lead. Cadmium water contamination causes nephrotoxicity. At high exposure levels, lead causes encephalopathy, cognitive impairment, behavioral disturbances, kidney damage, anemia, and toxicity to the reproductive system, particularly at high exposure levels; long-term exposure may cause bone damage as well (Wanna, 2016; Raghunandhan, 2016; Dorota, 2016).

Thus, disposal of contaminated water has always been a major environmental issue worldwide. Much effort has been devoted to finding effective methods for removing heavy metals from water and soil, such as chemical precipitation, oxidation, reduction, coagulation, ion exchange, reverse osmosis, solvent extraction, flocculation, membrane separation, filtration, evaporation, electrolysis, and adsorption. These methods have been used to remove and recover toxic contaminants from industrial effluents. Among these methods, adsorption has attracted much attention of researchers because of its simple operation, high removal rate, less secondary pollution, as well as low cost (Rouholah, 2016; Li, 2016).

Nanotechnology plays a key role in adsorption method. Nanomaterials are those that measure only 1–100 nm at least on one dimension and those constructed by them as basic unit structure; it ranged from the world of macro and micro, with specific properties different from other traditional solid materials, such as surface effect, small size effect, quantum size effect, macroscopic quantum tunneling effect, dielectric limit effect, etc. Traditional separation and enrichment techniques might have the disadvantages of low speed response, low sensitivity, and poor selectivity. However, when nanomaterials are applied in the rapid adsorption and enrichment of trace substances, all the above defects can be largely improved and ideal effect can be obtained, which could serve as a more ideal adsorption material for the analysis of trace elements (Kardar, 2016).

2 ADSORPTION PRINCIPLE AND THE INFLUENCING FACTORS OF METAL IONS BY NANOMATERIALS

With the decrease of the size of nanoparticles, the number of surface atoms increases rapidly, surface area, surface energy, and surfaces binding energy also increase, so do the surface area, surface energy, and surface binding energy. As the surface atoms are unsaturated and are easy to combine with other metal ions, nanomaterials have a strong adsorption ability for many metal ions and can reach adsorption equilibrium within a relatively short period of time. Because of the large surface area, the nanomaterials also have a high adsorption capacity.

As has been reported, the following five main factors affect the adsorption capacity and adsorption rate: solution PH, temperature, time, initial ion concentration, and surface area of the adsorbent (Dorota, 2016; Rouholah, 2016; Li, 2016; Kardar, 2016).

3 RESEARCH PROGRESS OF VARIOUS NANOMATERIALS FOR THE ADSORPTION OF HEAVY METAL IONS

3.1 *Metal oxide nanomaterials for the separation and enrichment of metal ions*

Vassilev et al. studied the properties of heavy metal ion adsorption of nano titanium dioxide in 1996, indicating its high adsorption capacity, rapid adsorption and desorption, and good reversibility (Vassileva, 1997). Furthermore, they also studied the adsorption performance of ceric oxide for heavy metal ions such as Cd^{2+}, Co^{2+}, Cu^{2+}, Mn^{2+}, Pb^{2+}, and Zn^{2+}, and found that when pH $\geq$ 7, quantitative adsorption results (Vassileva, 2001). They also studied the adsorption properties of zircon alba for 18 elements (Hang, 2003). Yiping Hang et al. carried out a series of experiments on nano oxides, such as the adsorption properties of titanium dioxide for Ga^{3+}, In^{3+}, and Tl^{3+}, showing that they can be quickly absorbed on titanium dioxide under optimized pH condition and it has been successfully applied to the determination of trace elements (Ga^{3+}, In^{3+}, and Tl^{3+}) in geological samples (Liang, 2000). They also studied the absorption ability of nano titanium dioxide for Cr^{6+}/Cr^{3+} and used it for speciation analysis of chromium in water samples. Results showed that the relative standard deviations (n = 9) of Cr^{6+} and Cr^{3+} found by this method were 3.6% and 4.2%, respectively (ρ = 2.0 mg/L) (Li, 2002).

The applications of titanium dioxide (TiO_2), zirconium oxide (ZrO_2), and aluminum oxide (Al_2O_3)

for the separation and enrichment of metal ions are shown in Table 1.

3.2 *Surface-modified oxide nanomaterials for the separation and enrichment of metal ions*

Liu Yan et al. synthesized titania sol by sol-gel method and impregnated it on silica gel to obtain nano titanium dioxide material. The results showed that at pH 8–9, the heavy metal ions can be quantitatively enriched, and the adsorption of metal ions can be freed by 0.5 mol/L HNO_3 solution. Wenming Xiong et al. loaded nano aluminum oxide with dimethyl amino benzylidene rhodanine and identified trace silver in copper standard samples by Flame Atomic Absorption Spectrometry (FAAS) method with a detection limit of 1.7 µg/L, which can be used for the determination of trace silver in actual samples (Santos, 2005). Lian et al. used dithizone to modify nano titanium dioxide and then separated the enriched trace elements Cr^{3+} and Pb^{2+}, whose detection limits were 0.38 and 1.72 µg/L, respectively (Hong, 2005). Zheng sulfonamide modified nano titanium dioxide enriched Cr^{3+} and Pb^{2+} in food and water.

The applications of nanomaterials of surface-modified oxide for the separation and enrichment of metal ions are shown in Table 2.

3.3 *Magnetic oxide nanomaterials for the separation and enrichment of metal ions*

Because of their good magnetic performance, Fe_3O_4 nanoparticles are commonly used to prepare magnetic fluids, which can be used to adsorb metal ions. Because of their large specific surface area, nanoparticles can be easily reunited, and the use of different surfactants to modify the magnetic particles has become the key point. Yonggang Zhao surface-functionalized nano Fe_3O_4, and the adsorption of Cr^{6+} reached equilibrium within 10 min.

Table 1. Application of metal oxide nanomaterials for the separation and enrichment of metal ions (Suleiman J S, Hu B, Pu X, et al., 2007; Fuentes G, et al., 2009; Xu J, Wang Y, Xian Y, et al., 2003).

Adsorbents	Analysis method	Analysis object	Sample
TiO_2	ICP-AES	Mo(VI), W(VI)	Geological sample
TiO_2	ICP-AES	Cd(II), Co(II), Zn(II)	Environmental sample
TiO_2	UV-VIS	In(III), W(VI)	Water
TiO_2	UV	Mn(II)	–
ZrO_2	ICP-AES	Zn(II), Cu(II), Ni(II)	Water
ZrO_2	UV	Cr(VI)	Water
ZrO_2	ICP-MS	Cr(III), Cr(VI)	Water
Al_2O_3	GFAAS	Pb(II), Cd(II)	Water
Al_2O_3	UV-VIS	Ge(IV)	–
Al_2O_3	FAAS	Pd(II)	Synthetic sample
MgO	FAAS	Cu(II), Ni(II), Zn(II)	Water
SiO_2	FAAS	Ag(I)	Water

Table 3 shows the application of nanomaterials of magnetic oxides for the separation and enrichment of metal ions.

3.4 Carbon Nanotube (CNT) for the separation and enrichment of metal ions

Since its discovery in 1991, with many excellent physical and chemical properties due to its special structure, CNT has become a hot research topic. Mustafa et al. studied the adsorption performance of Multi-Walled Carbon Nanotubes (MWCNTs) for Cu^{2+} with AAS method, and the results show that the detection limit was 1.46 µg/L. Some researchers studied the adsorption properties of carbon nanotubes for Pb^{2+}, Cr^{6+}, Cd^{2+}, and Cu^{2+}[]. Shao modified MWCNTs with PAAM and PDMA and applied it to the adsorption of Pb^{2+}, and the results show that MWCNT–PAAM has better adsorption ability than MWCNT–PDMA. Some applications of CNTs for the separation and enrichment of metal ions are shown in Table 4 (ETAAS refers to the Electrothermal Atomic Absorption Spectrometry).

3.5 Other nanomaterials for the separation and enrichment of metal ions

Many new types of nanomaterials such as nano Barium Strontium Titanate powder (BST), organic–inorganic hybrid nanomaterials, and composite materials have also been developed. Cui et al. used the modified SiO_2 to adsorb Cr^{3+}, Cu^{2+}, Fe^{2+}, and Pb^{2+}, whose adsorption capacities are 6.2, 18.6, 4.7, and 6.0 µg/L, respectively, when pH = 4, and the corresponding detection limits are 0.19, 1.27, 0.4, and 1.79 µg/L.

Table 2. Application of nanomaterials of surface-modified oxide for the separation and enrichment of metal ions (Liu, 2008; Ezoddin, 2010; Pu, 2005; Shabani, 2009).

Adsorbents	Reagent	Analysis method	Analysis object	Sample
TiO_2	Silica gel	DFAAS	Pd(II)	Water
TiO_2	Two methylamino benzylidene rhodanine	FAAS	Au(III)	Water
Al_2O_3	–	FAAS	Cd(II), Pb(II)	Water
Al_2O_3	Gallic acid	ICP-MS	Fe(II), Fe(III)	Water, hair
Al_2O_3	Lauryl sodium sulfate-1,10 orthophenanthrolene	FAAS	Cu(II), Cd(II)	Water
Al_2O_3	Diphenyl thiocarbazone	FAAS	Au(III), Pt (IV), Pd(II)	Geological sample

Table 3. Application of nanomaterials of magnetic oxides for the separation and enrichment of metal ions (Faraji, 2010).

Adsorbents	Reagent	Analysis method	Analysis object	Sample
Fe_3O_4	Capric acid	ICP-AES	Cd(II), Co(II), Cr(III), Ni (II), Pb(II), Zn(II)	Water
Fe_3O_4	Lauryl sodium sulfate	FT-ICP-AES	Hg(II)	Water
Fe_3O_4	Polyacrylic acid	ICP-MS	Mn(II), Co(II), Cu(II), Zn(II), Pb(II)	Water
Fe_3O_4	Lauryl sodium sulfate	FAAS	Mo(VI)	Wastewater
Fe_3O_4	Bismuth	ICP-AES	Cr(III), Cu(II), Pb(II)	Environmental sample

Table 4. Application of carbon nanotube for the separation and enrichment of metal ions (Wu, 2009; Afkhami, 2009).

Adsorbents	Reagent	Analysis method	Analysis object	Sample
MWCNT	–	ETAAS	V(V)	Water
MWCNT	–	FAAS	Au(III)	Mineral water
MWCNT	–	FAAS	Cu(II)	Water
MWCNT	–	FAAS	Mn(II)	Water
MWCNT	Al_2O_3	FAAS	Ni(II)	Water
MWCNT	$KMnO_4$	FAAS	Cu(II)	Water
MWCNT	–	FAAS	Cu(II), Co(II), Ni(II), Pb(II)	Water
Single CNT	–	ICP-MS	La(III), Gd(III), Yb(III)	Water
Single CNT	HNO_3	ICP-MS	Cu(II), Co(II), Pb(II)	Water

4 CONCLUSION

Recently, nanomaterials have attracted much attention, because of their good performances, and the application of metal ions has becomes more wide. However, there are still some aspects that need further research and innovation.

1. Current research in nanomaterials is little, as most researchers focused only on materials such as titanium dioxide and aluminum oxide. Multi-development of new materials in should be carried out in the future, only by this way we can prepare nanomaterials more economically, easily, and with higher adsorption efficiency. Doping other metals, metal oxide, two types of metal oxide composite material, and preparation of core–shell type, hollow ball of different forms of nanomaterials will be combined with the carbon nanotube and the metal oxide, or using different organic reagents of nanomaterials for surface modification and grafting to improve the selectivity and stability.
2. At present, most of the nanomaterials are limited to the laboratory synthesis, the synthesis conditions are strict, and cannot realize industrialization. Further studies on the preparation of nano materials need to be conducted, to achieve low-cost and high-yield synthesis technology.
3. We can detect the enrichment metal ions using a variety of technologies, so that the detection method may transit from the spectrophotometric method to a more advanced non-colorimetric instrument detection.
4. The samples are too single, and we can improve the universality of the samples such as food, geology, sediment, hair, and blood.

REFERENCES

Afkhami A, Norooz-Asl R. Removal, preconcentration and determination of Mo(VI) from water and wastewater samples using maghemite nanoparticles[J]. Colloids & Surfaces A Physicochemical & Engineering Aspects, 2009, 346(1–3):52–57.

Chunxiang Li, Yongchao Qin, Pei Liang, et al. Nano TiO$_2$ separate and enrich with ICP-AES method for the W/Mo in the geological sample[J]. Journal of Analytical Science, 2002, 18(3):186–189.

Dorota Kołodyńska, Marzena Gęca, Ievgen V. Pylypchuk, et al. Development of New Effective Sorbents Based on Nanomagnetite[J]. Nanoscale Research Letters, 2016, 11(1):1–10.

Ezoddin M, Shemirani F, Abdi K, et al. Application of modified nano-alumina as a solid phase extraction sorbent for the preconcentration of Cd and Pb in water and herbal samples prior to flame atomic absorption spectrometry determination[J]. Journal of Hazardous Materials, 2010, 178(1–3):900–5.

Faraji M, Yamini Y, Rezaee M. Extraction of trace amounts of mercury with sodium dodecyl sulphate-coated magnetite nanoparticles and its determination by flow injection inductively coupled plasma-optical emission spectrometry[J]. Talanta, 2010, 81(3):831–6.

Fuentes G, Viñals J, Herreros O. Hydrothermal purification and enrichment of Chilean copper concentrates. Part 2: The behavior of the bulk concentrates[J]. Hydrometallurgy, 2009, 95(1–2):113–120.

Hang, Yongchao Qin, Zucheng Jiang, et al. Nano TiO$_2$ separate and enrich, slurry sampling fluorination assisted ETV-ICP-AES direct determination of trace rare earth elements [J]. Journal of chemistry in Colleges and Universities, 2003, 24(11):1980–1983.

Hong Z, Chang X, Ning L, et al. Sulfanilamide-modified nanometer-sized TiO2 microcolumn for the enrichment of trace Cr(III) and Pb(II).[J]. Annali Di Chimica, 2005, 95(7–8):601–6.

Kardar Z S, Beyki M H, Shemirani F. Takovite-aluminosilicate@MnFe2O4 nanocomposite, a novel magnetic adsorbent for efficient preconcentration of lead ions in food samples.[J]. Food Chemistry, 2016, 209:241–247.

Li A, Lin R, Lin C, et al. An environment-friendly and multi-functional absorbent from chitosan for organic pollutants and heavy metal ion.[J]. Carbohydrate Polymers, 2016, 148.

Liu R, Liang P. Determination of trace lead in water samples by graphite furnace atomic absorption spectrometry after preconcentration with nanometer titanium dioxide immobilized on silica gel. J Hazard Mater 152:166[J]. Journal of Hazardous Materials, 2008, 152(1):166–71.

Ozdes, D., Gundogdu, A., Kemer, B., Duran, C., Senturk, H.B., Soylak, M. Removal of Pb(II) ions from aqueous solution by a waste mud from copper mine industry: equilibrium, kinetic and thermodynamic study, J. Hazard. Mater. 166 (2009) 1480–1487.

Pei Liang, Chunxiang Li. Nano TiO$_2$ separate and enrich with ICP-AES measurement method for the Cr(VI)/Cr(III) [J]. Journal of Analytical Science, 2000, 16(4):300–303.

Pu X, Hu B, Jiang Z, et al. Speciation of dissolved iron(II) and iron(III) in environmental water samples by gallic acid-modified nanometer-sized alumina micro-column separation and ICP-MS determination. [J]. Analyst, 2005, 130(8):1175–81.

Raghunandhan R, Chen L H, Long H Y, et al. Chitosan/PAA based fiber-optic interferometric sensor for heavy metal ions detection [J]. Sensors & Actuators B Chemical, 2016, 233:31–38.

Raval N P, Shah P U, Shah N K. Adsorptive removal of nickel(II) ions from aqueous environment: A review.[J]. Journal of Environmental Management, 2016, 179:1–20.

Rouholah Zare-Dorabei, Somayeh Moazen Ferdowsi, Ahmad Barzin, et al. Highly efficient simultaneous ultrasonic-assisted adsorption of Pb(II), Cd(II), Ni(II) and Cu (II) ions from aqueous solutions by graphene oxide modified with 2,2-dipyridylamine: Central composite design optimization[J]. Ultrasonics Sonochemistry, 2016:265–276.

Santos D P, Bergamini M F, Fogg A G, et al. Application of a Glassy Carbon Electrode Modified with Poly(Glutamic Acid) in Caffeic Acid Determination [J]. Microchimica Acta, 2005, 151(1):127–134.

Shabani A M H, Dadfarnia S, Dehghani Z. On-line solid phase extraction system using 1,10-phenanthroline immobilized on surfactant coated alumina for the flame atomic absorption spectrometric determination of copper and cadmium[J]. Talanta, 2009, 79(4):1066–70.

Suleiman J S, Hu B, Pu X, et al. Nanometer-sized zirconium dioxide microcolumn separation/preconcentration of trace metals and their determination by ICP-OES in environmental and biological samples[J]. Microchimica Acta, 2007, 159(3):379–385.

Vassileva E, Becker A, Broekaert J A C. Determination of arsenic and selenium species in groundwater and soil extracts by ion chromatography coupled to inductively coupled plasma mass spectrometry[J]. Analytica Chimica Acta, 2001, 441(1):135–146.

Vassileva E, Hadjiivanov K. Determination of trace elements in AR grade alkali salts after preconcentration by column solid-phase extraction on TiO2 (anatase) [J]. Fresenius Journal of Analytical Chemistry, 1997, 357(7):881–885.

Wanna Y, Chindaduang A, Tumcharern G, et al. Efficiency of SPIONs functionalized with polyethylene glycol bis(amine) for heavy metal removal[J]. Journal of Magnetism & Magnetic Materials, 2016, 414:32–37.

Wu P, Chen H, Cheng G L, et al. Exploring surface chemistry of nano-TiO2 for automated speciation analysis of Cr(III) and Cr(VI) in drinking water using flow injection and ET-AAS detection[J]. J.anal. at.spectrom, 2009, 24(8):1098–1104.

Xu J, Wang Y, Xian Y, et al. Preparation of multiwall carbon nanotubes film modified electrode and its application to simultaneous determination of oxidizable amino acids in ion chromatography[J]. Talanta, 2003, 60(6):1123–30.

Yakout A A, El-Sokkary R H, Shreadah M A, et al. Removal of Cd(II) and Pb(II) from wastewater by using triethylenetetramine functionalized grafted cellulose acetate-manganese dioxide composite.[J]. Carbohydrate Polymers, 2016, 148.

Advances in Energy, Environment and Materials Science – Wang & Zhou (Eds)
© 2017 Taylor & Francis Group, London, ISBN 978-1-138-03600-0

A study on the adsorption characteristics of straw semi-char

R.R. Xiao & W. Yang
College of Chemistry, Chemical Engineering and Materials Science, Zaozhuang University,
Zaozhuang Shandong, China

G.S. Yu
Key Laboratory of Coal Gasification of Ministry of Education, East China University of Science
and Technology, Shanghai, China

ABSTRACT: In this paper, the adsorption properties of pyrolysis semi-char possessing a porous structure are studied. The BET specific surface area, pore volume, and adsorbability of semi-char increased with the pyrolysis temperature, and the pore diameter reached the maximum at about 500°C. It was shown that biomass semi-char could adsorb tar and the optimum adsorption conditions during pyrolysis were: proportion of semi-char at 500°C, 5%; carrier gas flow rate, approximately 1.0 L/min; and stirring speed, 1500 rpm.

1 INTRODUCTION

Bioenergy is an important renewable energy. The use of biomass gasification technology can achieve zero CO_2 emissions, saving of conventional energy, and meet the requirements of country sustainable development (Dong et al. 2007, Yuan et al. 2005, Garcia et al. 1999). Pyrolysis was treated as the pretreatment technology during biomass entrained-flow gasification process innovatively. Solid biomass, whose energy density is low, can be converted to clean syngas by this technology. The gas can be used to generate electricity or as civil energy, which is a significant conversion way (Li et al. 2006). However, a lot of tar result during the pretreatment process. The composition of tar is very complicated, which contains 10000 types of organic matter (Jia et al. 2000). The generation of tar not only reduced the syngas production, but also could jam and corrode the equipment (Sun & Jiang 2006). Therefore, treatment of tar during pyrolysis is crucial.

Pyrolysis semi-char possesses favorable hydrophilicity and porous structure. In this study, it was used to adsorb tar innovatively. After adsorbing tar, semi-char can be used as gasification material

directly (Li et al. 2010). The adsorbability of semi-char and treatment of tar are studied in this paper.

2 EXPERIMENT

2.1 Material

Straw was selected as the experiment material. It was first crushed and screened to 40–80 mesh, then it was dried at 105°C for 3 h. Proximate and ultimate analysis results of biomass are shown in Table 1.

2.2 Pyrolysis experiment

Before the experiment, nitrogen went through the reactor for about 5 min first in order to discharge air in the reactor. Biomass was put into a tube furnace, which was heated by electric stove. Pyrolysis temperature was controlled and shown by the temperature controller. After pyrolysis, semi-char remained in the reactor until it was cooled down. Four washing bottles were used to treat tar. A certain proportion of deionized water and semi-char were loaded in the

Table 1. Proximate and ultimate analyses of biomass.

	Proximate analysis [%]			Ultimate analysis [%][a]					
	V_d	FC_d	A_d	C	H	N	O^b	S	HHV [MJ·kg^{-1}]
Straw	71.70	18.58	9.72	43.68	5.70	0.97	39.72	0.21	17.10

Note: (a) weight percentage on dry basis
(b) by difference

first bottle. The mixture was stirred with magnetic stirrer to accelerate the mass transfer and heat transfer processes. The experiment conditions are: carrier gas, N_2; final pyrolysis temperature, 400–700°C; and heating rate, 5°C/min.

2.3 Representation methods and technology

The amounts of C, H, O, N, and S in biomass were analyzed by Vario MACRO Elemental Analyzer of Elementar, Germany. Proximate analysis was conducted by the infrared and fast coal quality analysis instrument of 5E-MAC of Changsha Kaiyuan Instruments Co. LTD. The pore surface area (SBET), pore volumes, and average pore diameter measurements were carried out using a modified Micromeritics ASAP-2020 automated surface area analyzer.

3 ADSORPTION PROPERTIES OF SEMI-CHAR

Biomass semi-char was mainly from the cracking of cellulose and lignin. Unmodified hydrolysis lignin itself can be used as adsorbent (Li & Luo 2007). Adsorption properties of semi-char are studied in this paper.

3.1 BET analysis of semi-char

BET analysis results of semi-char are shown in Table 2. The BET specific surface area, pore volume, and pore size of semi-char were large, which provided large space for adsorbate condensing. The BET specific surface area and pore volume of semi-char increased with the pyrolysis temperature, and the increase amplitude was maximum at 500–600°C. Such porous structure suggested that semi-char have appreciable adsorptive capacities. Pore diameter reached the maximum at about 500°C. The increasing pyrolysis temperature resulted in semi-char decomposition and releasing volatile matters. The porous structure of semi-char was more obvious, which magnified its BET specific surface area, pore volume, and pore size. The shrinkage of the char

Table 2. Specific surface results of semi-char at different pyrolysis temperatures.

Temperature [°C]	BET Surface area [m²/g]	Volume [× 10⁻³ cm³/g]	Average pore size [nm]
400	5.22	1.12	54.56
500	16.60	3.91	82.30
600	53.59	1365.84	59.68
700	70.12	2051.21	41.85

and realignment of its structure led to the observed reduction of the average pore size at higher temperature. The lower average pore diameter of the semi-char at 700°C was consistent with its much larger surface area and pore volume.

3.2 Effect of temperature and chemical pretreatment on adsorption effect of semi-char

Straw pyrolyzed at 400, 500, 600, and 700°C to obtain semi-char, were denoted by A, B, C, and D, respectively. The semi-char at 500°C was soaked with phosphoric acid solution (50%) and saturated $CuSO_4$ solution for 72 h separately, and the samples were denoted by E and F, respectively. The adsorption test of semi-char use gas adsorption method. The computational formula of adsorption rate is:

$$K = \frac{m - m_0}{m_0} \times 100\% \qquad (1)$$

where K is adsorption rate; m_0 is semi-char quality before test; and m is semi-char quality after test.

The adsorption rate of semi-char on methanal and ammonia is shown in Table 3. The adsorption rate of methanal increased and the adsorption rate of ammonia decreased with the increase in pyrolysis temperature. This phenomenon showed that pyrolysis temperature affects the adsorption capacity of semi-char. The increasing pyrolysis temperature resulted in the enhancement of cracking degree of biomass and decreasing of oxygen-containing functional groups. The acid-type center was destroyed and the base-type center increased accordingly. Thus, the adsorption of methanal increased.

As is shown in Table 3, the adsorption rates of samples E and F on methanal were lower than sample B. However, the adsorption rates on ammonia were similar. It further showed that the adsorption of semi-char on methanal was relevant to the base-type center of semi-char. The base-type center of samples E and F reduced after chemical treatment and the acid-type center did not change. This illustrated that the surface of high-temperature semi-char was basic, and hence the adsorption capacity on acidic material was commendable.

3.3 Adsorption efficiency of semi-char on tar

On the basis of the above research, biomass semi-char was used to adsorb tar and the adsorption rate increased by mechanical agitation. The new method enhanced the adsorption rate of semi-char on tar and reduced the tar content of gas. The mixture of semi-char after adsorbing tar can be used to prepare coal–water slurry, which was the feed of biomass entrained flow gasification (Zhang 2008).

Table 3. Adsorption rate of different semi-char samples.

Samples	Adsorption rate of methanal [%]	Adsorption rate of ammonia [%]	Specific surface area [m²·g⁻¹]
A	5.22	4.00	5.22
B	5.94	3.59	16.60
C	6.84	3.02	53.59
D	8.66	2.19	70.12
E	3.87	3.62	2.12
F	3.56	3.38	3.17

The adsorption efficiency of tar is:

$$\eta = \frac{M_1}{M_1 + M_2} \times 100\% \qquad (2)$$

where η is adsorption efficiency of tar; M_1 is the adsorption weight of tar; and M_2 is the total weight of unadsorbed tar.

3.4 pH values of tar

The pH values of tar at different pyrolysis temperature are presented in Fig. 1. Tar is a type of acid liquid, whose pH value is low, and hence it contains a certain amount of organic acid such as formic acid and acetic acid.

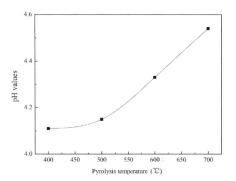

Figure 1. pH values of tar.

3.5 Influence of semi-char content

Semi-char samples (500°C) were added into a gas washing bottle (with deionized water) to adsorb condensed tar using mechanical agitator (1500 rpm). The carrier gas flow rate was 1 L/min. The adsorption efficiency as a function of stirring speed is presented in Fig. 2. The η values amplified first with the increase of semi-char content and reached maximum when the mass fraction of semi-char is about 5% and then it decreased. Thus, the semi-char proportion was 5% in subsequent experiments in order to treat tar ideally.

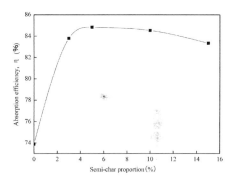

Figure 2. Effect of semi-char proportion on tar adsorption efficiency.

3.6 Effect of carrier gas flow rate

The effect of carrier gas flow rate on η values was studied at a stirring speed of 1500 rpm and semi-char temperature of 500°C. The results are shown in Fig. 3. The η values amplified first with the increase of carrier gas flow rate and reached the maximum value when carrier gas flow rate was about 1.0 L/min and then it decreased. At very high carrier gas flow rate, the flow velocity of the volatile matter taken out from the furnace was too fast in adsorption device and the residence time was too short. Therefore, the heat and mass transfer effect reduced. The condensable components of the volatile matter could not contact with

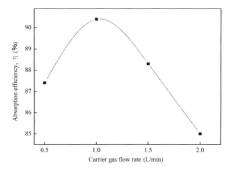

Figure 3. Adsorption efficiency at different gas flow rates.

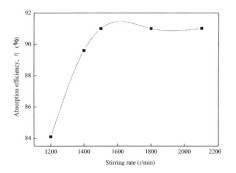

Figure 4. Adsorption efficiency at different stirring speeds.

adsorbent fully and effectively, which reduced the adsorption effect of semi-char on tar.

3.7 *Effect of stirring speed*

The effect of stirring speed on η values was studied at the carrier gas flow rate of 1.0 L/min and semi-char temperature of 500°C (5%). The results are shown in Fig. 4. The η values increased with the increase of stirring speed and reached a constant value eventually. The higher the stirring speed, more severe was the disturbance in the stirred tank and larger was tar and semi-char contact area. The mass transfer and heat transfer processes were strengthened. When the stirring speed reached a certain value, mass transfer and heat transfer processes achieved the best condition, and then kept stable relatively. The η values were unchanged at the stirring speed of 1500 rpm.

4 CONCLUSIONS

The porous structure of pyrolysis semi-char possesses favorable adsorbability. The BET specific surface area, pore volume, and adsorbability of semi-char increased with the pyrolysis temperature. The optimum conditions of adsorption of semi-char on tar are as follows: the semi-char proportion at 500°C, 5%; carrier gas flow rate, approximately 1.0 L/min; and stirring speed, 1500 rpm. The mixture of semi-char after adsorbing tar can produce coal–water slurry.

ACKNOWLEDGMENT

This work was subsidized by Zaozhuang independent innovation and achievements transformation projects (201502), which is gratefully acknowledged.

REFERENCES

Dong, Y. P., B. Deng & Y. Z. Jing (2007). A review of the research and development of biomass gasification technology in China. *Journal of Shandong University (Engineering Science) 37,*1–7.

Garcia, L., M. L. Salvador & J. Arauzo (1999). Catalytic steam gasification of pine sawdust, effect of catalyst weight/biomass flow rate and steam/biomass ratios on gas production and composition. *Energy & Fuels 13,* 851–859.

Jia Y. B., S. Y. Zhang & Z. H. Cheng (2000). Study on Tar Cracking in Pyrolysis or Gasification Process. *Coal Conversion 23,* 1–6.

Li J. F., D. H. Wang & Q. B. Xiang (2006). Application of biological material steal melt residue. *Environmental Sanitation Engineerin 14,* 1–4.

Li P., X. L. Chen & R. R. Xiao (2010). Study on the adsorption capacity of rice straw pyrolysis char. *Acta Energlae Solaris Sinica 31,* 947–950.

Li X. F. & X. G. Luo (2007). Progress of lignin in biochemical engineering. *Chemical Industry and Engineering Progress 26:*1722–1726.

Sun Y. J. & J. C. Jiang (2006). A Review of Measures for Tar Elimination in Biomass Gasification Processes. *Biomass Chemical Engineering 40,* 31–35.

Yuan Z. H., C. Z. Wu & L. Ma (2005). Principle and the use of biomass technology. *Beiijing: Chemical Industry Press,* 39–60.

Zhang W. (2008). The engineering research of biomass entrained flow gasification. *East China University of Science and Technology,* 10–50.

Decomposition of azo dye brilliant red B by radio frequency discharge plasma in aqueous solution

Jinxiu Wang & Songlin Zhang
College of Geography and Environment Science, Northwest Normal University, Lanzhou, P.R. China

Lei Wang
College of Environmental Science and Engineering, Xiamen University of Technology, Xiamen, P.R. China

ABSTRACT: Azo dye is widely used in dyeing and printing industry. Degradation of azo dye Brilliant Red B(ARB) by Radio Frequency Discharge Plasma (RFDP) in aqueous solution under different pH, •OH scavengers (methanol, isopropanol) and catalysts (Fe^{2+}, Fe^{3+}) was studied in this paper. Results showed that numerous active species such as •OH, H• and •O were produced in RFDP. Removal of ARB reached 77.89% after 1 hour discharge, and it would be better under acidic conditions. Hydroxyl radicals scavengers had an inhibition effect on ARB removal, while Fe^{2+} and Fe^{3+} had catalysis effect on it, with the sequence of $Fe^{2+} > Fe^{3+}$. The UV visible absorption spectroscopy illustrated that the main reason for the decolorization of ARB was the break of azo bond which was damaged by •OH radicals formed in the discharge process. The main degradation intermediates formed during the process were formic acid, acetic acid, oxalic acid and malonic acid.

1 INTRODUCTION

Color is one of the most important indicators of water pollution (Wang 2009). Dyes-containing wastewater is hardly treated well due to complicated structures and stable chemical properties of dyes (Li 1997). The main wastewater treatment methods for Azo dye suffer respective shortcomings. Adsorption method just transfers the pollutants from liquid to the surface of solid. Being toxic to microbes, biological method is inefficient (Lu et al. 2013). Flocculation method suffers low decoloring ratio and time-consuming. Therefore, development of new approaches for dye wastewater treatment has practical significance.

RFDP in aqueous solution has aroused extensive attentions from researchers because of its simple equipment, efficiency, no secondary pollution (Bruggeman & Leys 2009, Sun et al. 1999). Mukasa et al. (2009) used emission spectrum to study temperature of plasma electron and distribution of active group emission spectrum in underwater discharge plasma driven by 27.12 MHz frequency. Tatsuya et al. (2012) studied the characteristics of RFPD in a high-conductivity NaCl Solution. Discoloration of dyes by RFDP has been reported, Maehara et al. (1996) used radio frequency power supply with 13.56 MHz frequency to drive copper electrode in aqueous solution to produce plasma, which induced the degradation of methylene blue dye. As azo dyes constitute the largest class of dyes, ARB was chosen as a model pollutant in this study to provide information and reference data for removal of organic dye by RFDP in aqueous solution.

2 EXPERIMENTAL

2.1 Instruments and reagents

The main instruments used in this study are as follows: RSG500 radio frequency power supply (Changzhou Rishige electronics technology Co., Ltd), GDS-3154 Digital oscilloscope (Good Will Instrument Co., Ltd), Rigol RP1100D High-voltage probe (Rigol Technologies, Inc.), PINTEK PT-740 Current probe, S21–1 Constant temperature magnetic stirrer (Shanghai Sile Instrument Co., Ltd), Eutech Con510 Conductivity meter, Eutech pH meter (Singapore utility Instrument co., Ltd), Shimadzu UV-2550 ultraviolet and visible spectrophotometer, Shimadzu TOC-VCPH Total organic carbon analyzer, Dionex ICS-1100 ion chromatography.

2.2 Experiments and analysis methods

Experimental apparatus assembly consisted of radio frequency power supply with 13.56 MHz and 500 W maximum output power, matcher, plasma

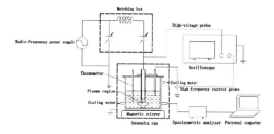

Figure 1. Experimental device.

generator, the cooling water system, thermometer, screening scan, digital oscilloscope, high-pressure bifurcation probe and current probe. The apparatus are illustrated in Figure 1. The electricity of plasma generator was tungsten wire with 1.0 mm diameter connected with coaxial cable. Discharge voltage and discharge current were measured by high pressure sensor and current sensor bifurcation respectively. The corresponding waveform and root mean square value were displayed and saved on the digital oscilloscope.

The intermediate products during degradation process were analyzed qualitatively by Dionex ICS-1100 ion chromatograph analyzer. All reagents were high-purity analytically pure and supplied by Aladdin. The ultrapure water was obtained from Millipore Elix5 + Synerg UV. H_2O_2 formed in the solution was measured by colorimetric method. Ferrous ion concentration was measured by adjacent dinitrogen spectrophotometry method.

3 RESULTS AND DISCUSSIONS

3.1 Effect of solution electrical conductivity on the breakdown voltage

Figure 2 shows the breakdown root mean square value of alternating-current (ac) voltage (RMS V) under different solution conductivity. Upon varying conductivity from 200 μS/cm to 1000 μS/cm, V_{break} decreased rapidly from 1450 V to 923 V, but it gradually decreases from 923 V to 736 V with conductivity varying from 1000 μS/cm to 2400 μS/cm, namely the lower the conductivity, the higher the breakdown voltage needed for the discharge. It indicated that solution conductivity played a vital role in the underwater discharge. The V_{break} decreased with increasing conductivity of solution. Since a high current density around the electrodes, air bubbles were formed around the electrodes and it was in series with the solution. Therefore, under the circumstance of constant conductivity, the effective voltage on the bubbles increased with the rising conductivity of the solution. It facilitated the breakdown of the bubbles (Sano & Wang 2001).

It means that more radio frequency power energy would be dissipated off in the form of thermal energy during discharge treatment at high solution conductivity (Ji & Zou 2012).

3.2 The typical emission spectra detected from RF underwater discharge

The hydroxyl radicals (•OH) and hydrogen atoms (H•) are considered to be the main active species that initiate the reaction in RFDP. But these views are indirect experiment speculation. In order to better study the reaction mechanism, the emission spectrometry was used to study these radicals. Figure 3 shows the typical emission spectra of ARB solution in RF underwater discharge at 110 W.

It can be seen from Figure 3 that the spectra were dominated by four kinds of bands. The emission peaks in ultraviolet region (200 ~ 400 nm), visible region (400 ~ 800 nm) and near infrared region (>800 nm) were observed. The four emission peaks in 262 nm, 283 nm, 289 nm and 309 nm were

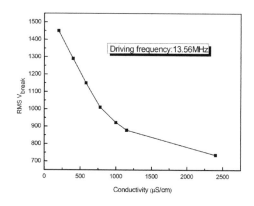

Figure 2. The breakdown voltage with the conductivity changes (input power: 110 W, pH$_0$: 7.0).

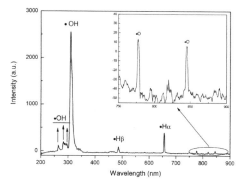

Figure 3. The emission spectra of RF underwater discharge (input power: 110 W, the initial conductivity: 500 μS/cm).

generated by hydroxide radicals which transited to ground state from electronic excited state (Boni et al. 2008). Hydrogen alpha emission peaks were located at 656 nm while hydrogen beta at 486 nm. Oxygen emission peaks were located at 777 nm and 844 nm. The spectra indicated that active radicals such as •OH, H• and •O were generated during the underwater discharge. These species can initiate degradation of organics in the solution (Liu et al. 2013).

3.3 *ARB removal during RF discharge process*

TOC is an important index to evaluate water pollution (Siepak 1999). Figure 4 shows the remaining ARB, TOC during the RFDP treatment. Decomposition of ARB proceeded smoothly when the solution containing ARB was subjected to RFDP.

It can be observed from Figure 4 that the remaining ARB and TOC decreased monotonically with increasing discharge time. After 60 min of discharge, approximately 77.89% of ARB, 69.91% of TOC disappeared. The disappearance of TOC was less than that of ARB, indicating that some organic intermediate products were formed during the discharge treatment.

3.4 *Effects of pH on ARB degradation*

pH is one of the most important factors affecting removal of ARB. Figure 5 shows the curve of removal rate and the discharge time of ARB solution under different initial pH conditions. It can be observed from Figure 5 that the degradation ration of ARB increased significantly with the decrease of pH_0. When initial pH values were 3.5, 5.5, the degradation ratios of ARB were 87.92% and 82.54% respectively. The degradation ratio of ARB with pH_0 9.0 and pH_0 11.0 was fundamentally the same as pH_0 7.5, reaching 77.89%. The degradation ratio of ARB under acid condition was better than that

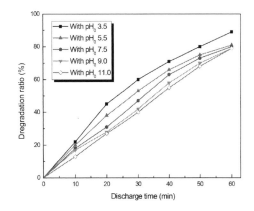

Figure 5.   Effect of the initial pH on ARB degradation (input power: 110 W, the initial conductivity: 200 μS/cm).

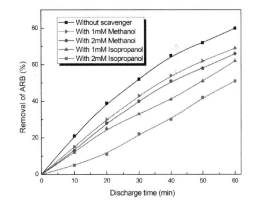

Figure 6.   Effects of organic additives on ARB removal during RFDP treatment (input power: 110 W, pH_0: 7.0, the initial conductivity: 200 μS/cm).

in neutral and alkaline conditions because stronger activity of radicals in acid condition resulted in a higher degradation ratio

Figure 6 shows the curve of pH value during RF discharge treatment. It can be seen from Figure 6 that the pH of ARB solution decreased monotonically, finally reached a stable value. The possible reason was that the formation of organic acid during discharge process made the pH value lower.

3.5 *Effects of the hydroxyl radical scavengers on ARB decolorization*

Two typical •OH scavengers, methanol and isopropanol as an example were discussed in this experiment. The reaction of methanol, isopropyl alcohol and hydroxyl radicals are as follows (Sun et al. 1999):

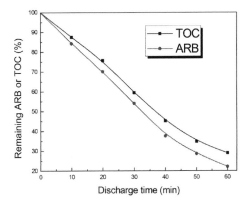

Figure 4.   Variation of TOC during RFDP treatment (pH_0: 7.0, the initial conductivity: 200 μS/cm).

$$\cdot OH + CH_3OH \rightarrow H_2O + \cdot CH_2OH$$

$$k_1 = 9.7 \times 10^8 \ mol^{-1} \ s^{-1} \tag{1}$$

$$\cdot OH + (CH_3)_2CHOH \rightarrow H_2O + (CH_3)_2COH$$

$$k_2 = 1.5 \times 10^9 \ mol^{-1} \ s^{-1} \tag{2}$$

Figure 6 shows the effects of •OH scavenger on ARB decolorization. It can be observed that the degradation ratio of the ARB significantly decreased in the presence of •OH scavengers. Isopropyl alcohol had a larger impact on the ARB decolorization. The inhibitory effect on ARB removal was more efficient when the concentration of •OH scavengers concentration was heavier, which indirectly proved that •OH scavengers played a vital role in the process of ARB degradation. As isopropyl alcohol was more efficient than methanol in consuming •OH scavengers (reaction (1) and (2)), the ARB removal was more efficient in the presence of isopropanol. The experimental results demonstrated that •OH was the primary species for the ARB degradation.

3.6 Effects of iron salts on the ARB degradation

The experimental results showed that Fe^{2+}, Fe^{3+} had catalytic effects on ARB degradation. As shown in Figure 7, the catalytic effect of molysite was enhanced with the increase of the concentration, and the catalytic effect of Fe^{2+} was better than that of Fe^{3+}.

In the presence of 1 mmol/L Fe^{2+} and Fe^{3+}, the degradation ration of ARB was 82.02% and 84.53% after 60 min discharge treatment respectively. The catalytic effect of Fe^{2+} was strong at beginning and decreased then, whereas the catalytic effects of Fe^{3+} gently increased with the rising time of discharge treatment. However, the reductive organic radicals produced in the reaction made Fe^{3+} revert to Fe^{2+}, enhancing the catalytic effect (Buxton & Greenstock 1998).

In order to further study the reaction mechanism, hydrogen peroxide (H_2O_2) generated in the process of reaction were studied. Experimental result demonstrated the concentration of H_2O_2 with ARB solution was lower than that of the solution without the ARB. In other words, H_2O_2 was consumed in the ARB removal. According to the Fenton reaction, It can be obtained from the experimental result that the catalytic effect of Fe^{2+} was more effective. The hydroxyl radicals had comparatively less effect on ARB decolorization because the reaction rate constant k of hydroxyl radicals and methanol was lower.

3.7 Ultraviolet-visible spectra variation

Figure 8 shows the ultraviolet and visible spectra of ARB solution with the condition that the initial concentration is 20 mg/L and pH_0 is 7.0 respectively. It can be observed that ARB solution had

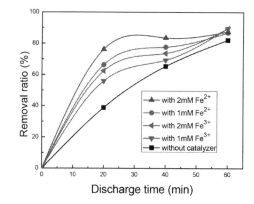

Figure 7. Effects of iron salts on ARB removal during RFDP treatment (input power: 110 W, pH_0: 7.0, the initial conductivity: 200 μS/cm).

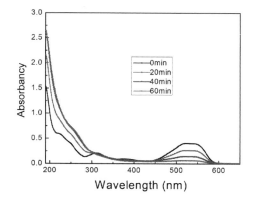

Figure 8. The Ultraviolet-visible spectral variation during RFDP treatment (input power: 110 W, pH_0: 7.0, the initial conductivity: 200 μS/cm).

three obvious characteristic absorption peaks at 514 nm, 562 nm in visible area and 312 nm in ultraviolet region. The absorption peaks at 514 nm and 562 nm were resonance absorption peaks of azo double bond and benzene, naphthalene ring on ARB molecules. The absorption peak at 312 nm was naphthalene ring absorption peak (Gao et al. 2003). With rising of the discharge time, the characteristic peak intensity decreased quickly at 514 nm and 562 nm, gently at 312 nm. It indicated that azo bond was broken firstly, followed naphthalene ring in the process of the ARB degradation. Absorption peaks in visible region almost completely disappeared after 60 min RF plasma treatment which illustrated that the break of big conjugate system of ARB resulted in decolorization (Tanaka et al. 2000).

Intermediate products, such as acetic acid, formic acid, oxalate acid and malonic acid were detected by ion chromatography. Figure 9 shows the intermediate products variation with the rising

296

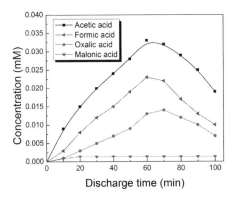

Figure 9. Variation of the intermediate product during RFDP treatment (input power: 110 W, pH_0: 7.0, the initial conductivity: 200 μS/cm).

discharge time. With longer discharge time, concentrations of these intermediate products increased gradually and decreased after 60 min discharge treatment except that malonic acid changed little. This indicated that intermediate products also can be degraded by RFDP treatment.

4 CONCLUSION

RFDP is an effective technique for ARB removal and its discharge breakdown voltage decreases with the increase of conductivity. The ARB decolorization ratio can reach 77.89% after 60 min of RFDP treatment. The ARB removal is relatively easier under acidic conditions (with pH_0 3.5). Radical scavengers have obvious inhibiting effects on ARB degradation. And the inhibitory effect of isopropyl alcohol is stronger than that of methanol. Fe^{2+} and Fe^{3+} all have a catalytic effect on the ARB degradation and the catalytic effect of Fe^{2+} is more efficient than that of Fe^{3+}. The hydroxyl radicals are the primary species for the ARB degradation during RFDP treatment. The optimal conditions for degradation of ARB are pH 3.5 and with iron salts as the catalyzer (Wang & Jiang 2008).

ACKNOWLEDGEMENTS

This work was supported by the National Science Foundation of China (51008262), the State Scholarship Fund of China (201406575032), the Natural Science Foundation of Fujian province, China (2015 J01651).

REFERENCES

Boni L.D., Andrade A.A. & Yamaki S.B. (2008). Two-photon absorption spectrum in diazoaromatic compounds. *J. Chem. Phys. Lett.* 463(4–6), 360–363.

Bruggeman P. & Leys C. (2009). Non-thermal plasmas in and in contact with liquids. *J. Phys. D.* 42(5), 53–61.

Buxton G.V. & Greenstock C.L. (1998). Critical review of rate constants for reactions of hydrated electrons, hydrogen atoms and hydroxyl radicals (•OH/•O⁻) in aqueous solution. *J. Phys. Chem. Ref. Data.* 17(2), 513–886.

Gao J, Liu Y, Yang W. (2003). Oxidative degradation of phenol in aqueous electrolyte induced by plasma from a direct Glow Discharge. *J. Plasma Sources Sci. Technol.* 12, 533–538.

Hichling A. (1971). Electrochemical processes in Glow Discharge at the gas-solution interface. *J. Mod. Aspects Electrochem.* 6, 329–373.

Ji L.L. & Zou S. (2012). Radio frequency underwater discharge operation and its application to Congo Red degradation. *J. Plasma Sources Sci. Technol.* 14(2), 111–117.

Li J. (1997). *Dyeing wastewater treatment.* Beijing, Chemical Industry Press.

Liu Y.J., Sun B. & Wang L. (2013). Emission spectroscopic study on radicals generated in liquid-Phase diaphragm Glow Discharge plasma. *J. Spectrosc. Spect. Anal.* 33(3), 790–793. (In Chinese)

Lu J., Yu Z.S. & Zhang H.X. (2013). Research progress in the microbial degradation of azo dyes. *J. Ind. Water Treat.* 33(1), 15–19.

Maehara T., Miyamoto I. & Kurokawa K. (1996). Degradation of Methylene Blue by RF Plasma in Water. *J. Plasma Chem. Plasma Process.* 28(4), 467–482.

Mukasa S., Nomura S. & Toyota H. (2009). Temperature distributions of radio frequency plasma in water by spectroscopic analysis. *J. Appl. Phys.* 106(11), 113–122.

Sano N. & Wang H. (2001). Nanotechnology: Synthesis of carbon 'onions' in water. *J. Nat.* 414(6863), 506–507.

Sengupta S.K., Singh R. & Srivastava A.K. (1988). A Study on non-faradaic yields of anodic contact Glow Discharge electrolysis using cerous ion as the •OH scavenger: an estimate of the primary yield of •OH radicals. *J. Indian J. Chem.* 37(6), 558–560.

Siepak J. (1999). Total Organic Carbon (TOC) as a sum parameter of water pollution in selected polish rivers (Vistula, Odra, and Warta). *J. Acta Hydrochimica Et Hydrobiologica,* 1999, 27(5), 282–285.

Sun B., Sato M. & Clements J.S. (1999). Use of a pulsed high-voltage discharge for removal of organic compounds in aqueous solution. *J. Phys. D.* 32(15), 1908–1915.

Tanaka K., Padermpole K. & Hisanaga T. (2000). Photocatalytic degradation of commercial Azo Dyes: *J. Water Res.* 34(1), 327–333.

Tatsuya A., Shinobu M. & Naoki H. (2012). Generation of radio frequency plasma in high-conductivity NaCl solution. *J. Jpn. J. Appl. Phys.* 51(10), 1–2.

Wang L. & Jiang X.Z. (2008). Plasma-induced reduction of Chromium (VI) in an aqueous solution. *J. Environ. Sci. Technol.* 42, 8492–8497.

Wang L. (2009). Aqueous organic dye discoloration induced by contact glow discharge electrolysis. *J. Hazard. Mater.* 171, 577–581.

Denitrification characters of an aerobic heterotrophic nitrification-denitrification Bacterium, *Pseudomonas pseudoalcaligenes* YY24

J.H. Sun & G.J. Wang
Key Laboratory of Tropical and Subtropical Fishery Resource Application and Cultivation, Ministry of Agriculture, P.R. China

J.H. Sun, M.M. Sun, J.H. Li, C.X. Chen, S.J. Zhou & X.T. Qiao
Tianjin Key Lab of Aqua-Ecology and Aquaculture, College of Fisheries, Tianjin Agricultural University, Tianjin, P.R. China

ABSTRACT: A bacterial strain YY24 with both aerobic denitrification and heterotrophic nitrification abilities was separated from a biofilter system of marine fish farm. Depending on the Identification with 16S rDNA sequence analysis, the sequence of isolate was 99% similar to P. pseudoalcaligenes. Aerobic denitrification and heterotrophic nitrification performance of YY24 was analyzed by measuring its growth and nitrogen removal efficiency. Under the conditions (C/N 10, 23°C and 120 rpm) with ammonium chloride (150 mg/L) the growth of YY24 entered logarithmic phase after 4 h, and reached stationary phase after 24 h. The removal ratio of Total Nitrogen (TN) and Total Ammonia (TAN) were 95.36% and 96.67% respectively at stationary phase. During the incubation, there was almost no nitrite and nitrate accumulation. The optimal conditions of nitrogen removal were as follows: pH ranged 5.5–8.5, the C/N ratio was above 10, temperature ranged from 23 to 30°C and the most suitable carbon source was sodium citrate.

1 INTRODUCTION

A conventional nitrogen removal system using heterotrophic denitrifiers bacteria and autotrophic nitrifying bacteria to convert nitrogenous compounds to nitrogen has been extensively used for waste water treatment (Zhao et al. 2010b). The processes of nitrogen removal have been reported in many studies, for example, anaerobic ammonium oxidation (Anammox) process (Van de Graaf et al. 1990; Strous et al. 1997), oxygen-limited autotrophic nitrification-denitrification (Oland) process (Kuai et al. 1998) and completely autotrophic nitrogen removal over nitrite (Canon) process (Sliekers et al. 2002). Disadvantages of such system were that the nitrogen removal process is more likely to consume and requires huge expanses of space to build anaerobic and aerobic tanks, separately (Khin et al. 2004). Bacterium with the capacity of aerobic denitrification and heterotrophic nitrification could convert ammonium into denitrification products through a single strain (Nishio et al. 1994). More recent studies have found that some bacteria, for example, *Alcaligenes faecalis, Thiosphaera pantotropha, Pseudomonas putida, Bacillus* sp. *Pseudomonas stutzeri, Providencia* rettgeri YL, *Rhodococcus pyridinivorans* CPZ24, *Halomonas* and *Diaphorobacter* sp. which both had the capacity to heterotrophic nitrification and had the notable ability of denitrification under aerobic conditions (Robertson et al. 1988; Patureau et al. 1997; Diep et al.1998; Su et al. 2001; Kim et al. 2005; Joo et al. 2005a; Khardenavis et al. 2007; Zhao et al. 2010a; Zhao et al. 2010b; Sun et al. 2012).

In a closed aquaculture system, inorganic nitrogen, including ammonia, nitrite and nitrate, would increased rapidly due to accumulation of uneaten feed, feces and metabolic wastes excreted by fish. Biofilter has been widely used for nitrogen removal in aquaculture system since bacteria and algae in biofilter contributed substantially to nutrient cycling, especially to the nitrogen cycle (Lei et al. 2003).

In the study, a bacterial strain with aerobic denitrification and heterotrophic nitrification abilities was separated from biofilter system in marine fish farm. This report aimed to identify it, and determine the nitrogen removal of YY24 at different environmental factors (ammonium concentration, shaking speed, C/N ratio and temperature). The purpose of it was to make sure its optimum conditions.

2 MATERIALS AND METHODS

2.1 *Medium for growth and storage*

The synthetic medium used was got ready for dissolving the following in 1 liters of distilled water: NH_4Cl 0.5; sodium citrate 5.66; trace element solution 50 mL; pH7.0. The trace element solution was composed of the following components (g/L): K_2HPO_4 5.0; $MgSO_4 \cdot 7H_2O$ 2.5; $FeSO_2 \cdot 7H_2O$ 0.05; $MnSO_4 \cdot 4H_2O$ 0.05; 1.5% salinity brine.

2.2 Bacterial isolation

From biofilter in water treatment system of Lida Marine Resources Development Co. Ltd. Tianjin, China, strain YY24 was isolated using the traditional bacterial separation method as follows: 5 g of biofilter sample was inoculated in medium, and then incubated at 120 rpm for 3 d at 23°C. The obtained inocula was plated in the Bromine methyl phenol Blue (BTB) medium (Takaya et al. 2003) and cultured at 23°C for 24 h. Bacteria turning BTB medium to blue was purified and inoculated in medium. An effective aerobic heterotrophic nitrifying-denitrifying bacterial isolate was screened depending on the 24 h removal rate of TAN and TN. Pure isolate was grown in a basal medium and kept in 25% glycerol, which was saved at −80°C for long time storage.

2.3 Bacterial identification

The molecular taxonomic status of the isolate was identified based on its partial 5'16S rDNA sequence comparison. The 16S rDNA was isolated, amplified and sequenced (Yingjun Co. Ltd. Beijing, China). The sequences were aligned on the Clustal_X 1.83 program (Thompson et al. 1997). A phylogenetic tree was built using MEGA software version 5 by the neighbor-joining method (Saitou et al. 1987) and further evaluated by boot-strap sampling (Felsenstein, 1985). The scale bar represents 0.02 fixed mutations per site, and boot-strap values were derived from 1,000 analyses. The phylogenetic tree was displayed using the Treeview (Arizona State University, MEGA 5, USA) program.

2.4 The aerobic denitrification and heterotrophic nitrification performance of strain YY24

Bacterium YY24 suspensions (5 ml) were cultivated in erlenmeyer flasks (250 ml) containing 100 ml basal medium with a shaking speed of 120 rpm at 23°C for 48 h. The cultures were collected for the subsequent analysis every two hours during the first six hours of incubation and every three hours after a relatively stable nitrification level with two replicate.

Growth of YY24 was detected by measuring the optical density (OD_{600} nm) using spectrophotometer (UVmini-1240, Tangxia electronic instruments and equipment Co. Ltd. Dong guan, China). pH was measured using pH meter (Mettler Toledo, China). Culture samples were centrifuged 10 min at 8,000 rpm, and supernatants were used to determine concentrations of ammonia-nitrogen (NH_4^+-N), Total Ammonia (TAN),

nitrate (NO_3^--N) Total Nitrogen (TN) and nitrite (NO_2^--N) (GB/T 17378.4-2007, China).

2.5 The independent effects of environmental factors on nitrogen removal of strain YY24

The effects of C/N ratio on nitrogen removal was studied by making the ratio of C/N to 6, 8, 10, 12, 14, 16, 18 and 20 and making ammonium chloride (150 mg/L) as the nitrogen source. To research the effects of carbon sources on nitrogen removal capability of strain YY24, sucrose, sodium citrate, glucose, acetate, lactose, methanol, sodium potassium tartrate tetrahydrate and soluble starch were used as carbon sources in medium. The effects of temperature were studied by making temperatures to 7, 15, 23, 30, 37 and 45°C. The effects of pH on the nitrogen removal efficiency were studied by making pH to 4.5, 5.5, 6.5, 7.5, 8.5, 9.5 and 10.5. In each experiments, bacterium YY24 was incubated in the same conditions (23°C, C/N 10, pH 7, sodium citrate) except the adjusted parameters. Cultures were sampled with two replicate at 24 h intervals to determine OD_{600}, pH, and contents of NO_3^--N, TAN, TN, NO_2^--N, and NH_4^+-N.

2.6 Statistical analysis

All data (means ± standard) are subjected to analysis using one-way ANOVA by SPSS 18.0 (SPSS Inc. Chicago, IL, USA). Differences among the means were tested by Duncan's multiple range tests, and the chosen level of significance was $P < 0.05$.

3 RESULTS

3.1 Identification of strain YY24

As shown in Fig. 1, strain YY24 and *P. pseudoalcaligenes* (NR 037000.1) were underneath the same set with 99% identity. YY24 was recognized as a strain belong to *P. pseudoalcaligenes* according to 16 s rDNA sequence analysis. This tree also showed a marked evolutionary divergence of strain YY24 from type strains being part of other groups such as *Alcaligenes, Pseudomonas, Bacillus, Rhodococcus, Arthrobacter* and *Paracoccus*.

3.2 The aerobic denitrification and heterotrophic nitrification performance of strain YY24

As shown in Fig. 2a, after a 4 h lag phase, strain YY24 entered exponential growth phase, and maximum growth rate appeared at 24 h. pH increased significantly with the growth of bacteria. NH_4^+-N concentrations decreased first and then increased with the prolongation of culture time, and it reached the minimum of 5.04 mg·L^{-1} at 24 h. The maximum removal ratio of TAN was 6.10 mg·L^{-1}·h^{-1} and the degradation ratio was 96.67% (Fig. 2b). Similar change was observed in TN, 95.36% of TN was eliminated at 24 h (Fig. 2c). Low concentrations of nitrate and nitrite were observed, with the decrease of TAN concentrations (Fig. 2d). TN concentrations in the media were also tested, and 95.36% of TN was eliminated from the media at 24 h, TN concentration decrease rapidly in the logarithmic phase (Fig. 2c).

3.3 C/N ratio, Temperature, Carbon source, pH affecting nitrogen removal ability of strain YY24

As shown in Table 1, the growth of strain YY24 improved with the increasing C/N ratio.

Ammonium removal percentage was C/N ratio-dependent and could reach up to 98% at C/N rate of 10. The TN removal efficiency reached the maximum at C/N ratio of 10. At the end of the experiment, pH in the medium was higher than the initial pH. As shown in Table 2, TN and ammonium removal rates increased first and then decreased with the increasing temperatures, and the maximum TN and ammonium removal rate appeared at 23°C. At the end of the experiment, pH in the medium at 7°C was lower than the initial pH while higher pH was observed at other temperatures. The maximal TN and ammonium removal rates were observed when sodium citrate was used as carbon source, and pH increased too (Table 3). At pH 7.5, there was the maximal TN

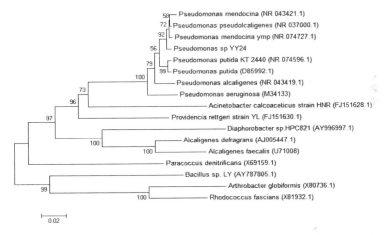

Figure 1. Phylogenetic tree based on partial 16S rRNA sequences showing phylogenetic relationships between strain YY24, other members of *Pseudomonas,* and heterotrophic nitrogen removal strains from other groups. (The tree was constructed using neighbor-joining method. Names of the different groups along with the accession numbers are shown in the parentheses. Bootstrap values (1000 replications) are indicated at the interior branches.)

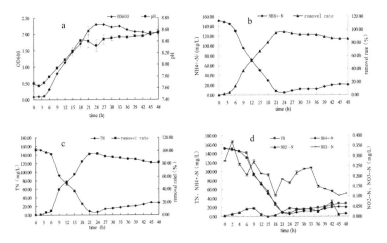

Figure 2. Changes in the growth, pH(a), nitrogen compounds and TN (b, c, and d)in medium inoculated with strain YY24 Values are represented as mean ± SE (n = 4).

Table 1. Effects of C/N ratio on the growth and nitrogen removal ability of *P. pseudoalcaligenes* strain YY24.

C/N	OD600	pH	TN removel ratio (%)	TAN removel ratio (%)	NO$_2^-$-N (mg/L)	NO$_3^-$-N (mg/L)
6	1.18 ± 0.05[a]	8.58 ± 0.01[b]	70.18 ± 1.24[d]	72.13 ± 0.04[a]	0.01 ± 0.00[ab]	0.12 ± 0.00[a]
8	2.02 ± 0.01[b]	8.40 ± 0.03[a]	91.45 ± 0.62[c]	93.30 ± 0.11[b]	0.01 ± 0.00[a]	0.17 ± 0.01[a]
10	2.36 ± 0.00[c]	8.30 ± 0.07[a]	98.08 ± 1.12[a]	98.74 ± 0.07[c]	0.01 ± 0.00[ab]	0.13 ± 0.01[a]
12	2.40 ± 0.04[c]	8.37 ± 0.02[a]	95.93 ± 0.37[b]	98.91 ± 0.04[cd]	0.01 ± 0.00[ab]	0.25 ± 0.00[b]
14	2.50 ± 0.01[e]	8.41 ± 0.04[a]	95.51 ± 0.01[b]	99.20 ± 0.04[e]	0.01 ± 0.00[ab]	0.14 ± 0.00[a]
16	2.38 ± 0.00[c]	8.38 ± 0.00[a]	96.51 ± 0.00[ab]	99.19 ± 0.02[e]	0.01 ± 0.00[b]	0.13 ± 0.00[a]
18	2.44 ± 0.05[ce]	8.39 ± 0.00[a]	97.91 ± 0.62[a]	99.11 ± 0.06[e]	0.01 ± 0.00[ab]	0.17 ± 0.01[a]
20	2.36 ± 0.01[c]	8.40 ± 0.01[a]	94.77 ± 1.37[b]	99.06 ± 0.12[de]	0.01 ± 0.00[ab]	0.25 ± 0.00[b]

Values are represented as mean ± SE (n = 4); Values in a row with different superscripts differ significantly (*P* < 0.05).

Table 2. Effects of temperature on the growth and nitrogen removal by *P. pseudoalcaligenes* strain YY24.

Tem	OD600	pH	TN removel ratio (%)	TAN removel ratio (%)	NO$_2^-$-N (mg/L)	NO$_3^-$-N (mg/L)
7	0.04 ± 0.00[a]	7.28 ± 0.01[a]	0.61 ± 0.44[a]	0.69 ± 0.17[a]	0.13 ± 0.11[e]	0.22 ± 0.01[d]
15	0.08 ± 0.01[a]	8.06 ± 0.01[b]	13.34 ± 0.27[b]	15.24 ± 2.32[b]	0.11 ± 0.11[d]	0.23 ± 0.00[c]
23	2.13 ± 0.02[c]	8.34 ± 0.09[c]	94.80 ± 0.49[e]	96.32 ± 1.15[d]	0.04 ± 0.11[b]	0.17 ± 0.01[a]
30	2.09 ± 0.01[c]	8.22 ± 0.07[c]	89.57 ± 0.52[f]	92.84 ± 0.35[d]	0.07 ± 0.11[c]	0.13 ± 0.01[b]
37	1.86 ± 0.02[b]	8.60 ± 0.06[d]	68.18 ± 0.13[d]	79.62 ± 1.10[c]	0.01 ± 0.11[a]	0.12 ± 0.02[b]
45	0.05 ± 0.00[a]	8.29 ± 0.03[c]	11.16 ± 0.49[c]	57.16 ± 1.11[b]	0.04 ± 0.04[b]	0.13 ± 0.02[e]

Values are represented as mean ± SE (n = 4); Values in a row with different superscripts differ significantly (*P* < 0.05).

Table 3. Effects of various carbon sources on the growth and nitrogen removal by *P. pseudoalcaligenes* strain YY24.

Carbon source	OD600	pH	TN removel ratio (%)	TAN removel ratio (%)	NO$_2^-$-N (mg/L)	NO$_3^-$-N (mg/L)
control	0.11 ± 0.00[d]	7.50 ± 0.00[e]	0.41 ± 0.05[a]	3.5 ± 0.02[b]	0.08 ± 0.01[b]	0.12 ± 0.00[a]
NaAc	0.43 ± 0.00[f]	7.21 ± 0.02[f]	15.72 ± 0.37[d]	17.53 ± 0.14[d]	0.01 ± 0.00[a]	0.12 ± 0.01[abc]
sodium citrate	2.15 ± 0.02[a]	8.37 ± 0.09[d]	92.8 ± 0.33[f]	94.32 ± 1.00[f]	0.00 ± 0.00[e]	0.13 ± 0.01[e]
soluble starch	0.05 ± 0.00[a]	7.46 ± 0.01[d]	1.25 ± 0.01[b]	1.94 ± 0.11[a]	0.08 ± 0.00[c]	0.22 ± 0.00[c]
potassium sodium tatrate	0.04 ± 0.00[e]	7.48 ± 0.02[a]	1.23 ± 0.06[b]	1.91 ± 0.05[a]	0.05 ± 0.02[d]	0.14 ± 0.00[abc]
Glucose	0.53 ± 0.05[ab]	7.26 ± 0.06[b]	20.98 ± 018[e].	22.36 ± 0.45[e]	0.05 ± 0.01[e]	0.13 ± 0.00[d]
sucrose	0.08 ± 0.00[a]	7.22 ± 0.04[d]	1.47 ± 0.13[b]	2.03 ± 0.06[a]	0.08 ± 0.00[d]	0.17 ± 0.01[f]
lactose	0.04 ± 0.00[c]	7.48 ± 0.02[c]	0.98 ± 0.02[ab]	1.69 ± 0.01[a]	0.05 ± 0.01[f]	0.25 ± 0.00[bc]
methanol	0.20 ± 0.00[b]	7.18 ± 0.02[d]	4.94 ± 0.04[c]	7.66 ± 0.07[c]	0.12 ± 0.00[e]	0.14 ± 0.00[ab]

Values are represented as mean ± SE (n = 4); Values in a row with different superscripts differ significantly (*P* < 0.05).

Table 4. Effects of pH on the growth and nitrogen removal by *P. pseudoalcaligenes* strain YY24.

pH	OD600	pH	TN removel ratio (%)	TAN removel ratio (%)	NO$_2^-$-N (mg/L)	NO$_3^-$-N (mg/L)
4.5	0.04 ± 0.00[a]	4.62 ± 0.00[a]	62.09 ± 0.16[c]	0.56 ± 1.10[c]	0.21 ± 0.00[c]	0.32 ± 0.01[e]
5.5	1.28 ± 0.08[c]	7.17 ± 0.09[b]	64.40 ± 0.55[d]	73.64 ± 0.07[d]	0.03 ± 0.01[a]	0.13 ± 0.01[ab]
6.5	1.47 ± 0.03[d]	8.10 ± 0.00[bc]	67.58 ± 0.20[e]	74.05 ± 0.07[d]	0.04 ± 0.00[a]	0.17 ± 0.00[a]
7.5	2.26 ± 0.02[f]	8.13 ± 0.02[bc]	93.40 ± 0.66[g]	95.10 ± 0.04[e]	0.03 ± 0.00[a]	0.13 ± 0.01[b]
8.5	1.61 ± 0.00[e]	8.14 ± 0.02[c]	74.03 ± 0.22[f]	75.58 ± 0.26[d]	0.08 ± 0.00[b]	0.22 ± 0.02[d]
9.5	0.43 ± 0.04[b]	8.42 ± 0.00[d]	16.77 ± 0.27[b]	17.74 ± 0.68[b]	0.21 ± 0.00[c]	0.23 ± 0.02[f]
10.5	0.04 ± 0.00[a]	9.45 ± 0.01[e]	1.39 ± 0.36[a]	2.78 ± 0.01[a]	0.09 ± 0.00[b]	0.22 ± 0.02[c]

Values are represented as mean ± SE (n = 4); Values in a row with different superscripts differ significantly (*P* < 0.05).

and ammonium removal rates (Table 4). By the end of this experiment, pH increased when the initial pH was below 7.5 while it decreased when the initial pH was above 7.5. The highest OD_{600} values appeared when the maximal ammonium removal rates were observed (Tables 1, 2, 3 and 4). Low concentrations of nitrate and nitrite were observed with the changes of C/N ratio, temperature, carbon source and pH (Tables 1, 2, 3 and 4).

4 DISCUSSION

The highest TAN removal ratio of 96.67% was obtained by strain YY24 when sodium citrate and ammonium acted as the unique organic carbon and nitrogen source, respectively. 91.44% of TAN was removed by *Pseudomonas putida* F6 at the condition of ammonium as the one and only nitrogen source (Liu et al. 2008). 98% of TN removel was obtained at the condition of succinate as the one and only carbon source at the C/N raio of 8 (Yuan et al. 2012). The nitrogen removal capacity of bacteria was dependent on bacterial species and culture conditions, especially carbon source and nitrogen source.

TAN concentrations decreased rapidly in the logarithmic phase, suggesting that heterotrophic nitrification occurred in the logarithmic phase. Similar results were reported in other research (Wang, 2006). However, Verstraete (1972) reported that nitrification process occurred in aging generally. The difference of nitrification time may be triggered by different strain. It would be beneficial for the determination of the successive supply time of bacteria in the practical application. Low nitrate and nitrite accumulations were observed in the whole experiment. However, the exact mechanism involved is still unclear. Some study indicated that rapid denitrification process could result in low nitrate and nitrite (Wang et al. 2007) and there was only a trace amount of either nitrate or nitrite in nitrogen removal process (Zhao et al. 2010b). The use of nitrate and nitrite during the ammonium removal process showed that the nitrogen removal pathway was aerobic denitrification-heterotrophic nitrification of stain YY24.

Lowest TAN removal rate appeared at C/N ratio of 6 because carbon source was deficient to remove TAN completely. At C/N ratio above and equal to 10, higher nitrogen removal rates were observed as compared to other studies (Joo et al. 2005b). C/N ratio of 10 was more applicable as compared to those above 10 because low supply of carbon source was need.

The carbon source utilization of heterotrophic microorganisms was species-specific. The maximal nitrogen removal rate was obtained when sodium citrate was used as organic carbon source for strain YY24. *Bacillus* sp. use glucose as the optimal carbon source (Kim et al. 2005). *P. pseudoal-*

caligenes strain YY24 utilized organic acids rather than sugar. Acetate was widely used as an organic substrate by some bacterial strains, for example, *Alcaligenes faecalis* 4, *Pseudomonas* sp. AS-1, and strain YL (Joo et al. 2005b; Robertson et al. 1988; Su et al. 2006; Taylor et al. 2009).

The optimum pH of YY24 was 7.5. *Providencia rettgeri* strain YL had high nitrogen removal capability at similar pH (Zhao et al. 2010). *Bacillus* MS30 achieved the maximal heterotrophic nitrification rate at weak basic environment of pH 7.5 (Mevel et al. 2000). The maximal values of nitrate reduction of *Bacillus* sp. and *Pseudomonas* sp. KW1 appeared at pH 7 (Patureau et al. 1997). The optimal pH of heterotrophic bacterium *Thiosphaera pantotropha* with high nitrogen removal ability was 8 (Robertson et al. 1988). It was suggested that neutral and weak basic environment were more geared to heterotrophic bacteria.

H^+ produced in nitrification process could lead to lower pH value, and the consumption of H^+ in the denitrification process could result in pH increase (Yang et al. 2009). The changes in pH were also associated with carbon sources (Yang et al. 2009). In present study, pH raised to 8.0–8.5 by end of experiment, which might be resulted from aerobic denitrification as well as organic carbon utilization.

5 CONCLUSIONS

P. Pseudoalcaligenes YY24 isolated exhibited predominant of simultaneous aerobic denitrification and heterotrophic nitrification capacities, and the optimum conditions for nitrogen removal were as follows: pH ranged 5.5–8.5, the C/N ratio was above 10, temperature ranged from 23 to 30°C and the best carbon source was sodium citrate. Strain YY24 could be promising and applicable for the wastewater treatment.

ACKNOWLEDGEMENTS

This trial was assisted by the National Agricultural Science and Technology Achievements Transformation Fund project (Grant No 2012GB2 A100021). This study was also supported by National Natural Science Foundation of China (31402313) and open fund of Tianjin Key Lab of Aqua-Ecology and Aquaculture.

ABBREVIATIONS

NO_2^--N: nitrite TAN: Total Ammonia
TN: Total Nitrogen NO_3^--N: nitrate
BTB: Bromine methyl phenol Blue
NH_4^+-N: Ammonia-nitrogen

REFERENCES

Diep C. N. Cam P. W. Vung N. H. & My N. H. (2009). Isolation of *Pseudomonas stutzeri* in wastewater of catfish fish-ponds in the Mekong Delta and its application for application for wastewater treatment. *Bioresource Technology*. 100 (16), 3787–3791.

Felsenstein J. (1985) Confidence limits on phylogenies: an approach using the bootstrap. *Evolution. 39 (4)*, 783–791.

GB/T 17378.4–2007, China Specifications for marine monitoring-Part 4: Seawater analysis. State General Administration of the People's Republic of China for Quality Supervision and Inspection and Quarantine, Beijing.

Joo H. S. Hirai M. & Shoda M. (2005a). Nitrification and denitrification in high-strength ammonium by *Alcaligenes faecalis. Biotechnology Letters. 27 (11)*, 773–778.

Joo H. S. Hirai M. & Shoda M. (2005b). Characteristics of ammonium removal by heterotrophic nitrification-aerobic denitrification by *Alcaligenes faecalis* 4, *Journal of bioscience and bioengineeering. 100 (2)*, 184–191.

Khardenavis A. A. Kapley A. & Purohit H. J. (2007). Simu ltaneous nitrification and denitrification by diverse *Diaphorobacter* sp. *Applied Microbiology Biotechnology. 77(2)*, 03–409.

Khin T. & Annachhatre A. P. (2004). Novel microbial nitrogen removal processes. Biotechnology Advances. *22 (7)*, 519–532.

Kim J. K. Park K. J. Cho K. S. Nam S. W. Park T. J. & Bajpai R. (2005). Aerobic nitrification denitrification by heterotrophic Bacillus strains. *Bioresource Technoogyl. 96 (17)*, 1897–1906.

Kuai L. & Verstraete W. (1998). Ammonium removal by the oxygen-limited autotrophic nitrification-denitrification system. *Applied & Environmental Microbiology. 64 (11)*, 4500–4506.

Lei P. & Jae-Yoon Jo. (2003). Performance of a foam fractionator in a Lab-scale seawater recirculating aquaculture system. *Journal Fish Science & Technology. 6 (4)*, 187–193.

Liu J. J. Wang P. & Wang H. (2008). One heterotrophic nitrification-aerobic denitrifying bacteria denitrification performance study. *Huan Jing Ke Xue Yan Jiu. (3)*, 121–125 [in Chinese].

Mével G. & Prieur D. (2000). Heterotrophic nitrification by a thermophilic Bacillus species as influenced by difierent culture conditions. *Canadian Journal of Microbiology. 46 (5)*, 465–473.

Nishio T. Yoshikura T. Chiba K. & Inouye Z. (1994). Effect of organic acids on heterotrophic nitrification by *Alcaligenes faecalis* OKK17. *Bioscience Biotechnology and Biochemistry. 58 (9)*, 1574–1578.

Patureau D. Bernet N. & Moletta R. (1997). Combined nitrification and denitrification in a single aerated reactor using the aerobic denitrifier *Commonas* sp. strain SGLY2. *Water Research. 31 (6)*, 1363–1370.

Robertson L. A. Van Neil E. W. Torremans R. A. & Kuenen J. G. (1988). Simultaneous nitrification and denitrification in aerobic chemostat at cultures of Thiosphaera pantotropha. *Applied & Environment Microbiology. 54 (11)*, 2812–2818.

Saitou N. & Nei M. (1987). The neighbor-joining method: a new method for reconstructing phylogenetic trees. *Molecular Biology and Evolution. 4 (4)*, 406–425.

Sliekers A. O. Derwort N. Gomez J. Z. Strous M. Kuenen J. G. & Jetten M. S. (2002). Completely autotrophic nitrogen removal over nitrite in one single reactor. *Water research. 36 (10)*, 2475–2482.

Strous M. Van Gerven E. Zheng P. Kuenen J. G. & Jetten M. S. M. (1997). Ammonium removal from concentrated waste streams with the anaerobic ammonium oxidation (ANAMMOX) process in different reaction configurations. *Water Research. 31 (8)*, 1955–1962.

Su J. J. Liu B. Y. & Liu C. Y. (2001). Comparison of aerobic denitrification under high oxygen atmosphere by *Thiosphaera pantotropha* ATCC 35512 and *Pseudomonas stutzeri* SU2 newly isolated from the activated sludge of a piggery wastewater treatment system. *Journal of Applied Microbiology. 90 (3)*, 457–462.

Su J. J. Yeh K. S. & Tseng P. W. (2006). A strain of *Pseudomonas* sp. Isolated from piggery wastewater treatment systems with heterotrophic nitrification capability in Taiwan. *Current Microbiology. 53*, 77–81.

Sun X. M. Li Q. F. Zhang Y. Liu H. D. Zhao J. & Qu K. M. (2012). A plant water heterotrophic nitrification aerobic denitrifying bacteria-system development and denitrification characteristics. *Wei Sheng Wu Xue Bao. 52* (6), 687–695 [in Chinese].

Takaya N. Catalan-Sakairi M. A. Sakaguchi Y. Kato I. Zhou Z. & Shoun H. (2003). Aerobic denitrificatlon bacteria that produce low levels of nitrous oxide. *Applied & Environmental Microbiology. 69 (6)*, 3152–3157.

Thompson J. D. Gibson T. J. Plewniak F. Jeanmougin F. & Higgins D. G. (1997). The Clustal_X windows interface: flexible strategies for multiple sequence alignment aided by quality analysis tools. *Nucleic Acids Research. 25 (24)*, 4876–4882.

Van de Graaf, A. A. de Bruijn P. Robertson L. Kuenen J. & Mulder A. (1990). Anoxic ammonium oxidation. In: *Proceeding of the 5th European Congress on Biotechnology*, C, Christiansen, L. Munck, and J. Villadsen(ed.), Munksgaard International Publisher, Copenhagen, 388–391.

Verstraete W. & Alexander M. (1972). Heterotrophic nitrification by *arthrobacter* sp. *Journal of Bacteriology. 110 (3)*, 955–961.

Wang L. B. Wan X. H. & Xu P. (2006). 4 strains of heterotrophic nitrification ability of nitrifying bacteria identification and preliminary study. *Shui Chan Yang Zhi. 27 (2)*, 147–151 [in Chinese].

Wang X. Ma Y. Peng Y. & Wang S. (2007). Short-cut nitrification of domestic wastewater in a pilot-scale A/O nitrogen removal plant. *Bioprocess Biosystems Engineering. 30 (2)*, 91–97.

Yang D. J. (2009). Intensive livestock farming research high ammonia nitrogen wastewater degradation bacteria. University of Shandong Agricultural, Shangdong, China [in Chinese].

Yuan M. D. & Xin Y. F. (2012). Screening and Identification of One Strain of Heterotrophic Nitrification-aerobic Denitrifying Bacteria and Its Denitrifying Activity. *Bei Hua Da Xue Xue Bao(Natural Science). 13 (3)*, 339–343 [in Chinese].

Zhao B. He Y. L. Huang J. Taylor S. & Hughes J. Joseph Hughes. (2010a). Heterotrophic nitrogen removal by *Providencia rettgeri* strain YL. *Journal of Industrial Microbiology Biotechnol. 37 (6)*, 609–616.

Zhao B. He Y. L. Hughes J. & Zhang X. F. (2010b). Heterotrophic nitrogen removal by a newly isolated *Acinetobacter calcoaceticus* HNR. *Bioresource Technology. 101 (14)*, 5194–5200.

Advances in Energy, Environment and Materials Science – Wang & Zhou (Eds)
© 2017 Taylor & Francis Group, London, ISBN 978-1-138-03600-0

An efficient thermo- and electrochemical method to produce hydrogen from coal using a molten NaOH reactor

L.Y. Zhu, B.H. Wang & H.J. Wu
College of Chemistry and Chemical Engineering Northeast Petroleum University, Daqing, China

L.N. Zhu & Y. Jiang
Daqing Chemical Research Institute of Petrochemical Research Institute of PetroChina Co. Ltd., Daqing, China

ABSTRACT: In this paper, a process of electrochemical splitting of coal for hydrogen production using the molten NaOH reactor is presented. The generation of hydrogen, an important new energy to replace fossil fuel, was enhanced up over a wide range of reaction temperature. With the increase of the temperature, the thermodynamic energy required to drive the coal splitting is decreased and improves the kinetics to constant hydrogen generation rate. This coal conversion is an endothermic process and could be mixed with solar thermal- and electrochemical process to realize the cooperative transformation of solar energy and fossil energy to hydrogen energy. The thermo- and electrochemical hydrogen generation using molten NaOH reactor can be operated at 360°C, which is a much lower temperature than that for the coal reforming process. A parametric study is conducted to determine the strong influence of temperature on hydrogen generation.

1 INTRODUCTION

With the depletion of oil and air pollution becoming an increasing concern, it is necessary to find a renewable energy to replace the current version. Hydrogen energy appears to be an effective way to provide better environment and sustainability (De Vries, Van Vuuren, and Hoogwijk 2007). A large variety of fossil energy sources can be used for hydrogen production. However, they are not necessarily environmentally friendly and emit a large amount of CO_2 in the hydrogen generation process. Together with the development of electrochemical technology, water electrolysis has attracted the attention of researchers as a promising approach to hydrogen production from fossil energy (Giddey, Kulkarni, and Badwal 2015). In 1979, Coughlin and Forooque proposed an environmentally friendly, clean, efficient, and less CO_2 emitting method using electrochemical splitting of coal for hydrogen (Baldwin et al. 1981).

As rapid decrease in the storage of energy has been closely related to an increase in the development of economy and the population, human beings are facing the most serious risk of energy and complications of environmental deterioration (GhaffarianHoseini et al. 2013). The way of utilization of fossil energy should be a relative environmentally benign manner. The combustion way of producing energy is inefficient and causes

pollution leading to a large amount of green house gas emission and global climate changes. The increasing problems of environmental effects promote the transformation of energy (Li and Fan 2008). All these emergency situations have led us to find new energy sources in an efficient way and minimized the harmful emissions, which are very essential in the energy promotion at reasonable use.

We have recently shown that the significant effects of temperature on the electrochemical process of coal can be applied to improve the hydrogen production from coal electrochemical splitting, especially the enhancement of energy transfer from the fossil fuel to new clean hydrogen energy. And the hydrogen generation rate is increased with a lower CO_2 emission using the electrochemical coal splitting under 360°C.

2 EXPERIMENTAL

The electrolysis apparatus used in electrochemical coal splitting experiment is shown in Figure 1. A high-purity nickel foil (99.99%, 2 cm^2) was used as the anode and cathode, which was covered with ceramic tube. The distance between anode and cathode was 2 cm. In this series of experiment, both the electrodes, two sheets of pure nickel sheets, are situated 5 mm below the surface of the electrolyte to facilitate H_2 evolution to minimize

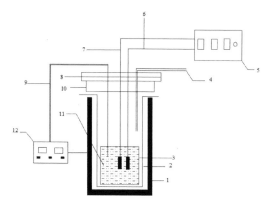

Figure 1. 1 Resistor furnace 2 Stainless steel sealed reactor 3 Corrosion-resistant corundum electrolytic cell 4 Gas conduit 5 DC-regulated power supply 6 Anode 7 Cathode 8 Heat preservation asbestos 9 K type thermocouple 10 hermetic seal 11 Molten NaOH 12 Temperature control box.

oxygen interaction with the cathode product. The coal after pretreatment (1 g) and the electrolyte (NaOH, 20 g) were put into the electrolysis cell (a Ø50*80 mm cell made by corundum). The electrolysis was carried out with constant current at 320°C, which is slightly higher than the melting point of NaOH, and lasted 6 h. The product gases formed at the anode and cathode were collected in measuring cylinders using the method of displacement of saturated sodium bicarbonate and analyzed with the multi-dimensional gas chromatography.

3 RESULTS AND DISCUSSION

Electrolysis of coals in alkaline media is an efficient way to produce hydrogen, other chemicals, and even fuels, and can reduce the ash and sulfur content of coal. Electrolysis of coal was proposed by Coughlin and Farooque, who recognized that coal was oxidized at anode, while protons are reduced to hydrogen at the cathode:

Anode:

$$C(s) + 2H_2O(l) \rightarrow CO_2(g) + 4H^+ + 4e^- \quad (1)$$

Cathode:

$$2H^+ + 2e^- \rightarrow H_2 \quad (2)$$

In aqueous solution, a number of oxidizing media were produced from water electrolysis on anode, including $\cdot OH_2$ and $\cdot OH$, which could also degrade the organic chemicals. On the other side, H^+ is formed into hydrogen at the cathode (Figure 2).

The hydrogen main comes from the water electrolysis in aqueous solution, but the same situation in molten NaOH reactor is different, the hydrogen is generated from water (Moisture in coal) electrolysis, NaOH electrolysis with a higher cell voltage, and coal dehydrogenation.

The thermo- and electrochemical process of pretreated coal in molten NaOH at different currents is investigated. Figure 3 shows the steady state of cell voltage for coal electrolysis in molten NaOH changes with time at different currents. Electrolytic experiment by different current density showed that the cathode showed steady activity and more organics were oxidized at a higher current value. The mass of coal decreased significantly after electrolysis from 2.1 g to 0.3 g at 360°C for a 0.4 A current in molten NaOH. According to the analysis of the products, 0.8599 g of solid coal is converted to gaseous products.

The product gases formed at the anode and cathode were collected and analyzed with multi-dimensional gas chromatography. The contents of gaseous products and the total gas volume were calculated by the data from GC analysis shown in Figure 4. The figure shows the efficiency of gas generation and the abundance of gas products.

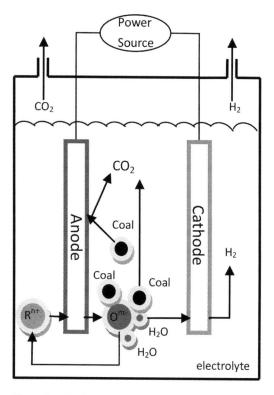

Figure 2. Hydrogen production from coal water slurry (Rn+/Om− are redox couple).

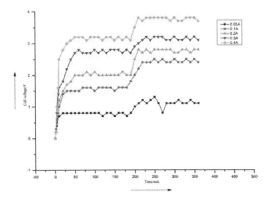

Figure 3. Change of cell voltage with time at different currents at 360°C.

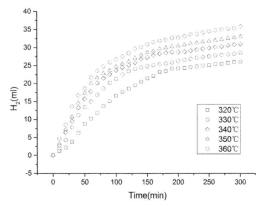

Figure 5. Effect of temperature on hydrogen evolution. Electrolysis of coal in molten NaOH was carried out at 0.4A and different temperatures.

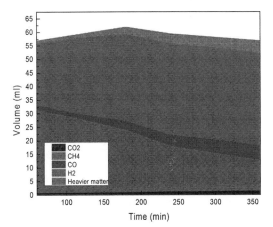

Figure 4. Volume of gases products at 360°C.

Table 1. Gas conversion and liquid conversion of the coal electrochemical oxidation at 0.4 A at different temperatures.

Temperature (°C)	Total conversion (%)	Gas conversion (%)	Liquid conversion (%)
320	83.1	45.1	38
330	84.53	45.03	39.5
340	85.65	42.85	42.8
350	86.74	41.14	45.6
360	87.3	38.5	48.8

From Figure 4, it can be concluded that the main components of gaseous products are hydrogen and methane. Meanwhile, the volume of methane decreased with time but that of hydrogen increased. According to the contents of gaseous products, the volume of carbon dioxide and carbon monoxide increased slightly with the time, but were not very high. In the system, the electrochemical oxidation could not oxidize the coal directly into CO_2 and CO, and the most possibility is to break up the organic matters to smaller organic molecules.

The effect of temperature on the thermo- and electrochemical oxidation of coal in molten NaOH was studied at 0.4 A. The generation of hydrogen increased with the temperature. At 0.4 A electrolysis current, increase of temperature from 320°C to 360°C showed an increase of hydrogen generation from 24.82 to 34.14 ml, which is presented in Figure 5. It is revealed that the temperature is a highly effective factor in the conversion of coal to

hydrogen by electrolysis. Higher temperature could reduce the activation energy and increase the kinetics of the electrolysis.

We observe that when the rate of the hydrogen evolution increased with the temperature, the rate of the total conversion and liquid conversion of coal is increased (Table 1). In particular, the rate of gas conversion decreases; in contrast to the gas conversion, the hydrogen evolution increases with the temperature, indicating the decrease in the rate of formation of competing parasitic products (carbon dioxide, carbon monoxide, oxygen).

In such a molten NaOH reactor, the temperature could increase by changing the electrolyte or mixing with other electrolyte, which indicates that the molten reactor could provide a higher temperature environment for the electrochemical oxidation of coal to increase the total conversion and the rate of hydrogen generation. In other words, increasing temperature with molten NaOH reactor is an efficient way to enhance hydrogen generation.

The electrochemical oxidation of coal using molten NaOH to produce hydrogen could be used for the solar energy conversion system (Figure 6).

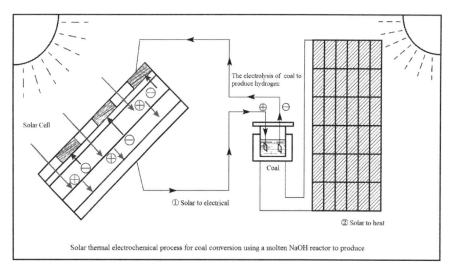

Figure 6.   Solar thermal electrochemical process for coal conversion using a molten NaOH reactor to produce hydrogen.

Solar-based energy is considered a green fuel for the 21st century, with much of the attention focused on the solar generation of electricity or fuel and will become the aims of the future lower carbon footprint of energy development strategy (De Vries, Van Vuuren, and Hoogwijk 2007). It is speculated that the conversion of solar energy to hydrogen energy is prerequisite for efficient energy upgrade from low-grade energy to high-grade energy (Licht 2009). The coal electrolysis process fully driven by solar energy to apply heat and electricity occurs at high Faradic efficiency of the anode oxidation and cathode reduction of coal with molten NaOH to hydrogen at a threshold electrolysis voltage.

4   CONCLUSION

The increase of hydrogen generation with increasing temperature provides an efficient low-energy pathway for electrochemical coal conversion process using molten NaOH. With the increase of hydrogen generation, the rate of formation of competing parasitic products (carbon dioxide, carbon monoxide, oxygen) decreases distinctly. This conversion process could be driven by solar energy to provide high temperature and meet the requirement of electricity, which is realized by combining and coupling the solar thermo- and electrochemical process to facilitate the transformation of fossil energy to hydrogen energy in a clean, efficient, and sustainable way.

ACKNOWLEDGMENT

The research is financially supported by the National Science Foundation of China (Grant No. 21376049, 21306022).

REFERENCES

Baldwin, Richard P, Keith F. Jones, Joseph T. Joseph, and John L Wong. 1981. Voltammetry and electrolysis of coal slurries and H-coal liquids. *Fuel* 60 (8):739–743.

De Vries, Bert JM, Detlef P. Van Vuuren, and Monique M. Hoogwijk. 2007. Renewable energy sources: Their global potential for the first-half of the 21st century at a global level: An integrated approach. *Energy policy* 35 (4):2590–2610.

GhaffarianHoseini, AmirHosein, Nur Dalilah Dahlan, Umberto Berardi, Ali GhaffarianHoseini, Nastaran Makaremi, and Mahdiar GhaffarianHoseini. 2013. Sustainable energy performances of green buildings: A review of current theories, implementations and challenges. *Renewable and Sustainable Energy Reviews* 25:1–17.

Giddey, S, A. Kulkarni, and SPS. Badwal. 2015. Low emission hydrogen generation through carbon assisted electrolysis. *International Journal of Hydrogen Energy* 40 (1):70–74.

Li, Fanxing, and Liang-Shih Fan. 2008. Clean coal conversion processes–progress and challenges. *Energy & Environmental Science* 1 (2):248–267.

Licht, Stuart. 2009. STEP (solar thermal electrochemical photo) generation of energetic molecules: A solar chemical process to end anthropogenic global warming. *The Journal of Physical Chemistry C* 113 (36):16283–16292.

Advances in Energy, Environment and Materials Science – Wang & Zhou (Eds)
© 2017 Taylor & Francis Group, London, ISBN 978-1-138-03600-0

Preparation and performance of Buton rock-modified asphalt waterproof coating

Su Zhao, Qian Wu, Fei Yu, Peng Ma & Dong-jie Guo
Institute of Material Science and Engineering, Shenyang Jianzhu University, Shenyang, China

ABSTRACT: Buton rock asphalt is a type of natural rock asphalt with excellent weather resistance. Replacement of part of matrix asphalt by Buton rock asphalt can not only improve the performance of the material but also save resources and expand the scope of application of natural asphalt. In this paper, we chose Maoming 90# asphalt as the matrix asphalt and Buton rock asphalt as the admixture to prepare Buton rock-modified asphalt waterproof coating. We obtain the ratio of raw materials by the orthogonal test: bentonite content, 10%; SBS content, 4%; water content, 10%; fluorocarbon emulsion content, 10%; and Buton rock-modified asphalt content, 66%. By investigating the effect of experimental conditions on the bonding strength of Buton rock-modified asphalt coating, we find that the optimum temperature is about 180°C and stirring rate is 800 rad/min. Heat resistance, adhesion, flexibility, and water resistance of the prepared Buton rock-modified asphalt coating are superior to the relevant standards to a great extent.

1 INTRODUCTION

The advantages of matrix asphalt are easy availability of raw materials and low price, thus it can be used for waterproof impermeability engineering. Currently, it is the most widely used waterproof material in impervious projects. However, because of the existence of its own hot cold crisp drip, poor weather resistance, and other shortcomings, the performance of the waterproof material is severely compromised. And the emergence of modified asphalt waterproof coating overcomes its own problems of matrix asphalt (Han, 2011; Gao, 2007; Wang, 2009).

Buton Rock Asphalt (BRA) is a type of natural asphalt produced in Buton Island, Indonesia. Different from other rock asphalts, it is formed by Paleozoic oil penetration into the rock, because of the constant movement, crush, and break of the crust, making the crude oil reserved in the deep crust continue to flock to the surface, rock asphalt substances under natural conditions, after thousands of years to form. Because of its perennial coexistence with the natural environment and long-term tolerance of harsh environment, its various properties are particularly stable, and it is usually with a very good weather resistance and other properties (Li, 2007).

In general, natural asphalt cannot be directly used as a type of asphalt, but can be modified with a small amount of matrix asphalt to form the modified asphalt, making the greatest use of its excellent performance. Numerous studies and engineering practices show that natural modified asphalt pavement with high service life, high

stability, high water resistance, high resistance to microbial attack, and high fatigue strength significantly improves and enhances the performance of asphalt pavement (Yan, 2009).

Although BRA has good weather resistance and is less expensive, it is mainly used in road construction and transportation aspects at present (Wang, 2008), and the study of the waterproof and anti-seepage engineering of buildings has not been reported yet. Therefore, in this paper, we take Maoming # 90 as matrix asphalt and BRA as admixture to prepare Buton rock-modified asphalt waterproof coating.

2 EXPERIMENT PART

2.1 Main raw materials

Maoming 90# asphalt; Indonesian BRA; Fluorocarbon resin emulsion, solid content not less than 45%; Bentonite, montmorillonite content more than 85%; linear SBS (Styrene–Butadiene–Styrene block copolymer).

2.2 Experimental method

1. Preparation of Buton rock-modified asphalt
   Weigh 100 parts of 90# asphalt, which is heated to 150°C, incorporated in 40% substitution of BRA, and high-speed stirring for 30 min, achieving uniform dispersion of BRA in the matrix asphalt. Then, the samples stand for 20 min under 30°C, finally producing Buton rock-modified asphalt.

Processing method of BRA: first, massive BRA is crushed, and 75 micron particle size of BRA powder is obtained. The parts by weight can be grams, kilograms, and other units of measurement.

2. Preparation of Bentonite paste

Weigh a certain amount of organic bentonite, adding an appropriate amount of hot water at 75°C (bentonite and water ratio of 1: 2), stirring evenly, standing for 3 h under 30°C, to obtain bentonite paste on standing.

3. Preparation of Buton rock-modified asphalt waterproof coating

Put 50–70% mass fraction of Buton rock-modified asphalt into the emulsifying machine, heated to 180°C, adding 4–12% of SBS, 10–15% of the bentonite paste, 5–15% of the fluorocarbon emulsion, 5–15% of water, stirring, making the bentonite paste and the SBS evenly disperse in asphalt, maintaining the temperature at about 180°C, stirring at 800 rad/min for about 30 min until the temperature is reduced to 30°C, to obtain Buton rock-modified asphalt waterproof coating.

2.3 *Performance testing*

Test the general properties of Buton rock-modified asphalt waterproof coating according to the provided method of JC/T408-2005 "Emulsified asphalt waterproof coating". Test the tensile properties with reference to GB/T19250. Test adhesion according to GB/T 1720.

Water absorption test:

At ambient temperature, the coating of samples is immersed in distilled water, taking out the samples after a certain time, sopping up the water, and weighing the weight. Calculate the water absorption rate of P (%) according to the Equation 1:

$$P(\%) = (W_2 - W_1)/W_1 \times 100\% \quad (1)$$

where W1 is the weight of the sample before immersion (g) and W2 is the weight of the sample after immersion (g).

3 RESULTS AND DISCUSSION

3.1 *Determination of composition content of Buton rock-modified asphalt waterproof coating*

Design orthogonal experiment according to the orthogonal table L9 (34) to determine the SBS, bentonite, water, and fluorocarbon emulsion of these four types of material mixed in proportion, and then determine the BRA content. Factors such as the levels of the test are shown in Table 1, and the results are shown in Table 2:

Analysis according to the data in Table 2 shows that the best solution should be D1B2A1C2: bentonite content, 10%; SBS content, 4%; water content, 10%; fluorocarbon emulsion content, 10%; and Buton rock-modified asphalt content, 66%.

3.2 *Effect of experimental conditions on the performance of Buton rock-modified asphalt waterproof coating*

3.2.1 *Effect of reaction temperature*

When other factors were constant, we investigated the impact of reaction temperature on the performance of modified asphalt waterproof coating. The test results are shown in Table 3. Figure 1 shows the effect of reaction temperature on bond strength of modified asphalt.

Table 1. Experimental factors and levels.

Test no.	A (SBS) %	B (Fluorocarbon emulsions) %	C (Water) %	D (Bentonite) %
1	4	5	5	10
2	8	10	10	12
3	12	15	15	14

Table 2. Factors and levels of composite-modified test and test results.

Test no.	A	B	C	D	Bond strength MPa
1	1	1	1	1	2.1
2	1	2	2	2	1.8
3	1	3	3	3	1.2
4	2	1	2	3	1.6
5	2	2	3	1	1.7
6	2	3	1	2	1.1
7	3	1	3	2	0.9
8	3	2	1	3	1.5
9	3	3	2	1	1.4
K1	5.1	4.6	4.7	5.2	–
K2	4.4	5	4.8	3.8	–
K3	3.8	3.7	3.8	4.3	–
K1/3	1.7	1.53	1.56	1.73	–
K2/3	1.46	1.67	1.6	1.27	–
K3/3	1.27	1.23	1.27	1.43	–
R	0.43	0.44	0.33	0.46	–

Note: K (K1, K2, K3) is the sum of bond strength for the same factor and the same level; K/3 (K1/3, K2/3, K3/3) is the average of K; R is the average difference between the maximum and the minimum values for the same factor and the different level.

Table 3. Effect of reaction temperature on the performance of the modified asphalt waterproof coating.

Reaction temperature				Surface drying time
°C	Low temperature flexibility	Heat resistance	Water imperm eability	h
120	Poor to excellent	Dripping	Water seepage	8
140	Poor to excellent	Flowing without dripping	No seepage	6
160	Poor to excellent	Not flowing, no slip	No seepage	6
180	Poor to excellent	Not flowing, no slip	No seepage	4
200	Fair	Not flowing, no slip	No seepage	2
220	Poor to fair	Not flowing, no slip	Seepage	2

Note: Low temperature flexibility (−20°C), water impermeability (0.3 MPa, 30 min).

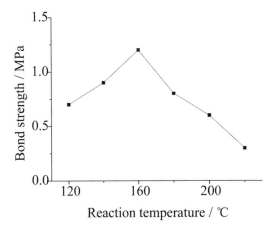

Figure 1. Effect of reaction temperature on the bond strength of modified asphalt.

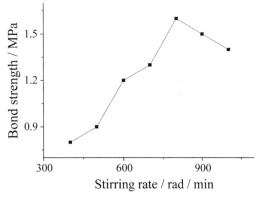

Figure 2. Effect of stirring rate.

Table 3 shows that, with the increase of temperature, asphalt aging becomes worse. When the temperature exceeded 200°C, the internal structure of asphalt was severely damaged (Zhang, 2012), and many of the basic properties of asphalt were lost rapidly, thereby reducing the bond strength. Therefore, in the process of preparing the modified asphalt, temperature control is very important.

By analyzing Figure 1, we can know that: when the reaction temperature was below 160°C, with the reaction temperature increasing, the bond strength of modified asphalt constantly increased, and the comprehensive performance was significantly enhanced. However, when the reaction temperature was higher than 160°C, with the reaction temperature increasing, the bond strength reduced; especially when the temperature was higher than 200°C, the declining rate of bond strength is high. The sample prepared after cooling had numerous cracks, and severe cracking and seepage occurred on detection. Finally, we comprehensively determined that the appropriate temperature was about 180°C in this experiment.

3.2.2 Effect of stirring rate

With other factors unchanged, we varied the stirring rate and studied its effect on the bond strength of modified asphalt coating. The results are shown in Figure 2:

By analyzing Figure 2, we could conclude that with the increase of stirring rate, the bond strength of the modified asphalt also increased. When the stirring rate was between 300 and 850 rad/min, the bond strength gradually increased, but when the stirring rate was more than 850 rad/min, the bond strength of modified asphalt decreased. With the increase of stirring rate, the components were more uniformly mixed, so that a cross-linked structure was formed more stably. And the active region, which was formed by "half polymerization effect" of BRA, was mainly composed of asphalt adhesion (Zhang, 2015). Furthermore, high-speed stirring made these "active regions" distribute more evenly in the system, so that the bond strength of modified asphalt was significantly increased. However, too high stirring speed would destroy the internal structure of the modified asphalt and decrease the bond strength. Therefore, the speed of stirring should not be higher than 850 rad/min. In this experiment, the stirring speed was 800 rad/min.

3.3 The related performance of Buton rock-modified asphalt waterproof coating

Take 66% of Buton rock-modified asphalt, 10% of bentonite, 4% of SBS, 10% of water, and 10% of fluorocarbon emulsion to prepare Buton rock-modified asphalt waterproof coating according to the experimental procedure of section 2.2 and test the performance according to the relevant standard of section 2.3.

3.3.1 General performance of waterproof coating
The test results are shown in Table 4.

Data from Table 4 show that the conventional performance of Buton rock-modified asphalt waterproof coating was in line with the relevant standards.

3.3.2 Tensile properties of waterproof coating
The tensile test results of Buton rock-modified asphalt coating are shown in Table 5.

As can be seen from Table 5, the tensile properties of Buton rock-modified asphalt waterproof coating were in line with the standards, and extensibility after heat treatment and ultraviolet treatment was also very good.

3.3.3 Water absorption and adhesion of modified asphalt waterproof coating
Water absorption and adhesion of waterproof coating is one of the key indices to evaluate a coating or coating system. Therefore, the determination of water absorption and adhesion of the coating has been widely concerned by the coating industry.

We tested the water absorption and adhesion of coating according to section 2.3. Test results are shown in Table 6.

Table 4. Conventional performance of buton rock-modified asphalt waterproof coating.

Project	Condition	Test results	Standard
Appearance	Observation by naked eye	Brown-black, dark brown, mushy at room temperature, dispersed in water without asphalt wire	No color difference, gels, caking, and obviously asphalt wire
Solid content %	$105 \pm 2°C$, Baking 30 min	53	≥ 45
Heat resistance	5 h	Without flowing, blistering, and slide at 200°C	90°C, not flowing, sliding, and dripping
Alkali resistance	$23 \pm 2°C$, 0.1% sodium hydroxide mixed with saturated calcium hydroxide	Good	No bubbles, cracks, peeling, chalking, softening, and dissolution
Bond strength MPa	$20 \pm 2°C$, 8 matrix method	0.73	≥ 0.30
Impermeability	Hydrostatic pressure was 0.3 MPa, 30 min	Impermeable	Impermeable
Flexibility	$-20°C$, around $\Phi 30$ mm rod	No cracks, fracture	No cracks
Surface drying time H	Standard conditions	0.25	≤ 8
Hard drying time h	Standard conditions	1	≤ 24

Table 5. Tensile test of Buton rock-modified asphalt waterproof coating.

Project	Conditions	Test results	Standard
Extensibility	Conventional treatment	16	≥ 6.0
	Heat Treatment, 70°C, 168 h	11	
mm	UV treatment, 500 W mercury lamp, 240 h	9	
	Alkali treatment, saturated calcium hydroxide solution, 168 h	9	

Table 6. Test results of water absorption and adhesion of waterproof coating.

Coating type	Water absorption (10 d) %	Water absorption (30 d) %	Adhesion level
Ordinary asphalt waterproof coating	15	20	3
Buton rock-modified asphalt waterproof coating	8	11	1

As is shown in Table 6, water absorption of Buton rock-modified asphalt waterproof coating was lower than that of ordinary asphalt waterproof coating, but its adhesion was higher than that of the ordinary asphalt waterproof coating. Therefore, the water resistance of Buton rock-modified asphalt waterproof coating is higher, and its performance is better than that of ordinary asphalt waterproof coating.

4 CONCLUSIONS

1. The ratio of raw materials of Buton rock-modified asphalt waterproof coating is determined by the orthogonal test: bentonite content, 10%; SBS content, 4%; water content, 10%; fluorocarbon emulsion content, 10%; and the mixing amount of Buton rock-modified asphalt, 66%.
2. By examining the impact of test conditions on bond strength of Buton rock-modified asphalt coating, we determine the optimum reaction temperature as 180°C and stirring speed as 800 rad/min.
3. As the results of the conventional performance testing of waterproof coating, tensile testing, and the test results of water absorption and adhesion show, heat resistance, adhesion, flexibility, and water resistance of the prepared Buton rock-modified asphalt coating are superior to the relevant standards.

REFERENCES

Hong-qing Han. 2011. Modified bentonite and its application. Inorganic chemicals industry 43 (10): 2–5.
Hai-hui Li. 2007. Rock asphalt Situation and Prospects. Friends of Science (10): 50–51.
Lei Yan. 2009. Natural asphalt Situation and Prospects. Henan Science and Technology 2 (10): 78–79.
Tong Gao & Chao Xu. 2005. Fluorocarbon emulsion modified bentonite emulsified asphalt waterproof coating. New building materials 10: 59–60.
Wei Wang. 2008. BRA modified asphalt mixture performance and research. Changsha University of Science and Technology.
You-peng Wang & Xiao-rong He. 2009. Palygorskite emulsified asphalt waterproof coating. New building materials 03: 66–68.
Yibo Zhang, Hongzhou Zhu, Guoan Wang, et al. 2012. Evaluation of low temperature performance for diatomite modified asphalt mixture. Advanced Materials Research 413:246–251.
Zhang H, Zhu C, Yu J, et al. 2015. Influence of surface modification on physical and ultraviolet aging resistance of bitumen containing inorganic nanoparticles. Construction & Building Materials 98:735–740.

Advances in Energy, Environment and Materials Science – Wang & Zhou (Eds)
© *2017 Taylor & Francis Group, London, ISBN 978-1-138-03600-0*

Pilot-scale hybrid biological fixed-film process for treating PVC wastewater

Junbo Zhou, Yang Liu & Zhaojian Zhang
Engineering Research Center for Polymer Processing Equipment, Ministry of Education, College of Mechanical and Electrical Engineering, Beijing University of Chemical Technology, Beijing, P.R. China

ABSTRACT: A biochemical treatment process of centrifugal mother liquid of Polyvinyl Chloride (PVC) was introduced as a pre-treatment for re-using the water for polymer production in a reactor. The pilot-scale experiment known as hybrid biological fixed-film process, which included two-stage biofilm treatment and two-stage filtration, was conducted in Zhongtai Chemical Co. Ltd. The results indicated that the centrifugal mother liquid of PVC after treatment could reach the water quality standards for further processing with COD_{cr} less than 50 mg/L, NH_3–N less than 5 mg/L, turbidity about 1 NTU, and pH 7~9.

1 INTRODUCTION

Polyvinyl Chloride (PVC) is one of the most commonly used thermoplastic materials worldwide (Sadat-Shojai, 2011), and data show that the annual demand of PVC is more than 40 million tons. The production process of vinyl chloride monomer can be divided into suspension, bulk, emulsion, and micro-suspension. Among them, the production of suspension PVC accounts for nearly 90% of the total PVC production (Zhang, 2003). PVC production process generates large amounts of centrifugal mother liquor, and the data show that the production of 1 ton of PVC needs 3–5 tons of deionized water (Li, 2001). Furthermore, the environmental problems of PVC wastewater should not be ignored.

It has been confirmed by many researchers that the fixed or suspended submerged biofilm systems are very effective and efficient in the removal of nitrogen and organic carbon by means of the attached growth biofilm (El-Shafai, 2013). Compared with the activated sludge process, the biofilm process shows several advantages such as stability and long retention time of microorganisms, which enable it to remove recalcitrant pollutants to reduce the surplus biomass (Liu, 2007). Because of the longer solid retention time of the biomass, the biofilm can operate at low hydraulic retention time, which results in less system footprint. However, the biofilm systems submerged by continuous flow have been proved efficient in nitrogen removal.

In conclusion, the biofilm process based on the up-flow submerged biofilm reactor is stable and flexible, and it represents a compact system with low sludge production and maintenance cost. Hybrid biological fixed-film process, which is used in the experiment, is a combined process based on biofilm reactor. The experiment aims at demonstrating the feasibility of biofilm system for industrial PVC wastewater preliminary treatment. In preparation for further treatment, such as membrane process for making effluent back to reactor, the effluent should meet the standard of circulation cooling water.

2 METHOD

2.1 *Microorganisms and culture conditions*

Microorganisms used for the HBF process had been pre-prepared as a dry powder mixed with a considerable amount of nutrients required for the activation. Powdered microorganisms (0.5 kg) were added to 250 kg of mixed liquor composed of tap water and PVC wastewater. The activation process was carried out without influent and effluent.

The mixed liquor went through an inner loop between the reactor and the buffer tank in a biological filming process after continuous aeration for 2 h in the tank. The biological film was produced when COD_{cr} was lower than 20 mg/L and the time span was about 24 h.

2.2 *Wastewater*

The PVC wastewater was directly obtained from suspension polymerization process of PVC resins in Zhongtai Chemical Co. Ltd. As is shown in Table 1, the wastewater contains Polyvinyl Alcohol (PVA), antifouling agent, heat stabilizer, terminating agent, and Vinyl Chloride Monomer (VCM),

Table 1. Composition of wastewater.

pH	CODcr (mg/L)	Conductivity (μs/cm)	NH$_3$–N (mg/L)	P (mg/L)
8 ~ 9	300 ~500	>250	10 ~ 15	<0.2

which might be toxic to microorganisms, and hence the wastewater shows weak biodegradability.

According to the composition and metabolic properties of cells, the growth and reproduction processes of microorganisms require a certain amount of nutrients. Considering the lack of phosphorus to maintain nutritional balance of the organism, H$_3$PO$_3$ (6 kg/month) is added to the wastewater (Li, 2008).

3 EXPERIMENTAL SETUP AND PROCEDURE

As shown in Fig. 1, PVC wastewater was first pumped into the HBF first-stage reactor after adding the necessary nutrients and adjusting the pH in the buffer tank T–01. The effluent from the upper overflow outlet of the HBF first-stage reactor was stored in tank T-02 before pumping into the HBF second-stage reactor. A filter was set up before the HBF second-stage reactor. Similarly to the HBF first-stage reactor, the effluent from the HBF second-stage reactor was stored in a same size tank T-03, then it went through the adsorber F-02, and finally met the standard for re-using or further process. As is shown in Fig. 1, the HBF reactor (overall dimensions: 1.046 m (width) × 0.523 m (length) × 4 m (height)) was filled with homogeneous quartz filler, and a gas–liquid distributor was installed right under the supporting layer. The filter and the adsorber, which were made of Fiber-Reinforced Polymer (FRP), had the same dimension (0.35 m (diameter) × 1.6 m (height)), but with different fillers. The filter was filled with natural quartz sand with an average diameter of 1 mm, while the adsorber was filled with granular activated carbon whose particle diameter was 4 mm. The HBF first-stage reactor set a discharge cycle of 72h and discharge span of 5 min with sewage valve in full open position while the HBF second-stage reactor used the air–water backwash with the same cycle time and span. During the air–water backwash process, the gas was first flushed at 18 L/m^2.s for 2 min, then water was added, rinsing the air and water for the next 3 min. The amount of backwater needs to be adjusted according to the quality of the effluent. The filter backwash cycle was set at 12h and backwashing time at 5 min. And the adsorber backwash cycle was set comparably at 240h and backwashing time at 5 min.

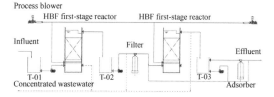

Figure 1. Pilot process flow diagram.

4 RESULTS AND DISCUSSION

The carbon-to-nitrogen ratio (C/N) of PVC wastewater was about 20, and the content of phosphorus was less than 1 mg/L.

The theoretical optimum C:N:P for microbial growth is 100:18:2. With reference to the optimum value, research shows that N, P may limit the water insufficient effect, and the situation is determined for the experiment. It is indicated that the lack of N, P in the PVC wastewater might limit the effect of sewage treatment, but it needs to be verified in the experiment.

4.1 Isolation and identification of bacteria

Two major microorganisms (*Bacillus subtilis* and *Bacillus amyloliquefaciens*) were analyzed from the on-site sampling detection through Polymerase Chain Reaction (PCR) and other biological methods, which could degrade organic substances in wastewater. *Bacillus subtilis* can secrete large amount of amylase, protease, and lipase in the breeding process, which can rapidly degrade organic substances such as starch, protein, and fat in wastewater. Under the combined effect of other microorganisms in the pool, the major part of organic substances were further decomposed into carbon dioxide and water, while a small portion was used to synthesize new cells. Finally, the wastewater was purified through those biological processes.

Bacillus subtilis in the form of spores can withstand a variety of harsh environment and inhibit matrix, and has been widely used in high concentrations for livestock and aquaculture water purification. Furthermore, *Bacillus subtilis* preparation works stably and can remove ammonia nitrogen, nitrite nitrogen, and CODcr and reduce the number of pathogens effectively.

The antibacterial substances of *Bacillus amyloliquefaciens* are mainly antimicrobial proteins, lipid peptides, and other polyketides. *Bacillus amyloliquefaciens* degrades butachlor and phenol in contaminated wastewater. In addition, it can dissolve anabaena blooms, and can thus be used to control eutrophication.

4.2 Removal efficiency of COD_cr

Experiments of PVC wastewater mainly aim to improve water quality for further processing such as membrane treatment. Therefore, adding little amount of substances to meet the demand of removing COD_{cr} to the acceptable capacity is considered very important.

Overall experimental process can be divided into four parts, as shown in Fig. 2. The biofilm start phase is before the first stage, in which there is no influent or effluent. The mixed PVC wastewater and tap water comprised a reflux circulation system between buffer tank and HBF reactor. The continuous influent was maintained constant at 150 L/h in the first stage. From these data, we can see that although the influent is volatile, the CODcr of the effluent in HBF second-stage reactor can stabilize at around 40 mg/L quickly, indicating that the fluctuations of PVC wastewater in the system are within the acceptable range, and the system still has great potential to improve its processing capacity.

After stabilization, the influent gradually increases to 600 L/h and the experiment attains the second phase. It can be seen from the chart that the effluent quality has deteriorated slightly with the influent growth before 450 L/h, but the efficiency of removing COD in HBF first-stage reactor remains relatively stable at the rate of 40%.

When the influent growth rate exceeded 450 L/h, the effect of the HBF first-stage reactor deteriorates significantly with effluent quality down to 220 mg/L and a lower COD removal rate of only 25%, indicating that when the hydraulic load was lower than 450 L/h, the sewage treatment capacity was most efficient under the current conditions.

In the third stage, the influent of HBF first-stage reactor is maintained at 400 L/h without reflux. After gradually increasing the influent of HBF second-stage reactor to 900 L/h, which equaled the reflux reaching 500 L/h, the efficiency of CODcr removal rate is the highest.

In the fourth stage, the effect of nitrogen and phosphorus on the treatment was further explored. Considering the fact that only a small amount of nitrogen and phosphorus (ammonia and nitrogen: 12 mg/L; phosphorus: 0.5 mg/L) was contained in the PVC centrifugal mother liquor, the effect of phosphorus on the biochemical reactor was first considered. Phosphoric acid (diluted to 2.89%) was added into the buffer tank T-01 (50 ml/day) as a phosphorus source. In this situation, the nitrogen content of the effluent was about 4 mg/L. As a result, the system did not lack nitrogen and needs to add nothing more. After adjustment, CODcr of the effluent dropped to below 50 mg/L and the removal rate was about 85%.

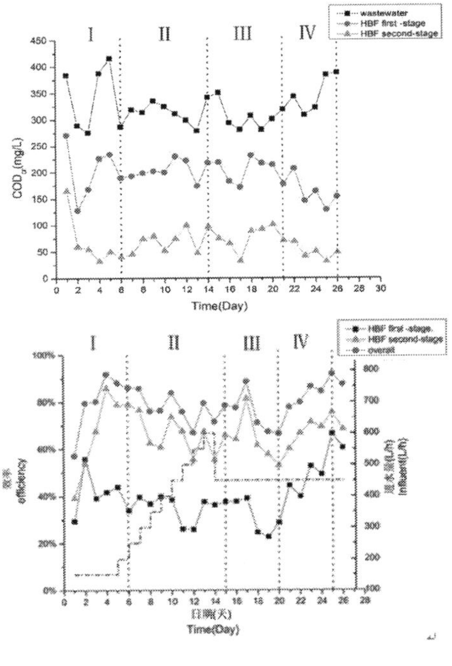

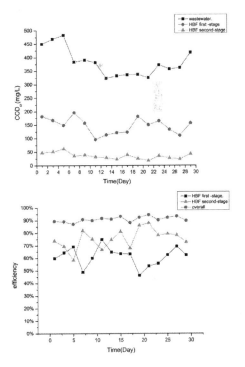

Figure 2.   Removal rate of COD_{cr} during the adjustment process over time.

Figure 3.   Removal rate of COD_{cr} during the repeated experiment over time.

To verify the effect of repeated industrial applications, the experiment was carried out again. The optimal operating parameters of the previous experiment were maintained, and the system entered the stable state quickly. It was found that the effect of the experiment is better than before through the statistical analysis in Fig. 3. The final removal rate of COD_{cr} maintained at about 90% and the efficiency of the HBF first-stage reactor increased dramatically in particular, considering only the environmental temperature.

The experiment was a natural cooling process. The environment temperature of the first experiment was lower than the second, and the temperature drop was obvious when the mother liquor flowed through the buffer tank. In addition, the influent temperature of HBF secondary-stage reactor was generally 1–3°C lower than that of the first-stage reactor, since no heating device was used in the process. Taking into account the general heat resistance of microorganisms, the best influent temperature was 30–35°C. According to the statistical analysis, the COD_{cr} removal rate is higher than 85%, showing the excellent adaptability of the process.

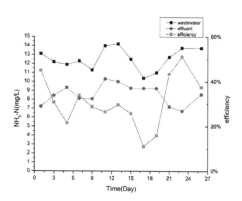

Figure 4. Removal of NH_3–N.

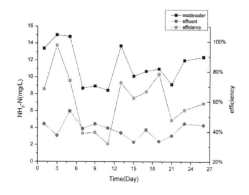

Figure 5. Removal of NH_3–N in contrast test.

4.3 Removal effect of nitrogen

Nitrogen is the necessary nutrient for the microorganisms in wastewater. The wastewater requires a certain proportion of carbonaceous compounds and nitrogen when microbes decompose the organic compounds, that is, C/N ratio. The high proportion of carbon and nitrogen and the high concentration of organic compounds will lead to aerobic heterotrophic bacteria predominating in the competition for oxygen with the nitrobacteria. Therefore, the process of nitrification is affected.

The PVC centrifugal mother liquid used for the experiment contains about 12 mg/L of ammonia and C/N is nearly 30, which is much larger than the optimum ratio mentioned above, but the result of the experiment is satisfactory. It can be seen from Figs. 4 and 5 that the effect of the contrast test is better than the initial value with ammonia less than 5 mg/L in the effluent.

5 CONCLUSION

The results of this pilot-scale experiment show that HBF process is quite suitable for industrial pre-treatment of the high-temperature PVC wastewater. The quality of effluent reaches the index of cooling circulating water, which is an essential precondition for advanced treatment.

Studies show that the system starts quickly and is quite stable against fluctuation of wastewater quality. Furthermore, the system shows strong ability to decompose COD_{cr} with minimal addition of other agents. In addition, the effect of NH_3–N decomposition is quite satisfactory under high carbon-to-ammonia nitrogen ratio.

REFERENCES

El-Shafai, S. A., & Zahid, W. M. (2013). Performance of aerated submerged biofilm reactor packed with local scoria for carbon and nitrogen removal from municipal wastewater. *Bioresource technology*, 143, 476–482.

Li Maoshuang, & Zhang Long. (2001). Studies on the treatment and reuse of PVC wastewater. *Qilu Petrochemical Technology*, 29(3), 211–214.

Li Y & Wu X. (2008). Development of production technology of PVC and its market analysis. China Chlor-Alkali, 10:1–3.

Liu, H., Yang, F., Wang, T., Liu, Q., & Hu, S. (2007). Carbon membrane-aerated biofilm reactor for synthetic wastewater treatment. *Bioprocess and biosystems engineering*, 30(4), 217–224

Sadat-Shojai, M., & Bakhshandeh, G. R. (2011). Recycling of PVC wastes.*Polymer Degradation and Stability*, 96(4), 404–415.

Zhang Qiang, & Liu Ying. (2003). Production and development of PVC production by suspension. *Polyvinyl Chloride*, 3, 26.

Advances in Energy, Environment and Materials Science – Wang & Zhou (Eds)
© 2017 Taylor & Francis Group, London, ISBN 978-1-138-03600-0

A study on the process of preparing dehydrated instant rice by enzymatic hydrolysis

Chunyu Xi, Mengni Cui & Wenyu Zhang
JILIN University, China

ABSTRACT: Compound enzyme is added with α-amylase and β-amylase to form dehydrated instant rice, which can change the structure of rice starch amylose and amylopectin content and ratio to inhibit the retrogradation degree of rice. We prepared α-amylase and β-amylase compound enzyme and determined its concentration rate, soaking time, and soaking temperature, which are the most important factors according to the main document. The best role conditions range of various factors were determined by the experiment, including enzyme concentration ratio, soaking time, and soaking temperature that can be set to three different levels, and the gelatinization degree was set as the y value. Finally, we designed the orthogonal experiment using the L_9 (3^4) form to analyze data, which can ensure the best role conditions of hybrid enzymes. In this paper, the best enzyme concentration ratio (beta/alpha), the optimum soaking temperature, and the optimum soaking time are found to be 9:1, 45°C, and 36 min, respectively.

1 INTRODUCTION

Rice is one of the major grain varieties in the world and about half of the population take rice as the staple food, as it can provide the basic ingredients including protein, carbohydrates, various vitamins, and minerals that humans need (Jin, 2002; Kang, 2007; Liu, 2008). After research and analysis for many years, it was widely accepted that rice starch, amylose starch, amylopectin proportion, fat content, protein content, and moisture content of rice may have an important influence on rice quality (YauN J N, 2013; Yu, 2012; Zhou, 2012).

Although the market has prospects for good development of instant rice, China's instant rice industry started late, and hence of the related technologies are not mature. A variety of instant rice have been developed due to product quality problems caused by the mature process. The self-heating instant rice relies on the heating layer and the water to heat the reaction of heating steamed rice, but there are many unstable factors due to the current self-heating technology, for example, sometimes the temperature is so low that it cannot meet the requirements; there should be a separate self-heating layer; the process cost is high; the promotion is very limited. Fresh instant rice tastes good, but needs to be heated by microwave with high hardware requirements, so it less convenient. However, return dehydration rate of instant rice is too high, and it tastes poor after heated by hot water, thus the development is restricted. Therefore, it is important to improve the production technology of dehydrated instant rice in order to get better quality.

The aim of our experiment is to solve the regenerative problem of dehydrated instant rice. According to the research, at present, the main methods to inhibit the retrogradation degree of rice include chemical emulsifiers and other food additives. The research by Qingjie Sun (2013) showed that the regenerative inhibitory effect is more significant by the conditions of β-amylase concentration of 0.111% and 41.4 min at 59.8°C. The study by ShuYi (2011) showed that when the proportion of cellulase and amylase is 1:4, the rehydration rate of instant rice is the highest. The research by Fengy Cai et al. (2013) showed that the retrogradation degree of steamed rice decreased when the additive amylase dose of 0.3 g/L, sucrose ester dose of 8G/L, and beta amylase dose of 2 g/L worked together in certain range

2 MATERIALS

2.1 *Rice*

5 kg Wuchang Rice
Execution standard: GB/T 19266
Production license number: QS230101021035

Table 1. Basic information of rice sample.

Entry	Content (/100 g)	Nutrient reference value (%)
Energy	139 kJ	17
Protein	7.3 g	12
Fat	0 g	0
Carbohydrate	75.7 g	25
Na	0 mg	0

2.2 α-amylase

G8290 Beijing Science & Technology Co. Ltd

Table 2. Basic information of α-amylase.

Name	Condition
pH	5.5–7.5
Temperature	50–70°C
Enzyme activity	10 W energy unit/g
BR	Food grade biochemical reagents

2.3 β-amylase

G8290 Beijing Science & Technology Co. Ltd

Table 3. Basic information of β-amylase.

Name	Condition
pH	5.5–7.5
Density	About 40 eyes
Water content	≤ 8%
Enzyme activity	10 W energy unit/g
BR	Food grade biochemical reagents

3 METHODS

First, put 20 g of rice into a beaker and mark its level with an accuracy of 0.1 g. Then, wash the rice with a weighed amount of distilled water, and soak it in 30 ml of distilled water and stir it for 10 s using a clean glass rod. Then, filter and drain the rice for three times. Dump 20 ml of prepared hybrid enzymes (Sidhu, 2009) into the beaker containing the rice sample. Use water bath pot as the heat source, set the specified temperature, and soak it for specified time. Pour the rice soaked into the rice cooker, and then pour 30 ml of distilled water for cooking. Then, switch on the power supply and start cooking. The rice should be finally discrete and dried when there is no white heart.

The cooked rice sample was tested by various methods such as Differential Scanning Calorimetry (DSC), by taking 2.0–2.5 mg of dry sample in a crucible covered by lid, pressing into the sample, and placing in the DSC equipment. The temperature was maintained between 30 and 200°C and the heating rate was 20°C/min.

4 RESOLUTION AND ANALYSIS

4.1 Single factor experiment

After the above experiment, we can conclude that the gelatinization degree of dehydrated rice

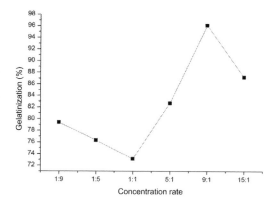

Figure 1. Degree of gelatinization of concentration rate of α-amylase and β-amylase.

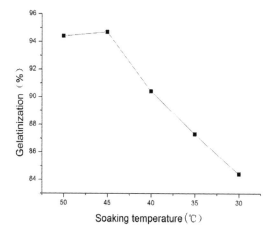

Figure 2. Degree of gelatinization of soaking temperature.

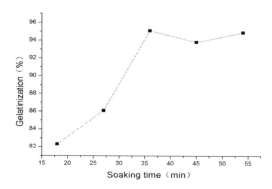

Figure 3. Optimal soaking period of α-amylase and β-amylase.

Table 4. Orthogonal factor level table.

Factor level	Enzyme concentration ratio (β/α)	Temperature (°C)	Time (min)
1	5:1	40	27
2	9:1	45	36
3	15:1	50	45

Table 5. Range analysis of orthogonal experiment design of anabiosis problem composite enzyme inhibition of dehydration convenient rice.

Number	A Enzyme concentration ratio (β/α)	B Temperature (°C)	C Time (min)	D Empty column	Y gelatinization (%)
1	1	1	1	1	83.89
2	1	2	2	2	90.91
3	1	3	3	3	87.05
4	2	1	2	3	93.23
5	2	2	3	1	96.45
6	2	3	1	2	95.31
7	3	1	3	2	91.05
8	3	2	1	3	93.03
9	3	3	2	1	94.56
K_1	87.28	89.39	90.74	91.63	
K_2	95	93.46	92.9	92.42	
K_3	92.88	92.31	91.52	91.1	
R	7.713	4.073	2.157	1.3	
Excellent level	A_2	B_2	C_2	D_2	

decreased with the increase of beta/alpha ratio from 1:9 to 1:1. Therefore, the best enzyme concentration ratio (beta/alpha) is 9:1. In accordance with this, the best soaking time and soaking temperature are 36 min and the 45°C, respectively.

4.2 Orthogonal experiment

The single factor experiments show that the factors affect each other, and hence we designed orthogonal experiment, selecting three factors, namely enzyme concentration ratio, soaking temperature, and soaking time, which should be set to three levels; and degree of gelatinization was the y value using the L9 (34) form (Wang, 2007).

It is evident from the test data of Table 5 that the most obvious factor among the primary and secondary factors A > B > C > D was soaking temperature, followed by enzyme concentration, soaking time, and the optimal combination $A_2B_2C_2$. In other words, when the enzyme concentration ratio (beta/alpha), soaking temperature, and soaking time were 9:1, 45°C, and 36 min, respectively, the gelatinization degree reached the highest level, and, at the same time, the retrogradation degree of dehydrated rice was the most optimum.

5 CONCLUSION

The experiment results showed that the composite enzyme prepared by alpha amylase and beta amylase and the specific process conditions of rice soaking can reduce the degree of retrogradation of rice starch, in addition to enhancing the organoleptic quality of dehydrated instant rice, which can avoid the "hard core" phenomenon. Analysis of orthogonal test shows that the optimal enzyme concentration ratio (beta/alpha), soaking temperature, and soaking time were 9:1, 45°C, and 36 min, respectively. Under this condition, the degree of gelatinization of the dehydrated instant rice is maximum (Fan, 1998).

REFERENCES

Ding Wenping, Wang Yuehui, Ding Xiaosen. Study on [J]. grain and feed industries rice starch gelatinization and retrogradation mechanism, 2013,43 (3): 11–21.

Fan J, Marks B P. Retrogradation kinctics of rice flours as influenced by cultivar[J]. Cereal Chem, 1998, 75(1): 153–155.

Feng Jin, The development trend of the [J]. food packaging, 2002, 34 (7): 36–37.

Kang East, Jinfeng He, Xichang Wang. Steamed Rice convenient production situation of our country and the research prospect of [J]. food industry, 2007, 32 (1): 40–42.

Kun LianTang. Starch pasting, aging characteristics and food processing [J]. Shaanxi Cereals & oils Science & technology, 2012, 9 (3): 26–29.

RuiWang, XiaojunMa. Study on the effect of [J]. amylases on aging Chinese instant Steamed Rice Journal of grain and oil, 2007, 36 (4): 14–17.

Shifeng Yu, Ying Ma. Research progress on rice quality influencing factors and its aging mechanism [J]. food industry, 2012, 1 (29): 285–288.

Sidhu, Xiong Liu. Response surface method optimize beta—amylase inhibitor glutinous amylopectin retrogradation process [J]. Food science and technology, 2009, 26 (3): 158–161.

Siming Zhao. Study on the characteristics and aging mechanism of rice starch [D]. Ph. D thesis, Huazhong Agricultural University, 2011.

Wei Liu, sun. Present situation and development trend of convenient rice market [J]. food and food industry, 2008, 15 (1): 3–5.

Wei Zhou, Yuanzhi Li. Study on the aging characteristics of instant rice [J]. modern food science and technology, 2012, 28 (5): 505–507.

YauN J N, Huang J J. Sensory analysis of cookedrice[J]. Food Quality and Prefernce, 2013, 34(7): 263–270.

Yi, Shu DiZhao. Enzymatic anti retrogradation Steamed Rice convenient drying technology research [J]. food industry, 2011, 16 (8): 28–29.

Yixuan Chen, Ying Zhang. The present situation of the development of rice anti aging [J]. Beijing agriculture, 2013, 28 (6): 119.

Advances in Energy, Environment and Materials Science – Wang & Zhou (Eds)
© 2017 Taylor & Francis Group, London, ISBN 978-1-138-03600-0

Effects of controlled release and stabilized fertilizers application on soil nitrogen nutrition and related enzyme activities of *Bougainvillea glabra* cultivation

Y. Shi, M. Zhao & M. Li
Hainan Tropical Ocean University, Sanya, China

ABSTRACT: The effects of two kinds of controlled release fertilizers (sulfur coated urea SCU, and polymer coated urea PCU) and a stabilized fertilizer (urea blended with nitrification inhibitor dicyandiamide, UD) application on soil nitrogen (N) nutrition and related enzyme activities of a flower shrub *Bougainvillea glabra* cultivation were studied with a pot experiment. The results showed that the total amount of inorganic N in SCU and PCU treated soil was significantly lower than that of urea-only (U) treated soil within the former 45 d after fertilization, while was significantly higher than that of U treatment within 45–120 d, because of the slow release of N nutrition in SCU and PCU treatments. The application of nitrification inhibitor dicyandiamide (DCD) slowed down the nitrification of NH_4, resulting in a higher NH_4^+–N and inorganic N content in soil compared with U treatment. Overall, the three kinds of soil enzyme activities (urease activity UA, nitrification potential NP, and nitrate reductase activity NRA) related to N transformation performed the order SCU > PCU > U, during the experimental period of 120 d, and DCD as a nitrification inhibitor significantly reduced the NP of soil.

1 INTRODUCTION

N is the main component of protein, nucleic acid, chlorophyll and many kinds of enzymes and vitamins in plants. Therefore, the level of N in soil directly affects the yield and quality of crops (Zhu et al, 2011). At present, with the huge demand for agricultural production, crop yield per unit area has been greatly improved, and the N in the soil has been unable to meet the needs of agriculture. People tend to supplement N nutrition by applying fertilizers to agricultural systems. The release of the quick acting fertilizer N usually disaccords with the law of crop nutrient need, resulting in excessive soil N nutrient and low utilization rate of fertilizer. The controlled release fertilizer can reduce the release rate of nutrient elements (González et al, 2015), and the stabilized fertilizer added with nitrification inhibitors can slow down the oxidation rate of NH_4^+–N in soil (Reiner and Rudolf, 2015). So the application of these two types of fertilizers can obviously reduce the loss of nutrient elements, decrease the greenhouse gas emissions, prevent the pollution of fertilizer application to the environment, and increase the utilization rate of fertilizers.

Soil enzymes play an important role in soil ecosystem, because it is a key participant in the material circulation and energy flow in soil, and is an essential factor of soil metabolism. The soil enzyme activity is closely related to soil physical and chemical properties and environmental conditions, so it is the early warning index and sensitive indicator of the changes of soil ecosystem (Bowles et al, 2014). Fertilization is one of the most profound agricultural measures to affect the soil quality evolution and its sustainable utilization. The contents of soil nutrients and the activities of enzymes in the soil could be significant affected by the application of different types of fertilizers.

The tropical and subtropical regions of southern China have the natural advantage of developing flower industry. Flower industry is one of the high investment, high technology, high yield, and high efficiency industries. During the growth and development of flower plants, it is necessary to use a variety of large, medium and micro elements. Therefore, it is necessary to apply fertilizers to meet the growth needs of the flowers. The flower fertilizers commonly used are organic fertilizers, nutrient water soaked with animal and plant residue, and inorganic fertilizers. It is easy to burn seedlings if the application of inorganic chemical fertilizers is not reasonable. Organic composts can not avoid the bad smell and sensory, and often cause seedling diseases and insect pests. Therefore, controlled release fertilizer and stable fertilizer may be the best choice for the development of flower industry. This is the reason why most of the controlled release fertilizer produced in the world is consumed in the market of flowers and lawn at present. The purpose of this study was to investigate the effects of controlled release fertilizer and stabilized fertilizer application on soil N nutrient and related enzyme activities of the shrub plant *B. glabra* cultivation. We hope that the results of the study can guide the management of soil nutrients and the choice of fertilizers for the shrub flowers cultivation.

2 MATERIALS AND METHODS

2.1 *Soil, plant and fertilizers*

In the present study, a paddy soil originated in granite latosol was used. The soil was collected from Miaolin Farm at Fenghuang town, Sanya, Hainan Province, China (18.305°N, 109.463°E), with a pH of 5.24 (soil:water = 1:2.5, w/v), organic matter of 18.65 $g \cdot kg^{-1}$, total N of 1.16 $g \cdot kg^{-1}$, NH_4^+–N of 6.2 $mg \cdot kg^{-1}$, NO_3^-–N of 17.4 $mg \cdot kg^{-1}$, and CEC of 6.37 $cmol \cdot kg^{-1}$. The fresh soil sample (0–20 cm) was collected and returned to the laboratory to remove impurities and residual roots, then pass through a sieve (2 mm) after naturally dried.

The plant material cultivated in this study was *B. glabra* Choisy, a kind of evergreen shrub belonging to the family Nyctaginaceae. The N fertilizers used in the present study were urea (containing N 46%), sulfur coated urea (containing N 35%), polymer coated urea (containing N 42%), and stabilized urea (containing DCD 5%). In addition, superphosphate and potassium sulfate were applied when the plant cultivation.

2.2 *Experimental design*

The pot experiment was carried out in the solar greenhouse of Hainan Tropical Ocean University from April to July in 2015. Five treatments were conducted: CK the control no N fertilization applied; U urea applied; SCU sulfur coated urea applied; PCU polymer coated urea applied; UD stabilized urea containing nitrification inhibitor DCD applied. Eighteen pots per treatment were prepared. The amount of N added to all the treatments was 200 $mg \cdot kg^{-1}$ except for the CK treatment, and the same amount of P (equivalent to 0.3 $g \cdot kg^{-1}$ P_2O_5) and K (equivalent to 0.2 $g \cdot kg^{-1}$ K_2O) fertilizers were applied in all the treatments. All the fertilizers were applied as base fertilizers when transplanted and other management measures were implemented according to the local triangle plum cultivation technique.

2.3 *Test items and methods*

The soil of subsample of three pots was taken after 15, 30, 45, 60, 90, and 120 days of transplant, respectively, to measure the concentration of NH_4^+–N and NO_3^-–N, the Urease Activity (UA), the Nitrification Potential (NP), and the Nitrate Reductase Activity (NRA). The content of inorganic N in soil was the sum of NH_4^+–N content and NO_3^-–N content. The soil UA was determined with urea-N residual method, the soil NP was determined with NO_2^-–N production method, and the soil NRA was determined with the Kandeler et al (1994) method.

3 RESULTS AND DISCUSSION

3.1 *Effects of controlled release and stabilized fertilizers application on soil N nutrition of B. glabra cultivation*

The contents of NH_4^+–N, NO_3^-–N and inorganic N in the soil of CK treatment with no chemical N fertilizer application fluctuated in a relatively low level of about 8, 12 and 20 $mg \cdot kg^{-1}$, respectively, within 120 d after fertilization and transplantation (Fig. 1). The content of NH_4^+–N in the soil of U treatment applied with urea decreased from nearly 200 $mg \cdot kg^{-1}$ to about 50 $mg \cdot kg^{-1}$ rapidly within 15–30 d, then decreased slowly. Accordingly, the content of NO_3^-–N in the soil of U treatment increased rapidly from about 30 $mg \cdot kg^{-1}$ to 100 $mg \cdot kg^{-1}$ (15–30 d), and then gradually reduced to the level of CK before the end of the experiment. During the period of experiment, the content of inorganic N in the soil of U was declined rapidly and steadily, from about 220 $mg \cdot kg^{-1}$ to 30 $mg \cdot kg^{-1}$. The contents of NH_4^+–N, NO_3^-–N and inorganic N in the soil treated with controlled release fertilizer (SCU or PCU) were significantly lower than the U treatment within 15–60 d, while obviously higher than that of U treatment after 60 days. The effect of PCU in general was slightly better than that of SCU. Due to the addition of nitrification inhibitor DCD in stabilized fertilizer, the content of NH_4^+–N in the soil of UD treatment was significantly higher than that of U treatment, while the content of NO_3^-–N was lower than the U treatment, and the total amount of inorganic N was slightly higher than that of U treatment. Overall, because the N nutrition released slowly in the controlled release fertilizers, and the release duration was prolonged, the amount of N in the soil treated with SCU and PCU was more balanced compared with U during the period of experiment of about 4 months. This was in favor of the flowers to absorb N nutrient from the soil continuously. The nitrification inhibitor DCD in stabilized urea retarded the oxidation of NH_4^+, resulting in a relatively low content of NO_3^- in soil, so the denitrification and leaching effects in the soil of UD were less than that of U, lead to a relatively higher

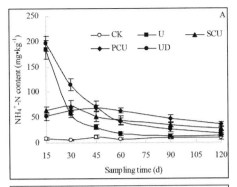

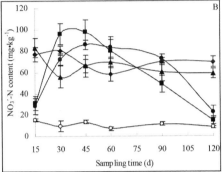

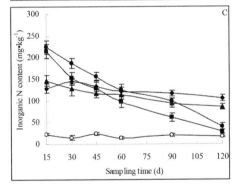

Figure 1. The dynamics of NH_4^+–N (A), NO_3^-–N (B), and inorganic N (C) contents in the soil of *Bougainvillea glabra* cultivation treated with different types of fertilizers.

content of inorganic N which could be utilized easily by plants.

The objective of efficient use of N fertilizer is to maximize the N utilization rate, minimize the NH_3 volatilization, NO_3^- leaching and denitrification, and promote the production-increase effect of N application (Qiao et al, 2012). The content of inorganic N can be used to characterize the level of available N to plants in the soil. The results of this study showed that the controlled release fertilizers and the stabilized fertilizer could effectively increase the content of inorganic N in the soil

during the period of the experiment of about 4 months, increasing the level of soil N supply and the utilization rate of fertilizer. In general, the purpose of agricultural production is harvest crops. Whereas, the ultimate goal of the flower cultivation industry is the lasting green and blooming period, the beautiful and strong plants and the simplified management procedure. The nutrient demand of different kinds of flowers generally have diverse peak. The purpose of applying controlled release fertilizers is to make the nutrient supply and plant demand synchronous. For the woody flowers, the demand for N nutrient in each stage is different from seedling stage to flowering stage. In this experiment, the growth of *B. glabra* was vigorous, and there was a relatively high demand for N fertilizer, especially in the prosperous period. The content of available nutrient in soil treated with controlled release fertilizers was basically consistent with the demand of *B. glabra*, while the content of N nutrient in urea treated soil was too high in the early stage of plant growth, but was too low in the middle and late stages. In the present experiment, soil treated with controlled release fertilizer increased inorganic N content markedly compared with fast-effective fertilizer in the middle and late stage of plant growth, gratified the demand of soil N in the critical period of the growth of plant, overcome the shortcomings of conventional fertilizer that the nutrient supply was insufficient in the late growth stage of *B. glabra*, and reduced the waste of fertilizer due to the excessive N in the early stage.

3.2 *Effects of controlled release and stabilized fertilizers application on soil enzyme activities of B. glabra cultivation related to N transformation*

As shown in Fig. 2, the UA, NP and NRA of CK treatment soil were all in a low level during the whole experimental period. The UA of the U treatment soil was significantly higher than other treatments in the first 15 d, and then decreased rapidly within 15–45 d, and then stabilized. The NP and NRA of U treatment soil did not fluctuate significantly during the experimental period. The UA of controlled release fertilizer (SCU or PCU) treatment soil was higher than the other treatments after 45 d, while the NP and NRA were higher than other treatments throughout. Compared with PCU, the soil treated with SCU had a higher NP and NRA, but had a lower UA. The NP of UD treatment soil was lower that the soil of other treatments during the whole experimental period, due to the addition of nitrification inhibitor DCD.

Soil enzyme is one of the important components of soil, which mainly comes from the root of higher plants and soil organisms. Soil enzyme directly

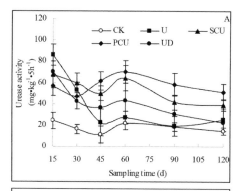

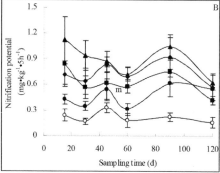

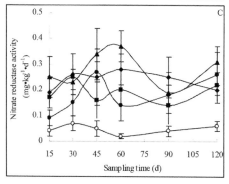

Figure 2. The dynamics of urease activity (A), nitrification potential (B), and nitrate reductase activity (C) of the *Bougainvillea glabra* cultivation soil treated with different types of fertilizers.

involves in the transformation of matter, nutrient release and fixation process, and is closely related to the soil nutrient. Soil urease is the only enzyme which has a significant effect on the transformation of urea, because the urea applied into soil can only be hydrolyzed in the presence of urease. NH_4 as the enzymatic reaction product of urease is one of the important N sources for plants, so the soil UA can be used to characterize the soil N supply capacity (Sanz-Cobena et al, 2008). The controlled release fertilizer slowed the dissolution of urea effectively, resulting in a certain level of urea in the

soil for a relatively long time, providing substrate for urease and enhancing UA by stimulating urease producing bacteria continuously. Because of the different coating materials of PCU and SCU, the nutrient release characteristics were diverse, lead to different contents of urea in soil and distinguishing stimulating effects to urease.

Soil NP is the ability of soil ammonia oxidizing microorganism to convert NH_4^+-N to NO_3^--N. The greater the soil NP is, the faster the transformation of NH_4^+ will be, the more the N leaching loss will be, and the larger the NO_3^- denitrification potential will be. Therefore, the reduction of soil NP can increase the utilization rate of N fertilizer and reduce the environmental pollution. The results of this study showed that nitrification inhibitor DCD combined with urea could effectively reduce the soil NP, slowed down the production rate /of soil NO_3^-, increased the content of NH_4^+-N and total amount of inorganic N in soil (Monaghan et al, 2013). The soil nitrate reductase is one of the soil important enzymes which involves in the process of reducing nitrate to nitrite (Abdelmagid et al, 1987). Because the substrate of nitrate reductase is NO_3^-, there is a close relationship between the level of NRA and the content of NO_3^--N in soil: Commonly, the soil NRA will be promoted with the increase of soil NO_3^--N content.

Moreover, fertilization changed the secretion characteristics of plant roots, and then altered the soil microbial population structure and a variety of enzyme activities. The nutrient released from controlled release fertilizer was slow, which could meet the need of nutrient at every stage of the plants, so it was still able to ensure the substrate concentration of soil enzyme catalytic reaction until the later stage of the experiment. Whereas, the common urea treatment reduced the utilization rate of N fertilizer, lead to the lack of nutrient supply in the late stage, being adverse to the enzymatic reaction, so a variety of enzyme activities were relatively low. In this study, SCU in the aspect of increasing the activity of soil enzyme was the best, which might be the contribution of the sulfur in the coated material of the controlled release fertilizer. Research have showed that a large amount of sulfur applied into the soil could affect the effectiveness of elements, acid-base balance, and the redox balance of the soil (Vvsr et al, 1988).

4 CONCLUSION

During the period of experiment, the content of inorganic N in the soil of U was declined rapidly and steadily, from about 220 mg•kg⁻¹ to 30 mg•kg⁻¹. Because the N nutrition released slowly in the controlled release fertilizers, and the release duration was prolonged, the contents of NH_4^+-N, NO_3^--N and inorganic N in the soil treated with controlled release fertilizer (SCU or PCU) were significantly lower than the U treatment within 15–60 d, while obviously higher than that of U treatment after 60 days. Because the nitrification inhibitor DCD in stabilized urea retarded the oxidation of NH_4^+, the content of NH_4^+-N in the soil of UD treatment was significantly higher than that of U treatment, while the content of NO_3^--N was lower than the U treatment, and the total amount of inorganic N was slightly higher than that of U treatment.

The UA of the U treatment soil was significantly higher than other treatments in the first 15 d, and then decreased rapidly within 15–45 d. The NP and NRA of U treatment soil did not fluctuate significantly during the experimental period. The UA of controlled release fertilizer (SCU or PCU) treatment soil was higher than the other treatments after 45 d, while the NP and NRA were higher than other treatments throughout. Compared with PCU, the soil treated with SCU had a higher NP and NRA, but had a lower UA. The NP of UD treatment soil was lower that the soil of other treatments during the whole experimental period, due to the addition of nitrification inhibitor DCD.

ACKNOWLEDGMENT

This work was financially supported by the Scientific Research and Trial Production Program of Sanya (2014 KS11), the Agricultural Science and Technology Innovation Support Program of Sanya (2014 NK27), the Program for Hainan Cultivated Land Improvement (HNGDg121501) and its ancillary project (2015PT28), and the Key Laboratory Support Program of Sanya (L1505).

REFERENCES

Abdelmagid, H. M., M. A. Tabatabai, & H. M. Abdelmagid (1987). Nitrate reductase activity of soils. *Soil Biol. Biochem. 19*, 421–427.

Bowles, T. M., V.Acosta-Martínez, & F. Calderón (2014). Soil enzyme activities, microbial communities, and carbon and nitrogen availability in organic agroecosystems across an intensively-managed agricultural landscape. *Soil Biol. Biochem. 68*, 252–262.

González, M. E., M. Cea, & J. Medina (2015). Evaluation of biodegradable polymers as encapsulating agents for the development of a urea controlled-release fertilizer using biochar as support material. *Sci. Total Environ. 505*, 446–453.

Kandeler, E., G. Eder, & M. Sobotik (1994). Microbial biomass, N mineralization and the activities of various enzymes in relation to nitrate leaching and root distribution in a slurry-amended grassland. *Biol. Fert. Soils 18*, 7–12.

Monaghan, R. M., L. C. Smith, & C. A. M. de Klein (2013). The effectiveness of the nitrification inhibitor dicyandiamide (DCD) in reducing nitrate leaching and nitrous oxide emissions from a grazed winter forage crop in southern New Zealand. *Agr. Ecosyst. Environ. 175*, 29–38.

Qiao, J., L. Yang, & T. Yan (2012). Nitrogen fertilizer reduction in rice production for two consecutive years in the Taihu Lake area. *Agr. Ecosyst. Environ. 146*, 103–112.

Reiner, R., & S. Rudolf (2015). The effect of nitrification inhibitors on the nitrous oxide (N_2O) release from agricultural soils-a review[J]. *J. Plant Nutr. Soil Sci. 178*, 171–188.

Sanz-Cobena, A., T. H. Misselbrook, & A. Arce (2008). An inhibitor of urease activity effectively reduces ammonia emissions from soil treated with urea under Mediterranean conditions. *Agr. Ecosyst. Environ. 126*, 243–249.

Vvsr, G., J. R. Lawrence, & J. J. Germida (1988). Impact of elemental sulfur fertilization on agricultural soils. I. Effects on microbial biomass and enzyme activities. *Can. J. Soil Sci. 68*, 463–473.

Zhu, X., W. Guo, & J. Ding (2011). Enhancing nitrogen use efficiency by combinations of nitrogen application amount and time in wheat. *J. Plant Nutr. 34*, 1747–1761.

Advances in Energy, Environment and Materials Science – Wang & Zhou (Eds)
© 2017 Taylor & Francis Group, London, ISBN 978-1-138-03600-0

Ultraviolet light–assisted synthesis of reduced graphene oxide–Bi_2WO_6 composites with enhanced photocatalytic water oxidation activity

J. Liu & Y. Li
School of Environmental Science and Engineering, North China Electric Power University, Baoding, China

J. Ke
School of Chemistry and Environmental Engineering, Wuhan Institute of Technology, Wuhan, China

ABSTRACT: A facile ultraviolet reduction method was reported in this study to couple hierarchical Bi_2WO_6 (BWO) flower-like with reduced Graphene Oxide (rGO), producing rGO/BWO composite. The prepared rGO/BWO composite exhibits enhanced ability of photocatalytic water oxidation ($177 \ \mu mol \cdot L^{-1} \cdot h^{-1} \cdot g^{-1}_{cat}$), which is 2.3 times higher than that of pure Bi_2WO_6 ($77 \ \mu mol \cdot L^{-1} \cdot h^{-1} \cdot g^{-1}_{cat}$) under simulated solar light irradiation. In addition, the photocatalytic oxidation performance for organic pollutant has been also evaluated. The coupled photocatalyst displayed improved activity, where almost 100% of RhB was degraded in 1 h, 1.9 times higher than that of pure BWO. It was found that the reduction of graphene oxide plays a key role in raising photo-induced charge carrier separation efficiency as electron collector through the interaction between reduced graphene oxidation and BWO. Thus, the strategy provides an efficient approach for the fabrication of graphene composites containing hierarchical ternary oxides.

1 INTRODUCTION

Photocatalytic production of hydrogen and oxygen by splitting water is one of the most challenges in solving energy demand and environment because of its great potential in converting solar energy to chemical energy (Wang 2015). To date, extensive studies have been conducted on binary metal oxides, such as TiO_2 (Yuan 2015, Zong 2011, Kiss 2014), WO_3 (Pilli 2013, Kalanur 2013), Co_3O_4 (Wang 2014, Deng 2015), and Bi_2O_3 (Ke 2017). Nevertheless, efficient large-scale production of hydrogen is hindered by the Oxygen Evolution Reaction (OER) due to high energy demand for formation of O = O bond deriving from two molecules (Xie 2015). In addition, valence bands of many binary metal oxides are typically consisting of O 2p, which are disadvantaged their applications in O_2 evolution. For ternary metal oxides such as $BiVO_4$ (Wang 2011, Zhang 2015) and Bi_2WO_6 (Sun 2014, Yan 2013), the bands of ternary metal oxides are comprised of atomic orbitals from more than one element, and thus the element ratios can finely tune the valence and the conduction bands as well as the band gap energy, which could make them promising for visible light driven photocatalysts. Among these ternary oxides, Bi_2WO_6 has attracted particular interest due to its suitable band gap 2.8 eV corresponding to a photo-response region

up to 450 nm (Chen 2015, Saison 2013, Qamar 2014). More importantly, the conduction band of Bi_2WO_6 is composed of the W 5d orbital; its valence band is formed by the hybridization of the O 2p and Bi 6s orbitals, which not only makes the VB largely dispersed, but also favors the mobility of photo-induced holes for specific oxidation reaction (Zhang 2014). However, the photocatalytic performance of pure Bi_2WO_6 is limited by the poor efficiency of charge carrier separation and utilization.

Recently, coupling the metal oxides with highly conductive graphene appears a promising road to overcome challenges in charge carrier transportation and photo-response range (Roy 2013, Williams 2013). It was demonstrated that layered graphene could enhance photo-generated electrons transportation in semiconductor particles owing to the abundance of delocalized electrons from the conjugated sp2 bonded carbon network (Xu 2011). In addition, graphene could darken the composites, hence extending their absorption to entire visible light region. Nevertheless, owing to existence of much functional groups like hydroxyl and epoxide groups on the basal planes and carboxyl groups at the edges, the electron transportation between metal oxide and graphene is hindered, which is harmful for photocatalytic performance of semiconductor based photocatalysts (Lee 2011).

Herein, we report the synthesis of a composite by coupling hierarchical Bi_2WO_6 flowers with reduced graphene sheets. The composite shows a markedly enhanced performance of producing O_2 from water under solar light irradiation.

2 EXPERIMENTAL SECTION

2.1 *Preparation of GO*

Layered GO was prepared by the developed Hummer's method (Zhu 2011).

2.2 *Preparation of BWO flowers*

The hierarchical Bi_2WO_6 (BWO) flowers were prepared by a facile hydrothermal route. In typical procedure, 2 mmol of $Bi(NO_3)_3 \cdot 5H_2O$ was dissolved into 40 mL of diluted nitrate acid (0.3 M). 1 mmol of $Na_2WO_4 \cdot 2H_2O$ was dissolved in 20 mL of purified water. And then the later solution was added into the former one drop by drop, stirred for 30 min. Then, 20 mL of NaOH solution (0.2 M) was dropwise added into the above mixed solution with vigorous stirring for 24 h. Finally, the mixed solution was transferred to 125 mL autoclave and kept at 160°C for 8 h. The pale yellow precipitation was obtained and washed by purified water and absolute ethanol for several times. The BWO flower sample was dried at 60°C for overnight.

2.3 *Preparation of rGO/BWO*

0.015 g of GO was added to 50 mL of absolute ethanol, sonicated for 1 h to make GO dispersed. And then 0.5 g of Bi_2WO_6 was added into the mixture and stirred for 30 min. After that, the mixed solution was transferred to the dark box with ultraviolet light irradiation for 5 h. The brown sample was centrifuged and washed by purified water for 3 times. The powder sample was dried under 60°C for overnight.

2.4 *Characterization*

The crystal phase of synthesized samples were measured by X-Ray Diffraction (XRD) using a Rigaku D/max25 system operated at 40 kV and 40 mA with Cu-Ka radiation ($\lambda = 1.5418$ Å) at a scan rate of 5° · min^{-1}. Nitrogen adsorption measurements at 77 K were performed using an ASAP2020 volumetric adsorption analyser. Scanning Electron Microscopy (SEM) was performed on a JEOL JSM-6360LV field emission microscope at an accelerating voltage of 15 kV. Raman spectra of the samples were recorded by ISA dispersive Raman spectroscopy using argon ion laser with 633 nm excitation. Fourier Transform Infrared Measurements (FTIR) were recorded on KBr pellets with a Bruker Tensor

27. Diffuse Reflectance Electronic Spectra (DRS) were measured with a Cary 4000 equipped with an integrating sphere accessory.

2.5 *Photocatalytic activity evaluation*

The photocatalytic reaction was processed in a black jacket reactor with a 300 W Xenon lamp as the simulated solar light. The photocatalytic performance was evaluated by oxidizing water molecules with $AgNO_3$ as an electron scavenger. In a typical procedure, 0.1 g of catalysts were added to 200 mL of solution including $AgNO_3$ (0.03 M) and La_2O_3 (0.2 g). Before irradiation, the suspensions were mixed under vigorous stirring for 30 min in the dark and degassed for removal of O_2 in solution and air. After that, the light source was lightened to trigger the water oxidation reaction. At the same time, the O_2 concentrations in the reactor were *in situ* taken by using a NEOFOX O_2 probe.

In addition, the catalyst was also used to test photocatalytic oxidation for the degradation of RhB. The concentrations of RhB and a catalyst were 10 mg/L and 0.5 g/L, respectively, and a high-pressure Xe-lamp (300 W, Philips) was used as a simulated solar source. Prior to the illumination, the RhB solution was mixed with catalysts and sonicated for 30 min in dark to establish the adsorption-desorption equilibrium. During photocatalytic process, 3 mL of the reaction solution was extracted for every 15 min and measured by UV-vis spectrophotometer (JASCOV-670).

3 RESULTS AND DISCUSSION

3.1 *Structure and morphology of the rGO/BWO*

The crystal phase structures of the as prepared samples were characterized by X-Ray Diffraction patterns (XRD). In Figure 1, the main diffraction patterns of rGO/BWO composite are similar to those of the pure BWO, characteristic of orthorhombic BWO (JCPDS No. 39-0256). The observed four apparent peaks match well with the (131), (200), (222), and (331) crystal planes of orthorhombic BWO, respectively (Zhang 2014). This result indicated that the strong ultraviolet irradiation did not damage the crystal phase of the BWO, which keeps the properties of BWO unchanged.

The morphologies of BWO and rGO-BWO were analyzed by Scanning Electron Microscopy (SEM) (in Figure 2). It is observed that BWO in Figure 2a mainly displays uniform hierarchical flower with a diameter of 2–3 μm. Regarding rGO-BWO (in Figure 2b), BWO flowers are kept and the graphene labeled by yellow line contact with the BWO flowers. This interface contact is beneficial for transportation of photo-generated electrons.

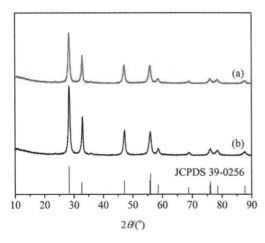

Figure 1.   XRD patterns of pure BWO (a) and rGO/BWO (b).

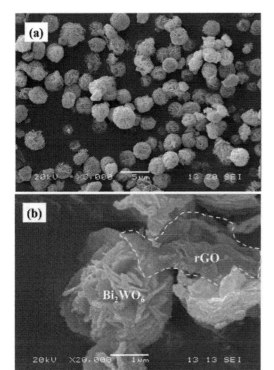

Figure 2.   SEM images of the BWO flowers (a) and rGO/BWO composite (b).

3.2   *Raman, FTIR, and optical properties*

In Figure 3, the strong peaks at 295, 707, and 796 cm^{-1} are ascribed to Bi-O and W-O vibration bands, respectively, which demonstrates the

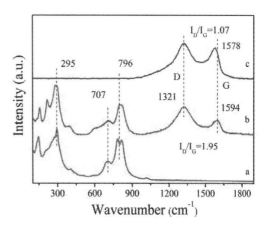

Figure 3.   Raman spectra of the BWO (a), rGO/BWO (b), and GO (c).

existence of BWO in the composite (Ma 2012, Li 2010). Meanwhile, the Raman spectrum of rGO/BWO composite displays two peaks at around 1319 cm^{-1} and 1594 cm^{-1} corresponding to D and G band, respectively. It is an apparent feature of layered graphene (Sun 2014, Lv 2014). In the case of pure GO, two prominent peaks appear at 1340 and 1582 cm^{-1}. It suggests that the ultraviolet reduction treatment has an apparent impact on the characteristic peaks of graphene. Besides, it is worth noting that the rGO/BWO composite exhibits obvious up-shift of G band peak compared with those of GO, which is a strong evidence for the chemical bonding between BWO and graphene (Ma 2013, Gao 2010). The previous studies reported that the n-type doping of grapheme leads to a downshift of G band, while the p-type doping results in an up-shift in comparison with pure graphene (Yeh 2014, Kudin 2008). Therefore, the Raman shift of the G band peak for the as-prepared rGO/BWO composite provides strong evidence for the accelerated charge transfer between rGO and BWO. Moreover, a higher ratio of D/G band intensity (1.95) is found in rGO/BWO in comparison with those in neat GO (0.906), thus indicating a likely higher structural integrity of graphene resulting from the preparation.

The FT-IR spectra of GO, BWO and rGO/BWO composite are shown in Figure 4. The representative peaks of GO are centered at 3431 cm^{-1} (O-H stretching vibration), 1719 cm^{-1} (C = O stretching vibration of COOH groups), 1403 cm^{-1} (tertiary C-OH stretching vibration), and 1032 cm^{-1} (C-O stretching vibration) (Rao 2015). Meanwhile, GO and rGO/BWO show absorptions at the same wavelength but with far lower absorption intensity, indicating the reduction of GO. The vibration band appearing at 1595 cm^{-1} clearly shows

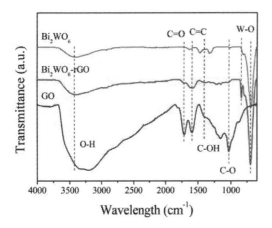

Figure 4. FTIR spectra of the BWO, rGO/BWO, and GO samples.

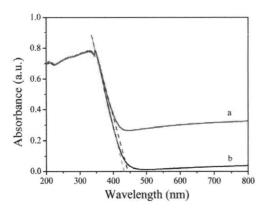

Figure 5. DRS of the as-obtained rGO/BWO (a) and BWO (b).

the graphene skeletal vibration (C = C), indicating its formation in the case of rGO/BWO. In addition, the vibration bands appear between 600 and 1000 cm^{-1}, which can be attributed to stretching vibration of the BWO. Two new absorptions at 698 and 840 cm^{-1} are associated with the stretching vibration of Bi-O and W-O-W bonds, respectively (Wang 2014). This result is consistent with the result of Raman, which demonstrates the formation of the rGO/BWO composite.

The photo-response property of the as-obtained samples was investigated. The measured results were displayed in Figure 5. The pure BWO flowers show photo absorption from UV to visible light rang (below 450 nm), which is attributed to the transition from the hybrid orbital of O 2p and Bi 6s to the W 5d orbital. While the rGO/BWO composite shows intense and broad background absorption in the visible light region, which was

similar with that of GO. The enhancement of absorption in the visible light region is ascribed to the introduction of black body properties, such as graphite-like materials (Wang 2014), indicating the existence of graphene in rGO/BWO. The big difference could also be seen from the color difference, where the BWO was yellow, the GO and rGO/BWO samples were both black. Because of the increased absorbance, a more efficient utilization of the solar energy can be obtained. Therefore, we can infer that the introduction of graphene in BWO flowers is effective for the visible-light response of the composite.

3.3 *Photocatalytic activity evaluation*

The pure BWO flowers and rGO/BWO composite were investigated for photocatalytic ability in terms of O_2 evolution from water with electron acceptor Ag^+ ions. In Figure 6, the concentration of O_2 evolution from H_2O after 3 h was 30 μmol/L for rGO/BWO, 2.1 times higher than the pure BWO (14.5 μmol/L). It is well known that light harvest, charge transfer and separation play important roles in the photocatalysis process. In general, valence band energy of BWO is more positive than the standard redox potential of H_2O/O_2 (1.23 eV vs. RHE), suggesting that the photo-generated holes of both BWO and rGO/BWO could theoretically oxidize H_2O to produce O_2 (Lee 2012). Furthermore, the rGO/BWO also showed enhanced photocatalytic performance for organic dye degradation compared with the composites of rGO and pure BWO (in Figure 7). The degradation efficiency of organic dye reaches to 100% in 60 min under solar light irradiation, which is 1.9 times higher than that of pure BWO. After 4 times recycling test, the

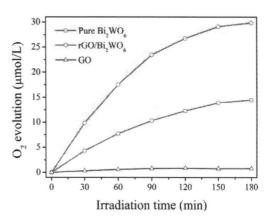

Figure 6. O_2 evolution curves of the BWO and rGO/BWO composite.

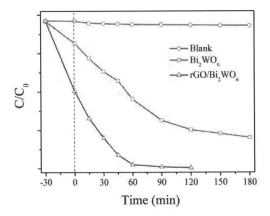

Figure 7. Degradation curve of RhB over the BWO and rGO/BWO under solar light irradiation.

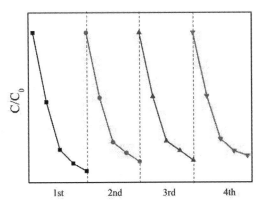

Figure 8. The recycling test of the rGO/BWO composite for RhB degradation.

rGO/BWO displays still good photocatalytic ability for RhB, as shown in Figure 8.

The significant enhancement could be ascribed to improvement in the light absorption range and reduction of the recombination of electron-hole pairs in BWO due to the presence of reduced graphene sheets in the composite. The GO firstly interacted with BWO by electrostatic interaction, and then the ultraviolet irradiation induced the formation of physical and chemical bonds between rGO and BWO. As a result, rGO can serve as an electron collector to efficiently separate the photo-generated electron-hole pairs, effectively lengthening the charge carrier lifetime. Furthermore, the extraordinary surface area of rGO provides more photocatalytic reactive sites not only on the catalyst surface but on the graphene sheet, which enlarges the reaction space markedly.

4 CONCLUSIONS

In this work, we successfully fabricated a hierarchical rGO/BWO composite by an ultraviolet assisted method for O_2 evolution from water and photocatalytic oxidation performance. The hierarchical BWO flower-like were firstly prepared by hydrothermal route, which shows uniform morphology with average diameter of 2–3 μm. The rGO/BWO displayed an enhanced O_2 production rate, 176 μmol·L^{-1}·h^{-1}·g$^{-1}_{cat}$, which is 2.3 times higher than that of pure BWO, 77 μmol·L^{-1}·h^{-1}·g$^{-1}_{cat}$. In addition to water splitting ability, the photocatalytic performance for RhB has been investigated, where RhB over the rGO/BWO composite was totally degraded in almost 60 min under solar light irradiation, 1.9 times higher than that of the pure BWO. The introduction of reduced graphene oxide plays a key role in raising photo-induced charge carrier separation efficiency as electron collector. Thus, the strategy provides an efficient approach for the fabrication of hierarchical ternary oxides composites containing graphene.

This work was supported financially by the National Natural Science Foundation of China (21507029, 21501138), the Natural Science Foundation of Hubei Province (2015CFB177), China Ministry of Education and the Fundamental Research Funds for the Central Universities (2016MS109).

REFERENCES

Chen, M. & Chu, W. 2015. Photocatalytic degradation and decomposition mechanism of fluoroquinolones norfloxacin over bismuth tungstate: Experiment and mathematic model. *Appl. Catal. B: Environ.* 168–169:175–182.

Deng, X. Bongard, H. Chan, C. & Tuysuz, H. 2016. Dual-templated Cobalt Oxide for photochemical water oxidation. *ChemSusChem* 9:409–415.

Gao, E. Wang, W. Shang, M. & Xu, J. 2011. Synthesis and enhanced photocatalytic performance of graphene-Bi$_2$WO$_6$ composite. *Phys. Chem. Chem. Phys.* 13:2887–2893.

Kalanur, S. Hwang, Y. Chae, S. & Joo, O. 2013. Facile growth of aligned WO$_3$ nanorods on FTO substrate for enhanced photoanodic water oxidation activity. *J. Mater. Chem. A* 1:3479–3488.

Ke, J. Liu, J. Sun, H. Hua, Y. Duan, X. Liang, P. Li, X. Tade, M. Liu, S. & Wang, S. 2017. Facile assembly of Bi$_2$O$_3$/Bi$_2$S$_3$/MoS$_2$ n-p heterojunction with layered n-Bi$_2$O$_3$ and p-MoS$_2$ for enhanced photocatalytic water oxidation and pollutant degradation. *Appl. Catal. B: Environ.* 200:47–55.

Kiss, B. Didier, C. Johnson, T. Manning, T. Dyer, M. Cowan, A. Claridge, J. Darwent, J. & Rosseinsky, M. 2014. Photocatalytic water oxidation by a pyrochlore oxide upon irradiation with visible light: rhodium substitution into Yttrium Titanate. *Angew. Chem. Int. Ed.* 53:14480–14484.

Kudin, K. Ozbas, B. Schniepp, H. Prudhomme, R. Aksay, I. & Car, R. 2008. Raman spectra of graphite oxide and functionalized graphene sheets. *Nano Lett.* 8:36–41.

Lee, G. Kim, K. & Cho, K. 2011. Theoretical study of the electron transport in graphene with vacancy and residual oxygen defects after high temperature reduction. *J. Phys. Chem. C* 115:9719–9725.

Lee, S. Carlton, C. Risch, M. Surendranath, Y. Chen, S. Furutsuki, S. Yamada, A. Nocera, D. & Shao-Horn, Y. 2012. The nature of lithium battery materials under oxygen evolution reaction conditions. *J. Am. Chem. Soc.* 134:16959–16962.

Li, Y. Liu, J. Huang, X. & Yu, J. 2010. Carbon-modified Bi_2WO_6 nanostructures with improved photocatalytic activity under visible light. *Dalton Trans.* 39:3420–3425.

Lv, H. Liu, Y. Hu, J. Li, Z. & Lu, Y. 2014. Ionic liquid-assisted hydrothermal synthesis of Bi_2WO_6-reduced graphene oxide composites with enhanced photocatalytic activity. *RSC Adv.* 4:63238–63245.

Ma, H. Shen, J. Shi, M. Lu, X. Li, Z. Long, Y. Li, N. & Ye, M. 2012. Significant enhanced performance for Rhodamine B, phenol and Cr(VI) removal by Bi_2WO_6 nanocomposites via reduced graphene oxide modification. *Appl. Catal. B: Environ.* 121–122: 198–205.

Ma, J. Meng, Q. Michelmore, A. Kawashima, N. Izzuddin, Z. Bengtsson, C. & Kuan, H. 2013. Covalently bonded interfaces for polymer/graphene composites. *J. Mater. Chem. A* 1:4255–4264.

Pilli, S. Janarthanan, R. Deutsch, T. Furtak, T. Brown, L. Turnerd, J. & Herring, A. 2013. Efficient photoelectrochemical water oxidation over cobalt-phosphate (Co-Pi) catalyst modified $BiVO_4$/1D-WO_3 heterojunction electrodes. *Phys. Chem. Chem. Phys.* 15:14723–14728.

Qamar, M. Elsayed, R. Alhooshani, K. Ahmed, M. & Bahnemann, D. 2014. Highly efficient and selective oxidation of aromatic alcohols photocatalyzed by nanoporous hierarchical Pt/Bi_2WO_6 in organic solvent-free environment. *ACS Appl. Mater. Interfaces* 7:1257–1269.

Rao, K. Sentilnathan, J. Cho, H. Wu, H. & Yoshimura, M. 2015. Soft processing of graphene nanosheets by glycine-bisulfate ionic-complex-assisted electrochemical exfoliation of graphite for reduction catalysis. *Adv. Funct. Mater.* 25:298–305.

Roy, P. Periasamy, A. Liang, C. & Chang, H. 2013. Synthesis of graphene-ZnO-Au nanocomposites for efficient photocatalytic reduction of nitrobenzene. *Environ. Sci. Technol.* 47:6688–6695.

Saison, T. Gras, P. Chemin, N. Chanéac, C. Duruphty, O. Brezová, V. Colbeau-Justin, C. & Jolivet, J. 2013. New insights into Bi_2WO_6 properties as a visible-light photocatalyst. *J. Phys. Chem. C* 117:22656–22666.

Sun, Z. Guo, J. Zhu, J. Ma, S. Liao, Y. & Zhang, D. 2014. High photocatalytic performance by engineering Bi_2WO_6 nanoneedles onto graphene sheets. *RSC Adv.* 4:27963–27970.

Wang, D. Li, R. Zhu, J. Shi, J. Han, J. Zong, X. & Li. C. 2012. Photocatalytic water oxidation on $BiVO_4$ with the electrocatalyst as an oxidation cocatalyst: essential relations between electrocatalyst and photocatalyst. *J. Phys. Chem. C* 116:5082–5089.

Wang, H. Lu, J. Wang, F. Wei, W. Chang, Y. & Dong, S. 2014. Preparation, characterization and photocatalytic performance of g-C_3N_4/Bi_2WO_6 composites for methyl orange degradation. *Ceram. Int.* 40:9077–9086.

Wang, Y. Wang, W. Mao, H. Lu, Y. Lu, J. Huang, J. Ye, Z. & Lu, B. 2014. Electrostatic self-assembly of $BiVO_4$-reduced graphene oxide nanocomposites for highly efficient visible light photocatalytic activities. *ACS Appl. Mater. Interfaces* 6:12698–12706.

Wang, Y. Zhou, T. Jiang, K. Da, P. Peng, Z. Tang, J. Kong, B. Cai, W. Yang, Z. & Zheng, G. 2014. Reduced mesoporous Co_3O_4 nanowires as efficient water oxidation electrocatalysts and supercapacitor electrodes. *Adv. Energy Mater.* 4:1400696.

Wang, T. & Gong, J. 2015. Single crystal semiconductors with narrow band gaps for solar water splitting. *Angew. Chem. Int. Ed.* 54:2–17.

Williams, K. Nelson, C. Yan, X. Li, L. & Zhu, X. 2013. Hot electron injection from graphene quantum dots to TiO_2. *ACS Nano* 7:1388–1394.

Xie, J. & Xie, Y. 2015. Transition metal nitrides for electrocatalytic energy conversion: opportunities and challenges. *Chem. Eur. J.* 21:1–12.

Xu, Z. Bando, Y. Liu, L. Wang, W. Bai, X. & Golberg, D. 2011. Electrical conductivity, chemistry, and bonding alternations under graphene oxide to graphene transition as revealed by in situ TEM. *ACS Nano* 5:4401–4406.

Yan, Y. Wu, Y. Yan, Y. Guan, W. & Shi, W. 2013. Inorganic-salt-assisted morphological evolution and visible light driven photocatalytic performance of Bi_2WO_6 nanostructures. *J. Phys. Chem. C* 117:20017–20028.

Yeh, T. Teng, C. Chen, S. & Teng, H. 2014. Nitrogen-doped graphene oxide quantum dots as photocatalysts for overall water splitting under visible light illumination. *Adv. Mater.* 26:3297–3303.

Yuan, Y. Ye, Z. Lu, H. Hu, B. Li, Y. Chen, D. Zhong, J. Yu, Z. & Zou, Z. 2015. Constructing anatase tio_2 nanosheets with exposed (001) facets/layered MoS_2 two dimensional nanojunction for enhanced solar hydrogen generation. *ACS Catal.* 6:532–541.

Zhang, G. Zang, S. Lan, S. Huang, Z. Li, C. Li, G. & Wang, X. 2015. Cobalt selenide: a versatile cocatalyst for photocatalytic water oxidation with visible light. *J. Mater. Chem. A* 3:17946–17950.

Zhang, N. Ciriminna, R. Pagliaro, M. & Xu, Y. 2014. Nanochemistry-derived Bi_2WO_6 nanostructures: towards production of sustainable chemicals and fuels induced by visible light. *Chem. Soc. Rev.* 43:5276–5287.

Zhang, Y. & Xu, Y. 2014. Bi_2WO_6: A highly chemoselective visible light photocatalyst toward aerobic oxidation of benzylicalcohols in water. *RSC Adv.* 4:2904–2910.

Zhu, M. Chen, P. & Liu, M. 2011. Graphene oxide enwrapped Ag/AgX (X = Br, Cl) nanocomposite as a highly efficient visible-light plasmonic photocatalyst. *ACS Nano* 5:4529–4536.

Zong, X. Xing, Z. Yu, H. Chen, Z. Tang, F. Zou, J. Lu, G. & Wang, L. 2011. Photocatalytic water oxidation on F, N co-doped TiO_2 with dominant exposed {001} facets under visible light. *Chem. Commun.* 47:11742–11744.

Advances in Energy, Environment and Materials Science – Wang & Zhou (Eds)
© 2017 Taylor & Francis Group, London, ISBN 978-1-138-03600-0

Improved sulfur resistance of cathode in PEMFC by the addition of Pd to the Pt/C catalyst

Yang Zhang, Jie Fu & Zixue Wang
College of Environmental and Chemical Engineering, Dalian Jiaotong University, Dalian, Liaoning Province, P.R. China

ABSTRACT: For ambient air compressed directly into the cathode as the oxidant, the performance of the Proton Exchange Membrane Fuel Cell (PEMFC) depends significantly on the quality of the air. Sulfur-containing compounds in the air will cause serious poisoning of the cathode catalyst. An easy and steerable method was used by adding transition metal Pd into the Pt/C catalyst to synthesize Pt-Pd bimetallic catalyst for improving sulfur resistance. Ethylene glycol was used as a reducing agent. The synergistic effect between two metals was shown in the X-Ray Diffraction (XRD) spectra. Na_2SO_3 was employed to test the catalyst poisoning by the method of Cyclic Voltammograms (CV) and potential scan. Compared with the curve with the Pt/C catalyst, the CV curve with the bimetallic catalyst was more stable after poisoning, and better recovery could be clearly seen after being poisoned. Owing to the addition of Pd, the sulfur resistance of the catalyst was greatly improved.

1 INTRODUCTION

The operating process of Proton Exchange Membrane Fuel Cell (PEMFC) needs ambient air, which contains various impurities (Yun et al. 2010). The air impurities, such as SO_2, NO_2, H_2S, and O_3, can influence the performance and the durability of PEMFC to some extent. Among these impurities, sulfur-containing compounds in the air, especially SO_2, are the most deleterious to PEMFCs (Imamura & Hashimasa 2006). According to Mepsted (2001) from Defense and Evaluation Research Agency, SO_2 even at low concentration could seriously affect the electrode performance. The concentration of SO_2 and NO_2 in the atmosphere should not exceed 0.5 ppm, when the impurities adsorption on the Pt catalyst occurred, and the active sites on the surface of the Pt were covered by the impurities (Morre et al. 2000). Thus, the O_2 adsorption on Pt surface and Oxygen Reduction Reaction (ORR) were hindered, which caused the decline of cell performance and this process was irreversible. Brosha et al. (2005) argued that SO_2 adsorption on the surface of Pt affected the charge transfer, blocking the proton conduction and caused the decay of the cell performance. Sulfur coverage of 14% caused a 95% loss in mass activity. When 37% of the Pt surface was covered with sulfur, the reaction pathway of the ORR on the Pt/C catalyst changed from a 4-electron process reaction to a 2-electron process reaction for peroxide (Garsany et al. 2007). The harm of SO_2 in the dry air to the cell performance was more apparent compared with the SO_2 in humid air. In addition, 0.1 ppm of SO_2 in the air had little impact on the performance of the fuel cell, and the performance was even more stable. While SO_2 concentration increased to 0.25 ppm, a decline in performance was found obviously. It was concluded that under the condition of the experiment, the initial concentration of SO_2 poisoning should be between 0.1 and 0.25 ppm (Brosha et al. 2005). Uribe et al. (2004) used a loading of ZnO-based compounds outside the device to remove H_2S in the air; when the H_2S concentration was at the level of 50 ppb, this method could ensure the cell performance without obvious attenuation in 1000 hours. However, this method needed additional auxiliary equipment and when the concentration of H_2S reached high level, and the H_2S absorption purification plant should be frequently replaced. Although sulfur poison effect of fuel cell is very serious, compared with CO poisoning reported in the literature, the solving method has not been widely reported.

The aim of the work presented in this paper is to produce a new kind of bimetallic catalyst with the transition metal Pd to improve the sulfur resistance of cathode catalyst in PEMFC.

2 EXPERIMENT

2.1 *Experiment principle*

Adding transition metal Pd into Pt catalyst can further improve the catalytic activity of ORR. Owing to the relatively lower price of Pd and its similar properties to Pt, the addition of Pd can change the electronic structure of the catalyst surface, decrease the surface of the band position, and reduce the oxygen adsorption of catalyst. Ascribe to the increasing active sites of ORR, the activity of catalyst for oxygen reduction increases accordingly. Through the method of crystal epitaxial growth in Pd particle surface preparation, Lim et al. (2009) produced out many Pt dendrites. Characterization results indicated that relative to the quality of Pt, the catalytic activity of oxygen reduction was 2.5 times higher than the Pt/C catalyst. The results indicated that the Pd alloy catalyst had good ORR activity.

2.2 *Process of Pt/C*

All experiment were performed on electrochemical workstation CHI 600C (Shanghai Chenhua Instrument Co. Ltd.). The gas was high-purity neat N_2 (99.997%). At first, 0.06 g of the carbon black XC-72 and 36 mL ethylene glycol were blended with ultrasonic oscillation, and then H_2PtCl_6 (0.74 mg/mL) glycol solution 6.86 mL was added. After ultrasonic oscillation and magnetic stirring evenly, ethylene glycol solution of NaOH (2 mol/L) was added until the pH reached 12. After stirring under protection of high purity nitrogen for 30 min, the suspension liquid was heated at 130°C for 3 hours with condenser pipe reflux maintaining access to nitrogen. After reaction, it needed to be washed five times by centrifugal separation. $AgNO_3$ was used to inspect chloride ion, and then vacuum drying was applied for 9 hours at 130°C. Therefore, the Pt/C catalyst was obtained.

2.3 *Process of Pt-Pd/C*

In this process, 0.06 g of the carbon black XC-72 and 36 mL ethylene glycol were blended with ultrasonic oscillation, and then 6.86 mL H_2PtCl_6 (0.74 mg/mL) glycol solution and 2.31 mL $PdCl_2$ (4 mg/mL) glycol solution were added. After ultrasonic oscillation and magnetic stirring evenly, ethylene glycol solution of NaOH (2 mol/L) was added until the pH reached 12. After stirring under protection of high-purity nitrogen for 30 min, the suspension liquid was heated at 130°C for 3 hours with condenser pipe reflux maintaining access to nitrogen. After reaction, it needed to be washed five times by centrifugal separation. $AgNO_3$ was used to inspect chloride ion, and then vacuum drying was applied for 9 hours at 130°C. Therefore, the Pt-Pd/C catalyst was obtained.

2.4 *Test of Pt/C*

A three-electrode system was employed for electrochemical testing. H_2SO_4 solution (0.5 mol/L) was used as an electrolyte. The Pt/C catalyst coated on the glassy carbon electrode was the working electrode, and Pt wire was the counter electrode. Calomel Electrode (SCE) was chosen as the reference electrode. Through the electrochemical workstation of CHI 600C, Cyclic Voltammograms (CV) test scanning potential interval was fixed from –0.2 to 1.2 V, with a scanning speed of 50 mV/s. The potential scan of ORR test interval was fixed from 0.5 to 1.5 V. Under protection with high-purity nitrogen, the curves of CV and ORR were recorded after five laps at the beginning and the end of the scanning. After the testing above, Na_2SO_3 measuring 0.05 mol/L was added into the H_2SO_4 solution used above. Test of CV and potential scan of ORR were repeated to find the difference of sulfur resistance between the catalysts of Pt/C and Pt-Pd/C. Finally, the Na_2SO_3 electrolyte was replaced by 0.5 M H_2SO_4 to test CV and ORR as the same. Trend of sulfur adsorption and sulfur desorption could be compared by the curves.

2.5 *Test of Pt-Pd/C*

Testing procedure of the Pt-Pd/C catalyst was the same as that in Section 2.4.

3 RESULTS AND DISCUSSION

3.1 *SEM and TEM characterization of Pt/C*

Figure 1 is the Scanning Electron Microscopic (SEM) image and Transmission Electron Microscopic (TEM) image of the Pt/C catalyst. It can be observed that the catalyst is sparse and uniform, but part of the catalyst is aggregate.

Figure 1. Magnified Pt/C catalyst by SEM and TEM, respectively.

Figure 2. Magnified Pt-Pd/C catalyst by SEM and TEM, respectively.

3.2 SEM and TEM characterization of Pt-Pd/C

SEM and TEM images of Pt-Pd/C catalyst are shown in Figure 2. Compared with the Pt/C catalyst, the active component of Pt-Pd/C metal particles increases, and the metal particles are better dispersed, and the distribution is very uniform. As the Pd and Pt crystal structure is very similar, we cannot identify the crystal phase through SEM phase filed. According to the characters of precious metal particle uniformity, it can be deduced that, in the precious metal particles, Pt was the crystal nucleus and Pd plating was the modified layer on the surface of Pt particles.

3.3 The spectra of energy-dispersive spectroscopy (EDS) of Pt/C catalyst

As shown in Figure 3, Pt can be easily found with abundant element of carbon.

3.4 The spectra of EDS of Pt-Pd/C catalyst

As shown in Figure 4, both Pt and Pd were detected. Furthermore, the amount of Pt and Pd was roughly the proportion of 3:1.

3.5 Characterization of XRD

The contrast of X-Ray Diffraction spectra (XRD) between Pt/C and Pt-Pd/C catalysts is shown in Figure 5. The degree of 2 Theta was founded at 23°, 39°, 46°, 67°, and 81° belonging to Pt metal crystal diffraction. The XRD result shows that all of the catalysts are face-centered cubic structure. It is worth noting that, because of the crystal peak of Pt-Pd/C relative to the Pt/C catalyst, it has slightly moved to the right. Peak had increased obviously

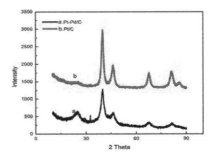

Figure 5. The XRD pattern of Pt/C and Pt-Pd/C catalyst.

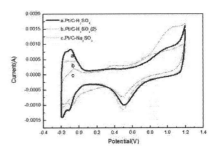

Figure 6. Pt/C catalyst cyclic voltammogram curves before and after poisoning.

compared with the Pt/C catalyst, which illustrates that Pd modification had a potential effect on the lattice surface of Pt, and therefore, it is likely to affect the catalytic activity of Pt.

3.6 CV Test of Pt/C catalyst with and without sulfur

The hydrogen peak area before poisoning was significantly greater than that after putting in Na_2SO_3 (Fig. 6). The smaller hydrogen peak area shows that the catalyst surface was partially covered by SO_3^{2-}. Therefore, the redox reaction of hydrogen was prevented, reducing the activity of the catalyst. The oxidation peak in high potential area increased which reflected that some of SO_3^{2-} was oxidized to SO_4^{2-} into the solution. Activity surface of the catalyst had partially recovered.

3.7 CV test of Pt-Pd/C catalyst with and without sulfur

The same variation with Pt-Pd/C catalyst can be seen in Figure 7. Pt-Pd catalyst surface was partially covered by SO_3^{2-}, hindering the hydrogen oxidation and oxygen reduction reaction. However, compared with the Pt/C catalyst, less peak area decline was found, which illustrates that the Pt-Pd/C bimetallic catalyst

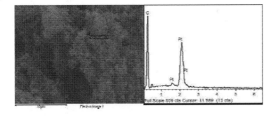

Figure 3. The spectra of EDS of Pt/C catalyst.

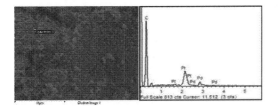

Figure 4. The spectra of EDS of Pt-Pd/C catalyst.

had reduced the poisoning of SO_3^{2-} ion directly on the catalyst, meanwhile reducing the adsorption energy on the surface of catalyst, keeping the activity of catalyst to the approximate original level. Therefore, the curve was more stable compared with Pt/C at the start or the end of poisoning.

3.8 *Comparison of two catalysts after poisoning*

As shown in Figure 8, the Pt-Pd/C catalyst was more active based on the equivalent Pt mass than the Pt/C catalyst after poisoning. Therefore, the resistance on sulfur was largely improved.

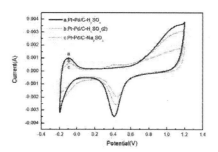

Figure 7. Pt-Pd/C catalyst cyclic voltammogram curves before and after poisoning.

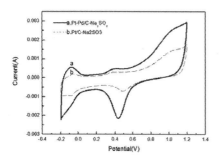

Figure 8. Comparison of Pt/C catalyst and Pt-Pd/C catalyst after poisoning by cyclic voltammetry.

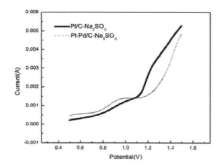

Figure 9. ORR graph of Pt/C and Pd-Pt/C catalyst after poisoning.

3.9 *ORR test*

The Pd modified Pt/C catalyst for Oxygen Reduction (ORR) had higher potential (Fig. 9), reflecting its oxygen reduction activity better than that of the Pt/C catalyst, when the catalyst poisoned at the same time.

4 CONCLUSION

In this paper, Pt-Pd/C bimetallic catalyst was produced by ethylene glycol reduction method. It can be concluded that the bimetallic catalyst with Pd has the better anti-sulfur performance. It means that, better performance and durability will be obtained when the fuel cells stack under sulfur pollution. CV test and potential scan of ORR showed that the Pt/C catalyst modified with Pd had higher activity ORR. To sum up, when sulfur oxide come into the bimetallic catalyst layer of the cathode, part of it will be absorbed by Pd, reducing the coverage of sulfur oxide on the surface of the Pt catalyst. Hence, excessive loss of catalyst activity is avoided. In other words, the ability of resistance on sulfur of the PEMFCs is improved.

ACKNOWLEDGMENTS

This research was financially supported by the education department of Liaoning province (L2013177), China.

REFERENCES

Brosha, E., F. Garzon, B. Pivovar, T. Rockward, J. Valeria & F. Uribe. (2005). Effect of Fuel and Air Impurities on PEM Fuel Cell Performance. Mid-Year DOE Fuel Cell Program Review.
Garsany, Y., Baturina, O. A., & Swider-Lyons, K. E. (2007). Impact of Sulfur Dioxide on the Oxygen Reduction Reaction at Pt/Vulcan Carbon Electrocatalysts. J. Electrochem. Soc. 154: B670-B675.
Imamura, D. & Y. Hashimasa. (2006). Effect of Sulfur-Containing Compounds on Fuel Cell Performance. ECS 212th Meeting.
Lim, B., M. Jiang, & P. H. C. Camargo. (2009). Pd-Pt Bimetallic Nanodendrites with High Activity for Oxygen Reduction. J. Science. 324:1302–1305.
Mepsted, G. (2001). Investigation of the effects of air contaminant of SPFC performance. Defense and Evaluation Research Agency.
Moore, J. M., P. L. Adcock, J. B. Lakeman, & G. O. Mepsted. (2000). The effects of battlefield contaminants on PEMFC performance. J. Power Sources. 85: 254–260.
Uribe, F., Adcock, P., & Bender, G. (2004). Electrodes for Hydrogen-Air PEM Fuel Cell. Los Alamos National Laboratory.
Yun, J. (2010). The research and development of fuel cell domestic and abroad. J. Modern Chemical Industry. 30(2): 106–109.

Author index